ADVANCES IN TRACE SUBSTANCES RESEARCH

Quantitative Methods
in
Aquatic Ecotoxicology

Quantitative Methods
in
Aquatic Ecotoxicology

Michael C. Newman

Savannah River Ecology Laboratory
The University of Georgia
Aiken, South Carolina

LEWIS PUBLISHERS
Boca Raton   Ann Arbor   London   Tokyo

Library of Congress Cataloging-in-Publication Data

Newman, Michael C.
    Quantitative methods in aquatic ecotoxicology / Michael C. Newman.
        p. cm. — (Advances in trace substances research)
    Includes bibliographical references (p.) and index.
    ISBN 0-87371-622-1
    1. Water—Pollution—Environmental aspects—Measurement. 2. Water—Pollution—Toxicology—
Measurement. 3. Aquatic organisms—Effect of water pollution on—Measurement. I. Title. II. Series.
    QH545.W3N49 1994   94-28619
    574.5'263—dc20   CIP

© 1995 by CRC Press, Inc.
Lewis Publishers is an imprint of CRC Press.

No claim to original U.S. Government works
International Standard Book Number 0-87371-622-1
Library of Congress Card Number 94-28619
Printed in the United States of America   2 3 4 5 6 7 8 9 0
Printed on acid-free paper

ADVANCES IN TRACE SUBSTANCES RESEARCH

Series Preface

The need to synthesize, critically analyze, and put into perspective the ever-mounting body of information on trace chemicals in the environment provided the impetus for the creation of this series. In addition to examining the fate, behavior, and transport of these substances, the transfer into the food chain and risk assessment to the consumers, including humans, will also be taken into account. It is hoped then that this information will be user-friendly to students, researchers, regulators, and administrators.

The series will have "topical" volumes to address more specific issues as well as volumes with heterogeneous topics for a quicker dissemination. It will have international scope and will cover issues involving natural and anthropogenic sources in both the aquatic and terrestrial ecosystems. To ensure a high quality publication, volume editors and the editorial board will subject each article to peer review.

Thus, **Advances in Trace Substances Research** should provide a forum where experts can discuss contemporary environmental issues dealing with trace chemicals that, hopefully, can lead to solutions resulting in a cleaner and healthier environment.

Domy C. Adriano
Editor-in-Chief

PREFACE

The Impetus for This Book

It is, therefore, urged without reason, as a discouragement to writers, that there are already books sufficient in the world; that all topics of persuasion have been discussed, and every important question clearly stated and justly decided; . . . [However], whatever be the present extent of human knowledge, it is not only finite, and therefore in its own nature capable of increase; but so narrow, that almost every understanding may, by a diligent application of its powers, hope to enlarge it. It is, however, not necessary, that a man forbear to write, till he has discovered some truth unknown before; he may be sufficiently useful, by only diversifying the surface of knowledge, and luring the mind by a new appearance to a second view of those beauties which it had passed over unattentively before.[1]

While browsing through the preface to Moriarty's book, *Ecotoxicology: The Study of Pollutants in Ecosystems*,[2] preparing to write this preface, I was struck by the similarity of our intentions. He suggests, and I concur, that an abundance of ecotoxicological data exists, but much of the data is insignificant. Considering myself an unwitting, albeit minor, contributor to this dilute data base and dismayed at the prospect of mediocrity as an inescapable theme in my professional career, I've dedicated considerable thought to factors contributing to this situation. Certainly, this topic is not trivial and, consequently, not characterized by practitioners lacking sufficient acumen or funding. For the most part, the professionals involved in this field are well trained, well intended, and well funded. Further, the prosaic argument that ecosystems are too complex to understand to any practical extent is inconsistent with the contrastingly rapid progress in disciplines such as molecular genetics, immunology, computer/information sciences, and physics, which deal with equally complex and less palpable subjects. Statistics demonstrating that the majority of publications in many "hard" sciences go practically uncited[3] provided me with a broader appreciation of the problem but no conceptual tools to improve a condition seemingly more severe in ecotoxicology than in many other fields.

I had resigned myself to the fact that all fields go through a presynthetic or descriptive phase before maturation. I assigned ecotoxicology (and its predecessors such as aquatic and wildlife toxicology) to this unsatisfying status until I stumbled on an article entitled "Strong Inference" by J. R. Platt.[4] Platt's arguments regarding qualities affecting the relative rates of advancement of various disciplines suggested that solidification of ecotoxicological principles and paradigms could be accelerated greatly by adapting a stronger inferential approach. In the process, we would also become better environmental stewards. One goal of this book is to encourage such an inferential approach.

Each chapter treats ecotoxicology as a scientific endeavor. As with all sciences, the focus will be "the organization and classification of knowledge on the basis of explanatory principles."[5] This treatment will develop a strong inferential and quantitative theme. Consequently, many aspects relevant to regulatory activities, e.g., standard methods and environmental legislation, will not be presented in balance with their importance. This omission should not be considered a mute assignment of the regulatory aspects of ecotoxicology to a status inferior to the scientific aspects. These critical topics are covered clearly and thoroughly in other volumes such as Rand and Petrocelli's *Fundamentals of Aquatic Toxicology*[6] and various publications of the U.S. Environmental Protection Agency (EPA).

The topics covered in this book are arranged in order of increasing ecological organization. This being the case, early chapters may contain information that some workers may not consider sufficiently high in biological organization to be considered pertinent to ecotoxicology, i.e., not community- or systems-level topics. I have expressed my objection to this artificial limitation elsewhere. "Few ecologists would disagree that progress in ecology would have been slowed by exclusion of all but community and system level research. It seems illogical to assume that growth in this new field of ecology would not be similarly compromised by such a restriction."[7] Regardless of the level examined, the intent in all chapters is a better understanding of the fate and effects of toxicants in ecosystems.

The Organization of the Book

This book, a quantitative treatment of the science of ecotoxicology, focuses primarily on aquatic systems but, with appropriate modification, many of the concepts and methods discussed can be applied to terrestrial systems. The first chapters consider fundamental concepts and definitions essential to understanding the fate and effects of toxicants at various levels of ecological organization as covered in the remaining chapters. Scientific ecotoxicology and associated topics are defined in Chapter 1. The historical perspective, rationale, and characteristics are outlined for the strong inferential and quantitative approach advocated in this book. The second chapter in the first section discusses the general measurement process. It considers methodologies for defining and controlling variance which could otherwise exclude valid conclusions regarding ecotoxicological endeavors. Ecotoxicological concepts at increasing levels of ecological organization are discussed in the second section (Chapters 3 through 7). Quantitative methods used to measure toxicant effects are outlined in each of these chapters. The final chapter summarizes the book with a brief discussion of ecotoxicological assessment.

REFERENCES

1. Murphy, A., ed. *The Works of Samuel Johnson, LL.D.* Vol. 1. (New York: George Dearborn Publishers, 1836).
2. Moriarty, F. *Ecotoxicology: The Study of Pollutants in Ecosystems.* (Orlando, FL: Academic Press, Inc., 1983).
3. Hamilton, D. P. "Publishing by—and for?—the numbers," *Science* 250:1331–1332 (1990).
4. Platt, J. R. "Strong inference." *Science* 146:347–353 (1964).
5. Nagel, E. *The Structure of Science. Problems in the Logic of Scientific Explanation.* (New York: Harcourt, Brace and World, Inc., 1961).
6. Rand, G. M., and S. R. Petrocelli, eds. *Fundamentals of Aquatic Toxicology: Methods and Applications.* (New York: Hemisphere Publishing Corp., 1985).
7. Newman, M. C., and A. W. McIntosh. "Preface." In *Metal Ecotoxicology: Concepts and Applications*, ed. M. C. Newman and A. W. McIntosh. (Chelsea, MI: Lewis Publishers, Inc., 1991).

Michael C. Newman
University of Georgia
Savannah River Ecology Laboratory
Aiken, South Carolina

ABOUT THE AUTHOR

Michael C. Newman

Michael C. Newman is a Research Associate Ecologist at the University of Georgia's Savannah River Ecology Laboratory. He received B.A. (Biological Sciences) and M.S. (Zoology) degrees from the University of Connecticut, and M.S. and Ph.D. degrees in Environmental Sciences from Rutgers University. He joined the faculty of the University of Georgia in 1983. His research interests include population level effects of toxicants, toxicity models, factors modifying bioaccumulation kinetics, and inorganic water chemistry. He has published more than 50 articles on these topics. He directed the development of UNCENSOR, a PC-based program for analyzing data sets containing "below detection limit" observations, which is presently used by more than 400 professionals. In 1991, with Alan W. McIntosh, he edited the book, *Metal Ecotoxicology: Concepts and Applications.*

Dr. Newman is also active in professional societies and teaching. He founded and was the first president of the Carolinas Chapter of the Society of Environmental Toxicology and Chemistry (SETAC), and serves on the National SETAC Awards Committee. He has organized special sessions at SETAC and other professional meetings including the First International SETAC Congress in Lisbon, Portugal. He has taught at the University of Connecticut, Rutgers University, the University of California–San Diego, and the University of Georgia.

ACKNOWLEDGMENTS

This volume was written during sabbatical leave from the University of Georgia. Work was supported by contract DE-AC09-76SROO-819 between the US Department of Energy and the University of Georgia. Completion of this volume would not have been possible without the support of the University of Georgia and US Department of Energy.

The hard work of Rose Jagoe as the SREL technical editor of this volume is greatly appreciated. She enhanced the quality of the book through careful assessment of mathematical accuracy and grammatical correctness in addition to her other numerous editorial tasks. Dawn Greene, John McCloskey, and Edith Towns provided excellent support with proofreading. Peter Landrum provided an excellent review of the entire manuscript. Sincere thanks to Drs. Margaret Mulvey, Joseph Pechmann, and Carl Strojan who reviewed many chapter drafts. The author is very grateful for the excellent reviews provided by the following people:

Philip M. Dixon, University of Georgia
Russell J. Erickson, U.S. EPA–Duluth
Karen A. Garrett, University of Georgia
Richard O. Gilbert, Battelle Memorial Institute
William L. Hayton, Ohio State University
Dennis R. Helsel, U.S. Geological Survey
Charles H. Jagoe, University of Georgia
Peter F. Landrum, NOAA Great Lakes Environmental Research Laboratories
Alan W. McIntosh, University of Vermont
Margaret Mulvey, University of Georgia
Joseph H. Pechmann, University of Georgia
John E. Pinder, III, University of Georgia
O. Eugene Rhodes, Jr., University of Georgia
Charles E. Stephan, U.S. EPA–Duluth
Carl L. Strojan, University of Georgia
Ward F. Whicker, Colorado State University

CONTENTS

Dedication

To my family, Peg, Ben, and Ian

Chapter 1

Introduction

We speak piously of taking measurements and making small studies that will 'add another brick to the temple of science.' Most such bricks just lie around the brickyard.[1]

I. ECOTOXICOLOGY AS A SCIENTIFIC DISCIPLINE

The [ecotoxicology] literature is both enormous and, in large part, trivial.[2]

Truhaut[3] is credited as the first to use the term *ecotoxicology* to define the "natural extension of toxicology, the science of the effects of poisons on individual organisms, to the ecological effects of pollutants."[2] Cairns and Mount[4] similarly defined ecotoxicology as "the study of the fate and effect of toxic agents in ecosystems," a phrasing that adds the study of the fate of pollutants but removes the word, science. It is puzzling to read these definitions if one accepts, as I do, that the aspirations of environmental scientists during the past 30 years have always been ecotoxicological in nature. What could be so lacking in the present body of knowledge to necessitate the implied reformation?

The impetus for defining this "new" approach seems to grow out of frustration with our continued inability to predict or, in many cases, concisely document effects at any but the lowest levels of ecological organization. Effective prediction remains an elusive goal, despite decades of sincere effort with ample funding. The frustration grows acute as the need for accurate description and prediction becomes more pressing.[5] Discomfort is invoked by statements such as Lederman's,[6] then President-Elect of the American Association for the Advancement of Science, that "understanding . . . ecological and environmental issues and providing guidance to policymakers" is one of the major tasks facing U.S. scientists today. Predictably, the ever-present banter about the relative virtues of applied versus basic science, standard versus nonstandard methods, field versus laboratory studies, and reductionist versus holistic approaches has taken on Babelian proportions. As the din increases, an attempt is at hand to regroup under a new standard, ecotoxicology.

Reformation provides the opportunity for great advances and equally great mistakes. This is particularly true in a melding of synthetic disciplines such as ecology and toxicology.[7] A series of insightful and timely papers providing much needed perspective (e.g., Cairns,[8,9] Cairns and Mount[4]) has been published. A contextual framework is also emerging for the application of ecotoxicological methods to environmental regulation and remediation (e.g., Connell,[10] Duffus,[11] Adams[12]). Still absent is a focused effort providing a scientific framework within this field. This absence of a solid scientific framework, in my opinion, has slowed progress during the past 30 years. Basic principles are left to be pondered as afterthoughts as legitimate and immediate needs for standardization or for information on the next of a seemingly endless number of new toxicants are satisfied. This opinion, seems to have been shared by Moriarty,[2] who stated in the last paragraph of his book, "I have tried to relate the problems of ecotoxicology to their ecological context. Failure so to do has led to much muddled thinking and to unreliable conclusions." More recently, Schwetz expressed a similar opinion that toxicology is "sometimes too much of an applied science. So most ideas come from other sciences."[5]

The goal of this volume is to contribute to the development of "an organization and classification of knowledge on the basis of explanatory principles"[13] for the science of ecotoxicology. The emphasis will be on detailing quantitative methods, because they lend themselves most readily to explicit formulation of conceptual models (hypotheses), falsification, and estimation of statistical confidence during the falsification process. However, it should not be forgotten while reading this volume that "the mathematical box is a beautiful way of wrapping up a problem, but it will not hold the phenomena unless they have been caught in a logical box to begin with."[1] General concepts in scientific logic will be discussed very briefly to aid in avoiding such logical errors. Further, explicit definitions fundamental to the discipline will be formulated to avoid confusion and ambiguity.

Another definition is required to distinguish the science of ecotoxicology from the impressive body of information fulfilling essential regulatory or monitoring needs. Ecotoxicology is the organization of knowledge about the fate and effects of toxic agents in ecosystems on the basis of explanatory principles. This definition is very similar to that of Jorgensen ("the science of toxic substances in the environment and their impact on the living organisms")[14] but emphasizes an ecosystem focus and several important qualities of scientific knowledge. The remainder of this chapter defines and clarifies the basic components of this definition: ecosystem, toxicant effect, toxicant fate, and the organization of knowledge based on explanatory principles.

II. TOXICANTS AND ECOSYSTEMS

Any ecosystem under study has to be delimited by arbitrary decision, but one has to remember always that the imposed boundaries are open. . . .[15]

The functional unit in ecology is the ecosystem. Many ecotoxicologists discuss only the biotic community residing in a defined area when dealing with "ecosystem" effects. However, an ecosystem includes the biotic community and its abiotic environment functioning together as a unit to direct the flow of energy and cycling of materials. The ecosystem approach embraces the concept of components functioning to maintain the system through a complex of feedback loops. Margalef[15] suggests that ecosystems are systems in which "individuals or whole organisms may be considered elements of interaction, either among themselves, or with a loosely organized environmental matrix."

It is important to remember that the ecosystem concept is an artificial construct used to frame concepts and hypotheses. It is not without limitations.[15,16] The ecosystem model should not be confused with reality despite its enormous usefulness. How closely the qualities of an operationally-defined ecosystem conform to those of the abstract one will be a function of many factors, including spatial and temporal scale, distinctness of system boundaries, and the specific qualities under study. For this reason, comparison of the qualities of an "impacted" ecosystem to those of an idealized ecosystem may be a worthwhile mental exercise, but it could not be used to conclude that an adverse effect has occurred. Appropriate comparisons would be to properties of a reference ecosystem or to the same ecosystem before to contamination.

III. TOXICANT EFFECTS IN ECOSYSTEMS

A. CLASSIFICATION BASED ON THE STRESS CONCEPT

Everybody knows what stress is and nobody knows what it is.[17]

1. Stress

Effects of toxicants on ecosystem components are often measured along a spectrum ranging from the molecular (e.g., induced proteins) to the whole-ecosystem level (e.g.,

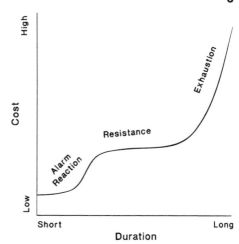

Figure 1 Cost associated with the three phases of Selyean stress.

a shift in system respiration:production or nutrient cycling). Most often, a measurement of "stress" is implied for any significant shift in a quality regardless of the ecological level examined. The measured quality may be a primary (e.g., modified level of a hormone directly influenced by the toxicant) or higher order (e.g., increased intensity of parasitic infection resulting from the generally debilitating effects of a toxicant) indicator of stress.

But what is stress? Hans Selye[17,18] developed the concept of stress and the associated general adaptation syndrome (GAS) as applied to individuals. "Stress is the state manifested by a specific syndrome which consists of all the nonspecifically induced changes within a biological system".[18] He stated quite clearly that stress is characterized by a specific syndrome in response to an external agent. It is a state achieved during any of a variety of activities, including many nondetrimental activities such as intense exertion. Further, it is a distinctive or specific suite of responses that are beneficial or compensatory. The associated general adaptation syndrome has three phases (Figure 1): the alarm reaction, adaptation (or resistance), and exhaustion components.[17] The alarm component involves immediate reactions, such as increased pulse rate and blood pressure. If the stressor continues exerting an effect on the organism, responses that "stimulate tissue-defense," such as enlargement of the adrenal cortex, occur. After a period of exposure to the stressor, the individual enters a characteristic exhaustion phase, indicating that the "finite adaptive energy" of the individual has been exhausted. With continued exposure to significant stress, the individual will be unable to maintain itself and will die. The concerted expression of specific mechanisms acts to regain or to resist deviation from homeostasis.

What is stress in the context of ecotoxicology? The rigorous definition originally given by Selye for individuals exposed to stressors has not been retained in studies of higher levels of ecological organization. The precise definition of stress depends on the level at which an effect is measured. Consequently, it is important to understand the various ecotoxicological meanings given to this concept in the literature.

Definitions vary, even at the individual level. Adams[12] compiled the following definitions of stress which focus on the level of the individual:

1. "the sum of all physiological responses that occur when animals attempt to establish or maintain homeostasis, the stressor being an environmental alteration and stress the organism's response,"
2. "adaptive physiological changes resulting from a variety of environmental stressors,"

3. "a diversion of metabolic energy from an animal's normal activities,"
4. "the sum of all the physiological responses by which an organism tries to maintain or reestablish normal metabolism in the face of chemical or physical changes,"
5. "alteration of one or more physiological variables to the point that long-term survival may be impaired," and
6. "the effect of any environmental alteration that extends homeostatic or stabilizing processes beyond their normal limits."

Examining population-level stress, including long-term, genetic consequences, the following definitions have been offered:

7. ". . . an environmental change that results in reduction of net energy balance (i.e., growth and reproduction). . . . Any reduction in production (somatic growth, reproduction or both) in response to an environmental change signifies reduced Darwinian fitness, and therefore represents a result of environmental stress,"[19]
8. ". . . an environmental condition that, when first applied, reduces Darwinian fitness; for example, reduces survivorship (S) and/or fecundity (m) and/or increases time (t) between life-cycle events,"[20]
9. ". . . anything which reduces growth or performance, it follows that, in a situation where a particular stress operates, there must be a reduction in fitness. . . . [if genotypes vary in fitness and stress is occurring consistently] evolutionary changes are to be expected,"[21]
10. "a recent anthropogenic change in the environment affecting a population's reproductive reserve or reducing its environmentally controlled abundance limit."[22]

At the community or ecosystem levels, the following definitions have been advanced:

11. ". . . . a detrimental or disorganizing influence. . . . negative responses to unusual external disturbances, or stressors of low probability to which a community of organisms is not preadapted,"[23]
12. ". . . an external force or factor, or stimulus that causes changes in the ecosystem, or causes the ecosystem to respond, or entrains ecosystematic dysfunctions that may exhibit symptoms,"[24]
13. "A stressor is any condition or situation that causes a system to mobilize its resources and increase its energy expenditure. Stress is the response of the system to the stressor. Responses to stressors may include adaptation or functional disorder,"[25]
14. ". . . a perturbation (stressor) applied to a system (a) which is foreign to that system or (b) which is natural to that system but applied at an excessive amount. . . ,"[26]

Esch and Hazen[27] provided the following definition attempting to cover all levels of ecological organization.

15. "The effect of any force which tends to extend any homeostatic or stabilizing process beyond its normal limit, at any level of biological organization."

Hoffman and Parsons,[28] although focusing on population genetics, gave a similar definition which covers all levels of organization.

16. ". . . the term "stress" [is used] to represent an environmental factor causing change in a biological system which is potentially injurious."

Careful review of these definitions suggests that "stress" is used to identify either (1) a response, (2) a characteristic or specific response, (3) an effect, or (4) an external factor causing a response or effect. In this treatment, the external factor is referred to as the stressor. The response or effect is stress. Inclusion of an effect that does not also constitute a response is contrary to the central theme of Selye's original concept. However, repeated omission of this theme in definitions necessitates the inclusion of nonresponse effects. It also necessitates establishing classes of stress that clarify its meaning when used in ecotoxicology.

Four qualities are present in these definitions regardless of the level of ecological organization. First, stress is a response to or effect of an external factor that is detrimental or disorganizing. Unlike the original concept of stress advanced by Selye, stress does not include a response to or effect of a nondetrimental factor. Second, the detrimental or disorganizing factor is atypical or present at atypical levels. The implication is that the system has not previously adjusted itself to the specific stressor to mediate its effects during predictable or highly probable exposures at a future time. Third, the system responds by or is characterized by a modification of energy flow or system structure. In the case of a response, the shift acts to maintain or reestablish some norm or homeostasis. Steady state is not an essential component of the response, although it is implied in several definitions. Fourth, temporal qualities are central to the concept of stress. Stress is a response to a recent stressor. In contrast to these four qualities common to the given definitions, the specific syndrome quality critical to Selye's stress concept has not been retained as an essential quality of stress at higher levels of organization.

With these common qualities identified, a clear, general definition of stress can be offered in the context of ecotoxicology: stress at any level of ecological organization is a response to or effect of a recent, disorganizing or detrimental factor. A stress response represents an effort to maintain or reestablish homeostasis, i.e., energy flow, material cycling, and/or system structure, to within a defined norm.

The following qualifiers are presented to differentiate between the various types of stress that can occur. The three categories of stress and three nonstress effects (Figure 2) defined here use Selye's original stress concept as a yardstick for comparison. Selye's theory is an arbitrary, yet commonly applied yardstick characterized by an enormous literature and refinement of ideas. This lends considerable justification to using Selye's concept as a touchstone.

1. Selyean stress is a specific or characteristic response to a recent disorganizing or detrimental factor. Its purpose is to maintain or reestablish homeostasis (i.e., energy flow, material cycling, and/or system structure) within a defined norm. It is not characterized by any previous adaptation to the stressor. The increase in pulse rate or blood pressure associated with physical exertion is typical of this category of stress. If the stressor continues eliciting a response for a longer period, other characteristics such as those already described for individuals exposed to a stressor may be expressed. Rapport et al.[24] described details of an ecosystem general adaption syndrome analogous to Selye's GAS. However, the many examples cited were suggestive of ecosystem effects like those described in the next category.

2. Preadaptive stress is similar to Selyean stress, but it is characterized by a previous adaptation to the stressor, i.e., the system has specific information with which to mediate the effect of the stressor. The adaptation may be recent and transient (e.g., acclimation) or long term and relatively permanent (genetic adaptation). These responses tend to be specific to the stressor. According to Rapport et al.,[24] this definition contains aspects of Selye's concept of eustress, a response to events that organisms or systems expect or anticipate. Induction of the P-450 monooxygenase system by polycyclic aromatic hydrocarbons is a response to a specific class of toxicants resulting from genetic adaptation. At the ecosystem level, preadaptive stress might be the type of response elicited by a regular or predictable stressor such as that associated with tides or seasonal fluctuations in soil moisture. To avoid confusion, it must be noted that this concept is not associated with the concept of genetic preadaptation.

3. Damage or distress is the adverse effect(s) of a stressor that is not a consequence of a system response. This category of effect is often defined as stress and, reluctantly, will be considered as stress here. At lower levels of organization, the term is usually used to denote cell, tissue, or organ damage. Rapport et al.[24] used "distress syndrome"

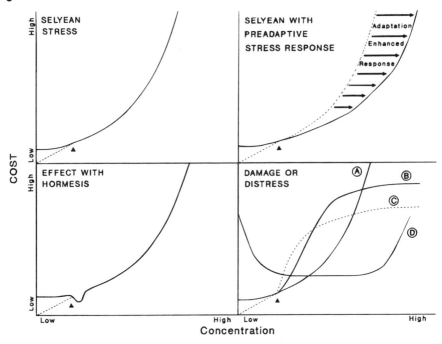

Figure 2 Examples of toxicant effects. The triangle denotes a possible threshold concentration. Below the threshold, the cost may be constant (solid line). Alternatively, the cost could increase with concentration (broken line). A simple Selyean and a Selyean effect with a superimposed preadaptation are illustrated in the top panels. An effect with hormesis is shown in the bottom left panel. Damage or distress effects can display a variety of trajectories. Examples shown here include Ⓐ zinc-induced gill damage,[53] Ⓑ DDT-induced fish mortality,[54] Ⓒ oxygen consumption by gill tissue exposed to cadmium,[55] and Ⓓ energy cost as influenced by salinity.[19] The curve shapes are products of the range of concentrations and units of effect as much as they are influenced by the type of effect.

when discussing this type of effect on ecosystems. The normal function or structure of a system is modified by a toxicant without involving an active response by the system. For example, an effect may be measured if sufficient numbers of cells are damaged in a target organ. Similarly, signs of ecosystem distress would include a reduction in species diversity, increased nutrient loss, or a shift in the productivity to biomass ratio.[24] In both cases, the effect is not an active response to the toxicant; it is an adverse consequence of intoxication.

2. Nonstress Effects

What are some other effects of toxicants in ecosystems? Several types of effects are outside the concept of stress as defined in section A.1.

1. A hormetic effect (Figure 2) is a stimulatory effect exhibited upon exposure to low levels of some toxicants or physical agents. It is not usually characterized by a toxicant-specific system response.[29] Although seemingly counterintuitive, the effect of a toxic substance at a certain level of exposure can seem beneficial. This general phenomenon is called hormesis. Southham and Ehrlich[30] defined it as "a stimulatory effect of subinhibitory concentrations of any toxic substance on any organism." Recent treatments of this topic include effects of cadmium on growth of wood ducks[31] and effects

of radiation.[32] A review of chemically induced hormesis was developed by Calabrese et al.[33]

2. A neutral effect is a measurable change that has no apparent impact (adverse or beneficial) on the system's overall qualities or probability of persistence. Although most measurable effects are likely to be positive or negative, it is illogical to reject the possibility of a neutral effect. Definition of this effect category can be particularly useful in the formulation of null hypotheses for statistical assessments of effect.

3. An ambiguous effect is a measured effect of undefined qualities relative to the degree of detriment/benefit, passivity, or preadaption. The present state of our understanding in ecotoxicology necessitates this category. Often effects measured at higher levels of organization fall into this category.

a. Balance between Beneficial and Adverse Effects

Our discussion has focused on a restricted portion of the range of toxicant concentrations. For many toxicants or physical agents, the response curve can assume a shape similar to D in the bottom right-hand corner of Figure 2. Below certain concentrations, the effect becomes increasingly detrimental. Familiar examples are associated with concepts such as Liebig's Law of the Minimum (e.g., phosphorus effects on crop yield) or Shelford's Law of Tolerance (e.g., salinity or temperature effects on marine species). An essential element can exert this pattern of effect on individuals. Odum's Push-Pull Model of Stress suggests that disordering effects at a certain level of stress can be beneficial at the ecosystem level also. Odum's discussion[34] of pulse stability in ecosystems presented such a beneficial effect in the context of a preadaptive stress. Lugo[25] used Odum's Push-Pull or Positive-Negative Effects Model to describe numerous ecosystem-level effects of stressors, including toxicant-associated effects.

3. Summary

The characteristics of effects based on the concept of stress are summarized in Table 1. Hypothetical diagrams of costs to a system with change in toxicant concentration are shown for each type of stress response in Figure 2. The responses are not necessarily exclusive of each other. For example, metallothionein induction may minimize cost at a low concentration of copper but, at a point of metallothionein saturation, "spill-over" of significant amounts of metal to other cellular fractions occurs (see Klaverkamp et al.[35]). At that point, the Selyean stress response may become increasingly important. Before metallothionein induction, a hormetic response could have occurred, and damage to kidney tissues could have occurred during exposure.

B. CLASSIFICATION OF EFFECTS BASED ON OTHER CRITERIA

Other classification systems of effects have been derived at various levels of biological organization. For example, a toxicant may be carcinogenic to an individual. At the population level, a toxicant may act as a destabilizing selection pressure. A brief discussion of some common systems follows, with each system discussed in detail in later chapters.

1. Temporal Context

Effects associated with individuals are frequently categorized in a temporal context, e.g., as "acute" or "chronic." These definitions are often used in discussions of toxicity testing methodologies. For example, an acute effect occurs immediately as a result of an intense exposure event. Casarett and Doull[36] defined acute effects as "those that occur or develop rapidly after a single administration of a substance." They defined chronic effects as "those that are manifest after an elapse of time." According to Rand and

Table 1 **Categories of ecotoxicological effect**

Category	Beneficial Response (B) Deleterious Consequence (D) Neutral Effect (N)	Specific Characteristic to Response?	Response Involves Preadaptation to Stressor	Examples
Selyean stress	B	Yes	No	Increased pulse rate and blood pressure with exertion
Preadaptive stress	B	Yes	Yes	Metallothionein induction
Hormetic response	B	No	No	Stimulation of algal population growth
Damage/distress	D	NA	NA	Thinning of eggshell due to DDT exposure; Decreased hunting efficiency of a predator due to intoxication
Neutral effect	N	NA	NA	Increase in toxicant concentration in a species with no adverse consequences
Ambiguous effect	?	?	?	

Petrocelli,[37] chronic effects "may occur when the chemical produces a deleterious effect as a result of a single exposure, but more often they are a consequence of repeated or long-term exposures." The difference between these two types of effects is a matter of degree. To illustrate this point, Casarett and Doull[36] explained that an "acute" exposure to a toxicant such as beryllium can produce an effect that will take some time to manifest itself. Finally, Suter et al.[38] briefly discussed interpretations of the term "chronic effect" to mean exposure over "greater than 10% of the organisms lifespan." He suggested that all life stages and processes must be exposed to detect chronic effects.

2. Lethality

Making the distinction between lethal and sublethal effects is also difficult.[2] Rand and Petrocelli[37] discussed death or failure to produce viable offspring in the context of lethal effects. Sublethal effects include deleterious behavioral or physiological changes. Unfortunately, it is often impossible to say whether a sublethal effect (diminution of predator avoidance behavior) would or would not result in death (lethal) for an individual within a natural ecosystem.

3. Site of Action

Most toxicological treatments (e.g., Casarett and Doull,[36] Rand and Petrocelli[37]) also distinguish between effects in the context of their sites of action. A systemic effect involves action on systems such as the central nervous, immune, or cardiovascular systems. A local effect occurs at the primary site of damage, such as a gill lesion caused by direct contact with the toxicant.

C. SUMMARY OF TOXICANT EFFECTS

Toxicant effects, including system responses, can be classified relative to the concept of stress. Six classes of effect (Selyean stress, preadaptive stress, hormetic response, damage/distress, neutral effect, and ambiguous effect) have been defined. Other classification schemes important in regulatory activities or traditional toxicology are based on time frame, lethality, or site of action criteria. Further, effects can be classified based on whether they exacerbate a preexisting condition, involve synergism with another agent, or represent an atypical reaction of a hypersensitive individual. Precise classifications of effects in the context of regulatory testing have been detailed in sources such as Sprague,[39,40] Buikema et al.,[41] Suter et al.,[38] and Weber et al.[42] The classifications that focus on lower levels of organization are discussed in Chapters 4 and 5.

What is the present state of knowledge relative to ecotoxicological effects? Figure 3 summarizes the present perception of the state of our knowledge of effects along the ecological spectrum of organization. Our ability to understand and assign causal relationships is best at the lower levels of organization and becomes precipitously poorer as the level of organization increases. The lower level effects are generally more sensitive (i.e., manifested at lower toxicant concentrations) than effects at higher levels of organiza-

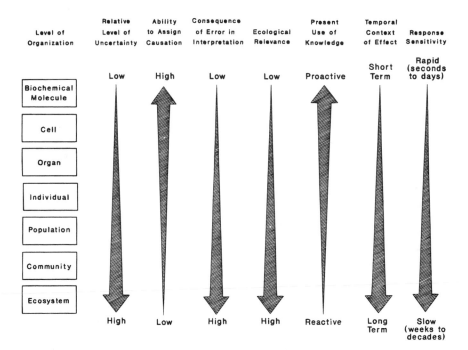

Figure 3 Features of ecotoxicological effects based on level of ecological organization. This figure is a composite derived from Figure 1 of Haux and Forlin,[56] Figure 1 of Adams et al.,[57] Figure 4 of Chapman,[43] and Figure 2 of Burton.[58]

tion. They often respond more rapidly to a toxicant. The associated advantage is that biochemical or physiological indicators can be used proactively (i.e., before irreversible or major ecological harm). When an ecosystem level effect is noted, it is used most often to document a degraded state. Such reactive documentation has no predictive value because the degradation has already occurred by the time an effect is seen. However, beyond documentation, this effect can be useful to establish a baseline for comparison as remedial actions are implemented. The responsiveness (rate at which the effect manifests itself after exposure to a toxic agent occurs) and the temporal context (duration of time that the effect will be significant after removal of the toxic agent) of an effect will be shorter and more rapid at lower levels of organization. Chapman[43] suggested that effects at lower levels tend to be more reversible than effects at higher levels of organization. At first glance, all these qualities seem to indicate that effects at lower levels of organization are superior to higher level effects as tools for managing ecosystem health. They respond quickly to change, are understood more readily, are assigned causation more easily, and are used more effectively before permanent or significant ecological degradation. Further, according to Moriarty,[2] "the immediate effects of pollutants are on individual organisms, by either direct toxicity or altering the environment. . . ."

Lower level responses are not superior overall to higher level responses, despite their virtues. Moriarty[2] continued his statement by saying that, although the immediate effects are at the individual level, ". . . the ecological significance, or lack of it, resides in the indirect impact on the populations of species." The goal of ecotoxicological stewardship is the protection of ecosystems, not biochemical moieties, physiological homeostasis, or even individuals. The probability of falsely assigning an adverse ecotoxicological effect is increased when higher level effects are neglected in favor of lower level responses.[43] Because our ability to relate lower level responses to ecosystem degradation is severely limited, it follows that the ecological relevance of a lower level response (e.g., a 50% decrease of total metal bound to metallothionein) is much more ambiguous than that of a higher level effect (e.g., a 50% decline in species richness).

IV. TOXICANT FATE IN ECOSYSTEMS

Rand and Petrocelli[37] aptly stated that toxicant fate is the "disposition of material in various environmental compartments (e.g., soil or sediment, water, air, biota) as a result of transport, transformation, and degradation." Concentration, distribution, speciation, and phase association of the toxicant in the various ecosystem components are considered, as well as toxicant sources and sinks (Figure 4). The toxicant will accumulate in both the biotic or abiotic components of the ecosystem being studied.

A. FATE IN BIOTIC COMPONENTS

In this volume, bioaccumulation is defined as the accumulation of a toxicant in or on an individual. Results of bioaccumulation studies are very often extrapolated to include population level implications. Studies may focus on taxonomic groups that accumulate relatively high concentrations of toxicants (e.g., metals in algae) or on groups more sensitive than others (e.g., DDT in nesting waterfowl). Alternatively, focus may be on assemblages physically associated with or functionally linked to components containing high concentrations of toxicant (e.g., infaunal species inhabiting contaminated sediments in the first case; scraper species grazing on metal-contaminated periphyton in the second case).

Accumulation is considered to include direct uptake from water through gills or epithelial tissues, and input from food via the gut. Usually, dietary sources include detritus and the input from grazing or predator-prey interactions. For this discussion, the concept of bioaccumulation explicitly includes accumulation of toxicants as a conse-

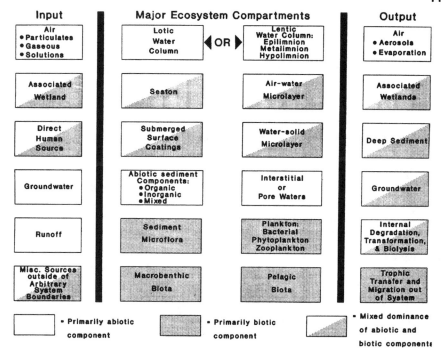

Figure 4 Compartments with significant ecotoxicological relevance to the fate of toxicants in aquatic ecosystems.

quence of parasite-host, parent-egg/embryo/juvenile, or general symbiotic interactions. Some of these exchanges may not be direct trophic interactions. Further, the accumulation of toxicants from items processed coincidentally with food items is also included as a component of bioaccumulation. Bioaccumulation is discussed in greater detail in Chapter 3.

As noted, the transfer of toxicant from one individual to another during trophic interactions is part of bioaccumulation. Biomagnification is said to occur if the concentration of toxicant increases during successive trophic transfers. This trophic transfer of toxicants is discussed in Chapter 7.

B. FATE IN ABIOTIC COMPONENTS

The intimate association of biotic and abiotic components of ecosystems is emphasized in Figure 4 using examples of significant inputs, outputs, and locations of toxicants in ecosystems. Indeed, any discussion of toxicant fate within ecosystems that ignores abiotic or biotic components will be incomplete. Not only are these components in close physical contact, they modify one another in ways to facilitate or inhibit exchange or transformation of toxicants. For example, bioturbation may enhance the movement of toxicants between sediments and overlying waters, and chemical speciation of toxicants in water influences bioaccumulation.

Physical and chemical mechanisms influence the fate of toxicants in ecosystem components and exchange between ecosystem components. Equilibrium partitioning may determine the distribution of organic pollutants. Redox reactions and dissolution/ precipitation or coprecipitation phenomena can strongly influence metal or radionuclide movement within ecosystems. Complexation, photolysis, and adsorption/desorption also can play major roles. Examples of important physical processes include sedimentation, atmospheric fallout, bulk movement of water, and bottom scouring.

V. ORGANIZATION OF KNOWLEDGE BASED ON EXPLANATORY PRINCIPLES

While ecologists have sought to elucidate the nature of these 'parallel effects' of [human-induced] stress on ecosystems, they have perhaps more often followed what Dyson identified as the predominant methodology of the biological sciences: a preoccupation with description of the diversity of phenomena, by and large to the exclusion of consideration of unifying themes.[24]

It has also been argued that ecology is largely ideographic (explains particulars) rather than nomothetic (derives universal laws). For a certain class of questions about ecosystems it is true that only particular explanations are needed (e.g., grass dominates this system because of frequent fires, a historical event) but the same is true in physics when known laws are applied to a given situation. This does not preclude the existence of laws.[44]

Well, there are two kinds of biologists, those who are looking to see if there is one thing that can be understood, and those who keep saying it is very complicated and that nothing can be understood . . . You must study the simplest system you think has the properties you are interested in.—Cy Levinthal[1]

A. INTRODUCTION

The ecotoxicological literature is replete with statements that a working knowledge of ecosystems is impossible because of ecosystem complexity, yet such statements are inconsistent with the rapid growth of knowledge in equally complex disciplines. A casual glance at the intricate charts of metabolic pathways should immediately cause such statements to be questioned. The complex of metabolic networks functioning within each individual is staggering but, taken *en masse,* our working knowledge of such systems is impressive. One can gain a similar impression from browsing through the molecular genetics literature.

What then is the reason for our limited knowledge base in ecotoxicology? I believe that a significant impediment to the growth of ecotoxicological knowledge lies in the approach by which such knowledge is sought, not in the complexity of the systems being studied. Although some approaches are adopted out of necessity, others are selected out of habit. Some have been learned from a parent discipline. Two significant examples are the predisposition for description in ecology (see above quotes from Loehle[44] and Rapport et al.[24]) and the propensity for single-species, endpoint toxicity tests transplanted from traditional toxicology. Other habits are more pervasive in life sciences (see quote from Levinthal[1]). Regardless of their origins, all habits are not equally fruitful; consequently, they should be subject to review, modification, replacement, or rejection. Based on this premise, the remainder of this chapter examines scientific habits worth considering as the field of ecotoxicology matures.

B. STRUCTURE OF SCIENTIFIC KNOWLEDGE
1. Historical Perspective

Chamberlin[45] suggested that the format of scientific inquiry has changed throughout history. Initially, knowledge was so limited that it seemed to be within the abilities of learned individuals to develop ruling theories that explained all phenomena. No sooner was a phenomenon presented than a ruling theory was used to explain it. This process built a large knowledge structure by repeated application alone. Chamberlin referred to this habit as precipitate explanation, the immediate and sufficient application of a theory to explain an observation. The ruling theory approach slowly was replaced by the working hypothesis approach. "Under the ruling theory, the stimulus is directed to the finding of facts for the support of the theory. Under the working hypothesis, the facts are sought for the purpose of ultimate induction and demonstration, the hypothesis being but a means for the more ready development of facts and their relations."[45] The working hypothesis approach questions theory but still retains a propensity toward precipitate explanation. When a theory or hypothesis is presented, it tends to be given favored status in testing. Chamberlin hypothesized that, although its roots are no longer considered valid, the habit of precipitate explanation continued into modern scientific inquiry. He suggested using the method of multiple hypotheses to minimize such bias. The method of multiple working hypotheses considers all potential hypotheses simultaneously. Equal amounts of effort are spent on the hypotheses; and, in this manner, the tendency to favor one hypothesis unintentionally is lessened.

Karl Popper[46,47] argued that scientific inquiry should test hypotheses by a formal process of falsification. No theory can be proven true, but it can be shown to be false through observation or experimental challenge. Repeated survival of a theory or hypothesis through a rigorous falsification process confers a favored status on it, and its strength is enhanced if it continues to be "corroborated by past experience." Regardless, it is never deemed true. Acknowledging that no scientist is totally objective, Popper referred to the testability of a hypothesis as "subject to inter-subjective testing." A good theory is one that can be tested by scientists, each of whom approaches it with his or her own prejudices. A process which Popper likened to natural selection occurs; the fittest theory survives.

The empirical basis of objective science has thus nothing 'absolute' about it. Science does not rest upon solid bedrock. The bold structure of its theory rises, as it were, above a swamp. It is like a building erected on piles. The piles are driven down from above into the swamp, but not down to any natural or 'given' base; and if we stop driving the piles deeper, it is not because we have reached firm ground. We simply stop when we are satisfied that the piles are firm enough to carry structure, at least for the time being.[47]

Survival of intersubjective testing alone does not constitute corroboration and consequent enhanced status for a theory. The testing must be rigorous. Some tests have greater powers to falsify based on logic alone. A classic illustration of this point involves Einstein's theory of relativity. The explanation of eccentricities in planetary orbits was less powerful in supporting Einstein's theory than the observation that light was attracted by gravity because it had more alternative explanations than the latter observation (see Popper,[46] pages 35–36, for more details of this example). The theory of relativity had a higher risk of rejection during the second test. Rousseau[48] suggested that avoidance of high-risk testing is one of three symptoms of "pathological science" (science without objectivity). A tradition of low-risk testing of theories in any field has an associated danger because "our habit of believing in laws is a product of frequent repetition."[46] The ability to separate dogma from paradigm becomes impaired if low-risk testing is common in a discipline.

Regardless of the logical power of a test, a test with insufficient measurement precision or accuracy is valueless in the process of falsification (see discussion of condensation bounds in Popper,[47] pages 123–127). It may even slow progress due to the confusing or ambiguous nature of associated conclusions. For this reason, Popper[47] advanced the opinion of "superiority of methods that employ measurements over purely qualitative methods," providing the impetus for the emphasis placed on quantitative methods in this book.

This brief discussion of the history of scientific inquiry will end here. Cessation does not suggest that Popper provided the capstone for scientific logic. Bayesian theory contests and extends many associated topics pertinent to our discipline. The interested reader is referred to Howson and Urbach[49] for an excellent presentation of this approach. Regardless, this discussion has identified three habits that should be avoided in ecotoxicology: precipitate explanation, low-risk hypothesis testing, and imprecise or inaccurate measurement.

2. Strong Inference

... strong inference is just the simple and old-fashioned method of inductive inference that goes back to Francis Bacon. ... The difference comes in their systematic application.[1]

As mentioned in the preface, Platt[1] observed that scientific disciplines progress at very different rates as a consequence of the general approach to extracting and organizing knowledge. Some lack a tradition of strong inference, e.g., rigorous hypothesis formulation and falsification procedures. Strong inference includes the well-known steps of hypothesis formulation, execution of experiments designed to falsify, alternate hypothesis generation, and continued testing until one hypothesis remains corroborated. However, Platt suggested that the distinction between disciplines lies in the value placed on rigorous and consistent application of such techniques.

Some fields focus too much on "surveys, taxonomic, design of equipment, systematic measurements and tables, theoretical calculations—all [of which] have their proper and honored place, provided they are parts of a chain of precise induction of how nature works."[1] He argued that such preoccupation is taught by example to students; it is not inherent to the field.

Some disciplines are characterized by a focused effort of falsification of hypotheses. Others show a meandering tendency toward precipitate explanation. Experiments are unintentionally designed to support favored hypotheses. He advocated using formal experimental inference methods ("the scientific method"), coupled with the method of multiple hypotheses, to minimize this problem.

He recommended the following practices to foster strong inference in a discipline.

1. Apply methods of inductive inference consistently and systematically.
2. Formulate hypotheses such that they are amenable to falsification. Use the "logic of exclusion" when possible.
3. State all reasonable alternate explanations of observations when presenting results.
4. Use the method of multiple hypotheses.
5. "... be explicit and formal and regular about it, to devote a half hour or an hour to analytical thinking every day, writing out a logical tree and the alternatives and crucial experiments explicitly in a permanent notebook."
6. After hearing a scientific explanation, ask two questions. Is there an experiment that could disprove it? What explanation does the present explanation exclude?

3. Selection of Hypotheses

Selection of hypotheses or theories is a subjective process. The Bayesian treatment presented by Howson and Urbach[49] supports this conclusion using the familiar "all ravens are black" argument. They explain that, if approached objectively, the possible number of theories to be tested equals the number of ravens in existence (n) raised to the number of possible color patterns (m) or n^m. Unless some level of subjective experience is used to select profitable hypotheses for testing, the falsification process would become impossibly cumbersome.

How are hypotheses selected? Several criteria have been advanced. Popper favors hypotheses that are easily falsifiable. Loehle[50] suggested that an optimum region (Medawar zone) exists relative to the tractability of the hypothesis and payoff for solving the problem. Loehle[44] recommended that ecologists be concerned with theory reduction during the process of selection. Theory reduction strives to explain one theory on the basis of another. This enhances parsimony and gives "two levels of explanation for the same phenomena." According to Loehle, topics that are particularly amenable to theory reduction are ecological diversity and succession. Linked also to parsimony is the application of Occam's Razor to selection of theories or hypotheses. Popper[47] favored simple hypotheses as ". . . they tell us more; because their empirical content is greater and because they are better testable." A strong, quantitative argument favoring parsimonious hypotheses can also be developed based on Bayesian theory.[51]

Chamberlin[45] advocated a process in which a series of likely hypotheses are considered equally and simultaneously. He observed that the movement from ruling theory to working hypotheses still remains biased toward a central hypothesis. The multiple working hypothesis approach remains subjective but lessens the tendency for the hypothesis to become "the controlling idea." "In developing the multiple hypotheses, the effort is to bring up into view every rational explanation for the phenomenon in hand and to develop every tenable hypothesis relative to its nature, cause or origin, and give all of these as impartially as possible a working form and a due place in the investigation. The investigator thus becomes the parent of a family of hypotheses; and by his parental relations to all is morally forbidden to fasten his affections unduly upon any one." Admittedly, phrases such as "rational explanation" and "tenable hypothesis" are permeated with subjectivity. Regardless, bias is lessened in the process of falsification, and a habit of thoroughly considering all hypotheses is fostered. The multiple working hypotheses approach has one additional advantage. It lessens the tendency to stop inquiry when a single "cause" is found. It increases the probability of detecting multiple or complex causes by evenly distributing effort between a set of hypothetical "causes". The approach fosters thoroughness as well as lessening bias.

VI. TOWARD STRONG INFERENCE AND CLEAR ECOLOGICAL RELEVANCE

In this chapter, the development of ecotoxicology as a science is strongly emphasized. The discipline is characterized by a strong need for prediction at all levels of organization, but it generally lacks sufficient knowledge for making such predictions, especially at higher levels of organization. This condition is unfortunate, as ecological relevance is highest for effects at the higher levels of organization.

Increased complexity was rejected as the explanation for our lack of understanding at higher levels. Instead, the opinion is forwarded that the approaches used in ecotoxicology especially at higher levels of organization, do not foster rapid growth of knowledge (Figure 5). A strong inferential approach is advocated to alleviate some of this difficulty. Strong inferential methods are applicable to all levels of organization, although powerful techniques such as random-assignment experiments or even quasi-experiments are logistically easier at lower levels. Criticisms based on the relative values of field versus laboratory, holistic versus reductionist, or standard versus nonstandard approaches become absurd if strong inference is an integral theme in all approaches. The decreased tractability of study at higher levels of organization or in the field must be counterbalanced by increased efforts and intensified inferential structure. Some difficulties can be offset to a degree by using the microcosm/mesocosm or large research team approaches such as those taken by Likens and co-workers.[52] Contrary to common belief, stronger inferential techniques are most valuable at higher levels of organization at which ecological rele-

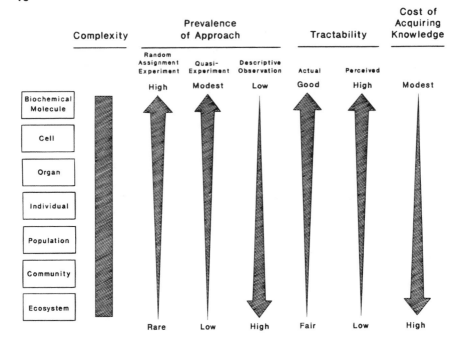

Figure 5 Elements of ecotoxicological knowledge and predominant approaches used to increase knowledge at each level of organization. This figure, used with Figure 3, qualitatively defines our present state of ecotoxicological inquiry. General approaches listed here include experiments with random assignment of experimental units to treatments, quasi-experiments (treatments, some measure of outcome, and experimental units present but not randomly assigned to treatments), and descriptive observation studies (see Cook and Campbell[59] for more details). All are valuable approaches but the values given to them relative to implying causal relationships are random assignment experiments > quasi-experiments > descriptive observation. The decrease in tractibility (ability to extract knowledge) with increasing level of organization is not a consequence of system complexity. It is a consequence of the predominant approaches customarily taken and differences in cost of inquiry at the various levels of organization.

vance is high but costs can be prohibitive. An analogy would be the intense effort invested in planning crucial, decisive experiments employing limited beam time on particle accelerators.

Strong inference is advocated as the most effective means of obtaining useful knowledge in the emerging and socially obligated field of ecotoxicology. It is characterized by the systematic and consistent application of standard methods of inductive inference. It includes the concept of multiple hypotheses as a means of minimizing subjectivity in selection of hypotheses. Methods for selecting hypotheses should be influenced by amenability to falsification and quantitative formulation, and parsimony including theory reduction. To be most effective, the measurement process must be demonstrably precise and unbiased. Hypotheses should be readily amenable to statistical analysis such that rejection criteria are defined in terms of probability.

Based on the materials discussed, 12 habits are suggested to enhance the rate at which ecotoxicological knowledge is acquired.

1. Be aware of and avoid the habit of precipitate explanation.
2. While recognizing the value of such approaches, avoid an excessive preoccupation

with "surveys, taxonomic, design of equipment, systematic measurements and tables, [and] theoretical calculation."

3. Systematically and consistently apply inductive inference techniques.
4. Favor experiments or observations with high risk of logical falsification.
5. Favor quantitative methods under rigorous precision/accuracy control.
6. Preferentially formulate hypotheses amenable to falsification.
7. Apply the principle of multiple hypotheses.
8. Favor hypotheses that are easily falsifiable.
9. Favor hypotheses that are tractable and have good probability of solving the problem.
10. Favor parsimonious hypotheses.
11. Favor hypotheses that enhance theory reduction.
12. Recognize negative results as a critically important component of the falsification process, not a consequence of the worker's failure to pick "the right question."

REFERENCES

1. Platt, J. R. "Strong Inference." *Science* 146:347–353 (1964).
2. Moriarty, F. *Ecotoxicology: The Study of Pollutants in Ecosystems.* (Orlando, FL: Academic Press, Inc., 1983).
3. Truhaut, R. "Ecotoxicology: Objectives, Principles and Perspectives." *Ecotoxicol. Environ. Saf.* 1:151–173 (1977).
4. Cairns, J., Jr., and D. I. Mount. "Aquatic Toxicology, Part 2 of a Four-Part Series." *Environ. Sci. & Technol.* 24(2):154–161 (1990).
5. Clemmitt, M. "Public, Private Health Concerns Spur Rapid Progress in Toxicology." *The Scientist* 6(4):1–7 (1992).
6. Lederman, L. M. *Science: The End of the Frontier?* A supplement to *Science:* (Washington, DC: AAAS, January 1991), 20.
7. Maciorowski, A. F. "Populations and Communities: Linking Toxicology and Ecology in a New Synthesis." *Environ. Toxicol. Chem.* 7:677–678 (1988).
8. Cairns, J., Jr. "Will the Real Ecotoxicologist Please Stand Up?" *Environ. Toxicol. Chem.* 8:843–844 (1989).
9. Cairns, J., Jr. "Restoration Ecology: A Major Opportunity for Ecotoxicologists." *Environ. Toxicol. Chem.* 10:429–432 (1991).
10. Connell, D. W. "Ecotoxicology—A Framework for Investigations of Hazardous Chemicals in the Environment." *Ambio* 16(1):47–50 (1987).
11. Duffus, J. H. "Interpretation of Ecotoxicity." In *Toxic Hazard Assessment of Chemicals.* M. Richardson, ed. (London: The Royal Society of Chemistry, Burlington House, 1986) 98–116.
12. Adams, S. M., ed. "Biological Indicators of Stress in Fish." *Am. Fish. Soc. Symp.* 8:1–8 (1990).
13. Nagel, E. *The Structure of Science. Problems in the Logic of Scientific Explanation.* (New York: Harcourt, Brace and World Inc., 1961).
14. Jørgensen, S. E. *Modelling in Ecotoxicology.* (New York: Elsevier, 1990) 11.
15. Margalef, R. *Perspectives in Ecological Theory.* (Chicago: The University of Chicago Press, 1968) 13.
16. Guttman, B. S. "Is 'Levels of Organization' a Useful Biological Concept?" *Bioscience* 26(2):112–113 (1976).
17. Selye, H. "The Evolution of the Stress Concept." *Am. Sci.* 61:692–699 (1973).
18. Selye, H. *The Stress of Life.* (New York: McGraw-Hill Book Company, 1956) 54.
19. Koehn, R. K., and B. L. Bayne. "Towards a Physiological and Genetical Understanding of the Energetics of the Stress Response." *Biol. J. Linn. Soc.* 37:157–171 (1989).

20. Sibly, R. M., and P. Calow. "A Life-Cycle Theory of Responses to Stress." *Biol. J. Linn. Soc.* 37:101–116 (1989).

21. Bradshaw, A. D., and K. Hardwick. "Evolution and Stress—Genotypic and Phenotypic Components." *Biol. J. Linn. Soc.* 37:137–155 (1989).

22. Shuter, B. J. "Population-Level Indicators of Stress." *Am. Fish. Soc. Symp.* 8:145–166 (1990).

23. Odum, E. P. "Trends Expected in Stressed Ecosystems." *Bioscience* 35:419–422 (1985).

24. Rapport, D. J., H. A. Regier, and T. C. Hutchinson. "Ecosystem Behavior Under Stress." *Am. Nat.* 125(5):617–640 (1985).

25. Lugo, A. E. "Stress and Ecosystems," in *Energy and Environmental Stress in Aquatic Systems,* eds. J. H. Thorp. and J. W. Gibbons. (Springfield, VA: National Technical Information Service CONF-771114, 1978) 62–101.

26. Barrett, G. W., G. M. Van Dyne, and E. P. Odum. "Stress Ecology." *Bioscience* 26(3):192–194 (1976).

27. Esch, G. W., and T. C. Hazen. "Thermal Ecology and Stress: A Case History for Red-Sore Disease in Largemouth Bass," In *Energy and Environmental Stress in Aquatic Systems,* ed. J. H. Thorp, and J. W. Gibbons, (Springfield, VA: National Technical Information Service CONF-771114, 1978) 331–363.

28. Hoffman, A. A., and P. A. Parsons. *Evolutionary Genetics and Environmental Stress.* (Oxford: Oxford University Press, 1991).

29. Stebbing, A. R. D. "Hormesis—The Stimulation of Growth by Low Levels of Inhibitors." *Sci. Total Environ.* 22:213–234 (1982).

30. Southam, C. M. and Ehrlich, J. "Effects of Extract of Western Red Cedar Heartwood on Certain Wood-Decaying Fungi in Culture." *Phytopathology* 33:517–524 (1943).

31. Brisbin, I. L., Jr., K. W. McLeod, and G. C. White. "Sigmoid Growth and the Assessment of Hormesis: A Case for Caution." *Health Phys.* 52(5):553–559 (1987).

32. Sagan, L. A. "What Is Hormesis and Why Haven't We Heard About It Before?" *Health Phys.* 52(5):521–525 (1987).

33. Calabrese, E. J., M. E. McCarthy, and E. Kenyon. "The Occurrence of Chemically Induced Hormesis." *Health Phys.* 52(5):531–541 (1987).

34. Odum, E. P. "The Strategy of Ecosystem Development." *Science* 162:262–270 (1969).

35. Klaverkamp, J. F., M. D. Dutton, H. S. Majewski, R. V. Hunt, and L. J. Wesson. "Evaluating the Effectiveness of Metal Pollution Controls in a Smelter by Using Metallothionein and Other Biochemical Responses in Fish," In *Metal Ecotoxicology Concepts and Applications,* ed. M. C. Newman and A. W. McIntosh. (Chelsea, MI: Lewis Publishers, 1991) 33–64.

36. Casarett, L. J., and J. Doull. *Toxicology: The Basic Science of Poisons.* (New York: Macmillan Publishing Co., Inc., 1975).

37. Rand, G. M., and S. R. Petrocelli. *Fundamentals of Aquatic Toxicology.* (New York: Hemisphere Publishing Corporation, 1985).

38. Suter, G. W., A. E. Rosen, E. Linder, and D. F. Parkhurst. "Endpoints for Responses of Fish to Chronic Toxic Exposures." *Environ. Toxicol. Chem.* 6:793–809 (1987).

39. Sprague, J. B. "Measurement of Pollutant Toxicity to Fish, I Bioassay Methods for Acute Toxicity." *Water Res.* 3:793–821 (1969).

40. Sprague, J. B. "Measurement of Pollutant Toxicity to Fish. III. Sublethal Effects and 'Safe' Concentrations." *Water Res.* 5:245–266 (1971).

41. Buikema, A. L., Jr., B. R. Niederlehner, and J. Cairns, Jr. "Biological Monitoring Part IV—Toxicity Testing." *Water Res.* 16:239–262 (1982).

42. Weber, C. I., W. H. Peltier, T. J. Norberg-King, W. B. Horning, II, F. A. Kessler, J. R. Menkedick, T. W. Neiheisel, P. A. Lewis, D. J. Klemm, Q. H. Pickering, E. L. Robinson, J. M. Lazorchak, L. J. Wymer, and R. W. Freyberg. *Short-term Methods*

for Estimating the Chronic Toxicity of Effluents and Receiving Waters to Freshwater Organisms, 2nd ed, EPA/600/4–89/001. (Cincinnati, OH: EMSL U.S. Environmental Protection Agency, 1989).

43. Chapman, P. M. "Environmental Quality Criteria. What Type Should We Be Developing?" *Environ. Sci. & Technol.* 25(8):1353–1359 (1991).
44. Loehle, C. "Philosophical Tools: Potential Contributions to Ecology." *Oikos* 51:97–104 (1988).
45. Chamberlin, T. C. "The Method of Multiple Working Hypotheses," *J. Geol.* 5:837–848 (1897).
46. Popper, K. R. *Conjectures and Refutations. The Growth of Scientific Knowledge.* (New York: Harper and Row, 1965).
47. Popper, K. R. *The Logic of Scientific Discovery.* (London: Hutchinson and Company, 1968).
48. Rousseau, D. L. "Case Studies in Pathological Science." *Am. Sci.* 80:54–63 (1992).
49. Howson, C., and P. Urbach. *Scientific Reasoning. The Bayesian Approach.* (La Salle, IL: Open Court Publishing Company, 1989).
50. Loehle, C. "A Guide to Increased Creativity in Research—Inspiration or Perspiration?" *Bioscience* 40 (2):123–129 (1990).
51. Jeffreys, W. H., and J. O. Berger. "Ockham's Razor and Bayesian Analysis." *Am. Sci.* 80:64–72 (1992).
52. Likens, G. E., F. H. Bormann, R. S. Pierce, J. S. Eaton, and N. M. Johnson. *Biogeochemistry of a Forested Ecosystem.* (New York: Springer-Verlag, 1977).
53. Hodson, P. V. *The Effect of Temperature on the Toxicity of Zinc to Fish of the Genus Salmo.* Ph.D. dissertation, University of Guelph, Ontario, 1974.
54. King, S. F. *Some Effects of DDT on the Guppy and the Brown Trout.* U.S. Fish and Wildlife Service Scientific Report 399. (Washington, DC: US Fish and Wildlife Service, 1962).
55. Dawson, M. A., E. Gould, F. P. Thurberg, and A. Calabrese. "Physiological Response of Juvenile Striped Bass, *Morone saxatilis,* to Low Levels of Cadmium and Mercury." *Chesapeake Sci.* 18(4):353–359 (1977).
56. Haux, C., and L. Förlin. "Biochemical Methods for Detecting Effects of Contaminants on Fish." *Ambio* 17(6):376–380 (1988).
57. Adams, S. M., K. L. Shepard, M. S. Greeley, Jr., B. D. Jimenez, M. G. Ryon, L. R. Shugart, and J. F. McCarthy. "The Use of Bioindicators for Assessing the Effects of Pollutant Stress on Fish." *Mar. Environ. Res.* 28:459–464 (1989).
58. Burton, G. A., Jr. "Assessing the Toxicity of Freshwater Sediments." *Environ. Toxicol. Chem.* 10:1585–1627 (1991).
59. Cook, T. D., and D. T. Campbell. *Quasi-Experimentation Design and Analysis Issues for Field Settings.* (Boston: Houghton Mifflin Company, 1979).

Chapter 2

The Measurement Process

Boswell: "Sir Alexander Dick tells me, that he remembers having a thousand people a year to dine at his house; that is reckoning each person as one, each time he dines there." Johnson: "That is about three a day." Boswell: "How your statement lessens the idea." Johnson: "That, Sir, is the good of counting. It brings everything to a certainty, which before floated in the mind indefinitely." Boswell: "But Omne ignotum pro magnifico est: one is sorry to have this diminished." Johnson: "Sir, you should not allow yourself to be delighted with error."
Boswell and Glover[1]

I. GENERAL

A. OVERVIEW

Rousseau[2] defined three symptoms of "pathological science" (the excessive loss of scientific objectivity). The first symptom is an aversion to crucial experiments that could disprove a favored theory. The second is a disregard for prevailing ideas and theories. Traditional theories are given inadequate consideration as the researcher becomes more and more enamored with a new discovery. These first two symptoms should seem familiar to the reader because they were discussed in Chapter 1 as low-risk testing and precipitate explanation. The third symptom was mentioned only briefly in that chapter. Rousseau suggested that the last symptom of pathological science often begins with an effect that is "at the limits of detectability or has very low statistical significance . . . Once the investigator [is] convinced that something new and important has been discovered, the fact that all of the parameters involved . . . are not under control is viewed as having little consequence . . . "[2] The improperly controlled or poorly understood measurement process acts as the seed from which increasingly biased behavior grows. Fortunately, the means of minimizing such misinterpretations of the results of most measurement processes are available. Several methods for controlling measurement difficulties in or defining the limitations of the measurement process are outlined here at the beginning of this book because, in their absence, later methods would be useless.

Although sometimes believed to be pertinent primarily to chemical analyses, the techniques described in this chapter for assessing such qualities as accuracy and precision are amenable to and necessary for any measurement process. For example, if a plankton net is towed, the measured numbers of individuals of each species from that tow have associated limits of detection. (This point will be discussed again in Chapter 7 regarding species abundance.) Further, questions regarding precision and accuracy of measurements must be answered before any meaningful data analysis. In this example, precision may be quantified with replicate tows of the same net or one tow of two identical nets coupled by a common yoke. Accuracy during enumeration could involve species "spikes" to a portion or aliquot of a sample.

B. NECESSITY OF CONTROLLED MEASUREMENT

It is difficult to find a sound reason why this crucial aspect of ecotoxicology had been given such low priority for many years. Although formal methods for implementing quality control have been available since the 1930s,[3] quality control was practiced informally with varying degrees of commitment until environmental regulations mandated otherwise. It remains underemphasized in college course work outside of statistics, applied chemistry, and engineering. Indeed, programs implemented to insure controlled measurement still elicit extreme responses. "Some of the scientists seem to consider quality assurance an insult to their professional integrity, an obstruction to 'real work' and an inference that scientists will cheat or falsify data or results."[4] Such an immoderate attitude may be remanent of concepts abandoned early in the development of science (see Chapter 1, Section V.B., or Chamberlin[5]). Regardless, intersubjective testing as practiced today requires clear documentation of measurement conditions.[6] Failure to do so inhibits progress and improvement of skills,[7,8] decreases the effectiveness of the decision-making process,[9,10] fosters self-delusion,[2] and inhibits our ability to detect the rare occasion of fraud.[11,12]

II. REGIONS OF QUANTITATION

A. OVERVIEW

Keith et al.[9] and Taylor[7] have provided lucid explanations of quantitation regions; consequently, their presentations have been condensed into this overview with only minor modifications. The interested reader should examine the original materials for more detail.

The certainty of a measured value can be gauged relative to the standard deviation (s_0) for samples with concentrations near 0, i.e., signals near the baseline noise of the measurement process. The relative uncertainty is often used for this purpose.[7]

$$\text{Relative Uncertainty (\%)} = 100 \, \frac{z\sqrt{2}}{N} \qquad (1)$$

where $z = z$ statistic at a confidence level of $100(1 - \alpha)$ %, and N = measured value of an observation expressed as a multiple of s_0.

For example, the relative uncertainty of a value three times larger than s_0 at a 95% confidence level is $100 \, [\, (1.960\sqrt{2})/3]$ or $\pm 92\%$. Such an observation has a measurement uncertainty nearly as large as its mean value. A more acceptable measurement uncertainty might be that associated with a $10s_0$ value, i.e., $100[(1.960\sqrt{2})/10]$ or $\pm 28\%$.

Estimation of the information required to calculate the relative uncertainty (mean value and s_0) can be accomplished in several ways. The sample signal (S_t) may be used to estimate the mean value if there is no significant blank signal (S_b). If the blank signal is measurable, then the difference between the sample and blank signals ($S_t - S_b$) may be used. Next, one of two methods can be used to estimate the standard deviation of the measurement process (s_0). As recommended by Taylor,[7] s_0 can be estimated by plotting the standard deviation of signal versus concentration for a series of replicate standards (or samples) including one set close to a concentration of 0. The Y-intercept of the plot is then used to estimate s_0. A minimum of three concentration levels with a total of at least seven measured signals is recommended for estimating s_0. Although this linear extrapolation method is preferred, s_0 can also be estimated with signals from replicates at one concentration near the limit of quantitation (see below). Taylor[7] recommends using a concentration close to $20s_0$ in this single-concentration approach.

Using the information generated, each measurement can now be assigned to a region of quantitation (see Figure 1). Again, the measured values are expressed as multiples of s_0. At $3s_0$ above the baseline signal (the limit of detection, LOD), the measured value is estimated to be within approximately $\pm 100\%$ of the "true" value at a 95% confidence level. Values less than the LOD are often reported as "below the detection limit" (<DL) or "not detected" (ND). Above the limit of detection but below $10s_0$ is the region of less-certain quantitation (also called the region of qualitative analysis by Currie).[13] In this region, a measurement is detectable but associated with a sufficiently large uncertainty (greater than approximately $\pm 30\%$) to render it semiquantitative. Above $10s_0$ (the

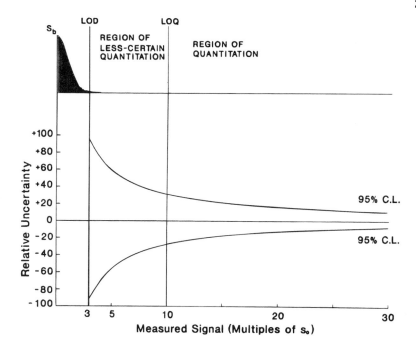

Figure 1 Regions of quantitation (modified from Keith et al.[9] and Taylor[7]).

limit of quantitation, LOQ) but below the point at which the linearity of the standard curve ends (limit of linearity, LOL) is the region of quantitation.

A major goal in design of quantitative procedures should be the generation of a data set with all observation values within the region of quantitation. Indeed, recommendations have been forwarded that only data in the region of quantitation should be used to make quantitative decisions.[9] Unfortunately, constraints such as available instrumentation, regulation-mandated methodologies, temporal variation in concentrations, and incorporation of control treatments or uncontaminated sites often produce data sets with observations in all or several measurement regions.[14]

B. DATA SETS WITH BELOW DETECTION LIMIT OBSERVATIONS
1. Definitions

Extending the above discussion, a straightforward definition of the limit of detection can be stated. The limit of detection is defined as $3s_0$ above the baseline signal. Below the limit of detection, the uncertainty of a measurement is approximately equal to or greater than the value itself. Note that specific applications may necessitate modification of this general definition, e.g., radiological methods assuming a Poisson distribution for signal noise. Estimation for nonlinear calibration methods,[15] comparison of instruments,[16] estimation using chromatographic techniques,[17] measurement of radioactivity,[13,18] and measurement of mixture analytes such as Aroclor 1242[19] are a few examples that might require some refinement of the above definition.

Further, several categories of limits of detection can be defined depending on the intended use of the resulting statistic. An instrument detection limit (IDL) measures the signal-to-noise ratio associated with the measurement equipment. The method detection limit (MDL) includes variation in all measurement steps leading to estimation of s_0. For example, the mass of sample used, extraction efficiency, and other procedural steps contributing to signal variation may be incorporated during estimation of a MDL.

2. Reporting

The limit of reporting (LR) of data may be the LOD or even the LOQ.[9,20] Most often, data sets are reported with all observations assigned numerical values except those below the LOD. Observations with values less than the LOD are noted as "below the LOD" (BDL, <DL, <LR or ND). The laudable intent of this practice is to report only values that are above the noise of the measurement process by a statistically defined amount. The resulting data set, which contains a subset of observations with no assigned numerical value below a certain point, is said to be censored. The data can be defined further as left-censored since censoring at the LOD omits low values from the left portion of the sample distribution.

Left censoring of data sets creates many problems. Censoring precludes use of valuable information by the decision maker. Fortunately, arguments against censoring are increasing in frequency.[21-25] According to these authors, it would be more effective to report values for all observations along with the associated measurement uncertainty. For example, the relative uncertainty could be estimated using Equation (1) if s_0 were reported. Then end users could examine and manipulate the data set as appropriate for their particular needs. For example, a rank order test may extract valuable information from a data set with a moderate proportion of observations below the LOD or LOQ.

3. Estimating Mean and Standard Deviation for Censored Data

a. Deletion and Substitution Methods

Regardless of sound arguments against censoring, left censoring of data sets remains a common practice. The most commonly used methods with censored data sets are deletion or substitution techniques.[25] In the deletion procedure, values below the reporting limit (LR) are not used to estimate mean and standard deviation. In the substitution procedure, some value such as 0, $\frac{1}{2}$LOD, or the LOD is substituted for all values below the LR when the mean and standard deviation are calculated. Although substitution of $\frac{1}{2}$LOD for a few LOD observations in a large data set may not cause major problems, the general application of these statistically unsubstantiated techniques can bias estimates of the mean and standard deviation. For example, deletion and substitution methods will produce biased estimates of the mean. In Figure 2A, two distributions with identical means but dissimilar standard deviations are used to illustrate this bias. The estimate of the mean for the distribution with the wider standard deviation will be biased upward more than that of the distribution with the narrower standard deviation if observations below the LOD are deleted, or if the LOD values are substituted for "<DL observations." If 0 or $\frac{1}{2}$LOD are substituted for the censored values, the mean estimated for the broader distribution will be smaller than that of the narrow distribution.

Similarly, standard deviations estimated with deletion or substitution methods are biased (Figure 2B). For example, deletion of or substitution of the LOD for censored values will bias the standard deviation for the distribution with the smaller mean more than that with the larger mean. Substitution of 0 will have the opposite effect.

b. Winsorized Mean and Standard Deviation

Gilbert[23] suggests that Winsorized estimates of mean and standard deviation can be used for censored data sets. The assumption is made that the data are distributed symmetrically. The "<DL" values are replaced by the smallest observation value above the LOD; however, the same number of largest values are replaced by the value of the next smallest observation. The mean and standard deviation are calculated for this modified data set. The simple, arithmetic mean for the modified data (Winsorized mean) is unbiased [Equation (2)]. However, Equation (3) is needed to provide an unbiased estimate of the Winsorized standard deviation.

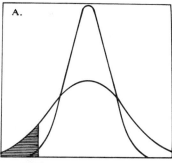

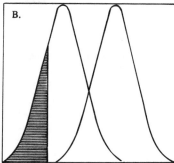

Figure 2 Problems associated with deletion or substitution techniques used to estimate means and standard deviations in left-censored data sets. Panel A shows two normal distributions with identical means but different standard deviations. Panel B shows two normal distributions with different means but identical standard deviations. The points censored (below the LOD) are within the shaded area to the left of each set of distributions. From Newman and Dixon[47] and reprinted from *American Environmental Laboratory,* Volume 2, page 27, (Copyright 1990 by International Scientific Communications Inc.).

$$\bar{x}_w = \frac{\sum_{i=1}^{n} x_i}{n} \tag{2}$$

$$s_w = \frac{s(n-1)}{v-1} \tag{3}$$

where s = standard deviation of the modified data set, n = the total number of observations, and v = the number of observations not modified.

Example 1
The following sulfate concentrations (mg/L) were measured during a routine water quality survey (sample size = 21, mean = 5.1, standard deviation = 2.1).

1.3 ("<2.5")	3.5	5.2	6.5	9.9
2.3 ("<2.5")	3.6	5.6	6.9	
2.6	4.0	5.7	7.1	
3.3	4.1	6.1	7.7	
3.5	4.5	6.2	7.9	

For this illustration, let's assume that the lowest two observations were censored, i.e., we have an embarrassing detection limit of 2.5 mg/L. The smallest two observations are replaced by the next largest value (<2.5 and <2.5 become 2.6 and 2.6). Next, the two

Example 1 *Continued*

largest values are replaced by the value for the next smallest observation (7.9 and 9.9 become 7.7 and 7.7). The modified data set is now the following.

2.6	3.5	5.2	6.5	7.7
2.6	3.6	5.6	6.9	
2.6	4.0	5.7	7.1	
3.3	4.1	6.1	7.7	
3.5	4.5	6.2	7.7	

The Winsorized mean ($\bar{x}_w$) is 5.08. The standard deviation of the modified data set (s) is 1.79. The number of observations (n) is 21, and the number of unmodified observations (v) is 17 (or $n - 4$). The Winsorized standard deviation (s_w) is estimated with Equation (3) to be [1.79 (21 − 1)]/(17 − 1) or 2.24. Rounding off, the Winsorized mean (5.1) for the censored data set is the same as that estimated for the uncensored data. The Winsorized standard deviation is slightly higher (2.2) than that estimated for the uncensored data (2.1).

This technique is useful if the distribution is symmetrical. In Example 1, the assumption was made that the distribution was normal. The technique can also be used for skewed data such as those from a log normal distribution if the observations are log transformed first. However, a backtransformation bias must also be estimated for log transformed data. Further discussion of Winsorized estimates can be found in Dixon and Massey,[26] Sokal and Rohlf,[27] Gilbert,[23] and Berthouex and Hinton.[28]

c. Probability Plotting

Perhaps the most straightforward method for estimating the mean and standard deviation of censored data sets is probability plotting. In Figure 3, the sulfate data from Example 1 are used to illustrate this approach. Again, the lowest two observations are assumed to be censored. The observations are ranked from smallest to largest. The cumulative percentage corresponding to each observation is then estimated and plotted against concentration. The mean is estimated from the concentration corresponding to the 50% cumulative percentage (P50). For the sulfate data, the estimate is approximately 5 mg/L (Figure 3). The standard deviation can be estimated using the concentrations associated with the 16th (P16) and 84th (P84) cumulative percentages, because roughly 68% of the central area under the normal curve is contained between these points. The standard deviation is estimated to be approximately 2.1 using Equation (5).

$$\bar{x}_{plot} = P50 \tag{4}$$

$$s_{plot} = \frac{P84 - P16}{2} \tag{5}$$

For log normal distributions, a similar approach using logarithmic-probability scale plots will generate acceptable estimates. Gilbert and Kinnison[21] provide Equations (6) and (7) to estimate the mean and standard deviation for samples from log normal distributions.

$$\bar{x}_l = \ln P50 \tag{6}$$

$$s_l = \ln\left[0.50\left[\frac{P50}{P16} + \frac{P84}{P50}\right]\right] \tag{7}$$

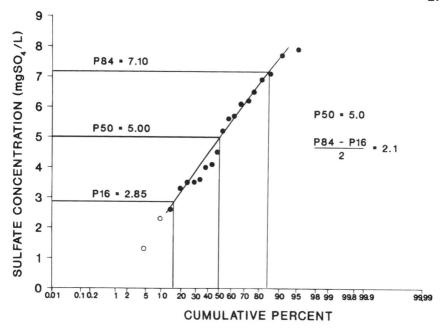

Figure 3 An example of estimating mean and standard deviation by probability plotting. The two lowest of the 21 sulfate concentrations may be less than the LOD. It is clear that, with the points above the censored values, a reasonable estimate of the mean and standard deviation can be obtained using concentrations corresponding to the 16, 50, and 84 cumulative percentages (see Example 1).

where P16, P50, P84 = concentrations from the ln-probability plot. Regardless of the plot scales used, this technique is characterized by a degree of subjectivity because visual curve fitting is involved. This can be avoided using regression techniques.

d. Distributional Methods

Although difficult with a complete data set, censoring makes assignment of an underlying distribution to a data set even more difficult. This is one of the major objections to censoring. However, several methods are applicable for estimating mean and standard deviation if the type of underlying distribution can be identified.

Maximum likelihood estimators (MLEs) were developed for a variety of distributions including the normal distribution.[29-32] With transformation and bias correction as noted here, these methods can be applied to data sets from log normal distributions. Cohen also developed MLEs for the Poisson,[33] Weibull,[34] and three-parameter gamma[35] distributions. The restricted MLE[31,32] as applied to normal or log normal data sets is presented here. Cases with small and large data sets are also considered, because the MLEs generated with few observations must be adjusted for a bias that becomes significant at small sample sizes.[36]

The restricted MLEs for the mean and standard deviation are the following:

$$\overline{X}_{MLE} = X_c - \lambda(\overline{x}_c - LOD) \tag{8}$$

$$s_{MLE} = \sqrt{s_c^2 + \lambda(\overline{x}_c - LOD)^2} \tag{9}$$

where x_c = the mean of observations above the LOD; LOD = limit of detection; s_c^2 =

variance of observations above the LOD; and λ = a statistic from Table 2 of Cohen[32] (Table 1 in the Appendix). Two values (h and γ) are needed to extract λ from Table 1 (Appendix). h is the number of observations below the LOD divided by the total number of observations. The γ is $s_c^2/(\bar{x}_c - \text{LOD})^2$.

Example 2

The sulfate data from Example 1 are used to generate MLEs. Again, assume that the lowest two observations are below the LOD. LOD = 2.5; $\bar{x}_c = 5.47$; $s_c^2 = 3.70$; $h = 2/21 = 0.095$; $\gamma = 3.70/(5.47 - 2.5)^2 = 3.70/8.82 = 0.42$.

Linearly interpolating from values given in Table 1 (Appendix), λ is estimated to be approximately 0.130162.

The $\bar{x}_{\text{MLE}}$ is estimated to be approximately $5.47 - 0.130162(5.47 - 2.5)$ or 5.08 mg/L. The s_{MLE} is approximately the square root of $3.70 + 0.130162 (5.47 - 2.5)^2$ or 2.20.

Because MLEs become increasingly biased as the number of observations in the data set decreases, estimates of bias provided by Schneider,[36] and Schneider and Weissfeld[37] may have to be applied to MLEs. These bias corrections are recommended if the number of observations is 20 or less.[37]

$$\bar{X}_{bc} = \bar{X}_{\text{MLE}} - \left[\frac{s_{\text{MLE}}}{n + 1}\right] B_x \tag{10}$$

where $\bar{X}_{bc}$ = the bias-corrected, MLE for the mean, and

$$B_X = -e^{2.692 - 5.439 \frac{n-k}{n-2k+1}} \tag{11}$$

$$S_{bc} = S_{\text{MLE}} - \frac{s_{\text{MLE}}}{n + 1} B_S \tag{12}$$

where s_{bc} = the bias-corrected MLE for the standard deviation, and:

$$B_S = -\left[0.3121 + 0.859 \frac{n - k}{n + 1}\right]^{-2} \tag{13}$$

where n = the total number of observations, and k = the number of observations below the LOD. Other techniques that assume a specific underlying distribution include order statistic,[38-41] regression on expected order statistics[25] and "fill-in" with expected values[42] methods. The order statistic methods are unbiased regardless of sample size but have higher mean square errors than the maximum likelihood methods.[14] When using order statistics techniques, pertinent tables such as those in Sarhan and Greenberg's publications[39-41] are designed for right-censored data sets. They must be converted to cope with left censoring. Alternatively, to avoid the associated tedium, the original data set can be multiplied by −1 to convert it to a right-censored data set amenable to use with these tables. After estimations are made, the sign of the mean is changed back to a positive number.

If the distribution is log normal, the above procedures can be used to estimate the mean and standard deviation but they are applied to the log transformed observation values. When doing this, it is important to realize that the backtransformed values are biased estimates of the arithmetic mean and standard deviation. Gilbert and Kinnison[21] described means of coping with this bias. Aitchison and Brown[43] provided the following equations to correct for the backtransformation bias.

$$\overline{X}_{lc} = e^{\overline{x}_l}\psi_n(s_l^2/2) \tag{14}$$

$$s_{lc}^2 = e^{2\overline{x}_l}[\psi_n(2s_l^2) - \psi_n[(n - 2)/(n - 1)s_l^2] \tag{15}$$

where s_l = standard deviation of the transformed values; $\overline{x}_l$ = mean of the transformed values; n = the number of observations; and $\psi_n(t)$ = a value with argument t extracted from Table A2 of Aitchison and Brown.[43] This table has been reproduced as Table 2 in the Appendix at the end of this book.

Table 2 in the Appendix covers a limited range of t values. The $\psi_n(t)$ can be estimated with the expansion of Equation (16).[44]

$$\psi_n(t) = 1 + \frac{n - 1}{n}t + \frac{(n - 1)^3}{n^2 2!}\frac{t^2}{n + 1} + \frac{(n - 1)^5}{n^3 3!}\frac{t^3}{(n + 1)(n + 3)} + \cdots \tag{16}$$

e. Robust Methods

Some methods work better than others if the underlying distribution from which the observations are taken is not known.[20,45] Helsel[46] recommends a modification of the probability plot method described above as the most robust method. For each observation above the LOD, a z score is estimated. Helsel's approach fits a regression line to the log transformed observation values above the LOD and their corresponding z scores. Next, the regression is used to predict "fill-in" values for the below-LOD values. All values including the predicted "fill-in" values are then backtransformed to arithmetic units. The mean and standard deviation are estimated using the data set that now includes "fill-in" values for the censored observations. Newman et al.[25] described a related method that uses the bias correction procedures of Aitchison and Brown[43] [Equations (14) and (15)] instead of avoiding bias by backtransforming observations before estimation of mean and standard deviation.

f. Programs Available for Mean and Standard Deviation Estimation

UNCENSOR[47] has been revised to facilitate the estimation of mean and standard deviation from left-censored data by all of the above-mentioned techniques except Winsorized estimates. Estimates are provided for normal and log normal distributions. For samples from log normal distributions, Equation (16) is expanded to the tenth term (i.e., the t in the last expansion is raised to the tenth power). Confidence intervals for estimates are provided for several procedures. Program and source code (PASCAL), and a detailed users manual are available at no cost from the author. Schneider[36] also provided pertinent FORTRAN code in an appendix of his book describing the techniques. Helsel[46] described estimation with the robust plotting method using commercially available software packages. He also offers FORTRAN code for dealing with multiple reporting limit data sets (data sets with more than one LOD). Recently, FORTRAN code and an excellent manual for MLE, regression, and the U.S. EPA's recommended delta-log normal methods were provided in an NCASI report by Berthouex and Hinton.[28]

4. Summary

Censoring is not recommended for data sets. When censoring does occur, deletion or substitution methods should not be used for estimating mean and standard deviation. A variety of techniques are available to estimate these univariate statistics. As outlined in

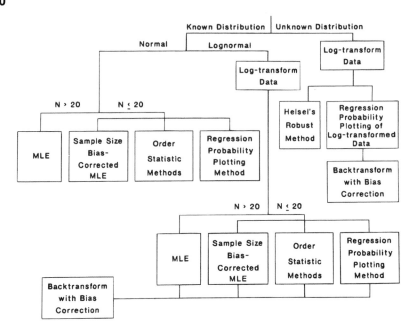

Figure 4 An outline of methods available for estimating mean and standard deviation for censored data sets.

Figure 4, they each have their advantages and disadvantages. When uncertain, it seems best to follow Helsel's recommendation that the robust techniques be used. Regardless of the technique selected, the original descriptions of the technique should be reviewed before application to familiarize oneself with the assumptions being made.

III. BLANK CORRECTION

A. OVERVIEW

Blank control is a critical part of most measurement processes. Failure to properly control or define blanks can lead to wasted resources and misappropriation of regulatory effort. For example, Settle and Patterson[48] indicated that pervasive lead contamination in the United States went essentially undetected due to poorly controlled blanks. More recently, Windom et al.[49] argued that U.S. Geological Survey NASQAN dissolved metals data used to assess trends in U.S. waters were suspect because analytical procedures were inadequate for blank control. Although Boatman[50] counters Windom and co-workers' specific conclusions about the U.S. Geological Survey data, he also expresses the opinion that most dissolved metals data are unreliable as a consequence of pervasive sample contamination.

Blank control can involve inexpensive[51] or very expensive[52] procedures. Consequently, careful consideration should be given to blank control at various steps in the measurement procedure to avoid easily eliminated problems that compromise data integrity or, at the other extreme, to avoid the accrual of unnecessary costs that limit the number of samples that can be analyzed.

A sequence of blanks may be used in the sampling and measurement process. An empty sample container may be filled in the field with water containing no analyte and then serve as a travel or sampling blank. A similar sampling blank may be generated during laboratory experiments. A solution containing no detectable analyte may be

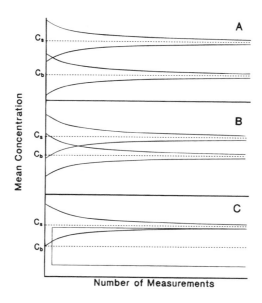

Figure 5 An example illustrating the influence of the relative magnitude of the blank and sample concentrations, and number of blank measurements. In Figure 5A, the 95% confidence intervals for the sample and blank separate as the number of measurements increases. Figure 5B shows that the number of measurements needed to get a similar separation increases as the sample mean value comes closer to that of the blank. Figure 5C demonstrates the difficulty encountered if the number of blank determinations remains constant at a small number. In Figure 5C, the confidence intervals do not separate as in 5B because the number of blank measurements is inadequate.

processed and analyzed along with a sample set to generate a procedural blank. A reagent blank may be used in calibration. Ideally, such a reagent blank is a solution containing everything that the sample contains (reagents and similar matrix) except the analyte being measured. Each of these blanks has its use in tracing sources of unacceptable contamination and quantifying acceptable contamination.

B. ESTIMATING THE EFFECTS OF THE BLANK ON THE MEASUREMENT PROCESS

General methods for calculating confidence intervals for differences between means[26] were used by Taylor[7,53] to estimate the 95% confidence interval for the true blank-corrected mean.

$$C_s = (C_m - C_b) \pm Z\sqrt{\frac{s_m^2}{m} + \frac{s_b^2}{b}} \tag{17}$$

where C_s, C_m, and C_b = the mean concentrations estimated in the sample after blank correction, in the original sample and in the blank respectively; s_m^2 and s_b^2 = the variances associated with the samples and blanks, respectively; m and b = the number of determinations of the sample and blank, respectively; and Z = Z statistic for $100\ (1 - \alpha/2)\%$ confidence interval (1.96 for a 95% confidence level).

Several factors influencing the effectiveness of blank correction can be illustrated with Equation (17) (Figure 5). First, the correctness of the estimated mean blank determination (C_b) is as important as that of the estimated mean sample value (Figure 5A).

Indeed, as the mean blank concentration may be subtracted from many samples, an argument could be made that its accurate estimation is more important than that of any single sample value. Also the relative magnitudes of the sample and blank values are important (Figure 5A versus 5B). As the sample value approaches that of the blank, its confidence interval overlaps more and more with that of the blank. Second, the variation about the blank can be as important as the magnitude of the mean blank value. Taylor argues convincingly from Equation (17) that as much effort should be made to estimate the blank value as is made to measure sample values in the trace range (ppm range) (Figure 5C). As the sample concentration approaches the blank (as analyses extend down toward the LOD with $s_m^2 \rightarrow s_b^2$), an equal number of blanks and samples are advocated. Regardless, Taylor[7] suggests the rule of thumb that blank correction should not exceed ten times the acceptable limit of error in trace analyses. He feels that 10% error in the blank is acceptable.

C. BLANK CONTROL CHARTS

The mean and standard deviation of the blanks associated with analyses should be monitored formally using the control chart methods described in the next section of this chapter. These methods use mean values and acceptable variation for "in control" analytical conditions as a basis for judging the quality of subsequent measurements. Relative to blanks, control charts with upper limits should suffice. This does not imply that a significant decrease in the measured blank value is not valuable information. A consistent drop in the blank below a previously defined level could indicate an unsuspected means of improving the measurement process, and reevaluation of procedures would then be made. It could also suggest inconsistencies in calculation methods such as those that may arise during the training of new personnel.

IV. ACCURACY AND PRECISION

A. ACCURACY

1. Overview

Accuracy is the closeness of a measured value to the "true" value. The "true" value may be obtained by adding a known spike to a sample or using an analyte-containing material with a matrix similar to that of the sample. Spiking allows the analyst to produce a material with a concentration similar to that of the sample to be analyzed. This has its advantages because accuracy estimates are concentration dependent. However, it is extremely difficult to develop a spiking methodology that incorporates the added analyte into the sample matrix such that it truly represents the analyte condition in the sample. For example, the digestion and analysis of a solid sample spiked with a metal solution would give an analyst very little information about the accuracy associated with the digestion. Alternatively, a standard material may be used that has the analyte incorporated into a matrix similar to that of the sample being studied. The standard material is supplied with a certified value and an acceptance interval for any analysis of the material. Both approaches have their place in improving and documenting accuracy.

2. Estimation of Accuracy

Accuracy is often expressed as percent recovery. If the sample has no background concentration to consider then the percent recovery is estimated with the simple formula,

$$P = 100 \frac{\text{MSC}}{\text{CSC}} \tag{18}$$

where MSC = measured spike concentration, and CSC = calculated spike concentration.

In this case, estimates of the mean and standard deviation of percent recovery data (P) are the familiar equations,

$$\bar{P} = \frac{\sum_{i=1}^{n} P_i}{n} \tag{19}$$

$$S_p = \sqrt{\frac{\sum_{i=1}^{n} (P_i - \bar{P})^2}{n - 1}} \tag{20}$$

where $\bar{P}$ = the mean P; P_i = the ith measure of P; and n = the number of measured P values.

If the spiked material has a background concentration, percent recovery is estimated with the following formula.

$$P = 100 \frac{MSC - UC}{CSC} \tag{21}$$

where UC = unspiked concentration.

An unbiased estimate of the mean percent recovery would be the same as given in Equation (19) for percent recovery without background concentration. However, an unbiased estimate of the variance of percent recovery data would be the following:[54]

$$S_p^2 = \left[\frac{\left(\bar{P} \frac{CV}{100} \right)^2}{n} \right] \left[\left(1 + \frac{1}{\frac{CSC}{UC}} \right)^2 + \left(\frac{n}{m \left(\frac{CSC}{UC} \right)^2} \right) \right] \tag{22}$$

where m = the number of measurements used to estimate UC; and CV = the coefficient of variation for the measurement process, i.e., standard deviation/mean. By taking the square root of this variance estimate, the standard deviation for P can be obtained when there is a significant background concentration.

As demonstrated by Provost and Elder[54] and reiterated here, it is important to realize that the variance of percent recoveries is not the same for samples with and without background concentrations. The variance associated with measurement of the background concentration must also be taken into consideration during calculations. When a significant background concentration is involved, the variance will be influenced by the ratio of the spike concentration to background concentration (CSC/UC). As the ratio decreases, the variance associated with the estimated mean percent recovery increases.

B. PRECISION
1. Overview
Precision is the effectiveness of a measurement process in producing similar results on repeated application. For example, precision of the measurements, 1.0, 2.5, 0.7 is inferior to that of 1.1, 0.9, 1.1. By analogy, precision would be the closeness of bullet holes in

a target to one another, whereas accuracy would be the closeness of a hole to the bulls eye of the target. A measurement process can be precise but inaccurate, e.g., a tight cluster of bullet holes at the very edge of the target. Although a process can also be imprecise but accurate (a widely spread cluster with a very minimal "average" distance from the bull's eye), imprecision very often spawns inaccuracy.

2. Estimation of Precision

Precision is estimated using replicate analyses. Mean and standard deviation may be estimated when two or more replicates are analyzed. Another estimator of precision, range may be calculated from replicate analyses. Most often the range is used for duplicate analyses. The range is the absolute value of the difference between the smallest and largest measurement in a set. The mean of a set of ranges is computed with Equation (23).

$$\overline{R} = \frac{\sum_{i=1}^{n} R_i}{n} \tag{23}$$

where R_i = the ith estimate of range; and n = the number of ranges estimated. The relationship between the range and the standard deviation depends on the distribution of the data. If the data can be assumed to be normally distributed, the standard deviation can also be estimated from the range.[3]

$$S = \frac{\overline{R}}{d_2\sqrt{n}} \tag{24}$$

where $\overline{R}$ = the mean range; and d_2 = a value obtained from Table 3 in the Appendix.

C. CONTROL CHARTS
1. Overview

Control charts traditionally have been used to monitor precision, accuracy, and blank acceptability of measurement processes. Average values and acceptable variation for these qualities are established when the measurement process is functioning properly ("in-control"). Checks against these "in-control" standards of quality are then made through time. There are two basic control chart approaches, Shewhart and cusum methods. Both are discussed briefly here. The interested reader is referred to Grant[3] for a more in-depth discussion of Shewart charts and to Van Dobben De Bruyn[55] for a similar treatment of cusum charts. General discussions of these methods relative to environmental sciences may also be obtained from the U.S. EPA.[56,57]

In general, an expected average value is estimated for the measurement process when it is "in-control." An estimate of the variation to be expected is also calculated. Critical limits are then set based on the probability of exceeding certain thresholds. Limits may include upper limits only or both upper and lower limits. For example, a control chart used to monitor a blank may not have a lower limit, because a blank significantly lower than normal is not viewed as a problem. Similarly, precision better than usual is not a problem. In contrast, the measurement accuracy as estimated by percent recovery can be either too low or too high.

In all these methods, the number of values used to calculate the mean and limits must be explicitly defined. The number of observation values used to compare an analytical session's results against the chart should also be the same.

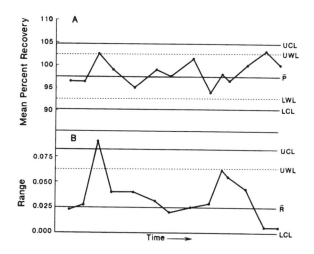

Figure 6 An example of Shewhart Chart Techniques. Figure 6A allows the analyst to monitor accuracy (see Example 3). Figure 6B allows precision to be estimated (see Example 5). Taken together, such accuracy and precision charts are called $\overline{X} - R$ charts.

2. Accuracy

a. Shewhart Charts

Taylor[7] describes two approaches to Shewhart charts. The first (property chart) uses individual property values, e.g., individual percent recovery values. The second ($\overline{X}$ chart) involves the mean of individual values (the mean of 6 percent recovery values generated during an analytical session for example). The $\overline{X}$ chart method is discussed here as it is less sensitive to spurious observations.

Accuracy may be estimated using mean percent recovery as mentioned above or using the mean value of a standard material measured in replicate along with the samples. In Figure 6A, a control chart with warning ($\pm 2S_p/\sqrt{n}$) and control limits ($\pm 3S_p/\sqrt{n}$) is constructed for percent recovery data. Roughly 95% (95.5%) of observed mean percent recoveries are expected to fall within the warning limits if the measurement process is controlled properly. More than 99% (99.7%) of all mean values generated by the "in-control" measurement process should be within the control limits. As outlined by Taylor[7] or Van Dobben De Bruyn,[55] rules for rejection of data may be generated such as the following.

1. If a value is outside the upper or lower control limits, the measurement process is judged "out-of-control."
2. If two successive points are outside the upper or lower warning limits, the measurement process is judged "out-of-control." Note that the two points must be outside the same (upper or lower) warning limit.
3. If there is an obvious trend, the system is "out-of-control." An example of such a trend would be seven consecutive points on one side of the mean.

The associated measurements are invalid if the process is judged "out-of-control." The system should be examined and the problem resolved. Quality control samples should be analyzed more frequently until it is certain that the problem has been resolved.

Control charts may also be used to suggest reexamination of the measurement process before it becomes "out-of-control." For example, a mean value beyond the warning limit or four points in a row on one side of the mean may prompt close examination of the process to catch any possible emerging problem.

Example 3

Twenty measurements were made of a standard material with certified lead concentration of 0.25 μg/g. Percent recovery (P_i) is used to estimate accuracy.

Measurement	$P_i(\%)$	$(P_i - \bar{P})^2$
0.27	108	112.36
0.22	88	88.36
0.21	84	179.56
0.31	124	707.56
0.23	92	29.16
0.23	92	29.16
0.25	100	6.76
0.24	96	1.96
0.25	100	6.76
0.23	92	29.16
0.27	108	112.36
0.28	112	213.16
0.20	80	302.76
0.25	100	6.76
0.22	88	88.36
0.26	104	43.56
0.22	88	88.36
0.23	92	29.16
0.24	96	1.96
0.26	104	43.56
Σ	1948	2,120.80

The mean percent recovery $(\bar{P})$ is estimated to be $\bar{P} = \Sigma P_i/n = 1948/20 = 97.4\%$.

According to Equation (20) the standard deviation for percent recovery values is estimated to be $s_p = \sqrt{2,120.80/(20 - 1)} = 10.6$. The standard deviation for the mean percent recovery is $s_p/\sqrt{n}$ or $10.6/\sqrt{20} = 2.4\%$

Shewhart chart limits can now be constructed using these estimates ($\bar{P}$ and its estimated standard deviation).

Upper Control Limit $= \bar{P} + 3(s_p/\sqrt{n}) = 97.4 + 3(10.6/\sqrt{20}) = 104.5\%$
Upper Warning Limit $= \bar{P} + 2(s_p/\sqrt{n}) = 97.4 + 2(10.6/\sqrt{20}) = 102.1\%$
Lower Warning Limit $= \bar{P} - 2(s_p/\sqrt{n}) = 97.4 - 2(10.6/\sqrt{20}) = 92.7\%$
Lower Control Limit $= \bar{P} - 3(s_p/\sqrt{n}) = 97.4 - 3(10.6/\sqrt{20}) = 90.3\%$

These results can now be used to construct an accuracy control chart for future lead analyses conducted under conditions identical with those used to generate these initializing data. Figure 6A shows the resulting chart. Estimates of mean percent recovery are plotted during any future analyses and assessed relative to the limits established above.

b. Cusum Charts

Cusum (cumulative sum) charts use all information generated before the present measurement session to evaluate quality. Deviations are summed through all previous measurement sessions, whereas, with Shewhart charts, initializing data are used to establish quality criteria. The change in slope of the cusum chart is used to monitor the quality of analyses. Cusum charts are more sensitive to consistent small deviations than Shewhart charts.[55]

In the U.S. EPA's *Handbook for Analytical Quality Control in Water and Wastewater Laboratories*[56] and *Region VI Laboratory Quality Control Manual*,[57] cusum charts for accuracy were developed using differences between observed and known values for a spike or standard sample. That approach is used here; however, alternate procedures have been described in Van Dobben De Bruyn.[55]

Cusum charts require defined α (probability of falsely judging the measurement process "out-of-control") and β (probability of falsely judging the measurement process "in-control") values. Values between 0.05 and 0.15 are recommended in the U.S. EPA manuals just noted. These values, along with an estimate of the standard deviation for the difference between the measured and known values, are used to estimate the cusum chart limits.

First, the variance of the differences (s_d^2) is estimated using Equation (25).

$$S_d^2 = \frac{\sum_{i=1}^{n} d_i^2 - \frac{\left[\sum_{i=1}^{n} d_i\right]^2}{n}}{n-1} \tag{25}$$

where d_i = the ith difference between the true and measured value; and n = the number of differences used in the estimation. With this estimate of the variance for the differences, the minimum (s_0^2) and maximum (s_1^2) acceptable variation are estimated.

$$s_0^2 = 0.64\, s_d^2 \tag{26}$$

$$s_1^2 = 1.44\, s_d^2 \tag{27}$$

The upper (UPPER (M)) and lower (LOWER (M)) control limits for M sets of differences to be plotted are the following.

$$\text{UPPER}(M) = \frac{2 \ln\left[\frac{1-\beta}{\alpha}\right]}{\frac{1}{s_0^2} - \frac{1}{s_1^2}} + M \frac{\ln\left[\frac{s_1^2}{s_0^2}\right]}{\frac{1}{s_0^2} - \frac{1}{s_1^2}} \tag{28}$$

$$\text{LOWER}(M) = \frac{2 \ln\left[\frac{\beta}{1-\alpha}\right]}{\frac{1}{s_0^2} - \frac{1}{s_1^2}} + M \frac{\ln\left[\frac{s_1^2}{s_0^2}\right]}{\frac{1}{s_0^2} - \frac{1}{s_1^2}} \tag{29}$$

Example 4

Again, the 20 measurements of the standard material with a certified concentration of 0.25 µg Pb/g are used. The numbers here are identical with those used in Example 3.

Example 4 *Continued*

Measurement	Measured − Certified Concentration (d)	d^2
0.27	0.02	0.0004
0.22	−0.03	0.0009
0.21	−0.04	0.0016
0.31	0.06	0.0036
0.23	−0.02	0.0004
0.23	−0.02	0.0004
0.25	0.00	0.0000
0.24	−0.01	0.0001
0.25	0.00	0.0000
0.23	−0.02	0.0004
0.27	0.02	0.0004
0.28	0.03	0.0009
0.20	−0.05	0.0025
0.25	0.00	0.0000
0.22	−0.03	0.0009
0.26	0.01	0.0001
0.22	−0.03	0.0009
0.23	−0.02	0.0004
0.24	−0.01	0.0001
0.26	0.01	0.0001
Σ	−0.13	0.0141

Using Equations (25) to (27):

$$
\begin{aligned}
s_d^2 &= [\Sigma d_i^2 - (\Sigma d_i)^2/(n)]/(n - 1) \\
&= [0.0141 - (-0.13)^2/(20)]/19 \\
&= 0.000698 \\
s_0^2 &= 0.64\, s_d^2 = 0.64(0.000698) = 0.000446 \\
s_1^2 &= 1.44\, s_d^2 = 1.44(0.000698) = 0.001005
\end{aligned}
$$

With s_0^2 and s_1^2, the upper and lower limits can be established for the cusum chart. For initial construction of the chart, they will be estimated for a range of M from 0 to 30 using Equations (28) and (29). Values for α and β are set at 0.15.

Upper (0) = (2ln((0.85/0.15))/(1/0.000446 − 1/0.001005) + 0(ln(0.001005/
        0.000446)/(1/0.000446 − 1/0.001005))
        = 0.00278

Upper(30) = (2ln((0.85/0.15))/(1/0.000446 − 1/0.001005) + 30(ln(0.001005/
        0.000446)/(1/0.000446 − 1/0.001005))
        = 0.02232

Lower(0) = (2ln((0.15/0.85))/(1/0.000446 − 1/0.001005) + 0(ln(0.001005/
        0.000446)/(1/0.000446 − 1/0.001005))
        = −0.00278.

Lower(30) = (2ln((0.15/0.85))/(1/0.000446 − 1/0.001005) + 30(ln(0.001005/
        0.000446)/(1/0.000446 − 1/0.001005))
        = 0.0168

Example 4 *Continued*

Figure 7 is the cusum chart resulting from these calculations. The cumulative sum of the differences squared (Σd_i^2) is plotted against M after each analytical session as shown. If the Σd_i^2 value goes above the UPPER (M) then the analysis is "out-of-control" and the associated measurements are invalid. The problem must be resolved before proceding with further analyses.

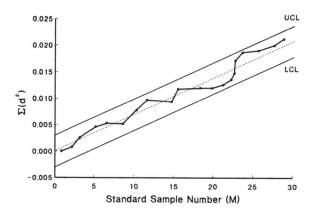

Figure 7 An example of a Cusum Chart for accuracy of a series of lead analyses (see Example 4).

3. Precision

a. Shewhart Charts

Precision charts (Figure 6B, for example) are often generated from duplicate analyses. The mean range and standard deviation about the mean range can be estimated using equations such as Equations (23), (24), and (25). However, the upper warning limit (UWL), and upper (UCL) and lower (LCL) control limits for the $\bar{R}$ can be fabricated as follows.

$$UCL = D_4\bar{R} \tag{30}$$

$$UWL = \bar{R} + \frac{2}{3}(D_4\bar{R} - \bar{R}) \tag{31}$$

$$LCL = D_3\bar{R} \tag{32}$$

The values for D_4 and D_3 depend on the number of measurements in the subgroup from which the range was derived. For ranges estimated from duplicate measurements ($n = 2$), $D_4 = 3.27$ and $D_3 = 0$. Values can be taken from Table 3 in the Appendix for range estimates from subgroup sizes of up to 20 replicates. Note that D_3 becomes greater than 0 when the replicate number increases to 6.

The control and warning limits for $\bar{R}$ charts are used like those for the $\bar{x}$ charts. However, there is no lower warning limit.

Frequently the $\bar{x}$ chart for accuracy and $\bar{R}$ chart for precision are combined into a composite graph such as Figure 6. Such a chart is called an $\bar{x} - R$ chart ("x bar $- R$ chart").

Example 5

Twenty duplicates of lead analyses for a set of similar samples are analyzed to give the following results.

	Duplicate 1	Duplicate 2	Range
	0.25	0.28	0.03
	0.25	0.22	0.03
	0.25	0.22	0.03
	0.27	0.28	0.01
	0.21	0.26	0.05
	0.31	0.28	0.03
	0.23	0.25	0.02
	0.23	0.22	0.01
	0.18	0.28	0.10
	0.17	0.16	0.01
	0.29	0.26	0.03
	0.37	0.35	0.02
	0.24	0.26	0.02
	0.29	0.31	0.02
	0.30	0.30	0.00
	0.28	0.31	0.03
	0.17	0.19	0.02
	0.30	0.27	0.03
	0.28	0.29	0.01
	0.29	0.29	0.00
Σ			0.50

The mean range ($\bar{R}$) is 0.50/20 = 0.025.

Using Equations (30), (31), and (32), the limits for the precision control chart can be estimated to be the following.

Upper Control Limit = $D_4\bar{R}$ = 3.27 (0.025) = 0.082
Upper Warning Limit = $\bar{R}$ + 2/3 ($D_4\bar{R} - \bar{R}$)
$$= 0.025 + 2/3 [3.27 (0.025) - 0.025]$$
$$= 0.063$$
Lower Control Limit = $D_3\bar{R}$ = 0(0.025) = 0.000

Figure 6B is the control chart resulting from these calculations. The mean range for each analytical session is plotted as shown and compared to the limits. For example, the third mean range indicates an "out-of-control" condition for analytical precision. The results of that session's work are invalid.

b. Cusum Charts

When ranges for duplicate analyses are used to estimate precision, the procedures just described for accuracy cusum charts can be used. The approach would be the same but the difference between the duplicate measurements would be used instead of the difference between the observed and "true" value of the samples.

V. VARIANCE STRUCTURE

During most measurement processes, variation is added to the final measured signal at several steps. Variation is introduced from field sampling to sample preparation to sample analysis. The variance structure for the measurement process can be written as $s^2_{total} = s^2_{sampling} + s^2_{preparation} + s^2_{analysis}$ assuming independence of variance components. Variance associated with individual steps in sampling (sampling strata, e.g., between sites, seasons, or treatments), sample preparation (aliquot selection, extraction, digestion), or analyses (replicate analyses) can be similarly defined.

Before any use of a data set, the variance structure of the measurement process should be understood. With proper experimental design, analysis of variance (ANOVA) methods can be used to estimate these variance components. As demonstrated in Examples 6 and 7, failure to define such structure can result in suboptimal data interpretation.

Example 6

In examining frequency distributions of essential and nonessential elements among individual fish, Pinder and Giesy[58] became concerned about the relative contributions of post sampling components of the measurement process to the overall distribution. In previous studies attempting to link distribution shape to essentiality of an element, these post sampling variance components were assumed to be insignificant relative to the variation between individuals. In their assessment, several muscle samples were taken from each of a large number of bass (*Micropterus salmoides*). Subsamples of each muscle sample were subjected to replicate analysis. The variance associated with differences among individual fish (σ^2_f), among subsamples for individual fish ($\sigma^2_{s(f)}$), and unexplained error, i. e., replicate analyses (σ^2_e) were estimated using nested analysis of variance (nested ANOVA). (See Sokal and Rohlf,[27] pages 271–320, for explanation and examples of these nested ANOVA techniques.)

Element	$s^2_f(\%)$	$s^2_{s(f)}(\%)$	$s^2_e(\%)$
Cd	0	34	66
Cu	35	0	65
Fe	42*	0	58
Hg	76**	4	20
Pb	39	13	48
Se	35**	0	65
Zn	65**	0	30

In this table, significant deviation of s^2 values are indicated by * for $\alpha = 0.05$ or by ** for $\alpha = 0.01$. Although only the s^2_f values were significant in these selected elements, the relative magnitudes of $s^2_{s(f)}$ and s^2_e to s^2_f suggest that, for these elements, the overall distribution in measured concentrations cannot be assumed to be generated by differences among individuals alone. The distribution could not be interpreted based on the assumption that variance between individuals alone controlled the type of distribution.

Example 7

In anticipation of a planned discharge containing small amounts of mercury into a southeastern U. S. stream, a biomonitoring survey was conducted of mercury concentrations in an endemic fish, the dusky shiner (*Notropis cummingsae*). Although this shiner occurs in distinct schools, the fidelity of individuals to a specific school and the extent of movement of a school within the watershed were unknown. Consequently, an effective sampling protocol could not be developed for monitoring mercury concentrations in fish from the stream above, at, and below the proposed point of discharge. After a preliminary survey indicating that a pooled sample of six individuals was sufficient to minimize variance between samples, the following survey was conducted.

Three regions (upstream, midstream at the site of the proposed discharge, and downstream) were sampled in the watershed during two seasons (October and immediately before

Example 7 *Continued*

spawning in April). Duplicate pooled samples from two schools at each of two sites within each region of the watershed were used to examine the variance structure. Because fish size was suspected to influence mercury concentration, a mixed ANOVA model was used to examine variance structure for average fish wet weight as well as mercury concentration in the pooled samples. (Again, see Sokal and Rohlf,[27] pages 271–320, for explanation and examples of these nested ANOVA techniques.) The components of variance estimated during this predischarge survey are tabulated here as percentages of the total variance for three variables.

Variable	Season(%)	Watershed Region(%)	Site(%)	School(%)	Error(%)
Wet weight	<1	28	<1	66	6
Mercury	20	6	2	5	67
Size-normalized mercury	24	22	11	1	42

There is a clear tendency for shiners of similar size to school together. This is important to note because mercury concentration is influenced by fish size. Sixty-six percent of the total variance in average shiner size was associated with among-school differences. The strong influence of watershed region on variance in average wet weight could easily be attributed to adult spawning movement upstream before spawning and juvenile movement downstream after hatching in feeder streams in upper reaches of the watershed. When the mercury concentrations are normalized to an average fish wet weight, the unexplained variance drops by 25% (from 67% to 42%). Once the contribution of average wet weight to mercury concentrations in pooled samples was taken into consideration, there seemed to be little motivation for sampling individual schools because only 1% of the total variance was attributed to among-school variation. Instead, the focus of the monitoring became the variation between regions with samples being taken at different sites within each region. Sampling was limited to one season when the least amount of size-dependent migration was occurring (October).

VI. SAMPLE SIZE

A. OVERVIEW

If too few individuals or too little material is taken for measurement, the confidence in the resulting measurements may be unacceptable for the intended purpose of the study. On the other hand, sampling too many individuals or too much mass can waste resources that might otherwise be available to address additional questions. For example, if the mean concentration of DDT in fish from a certain river were to be determined, it would be wise to ascertain the minimum number of individual fish to be sampled to obtain a reasonable estimate. If too few fish were sampled, the ability to estimate the mean concentration with a reasonable degree of precision could be lost. As a result, subsequent statistical analyses would be compromised. On the other extreme, if too many fish were sampled, resources could be wasted and sampling of additional locations on the river would be precluded. Further, if too small a piece of muscle tissue were taken from each fish, the resulting measurement might not accurately represent the concentration in fish muscle. Even if an adequate number of individuals were sampled, the nonrepresentative nature of the mass of muscle tissue used would compromise future statistical analyses.

Methods for estimating minimum sample size are described in this section. Gilbert[23] and Taylor[7] discussed determination of minimum number of individual samples or replicate analyses in more detail. The discussion below is essentially that of Gilbert.[23] Only techniques based on a prespecified relative error (| measured mean − actual mean |/

actual mean) will be outlined, although Gilbert[23] also detailed methods applicable if variance about the estimated mean or margin of error (| measured mean − actual mean |) are prespecified. Taylor[7] discussed means of incorporating monetary cost into decisions regarding the number of measurements to make at various steps in the process. Visman,[59] Ingamells,[60] Ingamells and Switzer,[6] Taylor,[7] and Wallace and Kratochvil[62] discussed methods for determining adequate sample weight in heterogeneous samples. This discussion draws heavily on these five publications.

B. NUMBER OF INDIVIDUALS TO SAMPLE
1. Overview
Stuart[63] illustrated the influence of sample number on estimation of population characteristics in the following way. Because the variance of the sample mean is the product of the population variance and $[1/n]$ $[(N − n)/(N − 1)]$, it follows that as n (the number of observations in the sample) increases toward N (the total number of individuals in the population), the variance of the sample mean approaches 0. The more samples taken, the better the estimate will be. If n is very small relative to N, e.g., $(N − n)/(N − 1)$ → 1, the term relating sample variance to that of the population approaches $1/n$. Two facts should be clear from this observation. First, the effectiveness of sample variance in estimating the population variance improves as sample number increases. Second, the improvement is minimal beyond a certain sample number. It follows that estimation of the minimum sample size aids in accurate estimation and effective use of analytical resources.

2. Minimum Number of Individuals if Analytical Error Is Neglible
In this approach, we assume that the magnitude of the analytical error is insignificant relative to that of the sampling error. However, the situation in which both are significant is discussed at the end of this section.

Gilbert[23] described a two-step approach to estimating the necessary number of samples. In his approach, the population variance (σ^2) need not to be known with assurance. The first step uses a z statistic to generate an initial estimate of the minimum sample number. In the second step, t statistics with $n − 1$ degrees of freedom are used iteratively to obtain the final estimate of sample number. If the number of degrees of freedom $(n − 1)$ is large, then the second step may not be necessary because the z statistic will approximate the t statistic.

The method described here assumes a previously specified, acceptable relative error $(d_r = |\overline{X} − \mu|/\mu)$. The estimate of the coefficient of variation required for this approach may be obtained during a preliminary survey or with an educated guess. (By a simple substitution of d or $|\overline{X} − \mu|$ for d_r and standard deviation for CV in Equations (33), (34), (35), and (36), the sample number based on a specified margin of error (d) can be estimated. In that case, the standard deviation is estimated during a preliminary survey or by best guess.)

If the size of the population (N) being sampled is small relative to the variance of the population (σ^2), the initial estimate of the required sample size is the following:

$$n = \frac{\left[\dfrac{z_{1-\alpha/2}\text{CV}}{d_r}\right]^2}{1 + \dfrac{\left[\dfrac{z_{1-\alpha/2}\text{CV}}{d_r}\right]^2}{N}} \tag{33}$$

However, if the size of the population is large relative to the variance of the population, then Equation (33) reduces to the following:

$$n = \left[\frac{z_{1-\alpha/2}CV}{d_r}\right]^2 \tag{34}$$

The initial estimate of minimum sample number (n) is made with previously defined α and d_r, estimated CV, and Equation (33) or (34). Using this initial n to define the associated degrees of freedom ($n - 1$) for the t statistic, the sample number is again estimated using either Equation (35) (if N is small relative to σ^2) or Equation (36).

$$n = \frac{\left[\dfrac{t_{1-\alpha/2,n-1}CV}{d_r}\right]^2}{1 + \dfrac{\left[\dfrac{t_{1-\alpha/2,n-1}CV}{d_r}\right]^2}{N}} \tag{35}$$

$$n = \left[\frac{t_{1-\alpha/2,n-1}CV}{d_r}\right]^2 \tag{36}$$

The new estimate of n is then used again in either Equation (35) or (36) to get yet another estimate. This process is iterated until no large changes occur in n between iterations.

Recently, it has been noted that the traditional approach often underestimates the minimum sample size.[64] Blackwood,[65] in commenting on implications in environmental sampling, attributed the bias to the neglect of assurance levels (the probability that the specified confidence interval (C.I.) e.g., 95% C.I., will have the specified interval width). Kupper and Hafner[64] provided formulas for generating better estimates (n_m^*) given an initial estimate (n_m) [from equation (34)], and assurance level ($1 - \gamma$). Table 4 in the Appendix gives such estimates for n_m values up to 100.

Example 8

Lead concentrations in a large population of green sunfish are to be estimated for a roadside pond. During a preliminary sampling, the analytical variance was determined to be insignificant relative to the variance among individual fish from this site. A coefficient of variation of approximately 0.47 was estimated during this initial survey. The minimum number of fish to be sampled for measurement of lead concentration in this population can be estimated assuming an acceptable relative error of 0.25 and an α of 0.10 ($z_{1-\alpha/2} = 1.645$).

Initial estimate of n from Equation (34):
$n = [(1.645*0.47)/0.25]^2 = 9.56$ or 10 individuals.
Now, with $n - 1$ or 9 degrees of freedom and an α of 0.10, a t statistic can be used in Equation (36) to refine the estimated sample number.
$n = [(1.833*0.47)/0.25]^2 = 11.88$ or 12 individuals
Using this new estimate of n (12), the calculation using Equation (36) is repeated. Each new estimate of n is used in further iterations until the estimates of n stabilize.
$n = [(1.796*0.47)/0.25]^2 = 11.40$ or 12 individuals
$n = [(1.796*0.47)/0.25]^2 = 11.40$ or 12 individuals
The minimum sample size of 12 individuals is selected for the survey.

If the initial estimate from Equation (34) is used in Table 4 to estimate the minimum sample size ($\alpha = 0.10$, $1 - \gamma = 0.90$), the estimate is 17. Note that the assurance level for the initial estimate ($1 - \gamma'$) was only 0.39.

3. Minimum Number of Replicate Analyses

Now, in this discussion, the variances associated with other measurement components (tissue subsampling, analyses) were assumed to be minimal. This certainly is not an assumption to be made lightly. During preliminary design, it is helpful to estimate the number of replicate measurements needed to obtain an acceptable relative error. The process is identical with the process described for estimating minimum number of individuals to sample from a population. The CV for the measurement process (not including sampling) is used along with a predetermined α and an acceptable relative error.

Example 9

Let's assume that, in Example 8, portions of a muscle sample were analyzed for lead. The resulting CV was 0.07. An α of 0.05 and relative error of 0.05 were designated as acceptable for the analytical process. The estimated number of replicate analyses was determined in the following steps.

Initial estimate using z statistic:
$n = [(1.96*0.07)/0.05]^2 = 7.53$ or 8 replicate analyses.
Iterations using t statistics:
$n = [(2.365*0.07)/0.05]^2 = 10.96$ or 11 replicate analyses.
$n = [(2.228*0.07)/0.05]^2 = 9.73$ or 10 replicate analyses.
$n = [(2.262*0.07)/0.05]^2 = 10.03$ or 11 replicate analyses.

Eleven replicate analyses would be needed to meet the tight relative error at an α of 0.05 for this survey. Kupper and Hafner's[64] corrections for underestimation may be applied here as well to obtain a better estimate.

---

4. Sample and Replicate Numbers Under Other Variance Structures

As illustrated very briefly in Section V, components of variance can be significant at several steps in the sampling and postsampling stages of the measurement process. Consequently, estimation of minimum sample number or replicate number can be considerably more involved than demonstrated here. Reference to methods associated with the most common situations are provided here.

Taylor[7] outlined methods for estimating sample and replicate sizes when both sampling and analytical variance are significant. He also presented methods of incorporating measurement cost into determinations of sample numbers as well as methods applicable when samples are taken from several design strata. Cochran[66] also discussed incorporation of cost in estimations and estimation of sample number for stratified data. He presented more detail on the techniques described above for determining sample number, including estimation during studies of proportions in populations. Gilbert[23] outlined methods for sample number estimation when measurements are correlated. (In the above-mentioned methods, measurements are assumed to be uncorrelated.) Correlated measurements may arise when measurements are taken at sites near one another over time.

C. WEIGHT (OR VOLUME) OF SAMPLE
1. Overview

Many populations sampled in ecotoxicology are not composed of discrete units such as individual organisms. For example, subsamples of weights or volumes are taken for materials such as soil, sediment, or water. The distribution of the analyte of interest in such a sample may be homogeneous or heterogeneous. If the distribution is heterogeneous, the discrete particles within it can be well mixed (Figure 8A) or poorly mixed (Figure 8B). For a heterogeneous sample, the variability between sample increments (aliquots or units of sample analyzed) decreases as the weight analyzed increases (Figure 8). Therefore,

A

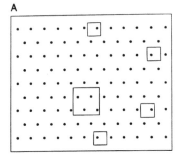

B

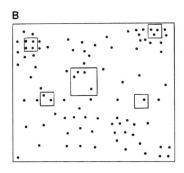

Figure 8 An illustration of sampling of heterogeneous samples. The sample may be well mixed (A) or poorly mixed (B). Means of estimating sample increments for such materials are described in the text.

it is important to analyze increments of sufficient weight to obtain an acceptable level of variability. But how large must the sample be to bring the variance to that acceptable level?

First, to answer this question, the variance associated with increment sampling must be estimated. To this end, the standard deviation of the overall measurement process is represented with the following equation.

$$s_{ov}^2 = s_s^2 + s_{ss}^2 + s_a^2 \tag{37}$$

where s_{ov}^2 = overall variance; s_s^2 = variance associated with increment sampling; s_{ss}^2 = variance associated with replicate measurements, and; s_a^2 = variance associated with a measurement. The s_s may then be estimated as described by Wallace and Kratochvil.[62] This is accomplished by taking the difference after the other variance components in Equation (37) are determined. First, s_{ss}^2 can be estimated by analyzing a homogeneous standard material to derive s_a^2 and subsamples of the material in question to derive s_{ss+a}^2. The s_a^2 is assumed to be similar for the reference material and the sample.

$$s_{ss}^2 = s_{ss+a}^2 - s_a^2 \tag{38}$$

If there is no suitable reference material, the combined s_{ss+a}^2 can be used in place of $s_{ss}^2 + s_a^2$. Alternatively, s_{ss+a}^2 and s_s^2 can be estimated using nested ANOVA methods as described in Example 9.[58]

The sampling variance is then estimated by analyzing a series of increments of sample to derive s_{ov} and solving for s_s^2 in Equation (37). A large s_s^2 suggests a significant variance between increments that must be controlled by estimating an acceptable minimum sample weight. A sample weight is estimated that is associated with a predetermined maximum sampling uncertainty.

2. Increment Weight (or Volume) for Well-Mixed Materials

A sampling constant (K_s or Ingamells' constant) is defined as "the weight of a single increment that must be withdrawn from a well-mixed material to hold the relative sampling uncertainty (RSD) to 1% at the 68% level of confidence."[62]

$$K_s = w\,RSD^2 \tag{39}$$

where w = weight; and RSD = relative standard deviation associated with analysis of samples of weight, w. In terms of the coefficient of variation as used previously, RSD = 100*CV. It is used primarily to characterize the material, not to determine sampling design.[60] Assuming that the sample is well mixed, the sample weight required to obtain a specified relative standard deviation can be estimated with K_s.

Example 10

The weight of a sample required to get a relative standard deviation of 5% for a well-mixed material with a K_s of 3.5 g is estimated.

$$K_s = w\,RSD^2$$
$$w = K_s/RSD^2$$
$$w = 3.5/5^2$$
$$w = 0.14 \text{ g}$$

If the sample is well mixed then K_s will be constant across different weight increments. If K_s is not constant then the material is not well mixed.

To determine if K_s is constant over a range of weights, estimates of s_s^2 are obtained for large (s_L^2) and small (s_s^2) weight increments. Wallace and Kratochivil[62] recommend that the ratio w_L/w_s be at least 10. The w_s should be kept as small as reasonable. If the sample is well mixed, then the following relationship should be true.

$$s_L^2 w_L = s_s^2 w_s \tag{40}$$

where w_s = weight of the small increments; and w_L = weight of the large increments. Wallace and Kratochvil[62] suggested that a large number of small and large increments be used to estimate s_L^2 and s_s^2. They recommended that the sample not be considered well mixed if the null hypothesis that $\sigma_L > (w_s/w_L)^{1/2}\sigma_s$ is rejected. A one-tailed F test with an α of 0.10 or lower is recommended.

3. Increment Weight (or Volume) for Poorly Mixed Materials

If the sample is not well mixed, the s_L and s_s are used to estimate the Visman's constant (degree of segregation). Wallace and Kratochvil[62] recommended 20 to 30 increments of each weight. Calculations begin with estimation of random- and segregation-associated variance in s_s^2.

$$s_s^2 = \frac{A}{wn} + \frac{B}{n} \tag{41}$$

where A = the random variance component; B = the segregation component; and n = the number of weight increments. Notice that the random variance component is lowered by increasing the number of increments analyzed or the weight of each increment. However, the variance associated with segregation in the sample is influenced only by the number of increments analyzed. This assumption that no correlation exists between increment size and variance due to segregation has been questioned in some instances. Wallace and Kratochvil[62] presented a detailed treatment of estimation when there is a correlation.

Using w_L, w_s, s_L^2, and s_s^2 estimated previously, A and B can be determined.

$$A = \frac{w_L w_s}{w_L - w_s} [s_s^2 - s_L^2] \tag{42}$$

$$B = s_L^2 - \frac{A}{w_L} = s_s^2 - \frac{A}{w_s} \tag{43}$$

An estimation of the degree of segregation (Z_s) can be determined with the above mentioned information.[59] As will be shown, Z_s is very useful in determining sample weights during design of sampling schemes.

$$Z_s = \sqrt{\frac{B}{A}} \tag{44}$$

The Z_s increases as the degree of segregation in the sample increases. A thoroughly segregated sample has a Z_s of 1. Taylor[7] suggested that it is invalid to use a single K_s when Z_s exceeds 0.05.

The optimum increment weight to be taken from the material (w_{opt}) was defined by Ingamells[60] with Equation (45). The optimum increment weight is the sample weight giving the best precision when a total weight of W_T is sampled from the field in n increments. Wallace and Kratochvil[62] stated that "when increment weights equal to A/B ... are collected ..., 50% of the sampling variance will arise from segregation regardless of the absolute values of A and B." Means of estimating weights for contributions of A to the overall variance other than 50% were discussed by Wallace and Kratochvil.[62]

$$w_{opt} = \frac{A}{B} = \frac{W_T}{n} \tag{45}$$

where W_T = the total weight of sample; and n = the number of increments analyzed.

Example 11

The following example is a modified version of one for molybdenum concentrations measured in cores of various lengths from an ore bed. To remain pertinent to ecotoxicology, measurements become molybdenum concentrations in contaminated sediments and core lengths become sample weights. Further details for this specific example are provided in Ingamells.[60]

Sixty-one samples are taken for weight increments of 0.3333 g (w_s) and 10.0000 g (w_L). The mean concentration of molybdenum was estimated to be 29.0000 mg/g. The s_s for the 61 w_s sample was 0.8380 and the s_L for the 61 w_L samples was 0.2450.

A. Using the small increment samples, K_s is estimated.

$$K_s = w\text{RSD}^2 = 0.3333(2.89) = 0.96 \text{ g}$$

Example 11 *Continued*

B. Is the sediment well mixed?

$$F_s = \sigma_L^2 / [(w_s/w_L)\,\sigma_s^2] \text{ as estimated by}$$
$$s_L^2/[(w_s/w_L)s_s^2]$$
$$= 0.2450^2/[(0.3333/10.0000)0.8380^2]$$
$$= 2.56$$

To extract $F_{\alpha(v1,v2)}$ from a table of F statistics, the degrees of freedom must be estimated for the denominator and numerator. Because both s_s^2 and s_L^2 were generated using 61 sample increments each, the degrees of freedom for both are equal to $n - 1$ or 60. $F_{0.05(60,60)}$ is approximately 1.53. Because $F_s > F_{0.05(60,60)}$, we conclude that the material is not well mixed.

C. What is the degree of segregation for the sediments?
To answer this question, A, B, and Z_s are estimated with Equations (43), (44), and (45).

$$A = 0.2210$$
$$B = 0.0379$$
$$Z_s = 0.41$$

D. What is the optimum increment weight for this sediment?
$W_{\text{opt}} = A/B = 0.2210/0.0379 = 5.85$ g.

VII.   OUTLIERS

A. OVERVIEW

Robert A. Millikan earned worldwide recognition and, in 1923, a Noble Prize, for work that included estimating the charge of an electron. More recently, he has earned an additional distinction. Although his conclusions and character remain unquestioned, his tendency to omit "bad" data has pushed him to the middle of the present-day debate about acceptable scientific behavior.[6,67,68] In his paper involving electron charge, he stated that data generated from all oil droplets were examined, but later review of his notebooks revealed that data from 49 of 140 droplets were excluded.[67] His criteria for rejection of all but the "good" data remain the focus of criticism.[6]

Advocating the other questionable extreme during the 1930s, Bessel published geodetic research in which no observations were rejected for any reason.[69] Clearly, protocol grounded in sound logic and statistics must be followed to avoid the obvious problems that could arise during the use of either of these two extreme approaches.

It is unfortunate that such extreme behaviors must be linked to two otherwise remarkable scientific careers. It was not due to a lack of available methods. Anscombe[69] indicated that formulation of statistical criteria for the rejection of outliers began in the 1850s. He also stated that outlier rejection methodologies were common by 1925. Regardless, such behaviors can be avoided by application of methods such as those described here.

Grubbs[70] suggested that the following points be considered before using a statistical method to reject or retain any suspect observations. First, if the individual making the measurement is skilled and is aware of a gross departure from the normal measurement process, then the associated observation should be rejected regardless of its "fit" to expectations. Second, before using a statistical method, one must be certain that the apparent extreme deviation from the expected distribution of observations does not result from assumption of an incorrect model. Perhaps the usual assumption of a normal distribution is invalid, and extreme values at one tail should be expected as in the case of a log normal distribution. Last, when a reason cannot be found for the anomalous observation, it should be reported, and how the observation was or was not used in interpreting results should be clearly stated.

B. SINGLE SUSPECT OBSERVATION

Several techniques are available for coping with outliers. The Winsorizing methods described for left-censored data sets could be used to avoid the effects of outliers on estimating mean and standard deviation. Dixon's test (see Dixon and Massey[26] and Grubbs[70]) is another frequently used technique for single outliers. Sokal and Rohlf[27] have given a detailed description of Dixon's test along with an example. However, Grubbs[70] suggested that the following method is better than Dixon's test for samples with single suspected outliers. Test criteria for high values and low values are given in Equations (46) and (47), respectively.

$$T = \frac{\text{outlier} - \bar{x}}{s} \tag{46}$$

$$T = \frac{\bar{x} - \text{outlier}}{s} \tag{47}$$

where $\bar{x}$ = mean of all observations; outlier = value of the suspected outlier, and s = standard deviation of all observations. The T value estimated from Equation (46) or (47) is then compared with critical values in Table 5 (Appendix).

Example 12

Mercury concentrations are measured in a series of fish samples. The following concentrations (μg/g wet weight) are generated: 0.052, 0.058, 0.084, 0.088, 0.092, 0.093, 0.124, 1.719 (mean = 0.289; standard deviation = 0.578).

The obvious question is then asked, "Is the highest value an outlier under the assumption of a normal distribution?" Equation (46) is used to estimate T because the suspected outlier is the largest observation.

$T = (1.719 - 0.289)/0.578 = 2.474$

From Table 5, a critical value for T (α = 0.05, one-tailed test) for a sample size of 8 is found to be 2.03. Because T exceeds this critical value, the suspected outlier is judged not to come from the same normal distribution as the other observations.

C. SEVERAL SUSPECT OBSERVATIONS

If several observations in a set of data are outliers, our ability to detect them may be compromised in tests such as those described for single outliers. Multiple outliers tend to mask each other in such tests. Rosner[71] developed a generalized extreme studentized deviate (ESD) procedure for examining outliers when many may be present. This generalized ESD many-outlier method can be used for data sets with single or multiple outliers when the sample size is between 25 and 500.

Rosner's general approach involves the following steps. A maximum number of possible outliers (k) is specified. Then k possible outliers are identified in the data using the following procedure. First, the mean ($\bar{x}_i$) and standard deviation (s_i) are estimated using all observations in the data set. Next, all observations (x_i values) are ranked by their R_i [Equation (48)]. An R_1 or the MAX ($|x_i - \bar{x}_i|/s_i$) is identified. This observation with the largest $|x_i - \bar{x}_i|/s_i$ is the first potential outlier. The suspect observation is now omitted before proceeding to the next step. The mean [Equation (49)] and standard deviation [Equation (50)] are estimated for the $n - 1$ remaining observations. They are used to estimate $|x_i - \bar{x}_i|/s_i$ for each of the remaining observations. A MAX ($|x_i - \bar{x}_i|/s_i$) is identified and designated as R_2. The process is continued and generates values for R for $n, n - 1, \ldots n - k + 1$ numbers of observations. The k observations associated with these R values are the suspect observations.

$$R_{i+1} = \frac{|x_i - \bar{x}_i|}{s_i} \tag{48}$$

where i = the number of observations omitted from the calculations, e.g., $i = 0$ for estimation of R_1, $i = 1$ for estimation of R_2, etc.

$$\bar{X}_i = \frac{1}{n - i} \sum_{p=1}^{n-i} x_p \tag{49}$$

where n = the number of observations in the original data set (including suspected outliers); and x_p = the value of observation p from 1 to $n - i$.

$$s_i = \sqrt{\frac{1}{n - i - 1} \sum_{p=1}^{n-i} (x_p - \bar{x}_i)^2} \tag{50}$$

Each of the R values corresponding to the k suspect observations are compared to tabulated critical values (Table 6 in the Appendix). Each critical value (λ_i) is extracted from this table using the associated total number of observations in the original sample (n), α, and $\ell + 1$ (the number of outliers that have been removed plus 1). There are no outliers if all R values are less than or equal the corresponding λ_i values. If some R_i values are greater than the corresponding λ_i values then there are ℓ outliers.

An example of Rosner's test is given below. Further discussion with another example can be found in Gilbert[23] (pp. 186–191).

Example 13

A sample of fish is analyzed for mercury concentration (μg/g wet weight). Several outliers are suspected. A maximum number of outliers is set at 5 and Rosner's techniques are applied.

			Concentration		
(x_i)	$R(n)$	$R(n - 1)$	$R(n - 2)$	$R(n - 3)$	$R(n - 4)$
0.052	0.479	0.476	0.591	1.261	1.317
0.052	0.479	0.476	0.591	1.261	1.317
0.058	0.464	0.456	0.550	1.130	1.171
0.065	0.448	0.432	0.503	0.978	1.000
0.077	0.419	0.391	0.423	0.717	0.707
0.084	0.402	0.367	0.376	0.565	0.537
0.088	0.393	0.354	0.349	0.478	0.439
0.092	0.383	0.340	0.322	0.391	0.341
0.093	0.381	0.337	0.315	0.370	0.317
0.093	0.381	0.337	0.315	0.370	0.317
0.094	0.379	0.333	0.309	0.348	0.293

Example 13 *Continued*

0.097	0.371	0.323	0.289	0.283	0.220
0.104	0.355	0.299	0.242	0.130	0.049
0.112	0.336	0.272	0.188	0.043	0.146
0.114	0.331	0.265	0.174	0.087	0.195
0.124	0.307	0.231	0.107	0.304	0.439
0.129	0.295	0.214	0.074	0.413	0.561
0.137	0.276	0.187	0.020	0.587	0.756
0.148	0.250	0.150	0.054	0.826	1.024
0.199	0.129	0.024	0.396	1.935	2.268
0.205	0.114	0.044	0.436	2.065	**2.415(R_5)**
0.210	0.102	0.061	0.470	**2.174(R_4)**	
0.793	1.286	2.044	**4.383(R_3)**		
1.390	2.707	**4.075(R_2)**			
1.719	**3.490(R_1)**				

R_1 = MAX ($|x_i - \bar{x}_i|/s_i$) for n = 3.490; $\bar{x}_i$ = 0.253, s_i = 0.420
R_2 = MAX ($|x_i - \bar{x}_i|/s_i$) for $n - 1$ = 4.075; $\bar{x}_i$ = 0.192, s_i = 0.294
R_3 = MAX ($|x_i - \bar{x}_i|/s_i$) for $n - 2$ = 4.383; $\bar{x}_i$ = 0.140, s_i = 0.149
R_4 = MAX ($|x_i - \bar{x}_i|/s_i$) for $n - 3$ = 2.174; $\bar{x}_i$ = 0.110, s_i = 0.046
R_5 = MAX ($|x_i - \bar{x}_i|/s_i$) for $n - 4$ = 2.415; $\bar{x}_i$ = 0.106, s_i = 0.041

The estimated R values for the complete data set are given in the second column (from the left). The R_1 is typed in bold at the bottom of that column. The corresponding observation value (1.719) is removed from the data set and the process is repeated. The results are tabulated in column three. R_2 is identified (bold) and the corresponding, suspect observation value (1.390) is removed before proceeding. The process is repeated again and again (columns 4, 5, and 6) until k (= 5) suspect observations are identified.

The five observations associated with R_1, R_2, R_3, R_4 and R_5 are tested as potential outliers. If $R_{i+1} > \lambda_{\ell+1}$ then the corresponding observation is an outlier. (Note that ℓ is the number of outliers with the maximum value for ℓ being k.)

i	x_i	$n - i$	$\bar{x}_i$	s_i	R_{i+1}	$\lambda_{\ell+1}$
0	1.719	25	0.253	0.420	3.490	2.82
1	1.390	24	0.192	0.294	4.075	2.80
2	0.793	23	0.140	0.149	4.383	2.78
3	0.210	22	0.110	0.046	2.174	2.76
4	0.205	21	0.106	0.104	2.415	2.73

The $\lambda_{\ell+1}$ values are taken from Rosner's table (Table 6 in the Appendix of this book) for values of n (total number of sample observations), α, and $\ell + 1$ (actual number of outliers plus 1). They are tabulated above in the extreme right column. There are three outliers (1.719, 1.390, and 0.793) as three observations have $R_{i+1} > \lambda_{\ell+1}$

Linear interpolation may be used in Rosner's table (Table 6 in the Appendix). The tables in Rosner[71] do not have values for samples with less than 25 observations. Rosner suggested that, when the sample has less than 25 observations, the very similar methods in Hawkins (ref. 72, Appendix 4) be employed. Gilbert[23] suggested that Dixon's test can be used but outlier masking may occur.

For the reader's convenience, FORTRAN code from Rosner[71] is given below. (Reprinted with permission from *Technometrics*. Copyright 1983 by the American Statistical Association and the American Society for Quality Control. All rights reserved.)

FORTRAN IV Program to Compute $R_1, \ldots, R_{NOUT}$ for an
Arbitrary Unordered Sample $X_1, \ldots, X_{NSAM}$

```
      SUBROUTINE WT (X,N,NSAM,
      NOUT, WTVEC,NI,Q)
C COMPUTE ESD OUTLIER STATISTICS
      DIMENSION X(N), WTVEC(N1), Q(N)
      II = 1
      SUM = 0.0
      SUMSQ = 0.0
      FN = 0.0
      DO 2 I = 1, NSAM
      Q(I) = 0.0
      SUM = SUM + X(I)
      SUMSQ = SUMSQ + X(I)**2
      FN = FN + 1.0
    2 CONTINUE
    1 SS = SUMSQ - (SUM**2)/FN
      S = (SS/(FN - 1.0))**.5
      XBAR = SUM/FN
      BIG = 0.0
      IBIG = 0
      DO 3 I = 1, NSAM
      IF (Q(I).EQ.1.0) GO TO 3
      A = ABS (X(I) - XBAR)
      IF (A.LE.BIG) GO TO 3
      BIG = A
      IBIG = I
    3 CONTINUE
      WTVEC (II) = BIG/S
      Q (IBIG) = 1.0
      II = II + 1
      IF (II.GT.NOUT) GO TO 999
      SUM = SUM - X (IBIG)
      SUMSQ = SUMSQ - X (IBIG)**2
      FN = FN - 1.0
      GO TO 1
  999 RETURN
      END
```

The following definitions are given for each variable in the code.

NSAM = the number of samples in the data set.
NOUT = the maximum allowed number of outliers,
N = the maximum size of any data set entered into this program,
$N1$ = the maximum number of outliers for any data set entered into this program,
X = "single precision input vector of dimension N, whose first NSAM elements represents the unordered data set to which this [program] is to be applied,"

WTVEC = "single precision output vector of dimension $N1$ whose first NOUT elements are $R1, \ldots, RNOUT$", and

Q = "single precision input vector of dimension N used internally. . . ."

VIII. SUMMARY

In this chapter, quantitative techniques for addressing the most common questions arising during the measurement process are presented. By no means should these techniques be viewed as sufficient to avoid illogical or invalid interpretation of quantitative data. The reader is directed to the references for more complete treatment of these and related topics. The reader is also referred to publications dealing with calibration curves and associated statistical methods,[73-76] bias in regression methods employing log transformation of the dependent variable,[77-81] closure problems in data sets with variables summing to a constant (for example, percentage distribution of a metal in various sediment phases),[82,83] concentrations as ratios and general treatment of ratios,[84] and completeness of water quality and analyses.[85-87]

In the first two chapters, the concept of scientific ecotoxicology has been developed and basic quantitative methods pertinent to ecotoxicology have been outlined. These chapters have provided a conceptual and quantitative foundation on which the remaining chapters are built. In the remaining chapters, quantitative methods are discussed at increasing levels of ecological organization. In the final chapter, these methods are summarized with a brief discussion of ecological assessment.

REFERENCES

1. Johnson, S. *The Complete Works of Samuel Johnson.* (Troy, NY: Princeton Pafraets Book Company, 1750, reprinted 1903).
2. Rousseau, D. L. "Case Studies of Pathological Science." *Am. Sci.* 80:54–63 (1992).
3. Grant, E. L. *Statistical Quality Control,* 3rd ed. (New York, NY: McGraw-Hill Book Company, 1964).
4. Zimmerman, S. W. "Quality Assurance and Science—Oil and Water?" *Energy Update* September 3–4 (1990).
5. Chamberlin, T. C. "The Method of Multiple Working Hypotheses." *J. Geol.* 5:837–848 (1897).
6. Ayala, F., R. M. Adams, M.-D. Chilton, G. Holton, D. Hull, K. Patel, F. Press, M. Ruse, and P. Sharp. *On Being a Scientist.* (Washington, D. C.: National Academy Press, 1989).
7. Taylor, J. K. *Quality Assurance of Chemical Measurements.* (Chelsea, MI: Lewis Publishers, Inc., 1987).
8. Rayl, A. J. S. "Misconduct Case Stresses Importance of Good Notekeeping." *The Scientist* November 11:18–19 (1991).
9. Keith, L. H., W. Crummett, J. Deegan, Jr., R. A. Libby, J. K. Taylor, and G. Wentler. "Principles of Environmental Analysis." *Anal. Chem.* 55:2210–2218 (1983).
10. Palca, J. "Get-the-Lead-Out Guru Challenged." *Science* 253:842–844 (1991).
11. Koshland, D. L., Jr. "Fraud in Science." *Science* 235:141 (1987).
12. Culliton, B. J. "Random Audits of Papers Proposed." *Science* 242:657–658 (1988).
13. Currie, L. A. "Limits of Qualitative Detection and Quantitative Determination." *Anal. Chem.* 40(3):586–593 (1968).

14. Newman, M. C., and J. E. Pinder, III. 1987. "Coping with Uncertainty: Limits of Detection, Limits of Quantitation and Nested Sources of Error." In *Proceedings of the AWWA Water Quality Technology Conference* (Denver: AWWA Water Quality Association, 1987) 509–532.

15. Schwartz, L. M. "Lower Limit of Reliable Assay Measurement with Nonlinear Calibration." *Anal. Chem.* 55:1424–1426 (1983).

16. Arellano, S. D., M. W. Routh, and P. D. Dalager. "Criteria for Evaluation of ICP-AES Performance." *Am. Lab.* August 1985:20–32 (1985).

17. Synovec, R. E., and E. S. Yeung. "Improvement of the Limit of Detection in Chromatography by an Integration Method." *Anal. Chem.* 57:2162–2167 (1985).

18. Donn, J. J., and R. L. Wolke. "The Statistical Interpretation of Counting Data from Measurements of Low-Level Radioactivity." *Health Phys.* 32:1–14 (1977).

19. Alford-Stevens, A. L. "Mixture Analytes. Procedures for Determining Detection Limits Should be Defined." *Environ. Sci. & Technol.* 21(2):137–139 (1987).

20. Helsel, D. R., and R. J. Gilliom. "Estimation of Distributional Parameters for Censored Trace Level Water Quality Data. 2. Verification and Applications." *Water Resour. Res.* 22(2):147–155 (1986).

21. Gilbert, R. O., and R. R. Kinnison. "Statistical Methods for Estimating the Mean and Variance from Radionuclide Data Sets Containing Negative, Unreported or Less-Than Values." *Health Phys.* 40:377–390 (1981).

22. Gilliom, R. J., R. M. Hirsch, and E. J. Gilroy. "Effect of Censoring Trace-Level Water-Quality Data on Trend-Detection Capability." *Environ. Sci. & Technol.* 18:530–539 (1984).

23. Gilbert, R. O. *Statistical Methods for Environmental Pollution Monitoring.* (New York: Van Nostrand Reinhold Co., 1987).

24. Porter, P. S., R. C. Ward, and H. F. Bell. "The Detection Limit. Water Quality Monitoring Data Are Plagued with Levels of Chemicals That Are Too Low to Be Measured Precisely." *Environ. Sci. & Technol.* 22:856–861 (1988).

25. Newman, M. C., P. M. Dixon, B. B. Looney, and J. E. Pinder, III "Estimating Mean and Variance for Environmental Samples with Below Detection Limit Observations." *Water Resour. Bull.* 25(4):905–916 (1989).

26. Dixon, W. J., and F. J. Massey, Jr. *Introduction to Statistical Analysis.* (New York: McGraw-Hill Book Company, 1969) 330–332.

27. Sokal, R. R., and F. J. Rohlf. *Biometry,* 2nd ed. (New York: W.H. Freeman and Company, 1981) 413–414.

28. Berthouex, P. M., and S. W. Hinton. *Estimating the Mean of Data Sets that Include Measurements Below the Limit of Detection,* NCASI Technical Bulletin 621. (New York, NY: National Council of the Paper Industry for Air and Stream Improvement, Inc., 1991).

29. Cohen, A. C., Jr. "Estimating the Mean and Variance of Normal Populations from Singly Truncated and Doubly Truncated Samples." *Ann. Math. Stat.* 21:557–569 (1950).

30. Cohen, A. C., Jr. "On the Solution of Estimation Equations for Truncated and Censored Samples from Normal Populations." *Biometrika* 44:225–236 (1957).

31. Cohen, A. C., Jr. "Simplified Estimators for the Normal Distribution When Samples Are Singly Censored or Truncated." *Technometrics* 1(3):217–237 (1959).

32. Cohen, A. C., Jr. "Tables for Maximum Likelihood Estimates: Singly Truncated and Singly Censored Samples." *Technometrics* 3(4):535–541 (1961).

33. Cohen, A. C., Jr. "Estimation of the Poisson Parameter from Truncated Samples and from Censored Samples." *J. Am. Stat. Assoc.* 3:158–168 (1954).

34. Cohen, A. C., Jr. "Multicensored Sampling in the Three Parameter Weibull Distribution." *Technometrics* 17(3):347–351 (1975).

35. Cohen, A. C., Jr., and Norgaard, N. J. "Progressively Censored Sampling in the Three-Parameter Gamma Distribution." *Technometrics* 19 (3):333–340 (1977).

36. Schneider, H. *Truncated and Censored Samples from Normal Populations.* (New York, NY: Marcel Dekker, 1986).

37. Schneider, H., and L. Weissfeld. "Inference Based on Type II Censored Samples." *Biometrics* 42:531–536 (1986).

38. Gupta, A. K. "Estimation of the Mean and Standard Deviation of a Normal Population from a Censored Sample." *Biometrika* 39:260–273 (1952).

39. Sarhan, A. E., and B. G. Greenberg. "Estimation of Location and Scale Parameters by Order Statistics from Singly and Doubly Censored Samples. Part I. The Normal Distribution up to Samples of Size 10." *Ann. Math. Stat.* 27:427–451 (1956).

40. Sarhan, A. E., and B. G. Greenberg. "Estimation of Location and Scale Parameters by Order Statistics from Singly and Doubly Censored Samples. Part II. Tables for the Normal Distribution for Samples of Size $11 \leq n \leq 15$." *Ann. Math. Stat.* 29:79–105 (1958).

41. Sarhan, A. E., and B. G. Greenberg. *Contributions to Order Statistics.* (New York, NY: John Wiley & Sons, Inc., 1962).

42. Gleit, A. "Estimation of Small Normal Data Sets with Detection Limits." *Environ. Sci. & Technol.* 19:1201–1206 (1985).

43. Aitchison, J., and J. A. C. Brown. *The Lognormal Distribution with Special Reference to Its Use in Economics.* (New York, NY: Cambridge University Press, 1957).

44. Finney, D. J. "On the Distribution of a Variate Whose Logarithm Is Normally Distributed." *J. R. Stat. Soc. Suppl.* 7:155–161 (1941).

45. Gilliom, R. J., and D. R. Helsel. "Estimation of Distributional Parameters for Censored Trace Level Water Quality Data. 1. Estimation Techniques." *Water Resour. Res.* 22 (2):135–146 (1986).

46. Helsel, D. R. "Less Than Obvious. Statistical Treatment of Data Below the Detection Limit." *Environ. Sci. & Technol.* 24 (12):1766–1774 (1990).

47. Newman, M. C., and P. M. Dixon: "UNCENSOR: A Program to Estimate Means and Standard Deviations for Data Sets with Below Detection Limit Observations." *Am. Environ. Lab.* 4/90:27–30 (1990).

48. Settle, D. M., and C. C. Patterson. "Lead in Albacore: Guide to Lead Pollution in Americans." *Science* 207:1167–1176 (1980).

49. Windom, H. L., J. T. Byrd, R. G. Smith, Jr., and F. Huan. "Inadequacy of NASQAN Data for Assessing Metal Trends in the Nation's Rivers." *Environ. Sci. & Technol.* 25(6):1137–1142 (1991).

50. Boatman, C. D. Comment on "Inadequacy of NASQAN Data for Assessing Metal Trends in the Nation's Rivers." *Environ. Sci. & Technol.* 25(11):1940–1941 (1991).

51. Jay, P. C. "Anion Contamination of Environmental Water Samples Introduced by Filter Media." *Anal. Chem.* 57:780–782 (1985).

52. Moody, J. R. "NBS Clean Laboratories for Trace Element Analysis." *Anal. Chem.* 54(13):1358A–1376A (1982).

53. Taylor, J. K. "Guidelines for Evaluating the Blank Correction." *J. Test. Eval.* 12(1):54–55 (1984).

54. Provost, L. P., and R. S. Elder. "Interpretation of Percent Recovery Data." *Am. Lab.* December 1983:57–63 (1983).

55. Van Dobben De Bruyn, C. S. *Cumulative Sum Tests in Theory and Practice.* (New York, NY: Hafner Publishing Company, 1968) 82.

56. U. S. EPA. *Handbook for Analytical Quality Control in Water and Wastewater Laboratories.* (Cincinnati, OH: Analytical Quality Control Laboratory, 1972).

57. U. S. EPA. *Region VI Laboratory Quality Control Manual,* 2nd ed. (Ada, OK: U. S. EPA Analytical Quality Control Program, 1972).

58. Pinder, J. E., III, and J. P. Giesy. "Frequency Distributions of the Concentrations of Essential and Nonessential Elements in Largemouth Bass, *Micropterus salmoides.*" *Ecology* 62(2):456–468 (1981).

59. Visman, J. "A General Sampling Theory." *Mat. Res. Stand.* 9:9–64 (1969).

60. Ingamells, C. O. "New Approaches to Geochemical Analysis and Sampling." *Talanta* 21:141–155 (1974).

61. Ingamells, C. O., and P. Switzer. "A Proposed Sampling Constant for Use in Geochemical Analysis." *Talanta* 20:547–568 (1973).

62. Wallace, D., and B. Kratochvil. "Visman Equations in the Design of Sampling Plans for Chemical Analysis of Segrated Bulk Materials." *Anal. Chem.* 59:226–232 (1987).

63. Stuart, A. *Basic Ideas of Scientific Sampling.* (New York, NY: Hafner Press, 1976).

64. Kupper, L. L., and Kerry B. Hafner. "How Appropriate Are Popular Sample Size Formulas?" *Am. Stat.* 43(2):101–105 (1989).

65. Blackwood, L. G. "Assurance Levels of Standard Sample Size Formulas." *Environ. Sci. & Technol.* 25(8):1366–1367 (1991).

66. Cochran, W. G. *Sampling Techniques,* 3rd ed. (New York, NY: John Wiley & Sons, 1977).

67. Sigma Xi. *Honor in Science.* (Research Triangle Park, NC: Sigma Xi, The Scientific Research Society, Inc., 1991).

68. Goodstein, D. "What Do We Mean When We Use the Term 'Science Fraud'?" *The Scientist* 3/2:11–12 (1992).

69. Anscombe, F. J. "Rejection of Outliers." *Technometrics* 2(2):123–147 (1960).

70. Grubbs, F. E. "Procedures for Detecting Outlying Observations in Samples." *Technometrics* 11(1):1–21 (1969).

71. Rosner, B. "Percentage Points for a Generalized ESD Many-Outlier Procedure." *Technometrics* 25(2):165–172 (1983).

72. Hawkins, D. M. *Identification of Outliers.* (New York, NY: Chapman and Hall, 1980).

73. VanArendonk, M. D., R. K. Skogerboe, and C. L. Grant. "Correlation Coefficients for Evaluation of Analytical Calibration Curves." *Anal. Chem.* 53:2349–2350 (1981).

74. Meyer, E. F. "Comments on Curve-Fitting Methods." *Anal. Chem.* 54:1878–1879 (1982).

75. Christian, S. D., and E. E. Tucker. "Least Squares Analysis with the Microcomputer. Part Five: General Least Squares with Variable Weighting." *Am. Lab.* 84: 18–32 (1984).

76. Helland, I. S. "On the Interpretation and Use of R^2 in Regression Analysis." *Biometrics* 43:61–69 (1987).

77. Beauchamp, J. J., and J. S. Olson. "Correction of Bias in Regression Estimates After Logarithmic Transformation." *Ecology* 54:1403–1407 (1973).

78. Duan, N. "Smearing Estimate: A Nonparametric Retransformation Method." *J. Am. Stat. Assoc.* 78 (383):605–610 (1983).

79. Koch, R. W., and G. M. Smillie. "Bias in Hydrologic Prediction Using Log-Transformed Regression Models." *Water Resour. Bull.* 22(5):717–723 (1986).

80. Newman, M. C. "A Statistical Bias in the Derivation of Hardness-Dependent Metals Criteria." *Environ. Toxicol. Chem.* 10:1295–1297 (1991).

81. Newman, M. C., and M. G. Heagler. "Allometry of Metal Bioaccumulation and Toxicity." In *Metal Ecotoxicology,* ed. M. C. Newman and A. W. McIntosh. (Chelsea, MI: Lewis Publishers, 1991) 91–130.

82. Johansson, E., S. Wold, and K. Sjodin. "Minimizing Effects of Closure on Analytical Data." *Anal. Chem.* 56:1685–1688 (1984).

83. Newman, M. C., and R. Mealy. "Multivariate Methods Used to Dissect Water Chemistry Data from Systems Receiving Thermal Effluents." In *Proceedings of the Southeastern Workshop on Aquatic Ecological Effects of Power Generation,* Report 124, ed.

K. Mahadevan, R. K. Evans, P. Behrens, T. Biffar, and L. Olsen. (Sarasota, FL: Mote Marine Laboratory, 1988) 165–185.

84. Doctor, P. G., R. O. Gilbert, and J. E. Pinder III. "An Evaluation of the Use of Ratios in Environmental Transuranic Studies." *J. Environ. Qual.* 9(4):539–546 (1980).

85. Johnsson, P. A., and D. G. Lord. *A Computer Program for Geochemical Analysis of Acid-Rain and Other Low-Ionic-Strength, Acidic Waters,* U.S. Geological Survey Water-Resources Investigations Report 87-4095. (Denver, CO: U.S.G.S. Books and Open-File Reports, 1987).

86. Peden, M. E., S. R. Bachman, C. J. Brennan, B. Demir, K. O. James, B. W. Kaiser, J. M. Lockard, J. E. Rothert, J. Sauer, L. M. Skowron, and M. J. Slater. *Development of Standard Methods for the Collection and Analysis of Precipitation,* EPA/600/4-86/024. (Cincinnati, OH: Environmental Monitoring and Support Laboratory, 1986).

87. U.S. EPA. *National Stream Survey Phase I: Quality Assurance Report,* EPA/600/4-88/018. (Las Vegas, NV: U.S. EPA, Environmental Monitoring Systems Laboratory, 1988).

Chapter 3

Bioaccumulation

Models are, for the most part, caricatures of reality, but if they are good, then, like good caricatures, they portray, though perhaps in distorted manner, some of the features of the real world.[1]

I. GENERAL

Toxicant bioaccumulation became a topic of public and scientific concern early in the 1950s. Rachel Carson's *Silent Spring* thrust the consequences of pesticide accumulation by wildlife species into the public's awareness. Contributing to this awakening were the tragic consequences of heavy metal bioaccumulation in food species so apparent after outbreaks of Minamata and Itai-itai disease. Accumulation of fission products in food species and humans from nuclear weapons testing also became an issue of concern.

Fortunately, much has been accomplished in the area of bioaccumulation during the past four decades. After World War II, considerable talent was focused on quantitative prediction of uptake and elimination of radionuclides by human and nonhuman species. An abundance of quantitative techniques grew from these efforts (see Whicker and Schultz[2] as evidence). Sentinel species were also developed to monitor other pollutants in marine and freshwater systems. Pharmacokinetic models developed in the late 1930s and synthesized in such sources as Atkins[3] and Wagner[4] were readily adopted by ecologists predisposed to compartment modeling.

Today, an impressive body of knowledge has been developed for toxicant bioaccumulation. Although this body of knowledge provides a sound foundation on which to build, much more remains to be learned. Indeed, fundamental issues remain unresolved despite this body of knowledge. For example, ambiguity still exists regarding the best means of expressing[5,6] and the major factors underlying[7] variation in toxicant concentrations among individuals from the same population. Also, as pointed out recently by Barron et al.,[8] many recent advances made in pharmacokinetic or toxicokinetic modeling have yet to be incorporated into ecotoxicological studies. It is the purpose of this chapter to outline techniques presently in use and those potentially useful in ecotoxicology.

II. MODELING BIOACCUMULATION: GENERAL APPROACH

Models of mathematically distinct compartments are the rule in ecotoxicology. Although these compartments are sometimes linked to specific organs (excretion by the kidneys) or tissue pools (loss from fatty tissues), more often than not they are treated as simply kinetically distinct (fast or slow elimination) compartments. Some speculation may be made to link the mathematical compartment to a physical compartment. In other studies, clear linkage to physical compartments is made. A good example of such effective linkage is the work of Lyon et al.[9] with metal clearance from the hemolymph of crayfish. The clearances of several metals from a fast compartment were correlated with protein ligand binding and conformed to predictions of the Irving-Williams series. Unfortunately, speculation can be based as much on preconceptions as on the direct results of the study. Consequently, it is critical for the reader to remember clearly the distinction between physical and kinetic compartments when interpreting reports of bioaccumulation modeling.

A. ELIMINATION
1. General

Barron et al.[8] have recently emphasized the distinctions among the terms depuration, clearance, and elimination although they are used frequently as synonyms in ecotoxicology. Depuration is the loss of a toxicant from the organism after it has been placed in an environment devoid of toxicant. This term is associated with a particular experimental design involving the transfer of organisms that have accumulated elevated levels of toxicant to an environment with no significant sources of the toxicant. In the design, much effort is made to ensure that toxicant lost to the environment is not available to be taken up again during the course of depuration. Clearance "is best interpreted as the rate of substance transferred, normalized to a concentration, and has units of flow (e.g. mL/h)." It can describe transfer of materials between compartments within the organism. It will be discussed later in the specific context of clearance volume-based models. In this text, elimination is the summation of metabolic, excretory, physiochemical, and radioactive decay processes resulting in the decrease of toxicant in the organism. (This definition adds physiochemical and radioactive decay to processes that Barron et al.[8] list as contributing to elimination.) Metabolic degradation of an organic toxicant would contribute to elimination under this definition as well as renal excretion. Radioactive decay could be included as part of elimination if the loss of a radioactive contaminant was being modeled. However, if the radionuclide is simply being used as a tracer in the experimental design, radioactive decay would not be incorporated as a component of elimination. Growth dilution, the apparent decrease in toxicant concentration resulting from organism growth, would not be considered a component of elimination. With growth dilution, the total amount of toxicant in the organism would not decrease as a consequence of growth. The concentration would decrease because the amount of tissue in which the toxicant was distributed increased.

Elimination occurs through a complex array of physical, chemical, and biological mechanisms. As mentioned briefly, oxidation or hydrolysis linked perhaps to an additional "synthetic" biotransformation[10] can render a nonpolar organic toxicant to a water-soluble form amenable to excretion. (Note that increased water solubility does not necessarily lead to enhanced elimination.) Other general processes including transport across gills, exhalation, secretion via the gallbladder or hepatopancreas, molting, excretion, or egg deposition have been noted in reviews by Whicker and Schultz[2] and Spacie and Hamelink.[11] Many such biological processes are not temporally constant, e.g., developmental changes in biliary excretion of methylmercury.[12] Physical mechanisms may also be important. Radioactive decay can be significant in modeling elimination of many short-lived radionuclides. Also, physical desorption of surface-bound toxicants[13] can contribute to elimination. Toxicants can be lost from plants by leaching, herbivore grazing, and leaf fall.[2]

Despite this remarkable diversity of elimination mechanisms, most quantitative treatments of elimination draw on a few mathematical models. Perhaps this is primarily a consequence of some major underlying principle, e.g., the predominance of apparent first-order kinetics. This is at least partially true and has greatly enhanced progress. Regardless, it is profitable to assume that habit or tradition also plays a role in the uniformity of approaches taken. Under this assumption, nontraditional approaches are presented in this chapter alongside generally applied methods. They are not presented as superior or inferior to traditional methods: they are presented as alternatives to be critically evaluated during model selection and development.

2. Reaction Order

During this discussion, assumptions are made regarding the apparent reaction order of elimination kinetics. Consequently, a brief discussion of reaction order is warranted.

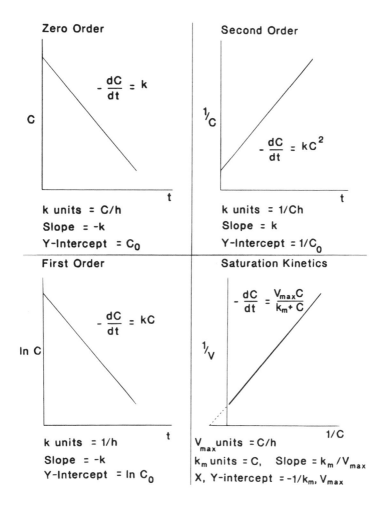

Figure 1 Zero-order, first-order, second-order, and saturation kinetics. Slopes and intercepts from linear plots such as those shown here are often used to estimate parameters describing the reaction kinetics.

In the case of a reaction involving only one reactant, reaction order refers to the power to which the reactant concentration is raised in the equation describing the reaction rate ($-dC/dt = kC^n$). Most reactions germane to this chapter are described by low-order (zero, first, or, rarely, second; $n = 0$, 1, or 2) or Michaelis-Menten saturation kinetics. Figure 1 summarizes these four reaction kinetics models. With zero-order kinetics, the reaction rate is independent of concentration, i.e., proportional to concentration raised to the 0 power ($C^0 = 1$). The rate is determined by C^1 with first-order kinetics. First-order kinetics are so pervasive in bioaccumulation modeling that they are assumed unless specified otherwise.[8] Second-order kinetics may involve a single reactant or two different reactants. In the first case, the reaction rate is determined by C^2. When two different reactants are involved (C_1, C_2), the order of the reaction is determined by the sum of their exponents. For example, if $-dC_1/dt = kC_1^1 C_2^1$ then the reaction order is $1 + 1$ or 2.

With many enzyme-catalyzed reactions, saturation kinetics (Michaelis-Menten kinetics) can be relevant. With Michaelis-Menten kinetics, the reactant (substrate, S) combines with the enzyme (E) to form a complex (ES). Next, the substrate in the complex is

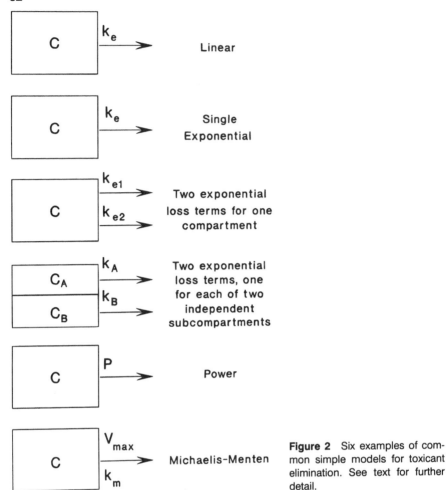

Figure 2 Six examples of common simple models for toxicant elimination. See text for further detail.

converted to product (*P*). The assumption is made that the complex formation step is reversible and the product formation step proceeds as a first-order reaction. However, the overall rate of reaction for $E + S \leftrightarrows ES \rightarrow E + P$ may shift from first order to zero order if the concentration of the reactant (*S*) increases beyond a certain point. With a finite amount of enzyme present, a concentration of substrate can be reached above which essentially all the available enzyme molecules are saturated. Beyond this concentration, the reaction will proceed at a rate independent of substrate concentration.

3. Linear Elimination Model

In the context of zero-order elimination kinetics, the rate at which the toxicant is eliminated from the organism (or compartment) would be independent of the concentration in the organism (Figures 2 and 3, linear). Such behavior could be modeled using Equation (1)

$$C_t = C_0 - k_e t \tag{1}$$

where C_t = concentration at time, t; C_0 = concentration at $t = 0$; and k_e = a constant with units of concentration/h.

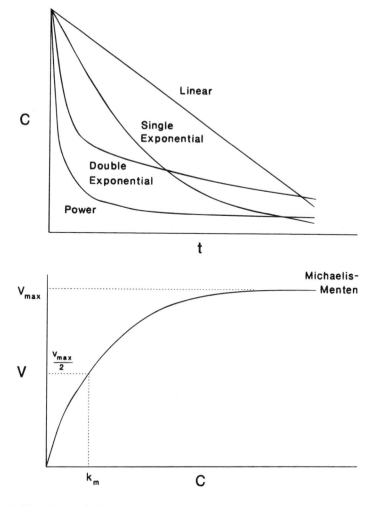

Figure 3 Illustrations of elimination kinetics associated with the models shown in Figure 2 and described in the text.

Extremely slow release of lead from bone can approximate zero-order kinetics when measured over a short time span. Schulz-Baldes[14] provides a detailed treatment of apparent linear kinetics for elimination of lead from the blue mussel, *Mytilus edulis*. As mentioned above, an enzyme-mediated mechanism under saturation conditions is another situation in which kinetics will display zero-order-like behavior.

4. Single (Mono-) Exponential Elimination Model

The most common model for elimination is based on first-order kinetics. In this case, the rate of elimination would be directly proportional to the concentration of toxicant in the organism at any time. A simple, first-order model with one pool of toxicant is described in Equation (2) and illustrated in Figures 2 and 3 (single exponential).

$$C_t = C_0 e^{-k_e t} \tag{2}$$

where k_e = elimination rate constant with units of 1/h.

The k_e in this equation describes the fractional change in concentration per hour. There are many examples of toxicant elimination displaying such monoexponential elimination. This model also describes simple radioactive decay kinetics; however, in that case, the decay rate constant is normally designated as λ, not k_e.

The time needed for the concentration to drop by a factor of 2 (one-half that present at the beginning of any time interval) is the elimination half-life or half-time. For the simple model described by Equation (2), it can be estimated by Equation (3).

$$t_{1/2} = \frac{\ln 2}{k_e} \tag{3}$$

Half-life estimates are derived most often by measuring loss at several intervals over time. However, design limitations can necessitate estimation based on two measurement times only, e.g., Steele et al.[15] and Scott et al.[16] It is important to realize that two-measurement methods are more sensitive to error than methods using several measurement intervals. Simulations by Phillips[17] suggested that selection of large time intervals and reduction of analytical error to a minimum can minimize the associated difficulties. Regardless, before using two-measurement methods, the reader should examine the work of Phillips.[17]

The mean lifetime of a particle in the compartment (τ) as defined by Equation (4) is $1.44t_{1/2}$.[2] This term is also called the mean residence time or turnover time.[18]

$$k_e = \frac{1}{\tau} \tag{4}$$

Example 1

Elimination of anthracene is measured in fish removed from a contaminated environment and placed in a laboratory tank receiving a continuous flow of uncontaminated water. The (fictitious) data set for elimination over 4 h is given in the first two columns of the table below. ($\ln C$ = natural logarithm of nanomoles of anthracene/g; ϵ = regression residual.)

Time (h)	Anthracene (nmol/g)	$\ln C$	Predicted $\ln C$	ϵ	ϵ^2
0	100	4.605	4.597	0.008	0.000064
1	53	3.970	3.917	0.053	0.002809
2	24	3.178	3.237	−0.059	0.003481
3	12	2.485	2.557	−0.072	0.005184
4	7	1.946	1.877	0.069	0.004761
				$\Sigma\,\epsilon^2$ =	0.016299

The natural logarithm of anthracene concentration ($\ln C$) and time (t) were fit using least-squares, linear regression to produce the following model: $\ln C = -0.680t + 4.597$. The associated correlation coefficient ($r^2 = 0.996$) indicated that more than 99% of the variation in $\ln C$ could be accounted for by the model. The elimination rate constant is estimated to be -slope (see Figure 1, first order) or 0.680/h. The $t_{1/2} = \ln 2/k_e = 0.693/0.680 = 1.02$ h.

According to accepted practices in the field, this linear equation can be backtransformed to arithmetic units to yield the following model, C (nmol/g) $= 99.19e^{-0.680t}$. Predictions of concentrations at any time are then made. However, unless there is no error in the

Example 1 *Continued*

model or the predicted median value is desired, predictions from such a backtransformed model will be biased. The bias arises from the fact that, although the arithmetic mean of ln Y is predicted accurately in the linear model, the arithmetic mean of Y is not accurately predicted in the backtransformed model. The median of Y (maximum likelihood estimate) is predicted in the backtransformed model. (See Beauchamp and Olson,[19] Sprugel,[20] Newman and Heagler,[21] and Newman[22] for discussion of this bias.) The magnitude of this bias will increase as the error variance of the model increases. The prediction bias can be estimated from the mean square error (MSE) if the residuals (ϵ) seem to be normally distributed as in the present example. The MSE is the sum of the ϵ_i^2 values divided by $n-2$ where n is the number of data pairs. The MSE is 0.016299/3 or 0.005433 in this example. The bias can be estimated using $e^{MSE/2}$ to be $e^{0.005433/2}$ or 1.0027. Any predicted value can be multiplied by 1.0027 to generate an unbiased prediction from the backtransformed model. For example, the initial anthracene concentration is estimated to have been the following.

$$\text{Biased estimate:} \qquad C = 99.19e^{-0.680t}$$
$$= 99.19e^{0}$$
$$= 99.19 \; nmol/g$$
$$\text{Unbiased estimate:} \qquad C_{\text{unbiased}} = 1.0027*99.19$$
$$= 99.46 \; nmol/g$$

The bias is relatively insignificant in this example but, in other cases, it may be important to consider.

If the residuals had not been normally distributed, the bias could have been estimated with a "smearing estimate."[21,23] The smearing estimate is $1/n(\Sigma e^{\epsilon_i})$. Each individual residual is used as the exponent for e and then summed. This sum is then multiplied by $1/n$ to obtain an estimate of the transformation bias.

In this example, the bias was insignificant, but this is often not the case. The bias should be estimated if models are fit using logarithmic transforms of the dependent variable and the results are used for predictive purposes. If the model is generated for descriptive rather than predictive purposes, the MSE should still be included in model presentation. Inclusion of the MSE will allow evaluation of the bias before future implementation for predictive purposes by other researchers. Finally, if the predicted median Y is acceptable or preferred, this bias correction is unnecessary.

5. Elimination Model with Two Loss Terms for One Compartment

If there are two components of elimination removing toxicant from the compartment (organism), Equation (2) can be modified to describe the resulting kinetics (Figure 2, two exponential loss terms for one compartment). The resulting time course of elimination would be similar to that of the single exponential in Figure 3.

$$C_t = C_0 e^{-(k_{e1}+k_{e2})t} \tag{5}$$

where k_{e1} and $k_{e2} = $ two distinct elimination rate constants. Equation (5) is useful for studying elimination of a radioisotope with significant radioactive decay. In that case, the decay rate constant (λ) is used in place of one of the elimination rate constants.

The elimination half-life of a compartment with more than one elimination component is defined as an effective half-life by Whicker and Schultz[2] and estimated by Equation (6).

$$t_{eff} = \frac{\ln 2}{k_{eff}} \tag{6}$$

where

$$k_{eff} = \sum_{i=1}^{n} k_{ei} \tag{7}$$

with k_{ei} being the ith k_e. Each elimination component has an elimination half-life defined by Equation (3).

Example 2

Cutshall[18] estimated depuration of ^{65}Zn from oysters removed downstream from the Department of Energy's Hanford Facility nuclear production reactors. The objective of the analyses was to define the kinetics of radioactive contaminant loss from a food species. Visually extracted data from Cutshall's Figure 1a were used in this illustration of modeling elimination kinetics. (It should be noted here and elsewhere that, when visually extracted data are used in examples, the data will be slightly different from the original data and, consequently, results will not be directly comparable with those in the original paper.)

What is the k_e for ^{65}Zn elimination? The following SAS program[24] fits these data to Equation (5) using nonlinear least-squares regression methods with and without weighting. It also fits the transformed data (ln ^{65}Zn activity) to the model [Equation (5)] using a linearized form, ln ^{65}Zn activity $= b*$day $+ a$ (see Figure 1, first-order kinetics). A λ of 0.00283 is used for ^{65}Zn.

```
DATA OYSTERZN;
    INPUT DAY ZINC @@;
    WNLIN2 = DAY**2;
    LZINC = LOG (ZINC);
    CARDS;
001 700 001 695 001 675 001 630 001 606 001 540 001 505 001 470
001 460 001 455 001 395 001 355 001 300 001 275 001 270 001 260
010 340 010 500 020 280 020 570 020 525 025 535 025 345 030 302
030 345 035 320 040 275 045 275 060 375 060 260 060 245 075 155
080 190 080 350 090 410 090 360 095 370 100 195 115 400 125 175
125 095 130 320 130 325 140 325 160 145 160 140 170 150 175 290
185 125 200 135 210 180 220 055 255 100 260 140 260 110 280 090
295 160 315 055 325 075 325 045 355 085 365 075 385 065 410 045
420 070 420 030 450 060 465 065 505 025 515 025 550 015 555 020
625 020
;
/* USING NONLINEAR REGRESSION WITH PARTIAL DERIVATIVES GIVEN.   */
/* INITIAL ESTIMATES AND BOUNDS ARE SELECTED BY CURSORY DATA    */
/* EXAMINATION. KE = THE ELIMINATION RATE CONSTANT, INITACT =   */
/* THE ACTIVITY OF 65-ZN IN THE OYSTER AT TIME = 0 DAYS, AND    */
/* 0.00283 IS THE DECAY RATE CONSTANT FOR 65-ZN                 */
/*                                                              */
/* FIRST - NONLINEAR REGRESSION WITH NO WEIGHTING               */
PROC NLIN;
    PARMS KE=0.004, INITACT=500;
    BOUNDS KE>0,
        1000>INITACT>0;
```

Example 2 *Continued*

```
    MODEL ZINC=INITACT*EXP ((− (KE+0.00283)) *DAY);
      DER. INITACT=EXP ((− (KE+0.00283)) *DAY);
      DER. KE=−INITACT*DAY*EXP ((− (KE+0.00283)) *DAY);
      OUTPUT OUT=PRED PREDICTED=PPCI RESIDUAL=RPCI;
RUN;
/* PLOTTING THE PREDICTED AND ORIGINAL ACTIVITY DATA VERSUS DAY */
/* AND THEN THE REGRESSION RESIDUALS VERSUS DAY           */
PROC PLOT;
    PLOT ZINC*DAY PPCI*DAY = "*"/OVERLAY;
    PLOT RPCI*DAY="*"/VREF=0;
RUN;
/* SECOND - NONLINEAR REGRESSION WITH DAY-SQUARED WEIGHTING  */
PROC NLIN;
    PARMS KE=0.004, INITACT=500;
    BOUNDS KE>0,
      1000>INITACT>0;
    MODEL ZINC=INITACT*EXP ((− (KE+0.00283)) *DAY);
      _WEIGHT_=WNLIN2;
      DER.INITACT=EXP ((− (KE+0.00283)) *DAY);
      DER.KE=−INITACT*DAY*EXP ((−(KE+0.00283))*DAY);
      OUTPUT OUT=PRED PREDICTED=W2PPCI RESIDUAL=W2RPCI;
RUN;
/* PLOTTING THE PREDICTED AND ORIGINAL ACTIVITY DATA        */
/*                                           VERSUS DAY  */
/* AND THEN THE REGRESSION RESIDUALS VERSUS DAY.            */
PROC PLOT;
    PLOT ZINC*DAY W2PPCI*DAY="*"/OVERLAY;
    PLOT W2RPCI*DAY="*"/VREF=0;
RUN;
/* NOW LEAST-SQUARES LINEAR REGRESSION OF LN 65-ZN VERSUS DAY */
/* WILL BE USED TO ESTIMATE KE. NOTE THAT, IN THIS MODEL, THE  */
/* DECAY RATE CONSTANT ISN'T INCLUDED. THEREFORE, THE ELIMINA- */
/* TION RATE CONSTANT IS THE SUM OF KE AND THE DECAY RATE CON- */
/* STANT FOR 65-ZN. TO DERIVE KE, THE DECAY RATE CONSTANT IS   */
/* SUBTRACTED FROM THE ESTIMATE FROM THE MODEL.               */
PROC GLM;
    MODEL LZINC=DAY;
    OUPUT OUT=LINEAR PREDICTED=PPCILIN RESIDUAL=RPCILIN;
RUN;
/* PLOTTING THE PREDICTED AND ORIGINAL ACTIVITY DATA        */
/*                                           VERSUS DAY  */
/* AND THEN THE REGRESSION RESIDUALS VERSUS DAY.            */
PROC PLOT;
    PLOT LZINC*DAY PPCILIN*DAY="*"/OVERLAY;
    PLOT RPCILIN*DAY="*"/VREF=0;
RUN;
```

The results of these calculations are illustrated in Figures 4 and 5. The unweighted, nonlinear regression model fit these data adequately (Figure 4A). The resulting estimates (± asymptotic standard error) of k_e and the initial ^{65}Zn activity were 0.00268 ± 0.00067/ h and 465 ± 20 pCi/g, respectively. However, examination of the regression residuals (Figure 4B) indicates that the assumption of independence between the variance of ^{65}Zn activity and day is incorrect. The variance decreases rapidly as duration of depuration increases.

Example 2 *Continued*

Weighting by some function of the independent variable (day) is one means of coping with such unequal variance or heteroscedasticity. For example, if the error variance increased linearly with an increase in time, a weighting of 1/day could be used. An observation toward the end of the time course would have less weight in the fitting than one at the beginning of the depuration process. In the present case, the error variance seemed to decrease rapidly as day increased and a weighting of day^2 was used. The commonly used weighting[25] of Y^2 or, in this case, pCi^2 was examined but was less effective than a weighting of day.^{2} (For a further discussion of weighting, please see Neter et al.,[26] pages 420–423, or Christian and Tucker[27]). Weighted, nonlinear regression results (k_e = 0.00244 ± 0.00035/h; initial ^{65}Zn activity = 454 ± 38 pCi/g) were slightly lower than those from the nonweighted regression.

Another means of fitting these data to Equation (5) would involve transformation of the dependent variable prior to linear regression. The ln ^{65}Zn activity was regressed against day to produce the following model: ln ^{65}Zn = − 0.0053 day + 6.07.

The results of the linear regression are given in Figure 5. The linear model fits these data adequately as indicated by the uniform distribution of the data points about the predicted line (Figure 5A) and the lack of pattern in a plot of regression residuals against time (Figure 5B). The absence of any obvious heteroscedasticity indicates the effectiveness of transformation as another means of coping with unequal variance with time.

As shown in Figure 1 (first order), the slope is used to estimate − k. The Y-intercept is the mean predicted estimate of ln ^{65}Zn activity at time = 0. Remember that the k includes the radioactive decay rate constant (λ = 0.00283) in this case. The k_e could be estimated by $k - \lambda$. The estimated k (±standard error) is 0.00531 ± 0.00024. The k_e is 0.00531 − 0.00283 or 0.00248. This estimate is very close to that obtained with the weighted, nonlinear regression methods.

The Y-intercept (6.07 ± 0.06) is the estimate of the mean predicted value of ln ^{65}Zn activity at time = 0. Normally, the value $e^{6.07}$ or 433 pCi/g dry weight would be used as the estimate of ^{65}Zn activity at time = 0. Again, this is not an estimate of the mean predicted value in arithmetic units. It was derived by backtransformation of the predicted mean of values expressed as logarithms. On backtransformation, the Y-intercept (predicted mean value at time = 0) becomes the predicted median value on an arithmetic scale, not the predicted arithmetic mean value. Consequently, the usual approach of predicting the mean initial concentration by backtransformation is biased. If the regression residuals are normally distributed as in this example, the bias can be approximated using the mean square error (The SAS program generated a MSE of 0.125 for this model). The bias is estimated to be $e^{MSE/2}$ or 1.06. The unbiased estimate of the mean ^{65}Zn activity at time = 0 is 1.06 * 433 or 460 pCi/g.

What are the effective and biological half-lives for ^{65}Zn in the oysters during depuration? The results of the linear regression procedures can be used in these estimations.

$$t_{eff} = \ln 2/k_{eff} = 0.693/0.00531 = 130.5 \text{ days}$$
$$t_{1/2} = \ln 2/k_e = 0.693/0.00248 = 279.4 \text{ days}$$

What is the mean lifetime of a particle as determined by k_e? τ = 1.44$t_{1/2}$ = 1.44 * 279.4 = 402.4 days

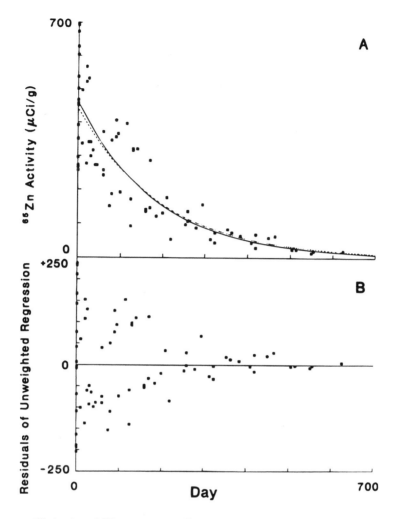

Figure 4 Elimination of ^{65}Zn from oysters.[18] Panel A shows the original data extracted from Ref. 18 and the results of nonlinear regression analyses. Both unweighted (solid line) and weighted (dashed line) regression results are shown. The clear decrease in the regression residuals from the unweighted regression is shown in panel B.

6. Elimination Model with Two Loss Terms for Two Subcompartments (Biexponential)

Other permutations on first-order elimination models are useful. If one is considering the loss from the whole body and elimination is taking place from two distinct subcompartments in the body, the following model is applicable (Figure 2, two exponential loss terms, one for each of two independent subcompartments; Figure 3, double exponential). As discussed in Whicker and Schultz,[2] a model with two subcompartments with exchange between subcompartments but elimination from only one will also display the same elimination kinetics as described here. An example of such a model would be a system with a storage subcompartment. In that case, the k_A and k_B have very different interpretations. (See pages 53–54 in Whicker and Schultz[2] and pages 44–49 in Atkins.[3])

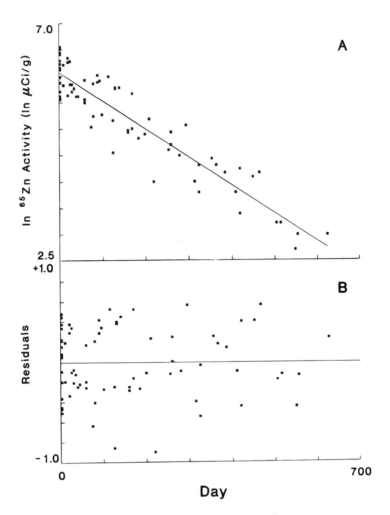

Figure 5 Elimination of ⁶⁵Zn from oysters.[18] Linear regression was used to analyze these data, and the results fit the transformed data nicely. Panel B shows the lack of any strong time dependence of the regression residuals.

$$C_t = C_A e^{-k_A t} + C_B e^{-k_B t} \qquad (8)$$

where C_A and C_B = concentrations in subcompartments A and B, respectively; and k_A and k_B = elimination rate constants (1/h) for subcompartments A and B, respectively.

Example 3

The elimination of methylmercury (^{203}HgCH$_3$) is measured in sunfish for 94 days. The ^{203}Hg activities (pCi/g) measured during the time course are tabulated below in the two left-hand columns. A plot of ln activity versus day showed a steeply sloped line initially then a distinct break to a shallower slope after 22 days of elimination. Two compartments within the fish were tentatively identified based on this plot. Backprojection or backstripping methods (Wagner,[6] pages 59–63) were used to obtain initial estimates of the rate constants and initial concentrations associated with these two subcompartments.

Example 3 *Continued*

Time (d)	Activity (pCi/g)	Predicted Activity	ε (Predicted Activity- Observed Activity) (pCi/g)
3	27,400	12,038	15,362
6	17,601	11,598	6,003
10	12,950	11,037	1,913
14	11,157	10,503	654
22	9,410	9,511	−101
31	**9,150**	8,507	
38	**6,060**	7,799	
45	**6,830**	7,151	
52	**7,270**	6,556	
59	**6,280**	6,011	
66	**5,950**	5,512	
73	**5,480**	5,053	
80	**5,310**	4,633	
87	**3,940**	4,248	
94	**3,310**	3,895	

In the first step of backstripping, data from the period during which the slow compartment dominated the kinetics (days 31 to 94 as printed in boldface) were used to estimate the associated C_B and k_B. The natural logarithm of activity was regressed against day for all data collected after day 22 to yield the following relationship $C = 12469e^{-0.0124day}$. This equation and the minor bias correction factor (1.002, MSE = 0.018) were used to generate predicted activities in this compartment through the entire elimination time course (column three, "Predicted Activity"). The residuals (measured activities as shown in column two minus the predicted activities in the slow compartment (B) as shown in column three) were estimated for days 3 through 22 and placed in column four. These residuals are estimates of the activity associated with the fast compartment during this period. The natural logarithms of these residuals were then regressed against day to yield $C = 34773e^{-0.2863day}$. The bias correction was very small for this compartment also (1.001, MSE = 0.0029). The random distribution of residuals from this second (fast component) regression model suggested that no further backstripping was warranted.

As estimates from backstripping can be quite misleading,[28] they are used here only as initial estimates for iterative, nonlinear regression analysis as implemented by SAS. (It is important to have reasonable initial estimates as iterative, nonlinear regression methods can converge on suboptimal "local solutions." Inaccurate estimators and poor model fit result from such solutions.) Initial estimates of C_A, C_B, k_A, and k_B from this backstripping technique were 34807, 12494, 0.2863, and 0.0124, respectively. The estimates of C_A, C_B, k_A and k_B (±asymptotic standard error) from the nonlinear regression model were 37,908 ± 4,824, 12,245 ± 858, 0.297 ± 0.044, and 0.012 ± 0.001, respectively. Roughly 76% of the activity was associated with the compartment displaying rapid elimination (37,908/(37,908 + 12,245)) and 24% was associated with the second compartment (12,245/(37,908 + 12,245)).

Remember that the k_A and k_B estimates include radioactive decay ($\lambda = 0.0145$ for ^{203}Hg). Consequently, 0.0145 must be subtracted from these estimates to determine the rate constants for biological elimination. It becomes immediately apparent upon doing so that the "slow" compartment actually had no detectable biological elimination. The loss was dominated by radioactive decay. In contrast, the first compartment had a large biological elimination rate constant of approximately 0.282.

Example 3 *Continued*

Considering the negligible elimination from the second compartment, this relationship for methylmercury elimination [Equation (8)] can be reduced to $C_t = C_A e^{-k_A t} + C_B$. The methylmercury in C_B was unavailable for elimination. With this new expression of the model, the limiting value of C_t would be C_B, not 0.

The SAS code used in these calculations was the following. Predicted results and associated residuals were also produced by this code.

```
DATA MERCURY;
  INPUT DAY PCI @@;
  LPCI = LOG(PCI)
  IF DAY>22;
  CARDS;
03 27400 06 17601 10 12950 14 11157 22 9410 31 9150 38 6060
45 6830 52 7270 59 6280 66 5950 73 5480 80 5310 87 3940
94 3310
;
/* RUNNING LINEAR MODEL FOR SLOW COMPARTMENT              */
PROC GLM;
  MODEL LPCI=DAY;
  OUTPUT OUT=EXPO PREDICTED=PEXPO RESIDUAL=REXPO;
RUN;
/* PLOTTING PREDICTED AND ORIGINAL ACTIVITIES VERSUS DAY  */
/* AND REGRESSION RESIDUALS VERSUS DAY                    */
PROC PLOT;
  PLOT LPCI*DAY PEXPO*DAY/OVERLAY;
  PLOT REXPO*DAY="*"/VREF=0;
RUN;
PROC PRINT;
VAR DAY PCI LPCI PEXPO REXPO;
RUN;
/* NOW FITTING "FAST" COMPONENT                           */
DATA FAST;
  INPUT DAY CPI @@;
  LPCI = LOG(PCI);
  CARDS;
03 15362 06 6003 10 1913 14 654 22 -101
;
/* RUNNING LINEAR MODEL FOR FAST COMPARTMENT              */
PROC GLM DATA=FAST;
  MODEL LPCI=DAY;
  OUTPUT OUT=FEXPO PREDICTED=PFEXPO RESIDUAL=RFEXPO;
RUN;
/* PLOTTING ORIGINAL AND PREDICTED ACTIVITIES VERSUS DAY  */
/* AND REGRESSION RESIDUALS VERSUS DAY                    */
PROC PLOT;
  PLOT LPCI*DAY PFEXPO*DAY="*"/OVERLAY;
  PLOT RFEXPO*DAY="*"/VREF=0;
RUN;
DATA MERCURY;
  INPUT DAY PCI @@;
  LPCI = LOG(PCI);
```

Example 3 *Continued*

```
    CARDS;
    03 27400 06 17601 10 12950 14 11157 22 9410 31 9150 38 6060
    45 6830 52 7270 59 6280 66 5950 73 5480 80 5310 87 3940
    94 3310
    ;
    /* NOW USING ESTIMATES FROM BACKSTRIPPING ABOVE TO FIT THESE   */
    /* DATA TO A NONLINEAR, DOUBLE EXPONENTIAL MODEL               */
    PROC NLIN DATA=MERCURY;
        PARMS KCB=0.0124 KCA=0.2863 CB=12494 CA=34807;
        BOUNDS 0 <KCB<1, 0<KCA<1, 5000<CB<20000, 20000<CA <45000;
        MODEL PCI=CA*EXP (−KCA*DAY)+CB*EXP (−KCB*DAY);
        OUTPUT OUT=DOUBLE P=PDOUB R=RDOUB;
    RUN;
    /* PLOTTING THE RESULTS OF THE NONLINEAR REGRESSION MODEL      */
    PROC PLOT;
        PLOT PCI*DAY PDOUB*DAY="*"/OVERLAY;
        PLOT RDOUB*DAY="*"/VREF=0;
    RUN;
```

But how many components should be stripped from a data set? There are several methods of estimating the appropriate number in addition to that used too frequently, the researcher's temperament and sense of optimism. Myhill's simulations[29] of two component elimination models suggest the following rules of thumb for data acceptability for stripping two components. If the ratio of the two k_e values is roughly 2 then acceptable convergence can be expected using iterative regression methods if the data error doesn't exceed 1%. It would require a very high quality data set to extract these two close components. For ratios of 4, 6, and 10, the data error should not exceed 5, 10, or 10%, respectively.

But what information is available for making the judgment between models of increasing numbers of components? Certainly, the MSE, plots of the original data's fit to the predicted model, and plots of regression residuals give some indication of the goodness of model fit as the number of exponential compartments increases. However, these do not provide an objective means of comparing models derived from data with a constant number of observation pairs but different numbers of estimated parameters. Barron et al.[8] suggest the methods of Boxenbaum et al.[25] or Akaike's information criterion[30] for this purpose. Akaike's information criterion (AIC) is the following:

$$\text{AIC} = n\left[\ln\left(\sum_{i=1}^{n} w_i(Y_i - Yp_i)^2\right)\right] + 2p \qquad (9)$$

where n = the number of data pairs fit to the model; Y_i = the Y value of the ith observation; Yp_i = the predicted Y value of the ith observation; w_i = weighting for the ith observation (1 for unweighted regression); and p = the number of parameters estimated by the model.

The AIC estimates the information content within the various candidate models. It assumes that the more complex model will contain more information and then estimates the consequent increase in the information. For example, if two models with very similar residual sums of squares are compared, the one with the least number of estimated parameters will be favored. The process of selecting the models with the lowest AIC is called minimum AIC estimation or MAICE. In formulating a modified AIC, Micromath Scientific Software[31] points out that the AIC is dependent on the magnitude of the data points. An alternative, Model Selection Criterion (MSC) based on the AIC is presented as a substitute for the AIC. Its value is independent of the magnitude of the data and, consequently, provides a "normalized" estimate of acceptability. They suggest that the following general judgments of model fit can be assigned to MSCs: <2 (unacceptable), 3 (marginal), 4 (typical of reasonably well-fit model), 5 (very good fit), >6 (exceptional and, perhaps, suspect). The MSC is estimated with Equation (10).

$$
\text{MSC} = \ln\left[\frac{\sum_{i=1}^{n} w_i(Y_i - \bar{Y})^2}{\sum_{i=1}^{n} w_i(Y_i - Yp_i)^2}\right] - \frac{2p}{n} \tag{10}
$$

Example 4

Using the program RSTRIP[31] to analyze the mercury data in Example 3, MSC values of 1.24, 4.09, 2.32, −0.49, and −0.78 were generated for one-, two-, three-, four-, and five-exponential components models, respectively. The MSCs supported our previous decision to extract two components from these data as the highest MSC was associated with the two-exponential component model. The MSC for the two-exponential model was within the range of "typical of a reasonably well-fit model." The results of the two-exponential model are used here to illustrate these calculations (assuming unweighted regression or weighting = $1/(Y_i^0 = 1)$.

Y	$Y-\bar{Y}$	$(Y-\bar{Y})^2$	Yp	$(Y-Yp$ or $\epsilon)$	$Wt\epsilon^2$
27,400	18,193.5	331,003,442.3	27,357.0	43.0	1,849.0
17,601	8,394.5	70,467,630.3	17,769.0	−168.0	28,224.0
12,950	3,743.5	14,013,792.3	12,808.0	142.0	20,164.0
11,157	1,950.5	3,804,450.3	10,952.0	205.0	42,025.0
9,410	203.5	41,412.3	9,472.8	−62.8	3,943.8
9,150	−56.5	3,192.3	8,462.4	687.6	472,793.8
6,060	−3,146.5	9,900,462.3	7,781.0	−1,721.0	2,961,841.0
6,830	−2,376.5	5,647,752.3	7,156.8	−326.8	106,798.2
7,270	−1,936.5	3,750,032.3	6,582.9	687.1	472,106.4
6,280	−2,926.5	8,564,402.3	6,055.2	224.8	50,535.0
5,950	−3,256.5	10,604,792.3	5,569.7	380.3	144,628.1
5,480	−3,726.5	13,886,802.3	5,123.1	356.9	127,377.6
5,310	−3,896.5	15,182,712.3	4,712.4	597.6	357,125.8
3,940	−5,266.5	27,736,022.3	4,334.6	−394.6	155,709.2
3,310	−5,896.5	34,768,712.3	3,987.1	−677.1	458,464.4
Σ		549,375,610.5			5,403,585.3

$$
\begin{aligned}
\text{MSC} &= \ln(549375610.5/5403585.3) - 2(4)/15 \\
&= \ln(101.67) - 8/15 \\
&= 4.62 - 0.53 \\
&= 4.09
\end{aligned}
$$

---

7. Power Elimination Model

Occasionally, a power function (see Figures 2 and 3, power) is used to best describe elimination, e.g., Whicker and Schultz[2] or Newman and McIntosh.[32]

$$C_t = C_1 t^{-P} \qquad (11)$$

where C_1 = concentration at day 1; and P = a constant. In such cases, a double-logarithm plot (ln C versus ln t) results in a straight line. Initially, such a model would seem to have no direct link to our previous discussion of reaction order. However, kinetics described with a power model could result from several simultaneous yet inseparable elimination components such as those associated with first-order kinetics. Whicker and Schultz[2] give several good examples of processes producing apparent power models. The simultaneous decay of a mixture of nuclear fission products can fit a power model although, when taken separately, each radionuclide involved displays exponential decay. Also the release of radionuclides from bone can be described with a power model. Whicker and Schultz[2] explain this observation by describing bone as a very heterogeneous "compartment" with an initial rapid loss of superficially bound radionuclide and then increasingly slower clearance from pools deeper within the bone.

Example 5

Newman and McIntosh[32] placed freshwater snails (*Physa integra*) from a contaminated reservoir in clean water under controlled laboratory conditions. During a 22-day period, snails were removed and analyzed for lead concentration. Visual inspection of the resulting depuration curve suggested a power model. The following code was used to fit these data.

```
DATA LEAD;
   INPUT DAY LEAD @@;
   LLEAD = LOG (LEAD);
   LDAY = LOG (DAY);
   CARDS;
0.16 41.0 0.16 31.0 0.16 25.3 1.00 30.5 1.00 22.7
1.00 22.0 2.00 13.5 2.00 15.0 2.00 16.5 4.00 10.0
4.00 12.0 4.00 15.0 8.00 08.7 8.00 13.0 8.00 15.0
12.0 08.0 12.0 09.0 12.0 12.0 16.0 08.0 16.0 10.5
16.0 12.0 22.0 12.5 22.0 10.6 22.0 05.5
;
/* USING LINEAR REGRESSION OF LN TRANSFORMED X AND Y       */
/* VARIABLES FOR POWER CURVE FITTING OF ELIMINATION        */
PROC GLM;
   MODEL LLEAD = LDAY;
OUTPUT OUT = POWER PREDICTED = PPOWER RESIDUAL = RPOWER;
RUN;
/* PLOTTING PREDICTED AND ORIGINAL LN CONCENTRATION        */
/* VERSUS LN DAY AND REGRESSION RESIDUALS VERSUS LN DAY     */
PROC PLOT;
   PLOT LLEAD*LDAY PPOWER*LDAY = "*"/OVERLAY;
   PLOT RPOWER*LDAY = "*"/VREF = 0;
RUN;
```

The resulting model (ln Pb = -0.272ln day $+ 3.001$) showed no pattern to its residuals and had a significant r^2 of 0.77 ($\alpha = 0.05$). The backtransformed model was Pb = 20.1 day$^{-0.272}$ with a prediction bias correction factor of 1.03. The slope and intercept had small standard errors of 0.031 and 0.064, respectively.

When a double-exponential model was imposed on these data as in Example 3, an adequate fit also resulted. The slow and fast compartment sizes were estimated to be 11.6($\pm$ 3.1) and 24.1($\pm$ 3.6), respectively. The k_e for the fast component was adequately

Example 5 *Continued*

estimated (0.737 ± 0.259/h). That for the slow compartment was near 0 and had a relatively large standard error (0.009 ± 0.018/h), suggesting negligible clearance from this compartment. Nearly 32% of the lead is essentially unavailable for elimination.

8. Michaelis-Menten Elimination Model

Finally, if Michaelis-Menten kinetics is warranted (Figures 1 to 3), the following equation defines the kinetics. As discussed in considerable detail in Piskiewicz,[33] such kinetics will conform to first-order when concentrations are below saturation and to zero-order when an excess of substrate is present. Spacie and Hamelink[11] discuss apparent Michaelis-Menten kinetics in data generated by Mayer[34] for DEHP elimination by fathead minnow.

$$C_0 - C_t + k_m \ln\left[\frac{C_0}{C_t}\right] = V_{max}t \tag{12}$$

where V_{max} = the maximum velocity of substrate conversion; and k_m = the substrate concentration at which the velocity of conversion is $V_{max}/2$. The rate of substrate conversion is defined by the equation,

$$v = \frac{V_{max}C}{C + k_m} \tag{13}$$

where v = the velocity of substrate conversion; and C = substrate concentration.

One of several linearizing plots is shown in Figure 1 [double-reciprocal or Lineweaver-Burk plot, Equation (14)]. Although the Lineweaver-Burk plot is the most commonly used transformation for deriving k_m and V_{max}, it is statistically the worst behaved transformation.[35] Other common linear plots based on other transformations include Eadie-Hofstee [v versus v/C, Equation (15)], Scatchard [v/C versus v, Equation (16)], and the least used but preferrable Woolf [C/v versus C, Equation (17)] plots.[35]

$$\frac{1}{v} = \frac{1}{V_{max}} + \frac{k_m}{V_{max}C} \tag{14}$$

$$v = V_{max} - \frac{k_m v}{C} \tag{15}$$

$$\frac{v}{C} = \frac{V_{max}}{k_m} - \frac{v}{k_m} \tag{16}$$

$$\frac{C}{v} = \frac{K_m}{V_{max}} + \frac{C}{V_{max}} \tag{17}$$

Relative to regression analysis, the Eadie-Hofstee (v versus v/C) and Scatchard (v/C versus v) plots have the complication that v (the measured rate of substrate conversion) is present in both the dependent and independent variables. Consequently, the assumption of insignificant error in the independent variable comes under question. Raaijmakers[35] discussed additional problems associated with such plots. Many of the same transformations employed are also used in modeling adsorption (see below). They are subject to the same biases discussed in detail by Kinniburgh.[36]

To incorporate Michaelis-Menten kinetics into elimination, zero-order kinetics can be used during that period of elimination when saturation is occurring; thereafter first-order kinetics can be used. The time at which the transition from pseudolinear kinetics to first-order kinetics (t^*) occurs was given by Wagner[4] as

$$t^* = \left[\frac{1 - \dfrac{1}{e}}{V_{max}} \right] C_0 + \frac{k_m}{V_{max}} \tag{18}$$

where e = base e (approximately 2.71828).

The slope of the pseudolinear phase of elimination (k_0^*) was also estimated by Wagner.[4]

$$k_0^* = V_{max} - \frac{k_m}{t^*} \tag{19}$$

Note that, because t^* depends on C_0, the slope is also dependent on initial concentration.

9. Comment on Interpreting Two Compartment Kinetics

As discussed in Section A.6, several compartment models can result in biexponential elimination kinetics. The one shown in Figure 2 (two exponential loss terms, one for each of two independent subcompartments) involves two independent subcompartments. (By strict definition, each of these subcompartments is actually a mathematical compartment. They were considered subcompartments only in the context of whole-body elimination kinetics.) Another compartment model resulting in identical overall elimination kinetics is shown in Figure 6. This model consists of two compartments with exchange between them. Elimination occurs only from the central compartment A. Such a model might be appropriate for modeling elimination of a toxicant subject to storage in the body. If elimination from the two compartments were monitored instead of overall elimination, very different curves would result (Figure 6). The elimination of toxicant from both compartments in the first model remains independent. However, when toxicant is introduced into compartment A in the second model, clearance becomes more complex.

The rate constants k_{AB}, k_{BA}, and k_{AO} can be calculated from the estimates of C_A, C_B, k_A, and k_B [Equation (8)] derived with methods discussed in Section A.6.[37]

$$k_{BA} = \frac{C_A k_B + C_B k_A}{C_A + C_B} \tag{20}$$

$$k_{AO} = \frac{k_A k_B}{k_{BA}} \tag{21}$$

$$k_{AB} = k_A + k_B - k_{BA} - k_{AO} \tag{22}$$

The half-life is estimated as $t_{1/2} = \ln 2/k_B$ for this model.[8]

The work of Holleman et al.[38] with ^{134}Cs elimination from reindeer is a good application of such a model. The model development is discussed in detail in their methods.

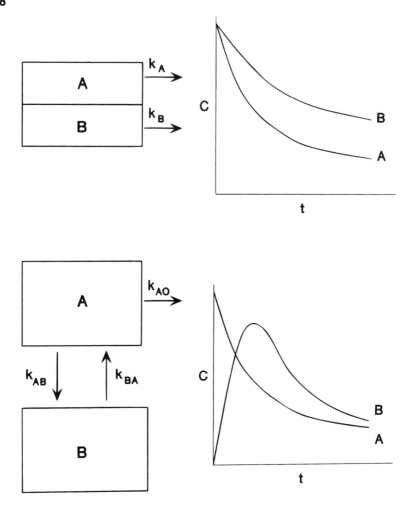

Figure 6 Examples of two-compartment models (with and without exchange between compartments) producing very different elimination kinetics from each compartment.

Another model very similar to that described by Equations (20) to (22) could also generate biexponential elimination kinetics. This model involves two interconnected compartments but elimination is from the peripheral second compartment (rate constant $= k_{BO}$) instead of the first compartment (rate constant $= k_{AO}$). In this case, the toxicant is introduced into the first compartment (A), exchanges with the second compartment (B), and is eliminated from this second compartment. The equations relating the biexponential parameters to constants for this model are the following.

$$k_{AB} = \frac{C_A k_A + C_B k_B}{C_A + C_B} \qquad (23)$$

$$k_{BO} = \frac{k_A k_B}{k_{AB}} \qquad (24)$$

$$k_{BA} = k_A + k_B - k_{AB} - k_{BO} \qquad (25)$$

A final two-compartment situation that can be modeled easily is one in which the toxicant is introduced into the first compartment (A) and passes (rate constant = k_{AB}) into the second compartment (B) from which it is eliminated (rate constant = k_{BO}). The k_A and k_B estimated from the biexponential elimination curve are estimates of k_{AB} and k_{BO}, respectively.

Derivations similar to Equations (20) to (25) can be determined for more complex models. For example, parameters estimated from an apparent triexponential elimination curve (three extracted elimination components) can be used to estimate constants for associated parameters for a model with storage in two compartments and elimination from a third, central compartment.[31,37] However, as the models become more complex, the importance of high-quality elimination data becomes more critical during parameter derivation.

B. ADSORPTION
1. General
Toxicant adsorption on surfaces such as those of algae,[13,39-42] periphytic microflora,[43] zooplankton,[44] fish,[45] and fish gills[46] plays a critical role in the initial stages of uptake. Adsorption ("the process through which a net accumulation occurs at the common boundary of two contiguous phases")[47] may involve weak, reversible bonding such as that associated with van der Waals forces (physical adsorption) or strong, irreversible bonding (chemical adsorption). (The terms adsorption and physical adsorption are sometimes used interchangeably[48] as in this treatment.) Although often visualized as simple ion exchange at specific binding sites, adsorption can include more complex processes. For example, adsorption of some ionic organic toxicants may not be driven by specific sites on the surface but, rather, by the lower affinity of nonpolar groups for the aqueous phase relative to the solid phase.[49] With chemical adsorption, desorption may be minimal.

Adsorption of a toxicant may result from a direct toxicant-surface interaction. However, adsorption may also be indirect, e.g., involve initial complexation with a ligand. For example, Sposito[47] lists the following metal-ligand-surface interactions that could occur. "1. The ligand has a high affinity for the metal and forms a soluble complex with it, and this complex has a high affinity for the adsorbent. 2. The ligand has a high affinity for the adsorbent and is adsorbed, and the adsorbed ligand has a high affinity for the metal. 3. The ligand has a high affinity for the metal and forms a soluble complex with it, and this complex has a low affinity for the adsorbent. 4. The ligand has a high affinity for the adsorbent and is adsorbed, and the adsorbed ligand has a low affinity for the metal." Such interactions may cause observed shifts in adsorption such as those associated with bacterial uptake of ^{241}Am in the presence of dissolved organic matter,[50] chromate accumulation by algae in the presence of various inorganic ligands,[40] and algal uptake of nickel in the presence of humic acids.[42] Complex ligand-toxicant interactions can also be important in estimating adsorption in tissues during pharmacokinetic modeling.[4]

To add further complications, other compounds or ions may compete for adsorption sites. For example, Group Ia metals, e.g., Li, Na, K, Rb, and Cs or oxyanions, e.g., AsO_4^{3-} and PO_4^{3-} may compete for sites. Sposito[47] (pages 129–132) and Stumm and Morgan[49] (pages 492–493) discussed adsorption of several metals varying in affinity for a solid adsorbent. Pertinent examples involving competition include pH effects on algal uptake of metals,[13,40,41] intermetal effects on algal uptake,[13] or hardness ion effects on metal toxicity.[46]

2. Common Forms of Freundlich and Langmuir Equations

Several methods have been developed to quantify adsorption based on the assumptions of a fixed number of binding sites, reversible adsorption, a homogeneous surface, and fixed adsorbate binding (the steric configuration of the bound toxicant doesn't change). The two most commonly used are the Freundlich [Equation (26)] and Langmuir [Equation (27)] isotherm equations.

$$\frac{X}{M} = KC^{1/n} \tag{26}$$

where X/M = the amount adsorbed/unit weight of adsorbent; C = the equilibrium concentration in the solution after adsorption has taken place; and K,n = derived constants.

$$\frac{X}{M} = \frac{abC}{1 + bC} \tag{27}$$

where a,b = derived constants.

The K and n constants for the Freundlich equation are often derived with the following linear transformation. The slope and intercept are used to estimate n and K, respectively.

$$\log \frac{X}{M} = \log K + \frac{\log C}{n} \tag{28}$$

The a and b constants for the Langmuir equation are similarly derived with the following linear equation.

$$\frac{C}{X/M} = \frac{1}{ab} + \frac{1}{a} C \tag{29}$$

First, the slope is used to estimate a. Next, b is derived with a and the intercept. Unlike the Freundlich isotherm equation which is wholly an empirical model, the Langmuir isotherm equation is derived from principles that allow direct physical meaning to be assigned to the associated constants. Equation (27) can be rewritten to conform to this theoretical context for the convenience of any reader who may wish to compare this discussion with the well-established literature on adsorption theory.

$$n = \frac{KCM}{1 + KC} \tag{30}$$

where n = amount adsorbed (mmol) per unit mass (g); C = equilibrium concentration in the aqueous phase (mmol/mL); M = the adsorption maximum (mmol); and K = the affinity parameter which is a measure of bond strength.

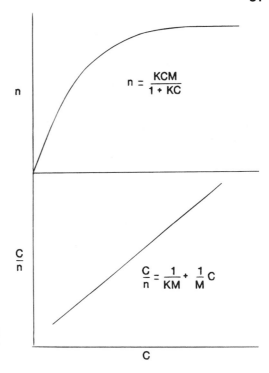

$$n = \frac{KCM}{1 + KC}$$

$$\frac{C}{n} = \frac{1}{KM} + \frac{1}{M} C$$

Figure 7 An adsorption model and its linear transformation. See text for details.

Figure 7 illustrates the shape of this model and its linear transform. The maximum number of binding sites can be estimated with Equation (30). The affinity parameters for various chemicals can be compared under different conditions. For example, Crist et al.[13] used this approach to estimate the relative affinity of alkali, alkali earth, and transition metals for algal cells. They also quantified the influence of pH on metal adsorption to these cells.

The traditional means of fitting adsorption data to the Langmuir isotherm model have been reexamined recently by Kinniburgh.[36] Table 1 summarizes the transformations examined.

Table 1 Linear transformations for the Langmuir Isotherm Model (from Table II of Kinniburgh[36]).

Transformation	Y versus X	K, M Estimates
Lineweaver-Burk	$1/n$ vs $1/C$	intercept/slope, 1/intercept
Reciprocal	C/n vs C	slope/intercept, 1/slope
Eadie-Hofstee	n vs n/C	-1/slope, intercept
Scatchard	n/C vs n	$-$slope, $-$intercept/slope

He found that unweighted, least-squares regression of these transformed data uniformly weighted low n observations more than high n observations. Further, the Eadie-Hofstee and Scatchard transformations had the experimentally measured n incorporated in the independent variable, leaving open the possibility of significant error in the independent variable. He also noted that, as mentioned previously in our discussion of Michaelis-Menten kinetics, the frequently used Lineweaver-Burk transformation behaves poorly during regression analysis. However, the Langmuir reciprocal behaved acceptably.

(It is reassuring to note that the Langmuir reciprocal transformation is the same as the Woolf transformation found by Raaijmakers[35] to be the best for fitting Michaelis-Menten kinetics.) Kinniburgh suggests that the Lineweaver-Burk, Langmuir reciprocal, Eadie-Hofstee, and Scatchard transformations fit data better (single pass, least-squares regression) when the following weightings were used n_i^4, n_i^4/C_i^2, n_i^2, and n_i^2, respectively. With these weightings, the regression results were closer to those derived by nonlinear regression methods. Example 6 is based partially on the data analysis of Kinniburgh[36] to illustrate his recommended procedures. Only the surface experiencing adsorption is changed to remain relevant to bioaccumulation.

Example 6
(Derived from Table III of Kinniburgh[36])
Adsorption of zinc from solution is measured for a freshwater, filamentous alga. The following data were generated. What are the K and M parameters for zinc adsorption?

n (mmol Zn adsorbed/g)	Equilibrium Zn Concentration (mmol/L)
0.075	0.030
1.40	0.069
1.95	0.118
2.51	0.166
3.03	0.217
3.53	0.270
4.02	0.325
4.41	0.388
4.79	0.453
5.22	0.512

Kinniburgh[36] outlined several linear transformations of the Langmuir equation and examined their abilities to fit this type of data. The recommended Langmuir reciprocal transformation with and without weighting will be used here as well as nonlinear regression. The following SAS code performs these calculations.

```
OPTIONS PS=58;
DATA ZINC;
   INPUT N C @@;
   WGT = (N**4)/(C**2); /* RECOMMENDED WEIGHTING              */
Y=C/N;
   CARDS;
0.75 0.030 1.40 0.069 1.95 0.118 2.51 0.166 3.03 0.217
3.53 0.270 4.02 0.325 4.41 0.388 4.79 0.453 5.22 0.512
;
/* USING NONLINEAR REGRESSION PROVIDING NO DERIVATIVES        */
/* AND NO WEIGHTING                                           */
PROC NLIN;
   PARMS K=3.00, M=9.00;
   BOUNDS K>0, M> 0;
   MODEL N = (K*C*M)/(1+K*C);
   OUTPUT OUT=PRED PREDICTED=PN RESIDUAL=RN;
RUN;
/* PLOTTING NONLINEAR REGRESSION DATA AND PREDICTIONS         */
/* VERSUS Concentration, AND REGRESSION RESIDUALS VERSUS      */
/* CONCENTRATION                                              */
PROC PLOT;
   PLOT N*C PN*C="*"/OVERLAY;
```

Example 6 *Continued*

```
    PLOT RN*C="*"/VREF=0;
RUN;
/* USING UNWEIGHTED LEAST-SQUARES LINEAR REGRESSION ON        */
/* LANGMUIR RECIPROCAL PLOT OF ADSORPTION DATA                */
PROC GLM;
    MODEL Y=C;
    OUTPUT OUT=LINEAR PREDICTED=PLIN RESIDUAL=RLIN;
RUN;
PROC PLOT;
    PLOT Y*C PLIN*C="*"/OVERLAY;
    PLOT RLIN*C="*"/VREF=0;
RUN;
/* USING WEIGHTED LEAST-SQUARES, LINEAR REGRESSION ON         */
/* LANGMUIR RECIPROCAL PLOT OF ADSORPTION DATA                */
PROC GLM;
    MODEL Y=C;
    WEIGHT WGT;
    OUTPUT OUT=WLINEAR PREDICTED=PWLIN RESIDUAL=RWLIN;
RUN;
PROC PLOT;
    PLOT Y*C PWLIN*C="*"/OVERLAY;
    PLOT RWLIN*C="*"/VREF=0;
RUN;
```

The nonlinear regression fit these data with very little pattern in the regression residuals. The estimates ($\pm$ asymptotic standard errors) of K and M were 2.10 ± 0.19 and 9.89 ± 0.52 mmol, respectively. (These are essentially the same estimates as given in Kinniburgh's Table IV.)

Although both linear regressions had high r^2 values (unweighted $= 0.966$, weighted $= 0.976$), residuals from the lowest C data points were large. A ∩-shaped pattern to the residuals was apparent when plotted against C. K and M were calculated from the unweighted regression results as follows (see Table 1):

$$K = \text{slope/intercept} = 0.1037/0.047 = 2.21$$
$$M = 1/\text{slope} = 1/0.1037 = 9.64 \text{ mmol}$$

Similarly, the estimates from the weighted regression were the following:

$$K = 0.1140/0.044 = 2.59$$
$$M = 1/0.1140 = 8.77 \text{ mmol}$$

These estimates are exactly opposite of those given in Table IV of Kinniburgh,[36] suggesting transcription error in that table. Contrary to the conclusion of Kinniburgh from these data, the unweighted regression produced estimates closer to those of the nonlinear regression model than the weighted regression.

3. Other Useful Adsorption Equations

Sposito[47] describes four classes of adsorption isotherms: S-, H-, L-, and C-curves for soils. It is reasonable to assume that adsorption onto biological surfaces may also deviate from the one class of adsorption isotherm models described above (L-curve). The reader is directed to Sposito[47] for a general discussion of other classes of adsorption isotherms and to the excellent article by Kinniburgh[36] for means of fitting data to more complex models.

C. ACCUMULATION MODELS: ONE COMPARTMENT

1. General

Uptake (the movement of a toxicant into or onto the organism) can occur directly from water, food, or sediments. Movement may be across the general integument, gills, pulmonary surfaces, or the gut. Initial phases of uptake may involve adsorption as described above. Uptake can vary tremendously depending on the specific species and environmental conditions. Indeed, uptake can vary even for individuals held under constant conditions. Changes may be a consequence of such processes as general physiological changes, modified calcium flux associated with insect molting, or fluctuating nonpolar organic uptake associated with seasonal lipid dynamics of molluscs.

According to Gordon,[51] there are six major mechanisms for movement of solutes across membranes: passive diffusion, active transport, facilitated diffusion or transport, exchange diffusion, pinocytosis, and solvent drag. Passive diffusion occurs as a movement of solute down a chemical, electrical, or activity gradient. Diffusion may involve "exchange diffusion" which is the simple exchange of ions of the same type across the membrane in both directions. Exchange diffusion involves a carrier molecule. Facilitated diffusion or transport also occurs down a chemical gradient but at a rate greater than predicted by passive diffusion alone. It involves a carrier molecule. Active transport requires energy to facilitate movement up a chemical gradient. Regulation of potassium and sodium (potassium higher in the cell than outside and sodium lower in the cell than outside) by a coupled ion pump (membrane-bound ATPase) is a good example of active transport.[52] Such a system is capable of saturation kinetics and requires energy as demonstrated by its failure upon metabolic inhibition. Elements such as cesium, which chemically mimics potassium, can be actively transported by such a mechanism. Another similar example is active calcium transport at low environmental concentrations.[53] Movement via pinocytosis involves enclosing material from outside the cell in a vacuole that then migrates across the cell to discharge its contents inside the organism. It requires energy. Solvent drag involves the movement of the solute in the direction of the solute movement. The relative importance of each mechanism will vary depending on many factors, including route of introduction (food versus water), the toxicant being studied, the external conditions, and the organism's state. Several mechanisms can work simultaneously or one can dominate uptake. Associated uptake dynamics can be influenced by energy requirements (active transport, facilitated diffusion, exchange diffusion, pinocytosis) or the possibility of saturation kinetics (active transport, facilitated diffusion, exchange diffusion).

Passive diffusion down an activity gradient is often the focus of model formulation for nonpolar organic toxicant uptake, e.g., Barber et al.[54] and Erickson and McKim.[55] It may also be significant in metal uptake under a variety of conditions.[1,56] Passive diffusion is described by Fick's Law [Equation (31)].

$$\frac{dS}{dt} = -DA\frac{dC}{dX} \qquad (31)$$

where dS/dt = the rate of movement across the membrane; D = the diffusion coefficient; A = the area across which diffusion is occurring; and dC/dX = the change in concentration across the membrane.

Although the focus during modeling is frequently on the concentration gradient (dC/dX) between the organism and the bulk external media, other aspects of this relationship are also critical. For example, the gill area available for uptake increases disproportionately with increases in body volume during growth.[57] (Such allometric relationships are normally described by power functions as discussed in more detail later in this chapter.)

Area available for uptake from food can also change as relative length of intestine can change disproportionately with size.[58] Thus, the change in A must be considered for uptake over periods of time during which growth can be significant. In the case of uptake via gills, the diffusion gradient is influenced by many factors including water flow across the gill and blood flow through the gills.[55,59] Such factors are often at the heart of physiologically based pharmacokinetic (PBPK) models such as those described later.

Uptake from food may be defined with the direct application of processes such as those described above if one considers a very simple system such as a gutless endoparasite exposed to a relatively uniform environment, e.g., ^{137}Cs bioaccumulation in liver flukes within white-tailed deer inhabiting a contaminated site. However, in most cases, uptake from food is more complex. For example, lead uptake by a snail scrapping and ingesting a heterogeneous mixture of microflora and associated abiotic debris can be extremely difficult to predict relative to uptake of dissolved lead from soft water.[43,60] Assessment of bioavailability, the degree to which a contaminant in a potential source is free for uptake, becomes very critical in such cases. Often an assimilation efficiency will be experimentally estimated based on the overall concentration in the food or some defined fraction of the food. Normalization of concentration can involve a major fraction suspected as being particularly available or unavailable for uptake.

Models of bioaccumulation range from very simple to highly complex. The following discussion provides a framework for such models beginning with simple compartment models. By far the most common models in use are those discussed initially as they provide adequate prediction for most users with minimum information on the system being modeled. However, complex models such as physiologically based pharmacokinetic models can provide in-depth understanding of the most significant processes controlling accumulation kinetics, estimation of target organ exposures, and enhanced extrapolation between species or toxicants.

2. Compartment Models

a. Rate Constant-Based Models

The elimination model formulations described to this point were based on rate constants. This type of model has been the mainstay of bioaccumulation modeling in aquatic toxicology. Discussion of rate constant-based models is expanded in this section to include uptake. However, an intentional omission in our discussions must be addressed before proceeding.

Although our discussion to this point has dealt solely with changes in concentrations, rate constant models can describe changes in mass of toxicant also. We have focused on concentrations because one is often concerned with the concentration in the organism relative to concentrations in its environment. Further, effects are more readily related to toxicant concentration within the organism or target organ than to amounts or mass of toxicant. Indeed, this is the reasoning behind the extensive use of concentrations in pharmacokinetics (see Wagner[4] or Greenblatt and Shader[10]). However, certain aspects of modeling concentrations must now be clarified to avoid confusion later.

Barron et al.[8] suggested that, if concentration units are used, they should be considered as amounts normalized to unit mass or volume. In fact, the rate constant model is described more accurately as a mixed model, i.e., one including rate constant and clearance components. This is an important point to make. Potential misinterpretation that may result from use of concentrations can be easily illustrated with one of our previous examples, elimination of methylmercury from two compartments (Example 3). Note that the C_A and C_B "concentrations" estimated for these two compartments would only be reasonable if the two compartments in question were of equal size (mass or volume assuming equal densities between compartments). However, if C_A and C_B are envisioned as the amounts in each compartment, this restriction is not pertinent. For this and related reasons, it is often easier to use mass in these equations if complex models with compartments of differing volumes are anticipated. Books by Atkins[3] and Whicker and Schultz[2] provide extensive model formulations based on amounts. Detailed examples of one-compartment model formulations involving transfer of amounts of materials are given in Willis and Jones[61] and Giesy et al.[62] Analogous examples of formulations involving two compartments are found in Goldstein and Elwood,[63] Holleman et al.,[38] and Konemann and Van Leeuwen.[64]

When dealing with concentration models involving multiple compartments, the volumes of each compartment must be estimated. The volume when multiplied by the concentration in the compartment gives an estimate of the amount of substance in the compartment. For example,[4] the total amount or dose (D_t) of a substance injected intravenously will be the sum of the amounts in all the compartments in the organism.

$$D_t = C_b V_b + \sum_{i=1}^{n} C_i V_i \qquad (32)$$

where C_b = the concentration in the blood; V_b = the estimated blood volume; C_i = the concentration in the ith compartment; and V_i = the volume of the ith compartment. It follows that the total volume in which the dose is distributed in the organism (V_t) can be defined.

$$V_t = V_b + \sum_{i=1}^{n} V_i \qquad (33)$$

One need only know the volumes and concentrations in the various compartments to define the distribution of mass within the organism.

A justifiable questioning of the above argument can be levied based on the previously stated fact that measured mathematical compartments are not *per se* linked to physical entities such as an organ or tissue type having an easily measured volume. If the compartments are not necessarily physical entities then how can compartment volumes be determined?

The apparent volume of distribution concept (see Greenblatt and Shader,[10] pages 8–12, 34–38) as used in pharmacokinetics can be adapted for this purpose. The apparent (or effective) volume of distribution (V_d) is the mathematically determined volume of a compartment in which the total amount of material is associated. The apparent volume is expressed in units of volume of a reference compartment. For example, the following procedure may be used to estimate the V_d for a material in the "blood" compartment. A known amount of drug or toxicant is injected intravenously and allowed to distribute itself in the blood. The blood concentration is then measured. The apparent volume of distribution of the blood (L) would be the total amount of the dose (mg) injected divided by the concentration (mg/L) measured in the blood after the material has been allowed to distribute itself within the blood only, i.e., V_d = dose/concentration. When several mathematical compartments are involved, each will have a V_d. These V_d values will be the apparent volumes expressed in units of volume of a reference compartment, blood in this case.

With this background, the models described in Figure 6 can now be further characterized by the addition of two V_d values, one for each of the two compartments. In the case of the two compartments with no exchange (Figure 6, top model), the V_d values can be estimated to be the following.

$$V_{dA} = \frac{D}{C_A} \tag{34}$$

where V_{dA} = the apparent volume of compartment A; C_A = the concentration in compartment A; and D = the amount of substance introduced into the organism.

$$V_{dB} = \frac{D}{C_B} \tag{35}$$

where V_{dB} = the apparent volume of compartment B; C_B = the concentration in compartment B; and D = as defined above.

In the second model shown in Figure 6 (dose administered to a central compartment, distributed to a peripheral compartment, and eliminated from the central compartment), estimation of apparent volumes can be performed using the following equations.

$$V_{dA} = \frac{D}{C_A + C_B} \tag{36}$$

$$V_{dB} = V_{dA} \left[\frac{k_{AB}}{k_{BA}} \right] \tag{37}$$

With constant exposure to the toxicant, the apparent V_d for the total organism at equilibrium ($V_{ss} = V_A + V_B$) was described by Barron et al.[8] as relating "the amount of chemical in the [organism] to its concentration in the [source] compartment at steady state . . . [It] expresses the affinity or capacity of the aquatic animal for a particular chemical in terms of the equivalent volume of exposure water holding the same quantity of chemical. [It] is viewed as the steady-state partitioning of compound between animal and water." It is expressed in units of volume/volume.

With the above refinements added to our discussion, we can now move to more involved models. Five of the most common single-compartment models of bioaccumulation are shown in Figure 8. The exponential model is identical with that described for single exponential elimination except a constant source (C_1) has been added with an uptake rate constant of k_u. The k_u has units of mL/(g*h). (Some authors will give units of 1/h for k_u under the assumption of equal densities for the source and tissue.)

Normally, uptake can be defined as conforming to first-order kinetics. However, as recommended by Spacie and Hamelink[11] and Barron et al.,[8] this assumption should be tested by exposing the organism to a range of concentrations. Barron et al.[8] recommended concentrations differing by at least fivefold. The following transformation of Equation (41) was recommended by Spacie and Hamelink[11] for this purpose.

$$C_t = k_u C_1 \left[\frac{1 - e^{-k_e t}}{k_e} \right] \tag{38}$$

where C_t = the concentration (amount normalized to mass in the compartment) in the compartment at time t; k_u = the uptake rate constant (mL/(g*h)); and C_1 = the concentration (constant) in the (assumed infinite) source. The bracketed term in Equation (38) is plotted against the various C_t values. A linear plot will result if C_1 and k_u are constant. (To verify first-order elimination kinetics, ln C_t/C_0 versus time for all elimination time courses regardless of C_0 should yield a single, straight line.)

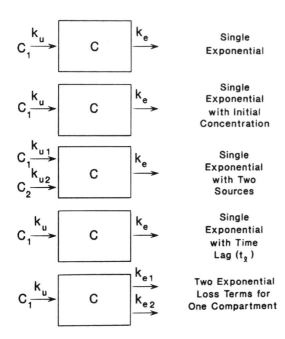

Single
Exponential

Single
Exponential
with Initial
Concentration

Single
Exponential
with Two
Sources

Single
Exponential
with Time
Lag (t_ℓ)

Two Exponential
Loss Terms for
One Compartment

Figure 8 Five common single compartment models for accumulation.

In this case of first-order kinetics, the single exponential elimination equation describing the change in concentration with time [Equation (39)] becomes Equation [40].

$$\frac{dC}{dt} = -k_e C \qquad (39)$$

$$\frac{dC}{dt} = k_u C_1 - k_e C \qquad (40)$$

The integrated form of Equation (39) is Equation (2). That of Equation (40) is the following:

$$C_t = \frac{k_u}{k_e} C_1 (1 - e^{-k_e t}) \qquad (41)$$

This model describes accumulation through time that eventually approaches steady state equilibrium with the source. The concentration in the compartment at steady state equilibrium (C_s) can be estimated.

$$C_s = \frac{k_u}{k_e} C_1 \qquad (42)$$

On division of both sides of Equation (42) by C_1, it becomes obvious that k_u/k_e is a measure of the distribution of toxicant between the source (C_1) and organism (C_s) at steady state equilibrium. As such, it has come to be referred to as the bioconcentration factor (BCF) when water is used as the source.

If, before the exposure being studied, the compartment has an initial concentration of toxicant (C_0) then the following modification of Equation (41) can be used.

$$C_t = \frac{k_u}{k_e} C_1[1 - e^{-k_e t}] + C_0 e^{-k_e t} \tag{43}$$

If there are two sources, Equation (41) can be modified to Equation (44).

$$C_t = \frac{k_{u1}C_1 + k_{u2}C_2}{k_e} [1 - e^{-k_e t}] \tag{44}$$

Equation (45) is pertinent if there is a time lag (t_l) between the beginning of exposure and initiation of uptake, e.g., induction of a transport mechanism or a delay in commencement of feeding.

$$C_t = \frac{k_u}{k_e} C_1[1 - e^{-k_e(t - t_l)}] \tag{45}$$

Finally, as an obvious extension of our discussion of elimination via two components from one compartment, accumulation under these conditions can be modeled with Equation (46).

$$C_t = \frac{k_u}{k_{e1} + k_{e2}} C_1[1 - e^{-(k_{e1} + k_{e2})t}] \tag{46}$$

It is important to remember that, although a common assumption, uptake need not be considered a first-order process. For example, Schulz-Baldes[14] describe linear accumulation kinetics ($C_t = kt + C_0$) for lead in blue mussels (*M. edulis*). Kolehmainen[65] provided an example of cyclic changes associated with seasonal changes in ^{137}Cs kinetics in bluegill (*Lepomis macrochirus*). Further, saturation of uptake mechanisms can be anticipated in many cases. Whicker and Schultz[2] (pages 74–78) provide an in-depth discussion of uptake models exhibiting exponential, linear, power, and cyclic time dependence.

Example 7
Ionic mercury accumulation in mosquitofish (*Gambusia holbrooki*) was measured over 6 days by Newman and Doubet.[66] The ambient concentration in water was 0.24 ng Hg/mL. Data for one of the fish studied were fit to Equation (41) using the following SAS code.

```
DATA MERCURY;
    INPUT DAY HG @@;
CARDS;
0 000 1 380 2 540 3 570 4 670 6 780
;
/* FITTING A SINGLE EXPONENTIAL ACCUMULATION MODEL        */
/* NOTE THAT THE HG CONC IN THE WATER WAS 0.24 NG/ML      */
PROC NLIN;
    PARMS KU=1000 KE=0.5;
```

Example 7 *Continued*

```
BOUNDS 0<KU, 0<KE<1;
MODEL HG=0.24 * (KU/KE) * (1-EXP(-KE*DAY));
OUTPUT OUT=MERCURY P=PHG R=RHG;
RUN;
```

The estimates of k_e and k_u ($\pm$ asymptotic standard error) were 0.59 ± 0.11/day and 1867 $\pm$ 242 mL/ (g*day), respectively. The BCF is estimated to be 1867/0.59 or 3164. The predicted concentration in the fish at equilibrium is (1867/0.59)*0.24 or 759 ng/g.

A more involved method of extracting accumulation parameters includes an exposure period during which organisms accumulate toxicant (elimination and uptake occurring together) followed by a depuration period during which the toxicant is eliminated in a toxicant-free environment (elimination only). Spacie and Hamelink[11] suggest that the accumulation portion of this exercise continue until the concentration in the organism approaches practical equilibrium. Practical equilibrium can be defined as the duration necessary to achieve concentrations 95% of those at final equilibrium ($t_{95\%} = -$ (ln $0.05/k_e$)). The elimination phase should continue for approximately three half-lives. However, there are cases with very slow elimination or very rapid uptake were accumulation to near equilibrium may not be practical or desirable.[67]

Example 8

Bioaccumulation of bromophos(O-(4-bromo-2, 5-dichlorophenyl) O, O-dimethyl ester) from water by the guppy (*Poecilia reticulata*) was measured by De Bruijn and Hermens.[68] Initially, fish were allowed to accumulate bromophos for 264 h (C_1 approximately 10.5 ng/L although it was 5.3 for the first several hours). The fish were then allowed to eliminate it in a bromophos-free environment. Concentrations (ng/g extractable fat) in the fish were assayed throughout the accumulation and elimination phases of the experiment. Data were visually extracted from De Bruijn and Hermens' Figure 1 to illustrate fitting of a single exponential bioaccumulation model with one source. Data are replotted in Figure 9 along with the results of the following regression analyses. The uptake and elimination rate constants were estimated simultaneously and then sequentially (k_e estimated from depuration phase data and then used in estimating k_u from the accumulation phase data) with these data. Although both nonlinear (weighted and unweighted) and linear regression on transformed data were used to estimate k_e for the sequential estimation method, only the linear regression estimate is presented here. The residuals from the linear regression model showed no obvious trends whereas those of the nonlinear regressions did.

```
DATA ACCUM;
    INPUT HOUR BRPHOS @@;
CARDS;
0.5 1900 001 3000 002 5200 004 6900 008 24000 024 50000
072 200000 144 400000 240 500000 264 500000
;
/* FITTING A SINGLE EXPONENTIAL ACCUMULATION MODEL WITH SIM- */
/* ULTANEOUS ESTIMATION OF KE AND KU. THE BROMOPHOS CONCEN-  */
/* TRATION WAS 10.5 NG/ML FOR ALL BUT THE INITIAL HOURS.     */
/* FISH CONC. WERE NG/G OF EXTRACTABLE FAT.                  */
PROC NLIN;
    PARMS KU=0.001 KE=0.01;
    BOUNDS 0<KU, 0<KE<1;
    MODEL BRPHOS=10.5 * (KU/KE) * (1-EXP(-KE*HOUR));
    OUTPUT OUT=BROMO P=PBRPHOS R=RBRPHOS;
```

Example 8 *Continued*

```
RUN;
DATA ELIMIN;
  INPUT HOUR BRPHOS @@;
  LBROMO=LOG (BRPHOS);
CARDS;
000 500000 012 450000 024 370000 048 290000 072 190000 096 150000
144 70000 216 21000 319 5000
;
/* NOW ESTIMATING KE FROM THE ELIMINATION DATA FOR LATER SUB- */
/* STITUTION INTO THE NONLINEAR REGRESSION (AS KE) FOR ESTI-   */
/* MATION OF KU.                                               */
PROC GLM;
  MODEL LBROMO = HOUR;
  OUTPUT OUT=BROMO2 P=P2BRPHOS R=R2BRPHOS;
RUN;
/* NOW USING THE ESTIMATED KE (0.01469 FROM THE PROC          */
/*                                          GLM ABOVE         */
/* IN THE ACCUMULATION MODEL AND SOLVING FOR KU.              */
PROC NLIN DATA = ACCUM;
  PARMS KU=0.001;
  BOUNDS 0<KU;
  MODEL BRPHOS=10.5* (KU/0.01469) * (1-EXP(-0.01469* HOUR));
  OUTPUT OUT=BROMO3 P=P3BRPHOS R=R3BRPHOS;
RUN;
```

The k_u and k_e ($\pm$ asymptotic standard error) derived simultaneously from the accumulation data were 345 $\pm$ 32 mL/(g fat*h) and 0.0052 $\pm$ 0.0010/h, respectively. Plots of predicted and original data, and plots of the regression residuals show good fit of the model. However, the k_e estimated from the elimination data (0.0147 $\pm$ 0.0002/h) was quite different from this estimate. The elimination data fit to the model was excellent ($r^2 = 0.998$) with no pattern to the residuals. When this estimate was used in the accumulation model, the k_u was estimated to be 644 $\pm$ 40 mL/(g fat*h). The residuals were lower than expected in the initial phases of accumulation but increased to above predicted values later.

A certain degree of covariance exists between k_e and k_u estimates when they are derived simultaneously. Such covariance can be avoided by the sequential approach described here. However, the simultaneous and sequential estimation methods should give similar results if nothing significant has changed between the accumulation and elimination phases of the experiment. The results shown above suggest that something had changed. The authors describe an initial (0 and 8 h analyses) concentration of bromophos of 5.3 ng/mL that increased to 10.5 ng/mL thereafter. Concentrations in the waters shown in their Figure 1 suggests that, for times less than 72 h, concentrations were less than this average value. The low concentrations initially in the water could have produced the lower-than-predicted pattern in the data residuals from the first portion of the accumulation model. Under this assumption, it would be best to use the estimate of k_e from the elimination phase and then estimate k_u using the sequential approach (k_e supplied to the nonlinear, regression model to derive k_u). The final estimates of k_e and k_u using this approach were 0.0147 $\pm$ 0.0002/h and 644 $\pm$ 40 mL/(g fat*h).

b. Clearance Volume-Based Models

Pharmacokinetics models often are formulated using the concept of clearance.[8,10,69] As defined above, clearance can be thought of as the volume (relative to that of a reference compartment) cleared per unit time (mL/h). If normalization to mass of tissue is employed in the model then the units are mL/(g*h). For example, if uptake of toxicant from water were being considered, then clearance would be interpreted as the volume of the toxicant source (water) cleared of toxicant per gram of organism every hour. If one is considering loss of a drug from an organ after intravenous injection, then clearance may be seen as the volume cleared from the organ (in terms of volume of a reference compartment such as the blood) in some unit of time. (As noted in Barron et al.,[8] uptake rate constants such as that described above are actually clearances, not proportional rate constants like elimination rate constants.)

Clearance (Cl) is defined in terms of apparent volume of distribution and rate constant as the following.

$$Cl = kV_d \tag{47}$$

For example, the clearance associated with elimination from one compartment would

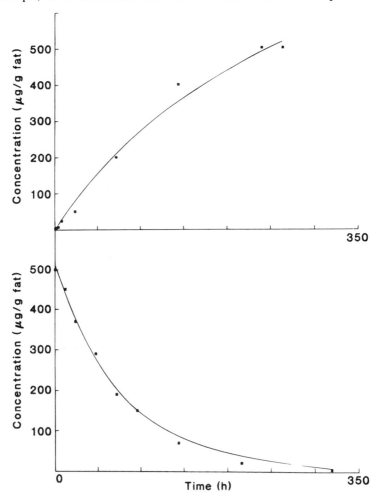

Figure 9 Fit of regression results to data from De Bruijn and Hermens.[68] Accumulation and elimination of bromophos from water in the guppy were fit in this example. See text for more details.

be k_e*V_d. By simple substitution into Equation (3), the $t_{1/2}$ in terms of clearance would be $(\ln 2*V_d)/Cl$.

The total clearance from the system being modeled is the sum of all the individual clearances.

$$Cl_t = \sum_{i=1}^{n} Cl_i \qquad (48)$$

where Cl_i = clearance from the ith compartment.

The units of the Cl_i values can be expressed relative to the external environment, e.g., a water compartment, or they may be expressed relative to an internal compartment. The internal compartment is often the blood. In the case of blood, the toxicant bound to ligands in the blood may be treated as yet another separate compartment. In that case, the reference compartment may become the free toxicant in the blood.

By substitution, these models can now be expressed in terms of clearance. Remember that V_d expresses the distribution of the toxicant between the reference compartment and the compartment in question. Landrum and Lydy[69] point out that, this being the case, the V_d is an estimation of the BCF. Because the BCF is estimated in the rate constant model as k_u/k_e then V_d estimates k_u/k_e. With this information and rearranging Equation (47), Equation (41) can be modified to a clearance volume-based model of accumulation.

$$C_t = V_d C_1 (1 - e^{-\frac{Cl}{V_d}t}) \qquad (49)$$

D. ACCUMULATION MODELS: SEVERAL COMPARTMENTS OR SOURCES

1. General

Accumulation compartment models can be expanded beyond the treatment described above to include multiple sources or several internal compartments. With enriched description comes the difficulties of deriving many parameters and ascertaining the sensitivity of the outcome to errors in each parameter estimate. Regardless, the richness of detail within such models can greatly enhance understanding of the system in question. Some of the simpler models are described here.

2. One Compartment with Uptake from Food and Water

Significant uptake from both food and water can occur for a wide range of toxicants.[14,70,72] The consequent accumulation can be accomodated in the model structure developed so far. Spacie and Hamelink[11] used Thomann's model[73] for accumulation of toxicant from two sources (food (C_f) and water (C_w)) to illustrate incorporation of a variety of factors into bioaccumulation models. To incorporate ingestion into the model, two terms were introduced, assimilation efficiency and weight-specific ration. The assimilation efficiency (α) is the amount of toxicant absorbed per amount of toxicant in the food. The weight-specific ration (R) is the amount of food provided per amount of organism present, e.g., 1 g food/100 g fish. Obviously, the feeding rate (f, g food/g fish) for free-ranging organisms can be used instead of R, e.g., Harrison and Klaverkamp.[74]

$$C_t = \frac{k_u C_w + \alpha R C_f}{k_e}(1 - e^{-k_e t}) + C_0 e^{-k_e t} \qquad (50)$$

Thomann's model also incorporates an initial concentration of toxicant [see Equation

(43)]. Although not included in Equation (50), the model was also expanded to include growth dilution. A growth rate constant was used based on the assumption of exponential growth during the period of exposure.

Each set of parameters associated with elimination, uptake from food, uptake from water, and growth would be estimated through a series of experiments such as those described above. The resulting parameters would then be incorporated into Equation (50). Assimilation efficiency from food is estimated relative to the entire animal here. If the model has several compartments then assimilation efficiency may be relative to the reference compartment, e.g., blood compartment in pharmacokinetics modeling of oral drug administration. Estimation of assimilation efficiency (or fractional absorption) in this context will be described in Section III.A.

3. Multiple Sources and Elimination Components

Generalizing to multiple uptake and elimination components from one compartment that has an initial concentration of toxicant, Whicker and Schultz[2] provide the model.

$$
C_t = \frac{\sum_{j=1}^{m} C_j k_{uj}}{\sum_{i=1}^{n} k_{ei}} \left[1 - e^{-\left(\sum_{i=1}^{n} k_{ei}\right)t} \right] + C_0 e^{-\left(\sum_{i=1}^{n} k_{ei}\right)t} \tag{51}
$$

4. Complex Multiple Compartment Models

More complex models range from those involving two compartments[75] to three compartments[76] to many compartments.[77,78] In many, the focus of uptake and elimination kinetics can move from mathematical compartments to specific organs or tissues. For example, Waiwood et al.[76] modeled zinc kinetics in lobster using the gills, hemolymph, hepatopancreas, and tail muscle as the major compartments. Such linkage of compartments to physical structures has the virtue of allowing direct physical interpretation of the associated parameters as well as leaving open the opportunity for incorporating physiologic and allometric relations into model development. Barron et al.[79] and Nichols et al.[80,81] provided good examples in which such models became physiologically based pharmacokinetic (or toxicokinetic) models (PBPK models). This genre of models is discussed in more detail below.

Important in many models is the internal degradation or conversion of the parent toxicant to metabolites. Such metabolism can involve first-order or saturation kinetics. If the parent toxicant concentration is high, metabolism may be described best by zero-order kinetics. Metabolites or chemical species can then be incorporated. Leversee et al.[82] and Foster and Crosby[83] provide good examples of metabolism of organic toxicants by aquatic biota. Walker[84,85] relates such degradation to elimination kinetics. Bioaccumulation of the herbicide, 3,5,6-trichloro-2-pyridinyloxy acetic acid as modeled by Barron et al.[79] is a good example of a clearance volume-based model incorporating metabolism.

Internal conversion may be important in modeling accumulation of inorganic toxicants as well as organic toxicants. A quick review of the complex, biologically mediated transformations of arsenic should make this obvious, e.g., Lunde,[86] Unlu and Fowler,[87] Edmonds and Francesconi,[88] and Zingaro.[89]

E. PHYSIOLOGICALLY BASED PHARMACOKINETIC MODELS
1. General

Compartment models derived from multiple exponential components are limited to two or three compartments by the associated data requirements.[80] For example, we have already discussed the high quality of data estimated necessary by Myhill[29] to strip two components from an elimination curve, i.e., data error less than 1% for a ratio of k_e values of 2. Further, because compartments are specific to the species and conditions of the experiment, extrapolation from the results is quite limited. Beyond two or three components, rate constant-based models are generally abandoned in favor of physiologically based models.

Physiologically based pharmacokinetic (PBPK) or physiologically based toxicokinetic (PBTK) models "define an organism in terms of its anatomy, physiology, and biochemistry."[80,81] This class of models, developed in the medical sciences, is not based on mathematically identified compartments. Rather, each compartment has an assigned physical meaning. Because of this linkage, physiological and allometric relationships can be incorporated. For example, accumulation can be linked to such processes as gill oxygen exchange,[90] bivalve filtration rates,[91] or temperature-dependent metabolic rate.[92] This being the case, extrapolation between similar species and conditions is accomplished more readily, and concentrations in target organs are assessed more clearly. Normally, simulation software such as SIMUSOLV,[93] Advanced Continuous Simulation Language (see Nichols et al.[79]) or TIME-ZERO[94] are used to develop such complex models.

Action of a tissue compartment on the toxicant can involve binding (storage) as described by adsorption in section B or elimination as described in section A. Elimination can be physical removal from the system such as urinary excretion or metabolism such as hepatic breakdown of organic toxicants. Kinetics can include saturation kinetics. Barron et al.[8] give the following equations as descriptions of tissues displaying only storage [muscle, Equation (52)] and tissues displaying metabolism also [liver, Equation (53)].

$$\frac{dC_m}{dt} = \frac{Q_m\left[C_i - \dfrac{C_m}{R_m}\right]}{V_m} \tag{52}$$

$$\frac{dC_h}{dt} = \frac{Q_h\left[\left(C_i - \dfrac{C_h}{R_h}\right) - Cl_h C_i\right]}{V_h} \tag{53}$$

where V_m, V_h = weight of muscle and hepatic compartments; R_m, R_h = blood/tissue partition coefficient; C_i = concentration entering the tissue; C_m/R_m, C_h/R_h = blood concentrations exiting the tissues; Q_m, Q_h = blood flow to the tissues; and Cl_h = clearance constant for hepatic compartment.

2. Uptake from Water

Athough some have involved metals, e.g., Part and Svanberg,[95] Van Der Putte and Part,[96] most PBPK models of toxicant accumulation have been developed for organic contaminants. Regardless of the class of toxicants being modeled, most models of accumulation via the gills begin with Fick's Law [Equation (31)]. For example, Hayton and Barron[97] define the rate of transfer across the gill with the following relationship.

$$\frac{dX}{dt} = D_m A K_m\left(\frac{C_0 - C_i}{d}\right) \tag{54}$$

where D_m = the diffusion coefficient (length^2/time); A = area of gill epithelium available for exchange; K_m = epithelium/water distribution coefficient; C_i, C_0 = concentration inside and outside of the gill; and d = the thickness of the epithelium.

Basic physiological and anatomical relationships are incorporated into the process of gill uptake to produce the relationship,

$$P = \left[\frac{d}{D_m A K_m} + \frac{h}{D_a A} + \frac{1}{K_b V_b} + \frac{1}{V_w} \right]^{-1} \tag{55}$$

where P = the uptake clearance (volume/time); D_a = the diffusion coefficient in water (length²/time); h = thickness of stagnant water layer on gill surface; K_b = blood/water distribution coefficient; V_b = effective blood flow through the gills; and V_w = effective water flow across the gills.

In this model, some parameters are related to the specific organism (d, h, A, V_b, V_w), whereas others (D_m, D_a, K_m, K_b) are linked to properties of the organic chemical in question. By defining these parameters for various organisms and organic chemicals our ability to predict bioaccumulation for untested species and chemicals is enhanced. Further, our ability to identify critical parameters is also enhanced.

This general approach is actively being developed for organic toxicants by such workers as Barron et al.,[79] Erickson and McKim,[55,59] Hayton and Barron,[97] and Nichols et al.[80,81] Although a seemingly powerful approach, it has yet to be applied in a focused manner for inorganic toxicants. Further, linkage of concentrations in various compartments including target organs to effect has lagged behind PBPK models used in mammalian toxicology. Such models which incorporate effect compartments in toxicological research are termed pharmacokinetic-pharmacodynamic models.[37]

Example 9

Because of the complexity of such models and the specialized software applied to model formulation, a detailed example would be difficult to provide. However, a general discussion of the model developed by Nichols et al.[80] could provide an appreciation for this approach. A more detailed description is given in the appendix of Nichols et al.[80]

Figure 10 is a visualization of this model. The organic toxicant enters the fish via the gills. This entry is controlled by a variety of physiological and anatomical features such as those described in Equation (55). The toxicant is distributed between several compartments including fat tissue, kidney, liver, richly perfused tissue, and poorly perfused tissue. Elimination occurs via saturation or first-order kinetics in the liver.

A large amount of information was needed to parameterize the model. Information on cardiac and respiratory function, metabolic and excretory rates, compartment volumes, and blood flow rates were estimated or derived from laboratory experiments with rainbow trout (*Oncorhynchus mykiss*).

---

3. Uptake from Food

General uptake from food has already been described in Section D.2 and Equation (50). However, complex PBPK models such as those derived for organic chemical bioaccumulation from water have not been derived for uptake from food. Bioenergetic considerations have been added to this general model by Pentreath.[98] Also, the ingestion of methylmercury was related to food intake for maintenance and growth, and growth

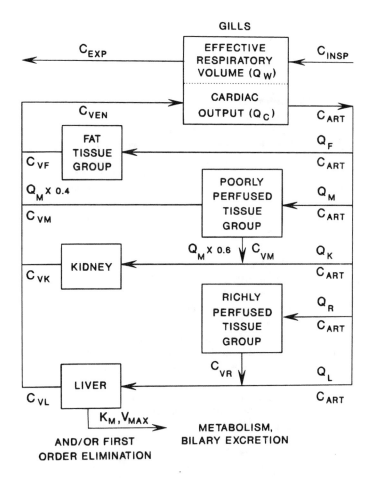

Figure 10 An example of a physiologically based pharmacokinetic model from Nichols et al.[80] An organic toxicant enters via the gills as determined by Q_w and Q_c. It is then distributed between several internal compartments and eliminated from the liver according to Michaelis-Menten kinetics. (Reprinted with permission from Academic Press.)

dilution was incorporated into the model. Intake of methylmercury was defined by Equation (56).

$$I_f = \phi\xi W^j + \phi \frac{1}{\epsilon} \frac{dW}{dt} \tag{56}$$

where ϕ = the concentration of methylmercury in the food; ξ = the maintenance food coefficient (size-dependent); W = the weight of the fish; j = allometric coefficient relating ξ to weight; and ϵ = efficiency of food utilization for growth.

The excretion efficiency (K) was also linked to fish weight.

$$K = a_2 W^{-b_2} \tag{57}$$

where a_2 and b_2 are derived parameters.

From these relationships, Equation (58) was derived to describe the change in methylmercury concentration with time as a function of feeding bioenergetics.

$$\frac{dC_t}{dt} = \phi\xi W_t^{j-1} + \phi\frac{1}{\epsilon}\frac{dW_t}{dt}\frac{1}{W_t} - \left[\frac{dW_t}{dt}\frac{1}{W_t} + \frac{a_2}{W_t^{b_2}}\right]C_t \qquad (58)$$

4. Growth and Allometric Considerations

In Equation (58), any of several growth models can be added in place of W_t. Pentreath selected the von Bertelanffy growth equation. [In Equation (50), exponential growth was assumed.] Other models such as the Richards growth equation[99] could be incorporated.

Allometric considerations (the effects of changes in size of the whole organism or an anatomical feature) are also critical to PBPK models. For example, the area available for absorption of a contaminant in food may change disproportionately with the volume or mass of an organism. Ribble and Smith[58] noted such a size-related shift in relative intestine length in fish. Size-specific metabolic rate shifts may also be critical in toxicant elimination and uptake.[21,100,101]

F. FUGACITY-BASED MODELS

Fugacity modeling can be directly related to the PBPK models described to this point. Gobas and Mackay[102] gave an excellent demonstration of this in their derivation of a bioaccumulation model for uptake via the gills. Fugacity-based models can enhance modeling as they allow expression of all amounts of toxicant in many potentially heterogeneous compartments in the same units. This enhances linkage to compartments outside of the organism.[103] Models dealing with gas, solid and liquid phases express all phase activities in terms of the gas phase fugacity.[104-106] Mackay[107] discusses this approach in detail in his recent book.

Fugacity is the "escaping tendency" of a substance from a phase.[102,103] This tendency for a substance to partition itself between phases is based on chemical potential or activity.[103] Fugacity is expressed in units of pressure (Pa). It (f) is related linearly to concentration (C) by the fugacity capacity (Z).[102]

$$C = fZ \qquad (59)$$

where C = concentration (mol/m^3); f = fugacity (Pa); and Z = the fugacity capacity (mol/(m^{3}Pa)).

When two phases are in equilibrium, the partition coefficient (P_{12}) will equal C_1/C_2 or Z_1/Z_2. Transport between phases is expressed with the transport constant, D. The rate of transport (mol/h) between two phases can be defined.

$$N = D(f_1 - f_2) \qquad (60)$$

where D = the transport constant (mol/(h*Pa)); f_1 = the fugacity of phase 1; and f_2 = the fugacity of phase 2.

In the context of the single compartment bioaccumulation model [Equation (41)],

$$k_e = D_o/V_o Z_o \qquad (61)$$

$$k_u = D_o V_o Z_w \qquad (62)$$

where D_o = the transport constant for the organism; V_o = the volume of the organism; Z_o = the fugacity capacity of the organism; and Z_w = the fugacity capacity of the water.

Gobas and Mackay[102] give the following fugacity-based model analogous to the rate constant-based [Equation (41)] and clearance volume-based [Equation (49)] models described previously.

$$C_t = C_w \left(\frac{Z_o}{Z_w}\right)(1 - e^{-\frac{D_{oi}}{V_o Z_o}}) \tag{63}$$

In a recent book,[107] Mackay described a more detailed, fugacity-based bioaccumulation model that incorporates uptake from food and water, fish metabolism, and growth. The associated BASIC program for personal computers was supplied with his book.

Example 10

As in Example 9, it would be impractical to provide a detailed derivation of a fugacity-based model here. Instead, a fugacity-based model derived by Mackay and Paterson[103] for bioaccumulation of styrene by ingestion and inhalation is depicted in Figure 11. It has the same general form as the PBPK model shown in Figure 10 but fugacity (f_i) and transfer parameters (D_i) are used to define the flow of contaminant within the organism.

III. MODELING BIOACCUMULATION: ALTERNATIVE APPROACHES

A. STATISTICAL MOMENTS APPROACH

Except for a very brief discussion of mean residence time estimation from half-life, models have been discussed in a solely deterministic context to this point. However, alternative models based on the distribution of contaminant molecules can also be formulated. In this statistical approach, the residence times within compartments can be estimated using only concentration versus time curves. Not only does this allow estimation without imposition of a specific model, it also allows estimation of statistics that are comparable between model types (mean residence time). Such an approach has been used very successfully in pharmacokinetics but rarely in ecotoxicology.[8]

Yamaoka and co-workers[108] brought statistical moments methods to pharmacokinetics. In their development of this approach, the first three moments (zero, first normal, and second central moments) of the plasma concentration versus time curve were used to estimate the mean and variance of drug residence time. These moments were defined by Equations (64) to (66)

$$\text{AUC} = \int_0^\infty C_p dt \tag{64}$$

$$\text{MRT} = \frac{\int_0^\infty t C_p dt}{\int_0^\infty C_p dt} \tag{65}$$

$$\text{VRT} = \frac{\int_0^\infty (t - \text{MRT})^2 C_p dt}{\int_0^\infty C_p dt} \tag{66}$$

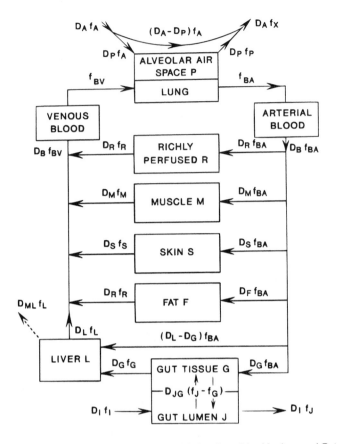

Figure 11 An example of a fugacity-based model developed by Mackay and Paterson[103] for styrene bioaccumulation. Styrene is taken up by inhalation and ingestion. Note that this model is very similar in form to the model in Figure 10. However, the units are changed to accomodate fugacity and transfer parameters. (Reprinted with permission from *Environmental Toxicology and Chemistry*. Copyright (1987) SETAC.)

where AUC = the area under the curve; MRT = the mean residence time; VRT = the variance of residence time; C_p = the plasma drug concentration; and t = time.

Because the C_p is usually observed for some finite, maximum time (T), the area under the right tail of the curve ($T \to \infty$) must be approximated. This is usually done assuming a terminal exponential component from T to ∞, e.g., $C_{p0}e^{-kt}$. The following unnormalized moments are used for this purpose.

$$S_0 = \int_0^T C_p dt + \frac{C_{p0}}{k} e^{-kT} \tag{67}$$

$$S_1 = \int_0^T tC_p dt + \left(\frac{C_{p0}}{k^2} + \frac{C_{p0}T}{k} \right) e^{-kT} \tag{68}$$

$$S_2 = \int_0^T t^2 C_p dt + \left(2\frac{C_{p0}}{k^3} + 2\frac{C_{p0}T}{k^2} + \frac{C_{p0}T^2}{k} \right) e^{-kT} \tag{69}$$

The integrated portions of Equations (67) to (69) can be estimated using the trapazoidal method and the plasma concentrations (C_{pi}) measured at n time intervals (t_i) up to T.

$$\int_{t_0}^{t_n} C_{pi}\, dt \approx \sum_{i=0}^{n-1} (t_{i+1} - t_i) \frac{(C_{pi} + C_{pi+1})}{2} \tag{70}$$

$$\int_{t_0}^{t_n} t C_{pi}\, dt \approx \sum_{i=0}^{n-1} (t_{i+1} - t_i) \frac{(t_i C_{pi} + t_{i+1} C_{pi+1})}{2} \tag{71}$$

$$\int_0^T t^2 C_{pi}\, dt \approx \sum_{i=0}^{n-1} (t_{i+1} - t_i) \frac{(t_i^2 C_{pi} + t_{i+1}^2 C_{pi+1})}{2} \tag{72}$$

Using Equations (67) to (69) with the substitutions given in Equations (70) to (72) for these integrated portions, the S_0, S_1, and S_2 are estimated. The area under the concentration-time curve (AUC) is estimated simply by S_0. The mean and variance of the residence time are estimated with Equations (73) and (74).

$$\text{MRT} = \frac{S_1}{S_0} \tag{73}$$

$$\text{VRT} = \frac{S_2}{S_0} - \left[\frac{S_1}{S_0}\right]^2 \tag{74}$$

If a specific model is assigned to the data set and an iterative regression method is used to fit a multiexponential model, these statistics can be easily estimated.

$$S_0 = \sum_{i=1}^{n} \frac{C_{p0i}}{k_i} \tag{75}$$

$$S_1 = \sum_{i=1}^{n} \frac{C_{p0i}}{k_i^2} \tag{76}$$

$$S_2 = \sum_{i=1}^{n} \frac{2C_{p0i}}{k_i^3} \tag{77}$$

where C_{p0i} = the concentration in the ith exponential component; k_i = the elimination rate constant for the ith component; n = the total number of conponents in the model.

Example 11

To illustrate the statistical moments approach, the elimination of anthracene from fish as presented in Example 1 (single exponential elimination model) will be used. In the previous example, one elimination component was obvious. The estimated initial concentration was 99.46 nmol/g and the elimination rate constant was 0.680/h. According to Equation (4) and associated assumptions, the mean residence time should be approximately 1.47 h. Now let's use the method of statistical moments to describe elimination.

i	Time (h)	Anthracene (nmol/g)	$t_{i+1} - t_i$	$(C_i + C_{i+1})/2$
0	0	100	1	76.5
1	1	53	1	38.5
2	2	24	1	18.0
3	3	12	1	9.5
4	4	7	—	—

Assuming no specific model except to estimate the area beyond the last sampling time (4 h), the calculations are the following.

From Equations (67) and (70):

$$S_0 = 1(76.5) + 1(38.5) + 1(18.0) + 1(9.5) + (99.46/0.680)e^{-0.680(4)}$$
$$= 142.5 + 9.6 = 152.1$$

From Equations (68) and (71):

$$S_1 = 142.5 + (99.46/0.680^2 + 99.46(4)/0.680)e^{-0.680(4)}$$
$$= 151.0 + 52.7 = 203.7$$

From Equations (69) and (72):

$$S_2 = 313 +$$
$$(2(99.6)/0.680^3 + 2(99.6)(4)/0.680^2 + 99.6(4^2)/0.680)e^{-0.680(4)}$$
$$= 313 + 309.6 = 622.3$$

Using Equations (73) and (74):

$$\text{MRT} = S_1/S_0 = 203.7/152.1 = 1.34 \text{ h}$$
$$\text{VRT} = S_2/S_0 - (S_1/S_0)^2 = 622.3/152.1 - (1.34)^2 = 2.3 \text{ h}$$

If the original, single exponential model is assumed then the calculations can be done using Equations (75) to (77) instead of Equations (67) and (72).

$$S_0 = 99.6/0.680 = 146.5$$
$$S_1 = 99.6/0.680^2 = 215.4$$
$$S_2 = 2(99.6)/0.680^3 = 633.5$$
$$\text{MRT} = S_1/S_0 = 215.4/146.5 = 1.47 \text{ h}$$
$$\text{VRT} = S_2/S_0 - (S_1/S_0)^2 = 633.5/146.5 - (215.4/146.5)^2 = 2.2 \text{ h}^2$$

The errors associated with the estimates of AUC, MRT, and VRT are dependent on t. The longer the period during which the concentrations are measured, the better the estimates. Yamaoka et al.[108] gave results of simulations (oral administration) indicating that measurement of concentrations until values drop to 5% of the maximum resulted in relative errors of 5, 10, and 40% for AUC, MRT, and VRT, respectively. When concentrations are measured to 1% of the maximum, relative errors become 1, 2, and 10% for AUC, MRT, and VRT, respectively.

With these estimates of AUC and MRT, clearance from the plasma (or blood, Cl_b) and volume of distribution at steady state (V_{ss}) can be determined.[8,31]

$$Cl_b = \frac{D}{AUC} \tag{78}$$

$$V_{ss} = Cl_b \, \mathrm{MRT} \tag{79}$$

Another potentially useful yet ignored application of this approach is the estimation of bioavailability. Yamaoka et al.[108] and Greenblatt and Shader[10] discussed the use of plasma AUC to estimate the bioavailability of orally administered drugs. The general approach involves estimation of AUC after intravenous injection (AUC_{iv}) and after oral administration (AUC_{oral}). The fractional absorption (f), a measure of bioavailability of orally administered chemicals, is estimated.

$$f = \frac{AUC_{oral}}{AUC_{iv}} \tag{80}$$

The f estimates the percentage of an orally administered drug absorbed and entering the blood relative to that in the blood via direct injection. Equation (80) assumes that the same dose was given by both routes. It may be easily changed to accomodate different doses.

$$f = \frac{D_{iv} \, AUC_{oral}}{D_{oral} \, AUC_{iv}} \tag{81}$$

where D_{iv} and D_{oral} = intravenous and oral doses.

The $\mathrm{MRT}_{gastric}$ and $\mathrm{VRT}_{gastric}$ can also be estimated by the differences between the oral and intravenous estimates, i.e., $\mathrm{MRT}_{oral} - \mathrm{MRT}_{iv}$ and $\mathrm{VRT}_{oral} - \mathrm{VRT}_{iv}$, respectively.[108]

The AUC values for various routes may be estimated using urinary excretion. The method is very similar to that described above. However, flow estimates are also incorporated into the calculations. The assumption is made that urinary excretion reflects the amount of drug in the plasma. Yamaoka et al.[108] and Greenblatt and Shader[10] provide details of this approach.

This approach can also be used for other routes of toxicant introduction and comparison of various toxicants entering by the same route. A common example in pharmacokinetics involves intramuscular versus intravenous administration. The approach is identical with that described for intravenous versus oral administration. Obviously, this approach has much to offer to estimation of bioavailability of toxicants in various environmental components. Unfortunately, it has been largely ignored.

B. STOCHASTIC MODELS

All the compartment models described in previous sections are based on the assumption that a deterministic system was being modeled. Constants and initial compartment concentrations were assumed to be invariant. All molecules within a compartment were assumed to have identical probabilities of leaving the compartment during any time interval. In reality, such assumptions are rarely if ever valid. For example, the elimination rate is a consequence of a multitude of processes including metabolic rate, gill ventilation, rate of urinary excretion, gut clearance, bile secretion, etc. Even for an individual, some distribution of elimination rate constants with an "average" value should be expected. If a group of individuals were used in a depuration experiment and subsets of the group were taken at time intervals for measuring concentrations, a distribution of initial concentrations between individuals must be expected except in the most fortuitous experiment. Further, all molecules of a contaminant are unlikely to have the same probablity of leaving a compartment during the course of depuration. Regardless, the bioaccumulation models used in ecotoxicology today are deterministic models based on these unlikely assumptions. For the reasons given above, Matis and co-workers[109-112] have advocated the wider use of stochastic models.

"A stochastic system has been defined as one whose output is uncertain and a stochastic model as one that is structured to account for at least part of the uncertainty [in stochastic systems]."[110] Stochastic models allow description of systems with heterogeneous compartments and stochastic flows between compartments. Whicker and Schultz's[2] description given earlier of radionuclide elimination from bone can be viewed as an example of a heterogeneous compartment. As mentioned, a series of exponential compartments could have been used instead of a power model, except that extraction of too many elimination rate constants would be required. Alternatively, a stochastic model based on a gamma distribution instead of multiple exponentials could describe this system with fewer parameter estimates. The gamma model would do so by incorporating a series of exponentials within its parameter estimates. It would also maintain a higher fidelity to the underlying mechanisms than a power model.

Let's use the explanation of Matis and Tolley[110] to illustrate a simple stochastic model. In this approach, the amount of a toxicant in a compartment experiencing elimination is modeled. The familiar form of the deterministic, monoexponential model is Equation (2). This model can be restated in the following form under the assumption that the mean value (μ_t) was being measured.

$$\mu_t = X_0 e^{-kt} \tag{82}$$

The model is now stochastic in its form. Although similar to that in Equation (2), this form predicts the mean concentration over time. It leaves open the possibility of random changes in the model.[112] Functions such as exponential, Weibull, gamma, Poisson, or negative binomial can be used to describe such uncertainty.

For example, k may be treated as a random variable. Matis and Tolley[10] provided details on a variety of such models with different distributions. Further, the probability of a molecule leaving the compartment within a time interval (the hazard rate) can be varied easily within the stochastic model format. A detailed example of modeling mercury accumulation in fish using stochastic models was also given in Matis et al.[112] A software package (KINETICA) for constructing stochastic models is also described. Perhaps with the increased availability of such software, stochastic models will become more common in ecotoxicology.

IV. INTRINSIC FACTORS AFFECTING BIOACCUMULATION

A. GENERAL

Factors intrinsic to the individual and extrinsic factors modify bioaccumulation. Intrinsic factors will be discussed briefly in the remainder of this chapter.

Intrinsic factors include the ability of individuals to transform, modify, or metabolize the toxicant. An obvious example is the internal transformation of organic toxicants such as benzo(a)pyrene[82] or p-nitrophenol.[83] However, transformations of inorganic toxicants may also be significant. For example, arsenic metabolism[86,88,89] especially relative to the role of arsenate as an analog in phosphate biochemistry[113,114] plays a critical role in bioaccumulation. Radionulcide bioaccumulation is strongly influenced by the biochemistry of chemical analogs, e.g., Morgan,[115] Blaylock and Frank,[116] Blaylock,[117] Rodgers,[118] and Hinton et al.[119] Less direct effects include such biochemical processes as calcium and phosphate effects on zinc uptake as controlled by calcium pyrophosphate granule formation[120] within the organism. Another intrinsic factor contributing to bioaccumulation is age[121,122] and associated developmental changes.[12] Sexual differences also exist.[123,124] Two other factors, individual size and lipid content are often expressed in quantitative terms and, as a result, are discussed in detail here.

B. LIPID CONTENT

Bioaccumulation of organic compounds is influenced by the lipid pool within individuals. Neely et al.[125] compared the lipophilic tendencies of eight organic compounds to the equilibrium bioconcentration factor (BCF or concentration in the organism/concentration in the water source) reported in fish. The partition coefficient between water and 1-octanol (K_{ow}) was used to estimate the lipophilic nature, and k_u/k_e [see Equation (42)] was used to estimate the BCF for each compound. The linear regression model describing this relationship was log BCF = 0.542log K_{ow} + 0.124. They stated that this relationship could be used to predict BCF for compounds with K_{ow} values within the tested range (log K_{ow} from 2.88 to 7.62). (Note that my reanalysis of these data[126] indicates a prediction bias of $10^{0.1173/2}$ or 1.14 (MSE = 0.1173) associated with these data). Veith et al.[127] and Mackay[128] produced similar linear relationships using more organic compounds with log K_{ow} values in the range of 1.0 to 7.05. Geyer et al.[129] found a strong relationship between log BCF and log K_{ow} for the marine mussel, M. edulis. Chiou[130] compared the K_{ow} with the partition coefficient of triolein (glyceryl trioleate, a lipid similar to triglycerides in animals) and verified the K_{ow} as a viable surrogate measure of organic contaminant lipophilicity. However, Figure 2 in Chiou's paper suggested that this relationship does deviate from linearity above a log K_{ow} of approximately 6. As suspected by Mackay,[128] the linearity of the relationship between log BCF and log K_{ow} did not persist with high K_{ow} compounds.

Connell and Hawker[131] discussed possible mechanisms associated with deviations from linearity in this relationship. They use the following models in their explanation. Equation (83) is a direct modification of Equation (41) using BCF instead of k_u/k_e in the equation. The BCF is linearly related to k_u and $1/k_e$. Equation (84) was proposed by Mackay and Hughes[132] based on the fugacity approach. Note that the inverse of k_e is linearly related to K_{ow} in this equation.

$$BCF = \frac{k_u}{k_e} (1 - e^{-k_e t}) \qquad (83)$$

$$\frac{1}{K_e} = \frac{V_l}{Q_w} K_{ow} + \frac{V_l}{Q_o} \qquad (84)$$

where V_l = the lipid phase volume in the animal; Q_w = the effective flow rate of water; and Q_o = the effective flow rate of octanol or organic matter between the water and lipid phase.

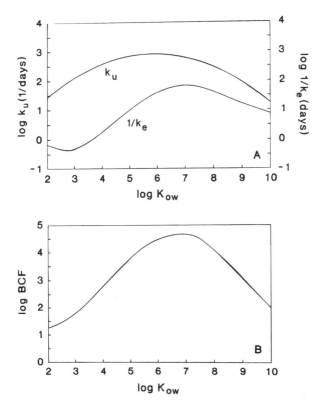

Figure 12 The effect of K_{ow} on the rate constants and BCF for organic toxicants (modified from Connell and Hawker[131]). Panel A shows the non-linear effect of log K_{ow} on both log k_u and log k_e, and panel B shows the consequent effect of log K_{ow} on log BCF.

Gobas et al.[133] suggested that uptake and elimination for organics with relatively low lipophilicity are controlled by membrane permeation. For those with higher lipophilicity (log K_{ow} > 3–4) permeation of membranes was rapid and, therefore, diffusion likely controls their uptake and elimination. Between log K_{ow} values of approximately 3 to 6, the diffusion control produces a linear relationship between log K_{ow} and log k_u or log k_e. (Note that, relative to internal compartments, cell permeability has little effect on compartmental modeling for very lipophilic compounds also.[133]) However, as lipophilicity increases (and K_{ow} increases), the k_u and k_e begin to decrease and the relationships become nonlinear (Figure 12A). This nonlinearity is thought to result from the decreased diffusion associated with the large molecular size of the highly lipophilic compounds. The result of these influences on k_e and k_u are shown in Figure 12B. The relationship between log BCF and log K_{ow} is linear between log K_{ow} of 3 to approximately 6.5 where diffusion controls uptake and elimination. The relationships deviate from linearity below approximately 3 as they are dominated by membrane permeation, not diffusion. Above a log K_{ow} of about 6.5, the log BCF decreases due to molecular size effects on diffusion rates. (Chiou's observations[130] that K_{ow} is less effective as a surrogate measure of fish lipid/water partitioning at high log K_{ow} values could also contribute to this trend.) Consequently, Connell and Hawker[131] fit polynomial equations to log BCF/log K_{ow} data for organic compounds with a very wide range of lipophilicity.

C. SIZE

The allometry (the study of size and its consequences) of bioaccumulation has been given less attention than warranted. This is surprising because size variability between individuals or samples from populations is one of the most predictable of characteristics influencing bioaccumulation. Allometric or scaling effects are also important in interspecies comparisons if species vary significantly in size.

The effect of size on toxic or therapeutic drug action has been clear to pharmacologists and physiologists since the 1920s, e.g., Campbell.[135] However, incorporation of scaling into environmental sciences did not begin with any intensity until the end of World War II. At that time, studies conducted with the emergence of nuclear weapons testing suggested that animal size was very important in determining body concentrations (or activities) of fission products. By the mid-1960s, a large body of literature on metal biomonitoring emerged to reinforce these conclusions. By the mid-1970s, Boyden[136,137] had adopted the power models used in physiological and anatomical allometry to quantitatively describe scaling effects on contaminant bioaccumulation. The scaling of body burden (amount of a substance/individual) for metals in shellfish was described with the equation,

$$B = aW^b \qquad (85)$$

where B = body burden (μg/individual); W = weight of the individual; and a and b = constants.

The relationship is most often transformed to facilitate least-squares, linear regression.

$$\log B = \log a + b \log W \qquad (86)$$

To express this relationship in terms of concentration, Boyden gave the following, simple conversion.

$$B' = \frac{B}{W} = \frac{aW^b}{W} = aW^{b-1} \qquad (87)$$

where B' = concentration (μg/g of tissue).

Equation (87) can also be transformed, as was Equation (85), to yield a linear relationship. However, because of the potential presence of W in both the independent (W) and dependent (B/W) variables, it is advisable to use Equation (86) instead of the analogous transformation of Equation (87) to fit this type of data.

Boyden[136,137] described three classes of relationships based on estimated b values (Figure 13). The two most common classes had b values of approximately 1 or lower. The first ($b \approx 0.75$) was characterized by increasing concentrations with decreasing animal size (weight). Boyden suggested that bioaccumulation was dominated by some process directly linked to metabolism because a b of approximately 0.75 was similar to the b often found for size-specific metabolic rate. The second class ($b \approx 1.00$) had concentrations independent of animal size. Boyden speculated that such a relationship might be linked directly to the number of binding sites available in the tissues. As animal size increases, the number of binding sites increases but the amount of element bound per gram of tissue remains constant. In the final and least common class ($b \approx 2.00$), the metal concentration increased with increasing size. Speculating, Boyden suggested that removal and avid binding in specific tissues or organs could result in such a relationship.

There are several statistical complications associated with the techniques described here. First, the potential prediction bias detailed earlier in this chapter must be considered. Second, as discussed by Boyden,[137] and Newman and Heagler,[21] using b may not be

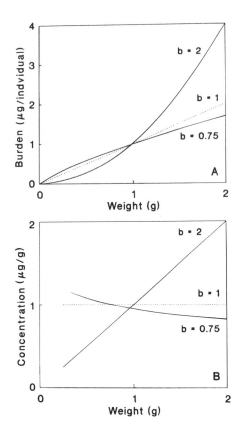

Figure 13 The effect of animal size (weight) on body burden (panel A, amount per individual) and concentration (panel B, amount per gram of tissue) of a toxicant. Three general curves with *b* values of 0.75, 1.00, and 2.00 are shown here to illustrate the various power models possible. See the text for more details.

valid. The use of predictive regression methods assumes no significant measurement error or inherent variability in the independent variable (weight); predictive regression techniques minimize variation about the dependent variable only. In many cases, this is a questionable approach as estimates of weight or size are often subject to significant error. For example, an average fish weight may be used when very small individuals of similar size are pooled to obtain sufficient tissue for analysis. In such a case, functional regressions that minimize the sum of the products of deviations of the independent and dependent variables from the regression line are more appropriate (see Newman and Heagler[21] and Ricker[138] for more details). Fortunately, if *b* and the correlation coefficient (*r*) are available from predictive regression results, the slope of the functional regression line (*v*) can be estimated by $v = \pm b/r$. The sign of *v* is the same as that of *b* and *r*.

Several conceptual complications have also been discussed regarding Boyden's speculation as described previously. Fagerström[101] argued that it is a mistake to compare *b* values from body burden or concentration with those from metabolic rates. The crux of his reasoning is that *b* values for states (burden or concentration) cannot be compared directly with the *b* value for a flux (metabolic rate). He used the accumulation of an indeterminant element[1] to demonstrate this point. For an indeterminant element, the following relationships are pertinent.

---

[1]An element is indeterminant if its concentration within the organism is directly proportional to that in the environment (source). It is determinant if its concentration in the organism remains constant over a range of environment concentrations, i.e., it is biologically controlled over a range of concentrations.

$$t_{1/2} \propto W^{1-b} \tag{88}$$

$$B' \propto W^0 \tag{89}$$

$$\theta \propto W^{1-b} \tag{90}$$

$$B \propto W^1 \tag{91}$$

where θ = metabolic turnover rate. The rate of elemental turnover (θ) will be directly proportional to the animal's metabolic rate. Contrary to Boyden's interpretations, Fagerström argues that a b value of 1, not 0.75, would suggest direct linkage to metabolic rate [Equation (91)].

Newman and Heagler[21] recently reanalyzed Boyden's data supplemented with more current data to reassess the general classification of relationships based on b (or v) values. They found no evidence supporting the three different classes of b values described by Boyden. Instead, continuous, skewed distributions of b (median = 0.80) and v (median = 0.85) values were evident. They also suggested that linear models may be more appropriate than power models for narrow allometry (description of relationships over a narrow range of sizes such as those often associated with populations). Power models may be more useful in the context of broad allometry, e.g., inter- or intraspecies comparisons with very wide ranges of animal sizes.

Several studies consider scaling effects on bioaccumulation kinetics, not observed body burdens or concentrations. It is clear that many allometric relationships could influence associated constants. For example, uptake by the gill assumes diffusion across a constant surface area and binding to a constant amount of tissue [see Equation (31)]. However, gill surface area changes disproportionately with animal size.[57,139] Many respiratory processes that directly affect gill uptake display clear allometric trends. The amount of gut surface available for uptake per unit body mass also changes disproportionately with size.[58] General body surface to volume ratio which changes with size can also influence uptake associated with adsorption processes.[140] Allometric relationships also may affect elimination constants. For example, renal elimination, number of nephra in the kidneys, urine output, heart beat, and gut beat are all linked disportionately to size in mammals.[141] Although not always the case,[142] enzyme concentrations such as those associated with contaminant degradation[85] display allometric relationships.

Anderson and Spear[143,144] successfully described the allometry associated with the simple one-compartment model constants [k_e and k_u in Equation (41)] using power models. Copper pharmacokinetics in the pumpkinseed sunfish (*Lepomis gibbosus*) gill were clearly defined with these models. But rainbow trout (*Oncorhynchus mykiss*) uptake showed only a slight scaling effect. For the sunfish but not the rainbow trout, these relationships were extended to include size-dependent toxicity. Newman and Mitz[145] also described size-dependent k_e and k_u for Zn accumulation in mosquitofish (*G. holbrooki*, formerly *G. affinis*) using simple power functions. Incorporating these power relationships into Equation (41), they demonstrated that b values may not be constant for a species/metal combination. This conclusion was contrary to Boyden's early speculation[136,137] and consistent with the results of many field studies, e.g., Cossa et al.,[146] Strong and Luoma,[147] and Lobel and Wright.[148] Using the same approach, clear power relationships were not found for Hg kinetics in mosquitofish.[66] Later, Newman and Heagler[21] speculated that metabolic rate and gill surface to total fish volume ratio may both play key roles in determining these relationships for Zn accumulation. However, no strong evidence supporting this speculation has been presented to date.

Rowland[134] and Hayton[149] also provided excellent and relevant treatments of allometry as related to PBPK modeling of mammals. Rowland's explanation began with the observation that many biological functions can be described by the power model [Equation (85) with B replaced by some biological function, $F(B)$]. The apparent volume of distribution and clearance are two such biological qualities pertinent to drug or toxicant pharmacokinetics which can be so described.

$$V_d = a_1 W^{b_1} \tag{92}$$

$$Cl = a_2 W^{b_2} \tag{93}$$

It follows by substitution that the $t_{1/2}$ can be expressed as a function of size-dependent V_d and Cl. [See Equation (47) and the text immediately following it for more information.]

$$t_{1/2} = \ln 2 \left[\frac{a_1}{a_2} \right] W^{b_1 - b_2} \tag{94}$$

As size decreases, $t_{1/2}$ will decrease if V_d decreases more rapidly than Cl. Smaller individuals will eliminate a substance faster than larger individuals.

Rowland[134] extended his treatment to include drug elimination kinetics from a monoexponential compartment. The concentration at any time (C_t) is determined by the dose administered (D), apparent volume (V_d), and elimination rate constant (k_e) as described in Equation (95).

$$C_t = \frac{D}{V} e^{-k_e t} \tag{95}$$

This follows directly from Equation (2) and rearrangement of Equation (34).

The concentration at any time can be determined as a function of the scaling of the apparent volume of distribution and clearance by substitution for V_d [Equation (92)], Cl [Equation (93)], and k_e [Equation (3) combined with Equation (94) into Equation (95)].

$$C_t = \frac{D}{a_1 W^{b_1}} e^{-\left[\frac{a_1}{a_2} \right] \left[\frac{t}{W_1^b - b_1} \right]} \tag{96}$$

Equation (96) can be used with nonlinear regression methods to fit size-dependent data to this model. However, to facilitate linear plotting, Rowland rearranges Equation (97).

$$\frac{C_t}{\frac{D}{W^{b_1}}} = \left[\frac{1}{a_1} \right] e^{-\left[\frac{a_1}{a_2} \right] \left[\frac{t}{W_1^b - b_2} \right]} \tag{97}$$

By taking the natural logarithm of both sides of Equation (97), a straight line relationship can be derived.

$$\ln \frac{C_t}{\dfrac{D}{W^{b_1}}} \text{ vs } \frac{t}{W^{b_1 - b_2}} \tag{98}$$

Rowland gave an example showing good fit to such a plot for methotrexate elimination from mice, rats, monkeys, dogs, and humans.

Except for ubiquitous data quality considerations mentioned previously, there is no reason to prevent these methods from being applied to ecotoxicology as well. The techniques can be modified to accomodate an initial concentration scenario when a dose scenario is not pertinent; substitution of body burden for dose would suffice. Incorporation of biexponential elimination would involve an expansion of Equation (96) similar to that with Equation (2) to Equation (8).

V. SUMMARY

Commonly used techniques for modeling bioaccumulation have been presented here along with several potentially useful alternative approaches. Examples illustrated the more important approaches, as well as the means of assessing the appropriateness of the models. The often blurry distinction between physical and kinetic compartments was made. The clearance-based and physiologically based pharmacokinetic models as well as the fugacity approach to modeling bioaccumulation were described. Stochastic formulations of these models were mentioned briefly. Model-free, statistical moments methods and quantitative methods of modeling effects of two intrinsic qualities of organisms (lipid content and animal size) were presented.

This treatment highlighted many subjects and approaches to provide the reader with a basic understanding of available methods. The reader is encouraged to explore the cited works listed in the references for a more comprehensive understanding of any particular method. A rich selection of techniques is available to the worker who invests the time to implement them effectively.

REFERENCES

1. Kac, M. "Some Mathematical Models in Science." *Science* 166:695–699 (1969).
2. Whicker, F. W., and V. Schultz. *Radioecology: Nuclear Energy and the Environment,* Vol. II. (Boca Raton, FL: CRC Press, Inc., 1982).
3. Atkins, G. L. *Multicompartment Models for Biological Systems.* (London: Methuen & Co. LTD, 1969).
4. Wagner, J. G. *Fundamentals of Clinical Pharmacokinetics.* (Hamilton, IL: Drug Intelligence Publications, Inc., 1979).
5. Pinder, J. E., III, and M. H. Smith. "Frequency Distributions of Radiocesium Concentrations in Soil and Biota." In *Mineral Cycling in Southeastern Ecosystems,* ed. F. G. Howell, J. B. Gentry, and M. H. Smith. (Augusta, GA: ERDA Symposium Series CONF-740513, 1975) 107–125.
6. Giesy, J. P., and J. G. Weiner. "Frequency Distributions of Trace Metal Concentrations in Five Freshwater Fishes." *Trans. Am. Fish. Soc.* 106(4):393–403 (1977).
7. Lobel, P. B., H. P. Longerich, S. E. Jackson, and S. P. Belkhode. "A Major Factor Contributing to the High Degree of Unexplained Variability of Some Elements Concentrations in Biological Tissue: 27 Elements in 5 Organs of the Mussel *Mytilus* as a Model." *Arch. Environ. Contam. Toxicol.* 21:118–125 (1991).
8. Barron, M. G., G. R. Stehly, and W. L. Hayton. "Pharmacokinetic Modeling in Aquatic Animals. I. Models and Concepts." *Aquat. Toxicol.* 17:187–212 (1990).

9. Lyon, R., M. Taylor, and K. Simkiss. "Ligand Activity in the Clearance of Metals from the Blood of the Crayfish (*Austropotamobius pallipes*)." *J. Exp. Biol.* 113:19–27 (1984).

10. Greenblatt, D. J., and R. I. Shader. *Pharmacokinetics in Clinical Practice.* (Philadelphia, PA: W. B. Sanders Co., 1985).

11. Spacie, A., and J. L. Hamelink. "Bioaccumulation." In *Fundamentals of Aquatic Toxicology,* ed. G. M. Rand, and S. R. Petrocelli. (New York, NY: Hemisphere Publishing Corp., 1985) 495–525.

12. Ballatori, N., and T. W. Clarkson. "Developmental Changes in the Biliary Excretion of Methylmercury and Glutathione." *Science* 216:61–63 (1982).

13. Crist, R. H., K. Oberhoiser, D. Schwartz, J. Marzoff, D. Ryder, and D. R. Crist. "Interactions of Metals and Protons with Algae." *Environ. Sci. & Technol.* 22(7):755–760 (1988).

14. Schulz-Baldes, M. "Lead Uptake from Sea Water and Food, and Lead Loss in the Common Mussel *Mytilus edulis.*" *Mar. Biol. (Berl.)* 25:177–193 (1974).

15. Steele, G., P. Stehr-Green, and E. Welty. "Estimates of the Biologic Half-Life of Polychlorinated Biphenyls in Human Serum." *N. Engl. J. Med.* 314:926–927 (1986).

16. Scott, D. E., F. W. Whicker, and J. W. Gibbons. "Effect of Season on the Retention of ^{137}Cs and ^{90}Sr by the Yellow-Bellied Slider Turtle (*Pseudemys scripta*)." *Calif. J. Zool.* 64:2850–2853 (1986).

17. Phillips, D. L. "Propagation of Error and Bias in Half-Life Estimates Based on Two Measurements." *Arch. Environ. Contam. Toxicol.* 18:508–514 (1989).

18. Cutshall, N. "Turnover of Zinc-65 in Oysters." *Health Phys.* 26:327–331 (1974).

19. Beauchamp, J. J., and J. S. Olson. "Correction for Bias in Regression Estimates after Logarithmic Transformation." *Ecology* 54(6):1403–1407 (1973).

20. Sprugel, D. G. "Correction for Bias in Log-Transformed Allometric Equations." *Ecology* 64(1):209–210 (1983).

21. Newman, M. C., and M. G. Heagler. "Allometry of Metal Bioaccumulation and Toxicity." In *Metal Ecotoxicology, Concepts and Applications,* ed. M. C. Newman and A. W. McIntosh. (Chelsea, MI: Lewis Publishers, 1991) 91–130.

22. Newman, M. C. "A Statistical Bias in the Derivation of Hardness-Dependent Metals Criteria." *Environ. Toxicol. Chem.* 10:1295–1297 (1991).

23. Koch, R. W., and G. M. Smillie. "Bias in Hydrologic Prediction Using Log-transformed Regression Models." *Water Res.* 22(5):717–723 (1986).

24. SAS Institute Inc. *SAS/STAT User's Guide,* Release 6.03 Edition. (Cary, NC: SAS Institute Inc, 1988).

25. Boxenbaum, H. G., S. Reigelman, and R. M. Elashoff. "Statistical Estimations in Pharmacokinetics." *J. Pharmacokinet. Biopharm.* 2(2):123–148 (1974).

26. Neter, J., W. Wasserman, and M. H. Kutner. *Applied Linear Statistical Models in Regression, Analysis of Variance, and Experimental Design.* (Homewood, IL: Richard D. Irwin, Inc., 1990).

27. Christian, S. D., and E. E. Tucker. "Least Squares Analysis with the Microcomputer. Part Five: General Least Squares with Variable Weighting." *Am. Lab.* February: 18–32 (1984).

28. Van Liew, H. D. "Semilogarithmic Plots of Data Which Reflect a Continuum of Exponential Processes." *Science* 138:682–683 (1962).

29. Myhill, J. "Investigation of the Effect of Data Error in the Analysis of Biological Tracer Data." *Biophys. J.* 7:903–911 (1967).

30. Yamaoka, K., T. Nakagawa, and T. Uno. "Application of Akaike's Information Criterian (AIC) in the Evaluation of Linear Pharmacokinetic Equations." *J. Pharmacokinet. Biopharm.* 6:165–175 (1978).

31. Micromath Scientific Software. *RSTRIP Polyexponential Curve Stripping/Least*

Squares Parameter Estimation, Version 4. (Salt Lake City, UT: Micromath Scientific Software, 1989).

32. Newman, M. C., and A. W. McIntosh. "Lead Elimination and Size Effects on Accumulation by Two Freshwater Gastropods." *Arch. Environ. Contam. Toxicol.* 12:25–29 (1983).

33. Piszkiewicz, D. *Kinetics of Chemical and Enzyme-Catalyzed Reactions.* (New York, NY: Oxford University Press, 1977).

34. Mayer, F. L. "Residue Dynamics of Di-2-ethylhexylphthalate in Fathead Minnows (*Pimephales promelas*)." *J. Fish. Res. Board Can.* 33:2610–2613 (1976).

35. Raaijmakers, J. G. W. "Statistical Analysis of the Michaelis-Menten Equation." *Biometrics* 43:793–803 (1987).

36. Kinniburgh, D. G. "General Purpose Adsorption Isotherms." *Environ. Sci. & Technol.* 20:895–904 (1986).

37. Gibaldi, M., and D. Perrier. *Pharmacokinetics,* 2nd ed., (New York, NY: Marcel Dekker, Inc., 1982).

38. Holleman, D. F., J. R. Luick, and F. W. Whicker. "Transfer of Radiocesium from Lichen to Reindeer." *Health Phys.* 21:657–666 (1971).

39. Fujita, M., and K. Hashizume. "Status of Uptake of Mercury by the Fresh Water Diatom, *Synedra ulna.*" *Water Res.* 9:889–894 (1975).

40. Stary, J., and K. Kratzer. "The Cumulation of Toxic Metals on Alga." *J. Environ. Anal. Chem.* 12:65–71 (1982).

41. Les, A., and R. W. Walker. "Toxicity and Binding of Copper, Zinc, and Cadmium by the Blue-Green Alga, *Chroococcus paris.*" *Water Air Soil Pollut.* 23:129–139 (1984).

42. Wang, H-K., and J. M. Wood. "Bioaccumulation of Nickel by Algae." *Environ. Sci. & Technol.* 18(2):106–109 (1984).

43. Newman, M. C., and A. W. McIntosh. "Appropriateness of Aufwuchs as a Monitor of Bioaccumulation." *Environ. Pollut.* 60:83–100 (1989).

44. Ellgehausen, H., J. A. Guth, and H. O. Essner. "Factors Determining the Bioaccumulation Potential of Pesticides in the Individual Compartments of Aquatic Food Chains." *Ecotoxicol. Environ. Saf.* 4:134–157 (1980).

45. McKone, C. E., R. G. Young, C. A. Bache, and D. J. Lisk. "Rapid Uptake of Mercuric Ion by Goldfish." *Environ. Sci. & Technol.* 5:1138–1139 (1971).

46. Pagenkopf, G. K. "Gill Surface Interaction Model for Trace-Metal Toxicity to Fishes: Role of Complexation, pH, and Water Hardness." *Environ. Sci. & Technol.* 17(6):342–347 (1983).

47. Sposito, G. *The Surface Chemistry of Soils.* (New York, NY: Oxford University Press, 1984).

48. Metcalf and Eddy, Inc. *Wastewater Engineering. Collection Treatment and Disposal.* (New York, NY: McGraw-Hill Book Co., 1972).

49. Stumm, W., and J. J. Morgan. *Aquatic Chemistry. An Introduction Emphasizing Chemical Equilibria in Natural Waters.* (New York, NY: Wiley-Interscience, 1970).

50. Giesy, J. P., and D. Paine. "Effects of Naturally Occurring Aquatic Organic Fractions on ^{241}Am Uptake by *Scenedesmus obliquus (Chlorophyceae)* and *Aeromonas hydrophila (Pseudomonadaceae).*" *Appl. Environ. Microbiol.* 33(1):89–96 (1977).

51. Gordon, M. S. *Animal Physiology: Principles and Adaptations,* 2nd ed. (New York, NY: The Macmillan Company, 1972).

52. Prosser, C. L. *Comparative Animal Physiology,* 3rd ed. (Philadelphia, PA: W. B. Saunders Company, 1973).

53. Hunn, J. B. "Role of Calcium in Gill Function in Freshwater Fishes." *Comp. Biochem. Physiol.* 82A(3):543–547 (1985).

54. Barber, M. C., L. A. Suarez, and R. R. Lassiter. "Modeling Bioconcentration of Nonpolar Organic Pollutants by Fish." *Environ. Toxicol. Chem.* 7:545–558 (1988).

55. Erickson, R. J., and J. M. McKim. "A Simple Flow-Limited Model for Exchange of Organic Chemicals at Fish Gills." *Environ. Toxicol. Chem.* 9:159–165 (1990).

56. Simkiss, K. "Lipid Solubility of Heavy Metals in Saline Solutions." *J. Mar. Biol. Assoc. U. K.* 63:1–7 (1983).

57. Hughes, G. M. "Measurement of Gill Area in Fishes: Practices and Problems." *J. Mar. Biol. Assoc. U. K.* 64:637–655 (1984).

58. Ribble, D. O., and M. H. Smith. "Relative Intestine Length and Feeding Ecology of Freshwater Fishes." *Growth* 47:292–300 (1983).

59. Erickson, R. J., and J. M. McKim. "A Model of Exchange of Organic Chemicals at Fish Gills: Flow and Diffusion Limitations." *Aquat. Toxicol.* 18:175–198 (1990).

60. Newman, M. C., and A. W. McIntosh. "Slow Accumulation of Lead from Contaminated Food Sources by the Freshwater Gastropods *Physa integra* and *Campeloma decisum*." *Arch. Environ. Contam. Toxicol.* 12:685–692 (1983).

61. Willis, J. N., and N. Y. Jones. "The Use of Uniform Labeling with Zinc-65 to Measure Stable Zinc Turnover in the Mosquitofish, *Gambusia affinis*—I. Retention." *Health Phys.* 32:381–387 (1977).

62. Giesy, J. P., J. W. Bowling, and H. J. Kania. "Cadmium and Zinc Accumulation and Elimination by Freshwater Crayfish." *Arch. Environ. Contam. Toxicol.* 9:683–697 (1980).

63. Goldstein, R. A., and J. W. Elwood. "A Two-Compartment, Three-Parameter Model for the Adsorption and Retention of Ingested Elements by Animals." *Ecology* 52(5):935–939 (1971).

64. Konemann, H., and K. Van Leeuwen. "Toxicokinetics in Fish Accumulation and Elimination of Six Chlorobenzenes by Guppies." *Chemosphere* 9:3–19 (1980).

65. Kolehmainen, S. E. "The Balances of ^{137}Cs, Stable Cesium and Potassium of Bluegill (*Lepomis macrochirus* Raf.) and Other Fish in White Oak Lake." *Health Phys.* 23:301–315 (1972).

66. Newman, M. C., and D. K. Doubet. "Size-Dependence of Mercury (II) Accumulation in the Mosquitofish, *Gambusia affinis* (Baird and Girard)." *Arch. Environ. Contam. Toxicol.* 18:819–825 (1989).

67. Hayton, W. L., personal communication, 1992.

68. De Bruijn, J., and J. Hermens. "Uptake and elimination Kinetics of Organophosphorous Pesticides in the Guppy (*Poecilia reticulata*): Correlations with the Octanol/Water Partition Coefficient." *Environ. Toxicol. Chem.* 10:791–804 (1991).

69. Landrum, P. F., and M. J. Lydy. "Toxicokinetics Short Course." Presented at 12th Meeting of the Society of Environmental Toxicology and Chemistry, Seattle, WA, November 3, 1991.

70. Preston, E. M. "The Importance of Ingestion in Chromium-5 Accumulation by *Crassostrea virginica* (Gmelin)." *J. Exp. Mar. Biol. Ecol.* 6:47–54 (1971).

71. Karlsson-Norrgren, L., and P. Runn. "Cadmium Dynamics in Fish: Pulse Studies with ^{109}Cd in Female Zebrafish, *Brachydanio rerio*." *J. Fish Biol.* 27:571–581 (1985).

72. Luoma, S. N., C. Johns, N. S. Fisher, N. A. Steinberg, R. S. Oremland, and J. R. Reinfelder. "Determination of Selenium Bioavailability to a Benthic Bivalve from Particulate and Solute Pathways." *Environ. Sci. & Technol.* 26:485–491 (1992).

73. Thomann, R. V. "Equilibrium Model of Fate of Microcontaminants in Diverse Aquatic Food Chains." *Can. J. Fish. Aquat. Sci.* 38:280–296 (1981).

74. Harrison, S. E., and J. F. Klaverkamp. "Uptake, Elimination and Tissue Distribution of Dietary and Aqueous Cadmium by Rainbow Trout (*Salmo gairdneri* Richardson) and Lake Whitefish (*Coregonus clupeaformis* Mitchill)." *Environ. Toxicol. Chem.* 8:87–97 (1989).

75. Blaylock, B. G., M. L. Frank, and D. L. DeAngelis. "Bioaccumulation of ^{95m}Tc in Fish and Snails." *Health Phys.* 42(3):257–266 (1982).

76. Waiwood, B. A., V. Zitko, K. Haya, L. E. Burridge, and D. W. McLeese. "Uptake

and Excretion of Zinc by Several Tissues of the Lobster (*Homarus americanus*)." *Environ. Toxicol. Chem.* 6:27–32 (1987).

77. Giblin, F. J., and E. J. Massaro. "Pharmacodynamics of Methyl Mercury in the Rainbow Trout (*Salmo gairdneri*): Tissue Uptake, Distribution and Excretion." *Toxicol. Appl. Pharmacol.* 24:81–91 (1973).

78. Skrable, K. W., G. E. Chabot, C. S. French, M. E. Wrenn, J. Lipsztein, and T. Lo Sasso. "Blood-Organ Transfer Kinetics." *Health Phys.* 39:193–209 (1980).

79. Barron, M. G., M. A. Mayes, P. G. Murphy, and R. J. Nolan. "Pharmacokinetics and Metabolism of Triclopyrbutoxyethyl Ester on Coho Salmon." *Aquat. Toxicol.* 16:19–32 (1990).

80. Nichols, J. W., J. M. McKim, M. E. Andersen, M. L. Gargas, H. J. Clewell, III, and R. J. Erickson. "A Physiologically Based Toxicokinetic Model for the Uptake and Disposition of Waterborne Organic Chemicals in Fish." *Toxicol. Appl. Pharmacol.* 106:433–447 (1990).

81. Nichols, J. W., J. M. McKim, G. J. Lien, A. D. Hoffman, and S. L. Bertelsen. "Physiologically Based Toxicokinetic Modeling of Three Waterborne Chloroethanes in Rainbow Trout (*Oncorhynchus mykiss*)." *Toxicol. Appl. Pharmacol,* 110:374–389 (1991).

82. Leversee, G. J., J. P. Giesy, P. F. Labdrum, S. Gerould, J. W. Bowling, T. E. Fannin, J. D. Haddock, and S. M. Bartell. "Kinetics and Biotransformation of Benzo(a)pyrene in *Chironomus riparius*." *Arch. Environ. Contam. Toxicol.* 11:25–31 (1982).

83. Foster, G. D., and D. G. Crosby. "Xenobiotic Metabolism of Nitrophenol Derivatives by the Rice Field Crayfish (*Procambarus clarkii*)" *Environ. Toxicol. Chem.* 5:1059–1070 (1986).

84. Walker, C. H. "Species Differences in Microsomal Monooxygenase Activity and their Relationship to Biological Half-lives." *Drug Metabl. Rev.* 7(2):295–323 (1978).

85. Walker, C. H. "Kinetic Models for Predicting Bioaccumulation of Pollutants in Ecosystems." *Environ. Pollut.* 44:227–240. (1987).

86. Lunde, G. "The Absorption and Metabolism of Arsenic in Fish." *Fiskeridir. Skr. Ser. Teknol. Unders.* 5(12):1–16 (1972).

87. Unlu, S. Y., and S. W. Fowler. "Factors Affecting the Flux of Arsenic through the Mussel *Mytilus edulis*." *Mar. Biol.* 51:209–219 (1979).

88. Edmonds, J. S., and K. A. Francesconi. "Isolation and Identification of Arsenobetaine from the American Lobster *Homarus americanus*." *Chemosphere* 10(9):1041–1044 (1981).

89. Zingaro, R. A. "Biochemistry of Arsenic: Recent Developments." In *Arsenic: Industrial, Biomedical, Environmental Perspectives,* ed. W. H. Lederer and R. J. Fensterheim. (New York, NY: Van Nostrand Reinhold Company, Inc., 1983).

90. Landrum, P. F., and C. R. Stubblefield. "Role of Respiration in the Accumulation of Organic Xenobiotics by the Amphipod *Diporeia* sp." *Environ. Toxicol. Chem.* 10:1019–1028 (1991).

91. Watkins, B., and K. Simkiss. "The Effect of Oscillating Temperatures on the Metal Ion Metabolism of *Mytilus edulis*." *J. Mar. Biol. Assoc. U.K.* 68:93–100 (1988).

92. Rose, K. A., R. I. McLean, and J. K. Summers. "Development and Monte Carlo Analysis of an Oyster Bioaccumulation Model Applied to Biomonitoring Data." *Ecol. Modell.* 45:111–132 (1989).

93. Neely, W. B., G. E. Blau, and G. L. Agin. "The Use of SIMUSOLV to Analyze Fish Bioconcentration Data." *Chemometrics Intelligent Lab. Syst.* 1:359–366 (1987).

94. Kirchner, T. B. *TIME-ZERO. The Integrated Modeling Environment Reference Manual.* (Ft. Collins, CO: Quaternary Software, Inc., 1989).

95. Part, P., and O. Svanberg. "Uptake of Cadmium in Perfused Rainbow Trout (*Salmo gairdneri*) Gills." *Can. J. Fish. Aquat. Sci.* 38:917–924 (1981).

96. Van Der Putte, I., and P. Part. "Oxygen and Chromium Transfer in Perfused Gills

of Rainbow Trout (*Salmo gairdneri*) Exposed to Hexavalent Chromium at Two Different pH Levels." *Aquat. Toxicol.* 2:31–45 (1982).

97. Hayton, W. L., and M. G. Barron. "Rate-limiting Barriers to Xenobiotic Uptake by the Gill." *Environ. Toxicol. Chem.* 9:151–157 (1990).

98. Pentreath, R. J. "The Accumulation of Mercury from Food by the Plaice, *Pleuronectes platessa* L." *J. Exp. Mar. Biol. Ecol.* 25:51–65 (1976).

99. McCallum, D. A., and P. M. Dixon. "Reducing Bias in Estimates of the Richards Growth Function Shape Parameter." *Growth Dev. Aging* 54:135–141 (1990).

100. Reichle, D. E. "Relation of Body Size to Food Uptake, Oxygen Consumption, and Trace Element Metabolism in Forest Floor Arthropods." *Ecology* 49(3):538–542 (1967).

101. Fagerström, T. "Body Weight, Metabolic Rate, and Trace Substance Turnover in Animals." *Oecologia (Berl.)* 29:99–104 (1977).

102. Gobas, F. A. P. C., and D. Mackay. "Dynamics of Hydrophobic Organic Chemical Bioconcentration in Fish." *Environ. Toxicol. Chem.* 6:95–504 (1987).

103. Mackay, D., and S. Paterson. "Fugacity Revisited. The Fugacity Approach to Environmental Transport." *Environ. Sci. Technol.* 116(12):654A–660A (1982).

104. Jorgensen, S. E. *Modelling in Ecotoxicology.* (New York: Elsevier, 1990).

105. Pankow, J. F. *Aquatic Chemistry Concepts.* (Chelsea, MI: Lewis Publishers, Inc., 1991).

106. Paasivirta, J. *Chemical Ecotoxicology.* (Chelsea, MI: Lewis Publishers, Inc., 1991).

107. Mackay, D. *Multimedia Environmental Models. The Fugacity Approach.* (Chelsea, MI: Lewis Publishers, Inc., 1991).

108. Yamaoka, K., T. Nakagawa, and T. Uno. "Statistical Moments in Pharmacokinetics." *J. Pharmacokinet. Biopharm.* 6:547–558 (1978).

109. Matis, J. H., and T. E. Wehrly. "Stochastic Models of Compartmental Systems." *Biometrics* 35:199–220 (1979).

110. Matis, J. H., and H. D. Tolley. "On the Stochastic Modeling of Tracer Kinetics." *Fed. Proc.* 39(1):104–109 (1980).

111. Matis, J. H., T. E. Wehrly, and K. B. Gerald. "The Statistical Analysis of Pharmacokinetic Data." In *Lecture Notes in Biomathematics. Tracer Kinetics and Physiologic Modeling Theory and Practice.* ed. S. Levin. (Berlin: Springer-Verlag 1983) 1–58.

112. Matis J. H., T. H. Miller, and D. M. Allen. "Stochastic Models of Bioaccumulation." In *Metal Ecotoxicology, Concepts and Applications,* ed. M. C. Newman and A. W. McIntosh. (Chelsea, MI: Lewis Publishers, 1991) 171–206.

113. Benson, A. A., and R. E. Summons. "Arsenic Accumulation in Great Barrier Reef Invertebrates." *Science* 211:482–483 (1981).

114. Morris, R. J., M. J. McCartney, A. G. Howard, M. H. Arbab-Zavar, and J. S. Davis. "The Ability of a Field Population of diatoms to Discriminate Between Phosphate and Arsenate." *Mar. Chem.* 14:259–265 (1984).

115. Morgan, F. "The uptake of Radioactivity by Fish and Shellfish. I. ^{134}Caesium by Whole Animals." *J. Mar. Biol. Assoc. U.K.* 44:259–271 (1964).

116. Blaylock, B. G., and M. L. Frank. "Distribution of Tritium in a Chronically Contaminated Lake." In *Behavior of Tritium in the Environment.* (Vienna: International Atomic Energy IAEA-SM-232/74, 1979) 247–256.

117. Blaylock, B. G. "Radionuclide Data Base Available for Bioaccumulation Factors for Freshwater Biota." *Nucl. Saf.* 23(4):427–438 (1982).

118. Rodgers, D. W. "Tritium Dynamics in Juvenile Rainbow Trout, *Salmo gairdneri.*" *Health Phys.* 50(1):89–98 (1986).

119. Hinton, T. G., F. W. Whicker, J. E. Pinder III, and S. A. Ibrahim. "Comparative Kinetics of ^{47}Ca, ^{85}Sr and ^{226}Ra in the Freshwater Turtle, *Trachemys scripta.*" *J. Environ. Radioact.* 16:25–47 (1992).

120. Jones, A. R., M. Taylor, and K. Simkiss. "Regulation of Calcium, Cobalt and Zinc by *Tetrahymena elliotti.*" *Comp. Biochem. Physiol.* 78A(3):493–500 (1984).

121. Bache, C. A., W. H. Gutenmann, and D. J. Lisk. "Residues of Total Mercury and Methylmercuric Salts in Lake Trout as a Function of Age." *Science* 172:951–952 (1971).

122. Williamson, P. "Variables Affecting Body Burdens of Lead, Zinc and Cadmium in a Roadside Population of the Snail *Cepaea hortensis* Muller." *Oecologia (Berl.)* 44:213–220 (1980).

123. Hirayama, K., and A. Yasutake. "Sex and Age Differences in Mercury Distribution and Excretion in Methylmercury-Administered Mice." *J. Toxicol. Environ. Health* 18:49–60 (1986).

124. Nicoletto, P. F., and A. C. Hendricks. "Sexual Differences in Accumulation of Mercury in Four Species of Centrachid Fishes." *Can. J. Zool.* 66:944–949 (1988).

125. Neely, W. B., D. R. Branson, and G. E. Blau. "Partition Coefficient to Measure Bioconcentration Potential of Organic Chemicals in Fish." *Environ. Sci. Technol.* 8(13):1113–1115 (1974).

126. Newman, M. C. "Regression Analysis of Log-transformed Data: Statistical Bias and Its Correction." *Environ. Toxicol. Chem.* 12:1129–1133 (1993).

127. Veith, G. D., D. L. DeFoe, and B. V. Bergstedt. "Measuring and Estimating the Bioconcentration Factor of Chemicals in Fish." *J. Fish. Res. Board Can.* 36:1040–1048 (1979).

128. Mackay, D. "Correlation of Bioconcentration Factors." *Environ. Sci. Technol.* 16:274–278 (1982).

129. Geyer, H., P. Sheehan, D. Kotzias, D. Freitag, and F. Korte. "Prediction of Ecotoxicological Behavior of Chemicals: Relationship between Physico-chemical Properties and Bioaccumulation of Organic Chemicals in the Mussel *Mytilus Edulis.*" *Chemosphere* 11(11):1121–1134 (1982).

130. Chiou, C. T. "Partition Coefficients of Organic Compounds in Lipid-Water Systems and Correlations with Fish Bioconcentration Factors." *Environ. Sci. Technol.* 19:57–62 (1985).

131. Connell, D. W., and D. W. Hawker. "Use of Polynomial Expressions to Describe the Bioconcentration of Hydrophobic Chemicals by Fish." *Ecotoxicol. Environ. Saf.* 16:242–257 (1988).

132. Mackay, D., and A. I. Hughes. "Three Parameter Equation Describing the Uptake of Organic Compounds in Fish." *Environ. Sci. Technol.* 18:439–444 (1984).

133. Gobas, F. A., P. C. Opperhuizen, and O. Hutzinger. "Bioconcentration of Hydrophobic Chemicals in Fish: Relation with Membrane Permeation." *Environ. Toxicol. Chem.* 5:637–646 (1986).

134. Rowland, M. "Physiologic Pharmacokinetic Models and Interanimal Species Scaling." *Pharmacol. Ther.* 29:49–68 (1985).

135. Campbell, F. L. "Relative Susceptibility to Arsenic in Successive Instars of the Silkworm." *Gen. Physiol. Notes* 9(6):727–733 (1926).

136. Boyden, C. R. "Trace Element Content and Body Size in Molluscs." *Nature* 251:311–314 (1974).

137. Boyden, C. R. "Effect of Size upon Metal Content of Shellfish." *J. Mar. Biol. Assoc. U.K.* 57:675–714 (1977).

138. Ricker, W. E. "Linear Regression in Fishery Research." *J. Fish. Res. Board. Can.* 30:409–434 (1973).

139. Muir, B. S. "Gill Dimensions as a Function of Fish Size." *J. Fish. Res. Board. Can.* 26:165–170 (1969).

140. Smock, L. A. "Relationships between Metal Concentrations and Organism Size in Aquatic Insects." *Freshwater Biol.* 13:313–321 (1983).

141. Adolph, E. F. "Quantitative Relations in the Physiological Constitutions of Mammals." *Science* 109:579–585 (1949).

142. Somero, G. N., and J. J. Childress. "A Violation of the Metabolism-size scaling paradigm: Activities of Glycolytic Enzymes in Muscle Increase in Larger-size Fish." *Physiol. Zool.* 53(3):322–337 (1980).

143. Anderson, P. D. and P. A. Spear. "Copper Pharmacokinetics in Fish Gills.—I. Kinetics in Pumpkin Seed Sunfish. *Lepomis gibbosus,* of Different Body Sizes." *Water Res.* 14:1101–1105 (1980).

144. Anderson, P. D., and P. A. Spear. "Copper Pharmacokinetics in Fish Gills—II. Body size relationships for Accumulation and Tolerance." *Water Res.* 14:1107–1111 (1980).

145. Newman, M. C., and S. V. Mitz. "Size Dependence of Zinc Elimination and Uptake from Water by Mosquitofish *Gambusia affinis* (Baird and Girard)," *Aquat. Toxicol.* 12:17–32 (1988).

146. Cossa, D., E. Bourget, D. Pouliot, J. Piuze, and J. P. Chanut. "Geographical and Seasonal Variations in the Relationship Between Trace Metal Content and Body weight in *Mytilus edulis.*" *Mar. Biol. (Berl.)* 58:7–14 (1980).

147. Strong, C. R., and S. N. Luoma. "Variations in the Correlation of Body Size With Concentrations of Cu and Ag in the Bivalve *Macoma balthica.*" *Can. J. Fish. Aquat. Sci.* 38:1059–1064 (1981).

148. Lobel, P. B., and D. A. Wright. "Total Body Zinc and Allometric Growth Ratios in *Mytilus edulis* Collected from Different Shore Levels." Mar. Biol. *(Berl.)* 66:231–236 (1982).

149. Hayton, W. L. "Pharmacokinetic Parameters For Interspecies Scaling Using Allometric Techniques," *Health Phys.* 57:159–164 (1989).

Chapter 4

Lethal and Other Quantal Responses to Stressors

Prior to the Renaissance period and extending well into that period, . . . the art of poisoning [was brought to] its zenith. . . . The record of the city councils of Florence and particularly the infamous Council of Ten of Venice contain ample testimony of the political use of poisons. Victims were named, prices set, contracts recorded, and, when the deed was accomplished, payment made. The notation 'factum', often appeared after the accomplishment of its transaction.[1]

I. GENERAL

The Toxic Substances Control Act (TSCA), Resource Conservation and Recovery Act (RCRA), Federal Insecticide, Fungicide and Rodenticide Act (FIFRA), Water Pollution Control Act (WPCA), and National Environmental Policy Act (NEPA) all require that toxicity be addressed for substances that may find their way into the environment. Unlike decisions based on mammalian toxicology, a discipline in which much basic research had been completed before enactment of such legislation, decisions associated with aquatic toxicology rely heavily on knowledge developed primarily within a regulatory framework.[2] This has the advantage of focusing effort in the most efficient way possible to address very real regulatory needs. However, it also imposes a structure on the associated knowledge base that is decidedly not scientific, i.e., not organized on the basis of explanatory principles (see Chapter 1). Also, there is an inherent delay in incorporating new information because the regulatory framework is built on the best information available at the time of enactment (P. Landrum, personal communication, 1994). Given the immediacy of our needs during the past several decades, this regulatory structure is reasonable and necessary. However, for the sustained growth of knowledge in aquatic ecotoxicology, a framework based on explanatory principles must emerge also. This chapter discusses the area of aquatic ecotoxicology with the strongest regulatory influence and, consequently, most in need of occasional reexamination.

Some toxicological consequences, such as the macabre one quoted above, are unambiguously "factum." Others are more difficult to demonstrate and, consequently, are prone to divergent interpretations. It is the purpose of this chapter to outline and illustrate the use of candidate methods for quantifying lethal effects. Both well-established and alternate approaches are discussed without regard to their acceptability for regulatory activities. Finally, the relative progress of regulatory and scientific ecotoxicology as applied to methods of quantifying lethal stress are contrasted briefly.

II. DOSE-RESPONSE AT A SET ENDPOINT

A. GENERAL

Toxic response is a function of both the intensity (e.g., concentration or dose) and duration of exposure. Some techniques for measuring dose-response relationships control the duration of exposure and vary dose. Such endpoint techniques are defined herein as those for which a response is assessed at a final set time (endpoint). With many toxicity tests, the response is a quantal response (an all-or-none response), e.g., dead or alive, unresponsive to prodding or responsive, immobile or mobile. Operationally defined responses such as the last two are used when death is difficult to score. Often the

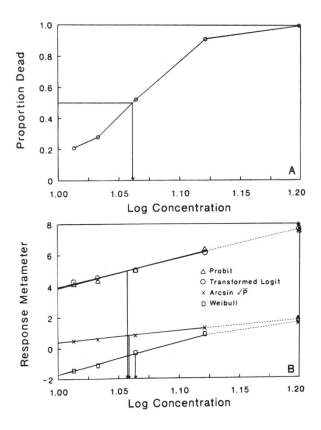

Figure 1 Sodium chloride toxicity to mosquitofish (*Gambusia holbrooki*). Panel A shows the sigmoidal plot of proportion dead versus log exposure concentration. Various linearizing metameters are used for these same data in Panel B. Response metameters for the probit, transformed logit, arcsine square root, and Weibull functions were extracted from Table 7 in the Appendix.

proportion (*P*) of the total number of individuals that are dead after a certain duration of exposure is used to summarize response in endpoint-based toxicity tests.

B. METAMETERS OF DOSE AND RESPONSE
The frequency distribution of individual tolerances within an exposed population is often skewed. Consequently, mathematical transformations of the original measurements are commonly made to normalize the distribution of tolerances.[3] For example, insect populations with subsets of individuals displaying extreme tolerances to a pesticide would be studied using logarithms of the exposure doses, not the arithmetic doses.

Measurements or measurement transformations used in the analysis of biological tests are referred to as metameters.[4] The logarithm of the exposure dose or concentration is the most common dose or concentration metameter.[5] A variety of effect metameters (metameters for toxicant effect) are used in endpoint techniques. Probit, logit, and arcsine of the square root transformations of *P* are some of the most common. They are used to make linear the characteristically sigmoid cumulative response curve (Figure 1A) resulting from plotting response (*P*) versus log dose or log concentration.

The probit of *P* versus log dose (or log concentration) is the most common pair of transformations. The probit was introduced by Bliss[6] using the concept of individual tolerance (measured as the smallest dose needed to kill a particular individual). When a particular dose is administered to many individuals in a population, those individuals with tolerances less than or equal to that dose will die. If groups of individuals from a population are exposed to a range of doses, a skewed frequency distribution of the proportion succumbing appears. Bliss[6] suggested that logarithms of lethal doses be used to produce a normal distribution from such data. To obtain a straight line, the normal equivalent deviations or NED (the proportion dying at each dose (P), expressed in terms of standard deviations from the mean of a normal distribution) could be plotted against the logarithm of dose. The resulting curve is a linear form of the cumulative normal distribution for mortality. Bliss viewed the negative values of NEDs below the mean as inconvenient. Consequently, he added 5 to the NED to make the occurrence of negative values rare. The effect metameter resulting from the addition of 5 to the NED is the probit or "probability unit." It must be mentioned that the advantage of adding 5 to the NED is debatable (see Finney,[7] page 358). Only the pervasive use of probits instead of NEDs in toxicology argues for continued use for the sake of uniformity. Use of the probit carries with it the assumption of a normal model [Equation (1)] for the log dose-response relationship. The Normal Cumulative Distribution Function model is:

$$F(x) = \frac{1}{\sqrt{2\pi}\sigma} \int_{-\infty}^{x} e^{-(x-\mu)^2/2\sigma^2} dx \tag{1}$$

where μ and σ are the mean and standard deviation, and $F(x)$ = theoretical proportion corresponding to a concentration metameter (P is the empirical proportion).

$$\text{Probit} = \text{NED} + 5 \tag{2}$$

The probit or the NED can be obtained from tables such as Table 7 in the Appendix, a table of probabilities for the normal curve, or by using special functions found in many statistical computer packages.

Berkson[8] advocated the use of logits ("log odds units") instead of probits. Use of the logit carries with it the assumption that a logistic model describes the log dose-response curve, not a normal model.[5] Berkson[8] argued for use of the logit metameter instead of the probit by pointing out that the probit transformation had no theoretical advantage over the logit and, unlike the logit, required a statistical table. Today, the requirement of a table is not generally a problem, and the theoretical underpinnings for use of the logistic model remain no clearer than those for the normal model. Indeed, the lack of any theoretical superiority of either of these two transformations was common knowledge soon after their introduction.[5] Although the logistic model can be related to the Hill equation for enzyme kinetics,[9] it is used empirically to fit log dose-response data. Both metameters simply work and give similar results (Figure 1B).

$$\text{logit} = \ln\left[\frac{P}{1 - P}\right] \tag{3}$$

A convenient form of the logit is the following (see Finney,[7] page 374, Equation 18.1.23). This transformation brings the logit values very close to those of the probit transformation except at the extreme ends of the log dose-response curve.

$$\text{Transformed logit} = \frac{\text{logit}}{2} + 5 \tag{4}$$

Both the logit and transformed logit can be found in tables such as Table 7 in the Appendix.

The third common response metameter is the arcsine, angle sigmoid, or angular transformation.[10]

$$\text{arcsine transform} = \text{arcsine}\sqrt{P} \tag{5}$$

Note that the arcsine may also be denoted as $\sin^{-1}$. The inverse sine of x is the value (angle) of y that satisfies $x = \sin y$, i.e., y is the angle whose sine is x.

Gaddum[5] suggested that the arcsine metameter not be used for proportions outside the range of 0.20 to 0.80. Using this metameter has the advantage of making the variance for the metameter constant[7] and, therefore, any weight factor used in fitting will not vary with the value of P.[5,7] Finney[7] suggested that this metameter is now less appealing than the probit or logit because of the easy accessibility of computers. No theoretical model has been presented for its use in modeling the dose-response relationships.

Other less commonly used transformations are available. For example, Finney[7] listed the Wilson-Worcester, Cauchy-Urban, and "linear" (or rectangular) sigmoid models. More recently, the Weibull model has been suggested for this purpose.[9,11] The Weibull model [Equation (7)] is a flexible generalization of the exponential model [Equation (6)]. It may have positive or negative skewness depending on the shape parameter (λ).[12,13]

The model for the Exponential Cumulative Distribution Function is:

$$F(x) = 1 - e^{-kx} \tag{6}$$

where k is greater than 0. The model for the Weibull Cumulative Distribution Function is:

$$F(x) = 1 - e^{-(\alpha x)^{\lambda}} \tag{7}$$

where α and λ are greater than 0. The α and λ are called the scale and shape parameters, respectively. The Weibull model reduces to an exponential model when the shape parameter is 1.

Although used to fit log dose-response data empirically, the Weibull model can also be linked to models of carcinogen action.[11] For example, when λ is an integer, the Weibull is analogous to a multiple-hit model. The Weibull transformation (U) defined by Equation (8) can be plotted against ln of dose to yield a straight line.

$$U = \ln(-\ln(1 - P)) \tag{8}$$

An example of such a plot is given in Figure 1B.

Example 1
What are the NED, probit, logit, transformed logit, angular transform, and Weibull metameter values for 50%, 16%, and 84% mortalities?

All values can be extracted from Table 7 in the Appendix. Alternatively, they may be generated with special functions and code such as the SAS code, Ned = PROBIT(P), Prbt = PROBIT(P) + 5, Logit = LOG(P/(1 − P)), T. logit = Logit/2 + 5, Angular = ARCSIN(SQRT(P)), and Weibull = LOG(−LOG(1 − P)), respectively.

Proportion	NED	Probit	Logit	T. logit	Angular	Weibull
0.50	0.000	5.000	0.000	5.000	0.785	−0.367
0.16	−0.994	4.005	−1.658	4.171	0.412	−1.747
0.84	0.994	5.994	1.658	5.829	1.159	0.606

Example 1 *Continued*

Note that the angular transform is expressed in radians here. If the desired transformation units are degrees then the following conversion is applied. One radian = 180°/π or approximately 57.29578°. The angular transform of 0.50 is 0.7854 radians*57.29578 degrees/radian or 45°.

C. THE LC50

Trevan[14] was the first to suggest using the LD50 or lethal dose killing 50% of the exposed organisms as a statistically reliable estimate of toxic effect instead of the more variable lethal threshold dose.[5] For animals exposed to concentrations in various media rather than doses, an LC50 (lethal concentration killing 50% of the exposed organisms) is estimated. For nonlethal or ambiguously lethal endpoints such as unresponsiveness to prodding or immobility of an invertebrate, the calculated concentration affecting 50% of the exposed organisms is often called the effective concentration (EC50). Graphical estimation of the LD50, LC50, or EC50 can be achieved using any of the response metameters (Table 7 in the Appendix) described to yield similar answers. For example, the values of probit (5.00000), transformed logit (5.00000), arcsine of the square root (0.78540), and Weibull (−0.36651) for $P = 0.50$ in Figure 1B correspond to concentration metameters of approximately 1.058, 1.057, 1.059, and 1.064, respectively. In turn, the antilogarithms of these concentration metameters correspond to 96-h LC50s of approximately 11.43, 11.40, 11.46, and 11.59 g NaCl/L.

1. Litchfield-Wilcoxon Method

The graphical method described is incomplete because it fails to give a confidence interval for the estimated LC50 or EC50. Litchfield and Wilcoxon[15] outlined a straightforward semigraphical procedure for estimating the LC50 and its 95% confidence interval. The following steps summarize their approach.

Step 1
On log-probability paper (or plotting the log concentration and probit of the proportion responding on arithmetic paper), produce a preliminary plot of the data (*P*'s versus the corresponding exposure concentrations). Because 0% and 100% mortality data cannot be plotted in this manner, omit such data from the plot. Draw a preliminary line through the points, "particularly those in the region of 40 to 60 per cent effect." Using this preliminary line, cull away points that fall below a predicted value of 0.01% or above a predicted value of 99.9%. Now draw a final line through the data points.

Step 2
Using the final line, extract the concentrations corresponding to the 16%, 50%, and 84% mortality (LC16, LC50, and LC84).

Step 3
Calculate a slope function (*S*):

$$S = \frac{\dfrac{LC84}{LC50} + \dfrac{LC50}{LC16}}{2} \qquad (9)$$

Step 4
Determine the total number of individuals (*N'*) tested between the 16% and 84% responses predicted by the line. Then calculate the 95% confidence interval (C.I.) as follows.

$$f_{LC50} = S^{\frac{2.77}{\sqrt{N'}}} \tag{10}$$

Upper limit of 95% C.I.:

$$\text{Upper} = LC50 \cdot f_{LC50} \tag{11}$$

Lower Limit of 95% C.I.:

$$\text{Lower} = LC50/f_{LC50} \tag{12}$$

Example 2

Newman and Aplin[16] exposed mosquitofish (*Gambusia holbrooki*) to a series of NaCl concentrations for 96 hours. The resulting data are tabulated below. What are the estimates of LC50 and its 95% C.I.?

[NaCl](g/L)	Log of [NaCl]	Numbers of Dead/Total	Proportion Dead	Probit
20.1	1.303	77/77	1.00	>7.576
15.8	**1.199**	**78/78**	**1.00**	**>7.576**
13.2	**1.121**	**69/76**	**0.91**	**6.341**
11.6	**1.064**	**40/77**	**0.52**	**5.050**
10.8	**1.033**	**22/79**	**0.28**	**4.417**
10.3	**1.013**	**16/76**	**0.21**	**4.194**
0.016	−1.796	0/79	0.00	<2.424

Optimum scaling of these or any data is difficult using log-probability paper. Such paper is not always available and is printed in 1, 2, or 3 cycles on the logarithmic scale. Regardless of the selection made for these data, the points will tend to be compressed into one small portion of the paper if the toxicant concentration range is narrow. Further, it is difficult to place points accurately between lines on such paper because the eye tends to place points with an arithmetic bias. Because use of such paper is analogous to the probit approach discussed and arithmetic paper can be easily used to scale the probit-log concentration data properly, these data will be graphed as the probit and log concentration metameters instead.

Figure 2 depicts these data. Only the boldface values in the table were plotted. The LC16, LC50, and LC84 can be extracted from this graph by taking the antilogarithm of the log concentration corresponding to the appropriate probit values for these percentages. Extract the probit values from Table 7 in the Appendix. They are 4.00554, 5.0000, 5.99446, respectively. (If you review the derivation of the probit, you might be confused by minor inaccuracies here. The probit for ±1 standard deviation should be 4.00000 and 6.00000, not 4.00554 and 5.99446. These slight inaccuracies arise because 16% and 84% are the "nearest whole percentage" estimates of ±1 standard deviation from the mean.)

$$LC16 = 10^{1.009} \text{ or } 10.21$$
$$LC50 = 10^{1.058} \text{ or } 11.43$$
$$LC84 = 10^{1.107} \text{ or } 12.79$$

$$S = (LC84/LC50 + LC50/LC16)/2$$
$$= (12.79/11.43 + 11.43/10.21)/2$$
$$= 1.12$$

Example 2 *Continued*

$$N' = 76 + 79 + 77 = 232$$

$$\begin{aligned}
f_{LC50} &= S^{2.77/\sqrt{N'}} \\
&= 1.12^{2.77/\sqrt{232}} \\
&= 1.12^{0.182} \\
&= 1.021
\end{aligned}$$

95% C.I. for the LC50:

$$\begin{aligned}
LC50*f_{LC50} &= 11.43*1.021 \\
&= 11.67
\end{aligned}$$

$$\begin{aligned}
LC50/f_{LC50} &= 11.43/1.021 \\
&= 11.19
\end{aligned}$$

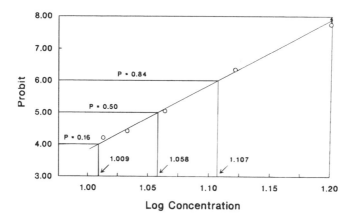

Figure 2 Sodium chloride toxicity data for mosquitofish using the probit-log concentration transformations. The probit values for the 0.16, 0.50, and 0.84 mortality proportions are used to estimate the LC50 and its 95% confidence interval (see Example 2, Litchfield-Wilcoxon method).

2. Maximum Likelihood Method (Normal, Logistic, and Weibull Models)

The Litchfield-Wilcoxon approach, being a semigraphic technique will yield slightly different answers depending on the person executing the procedures. A more formal approach to the estimation of LC50 is desirable.

The first approach to come to the reader's mind is probably least-squares linear regression. This method fits data to a model by minimizing the sum of the squares of the deviations between the observed and predicted values of Y. Indeed, such methods including weighted regression methods can be used to fit this type of data. Values can be predicted within the range of the dose metameter values used in the regression. Examples of such methods for unweighted and weighted regression are provided in Gad and Weil,[17] pages 94–98, and Christensen and Nyholm.[9] However, such an approach cannot easily include 0% and 100% responses if they occur. To get around this problem, Berkson's substitutions of $1/2n$ and $1 - 1/2n$ could be used[8] (n = the number of individuals in the exposure treatment associated with the P values of 0.00 or 1.00). Bliss[6] provided a more involved means of coping with 0% and 100% mortality treatments. However, this "fudging"[18] can compromise results. Armitage and Allen[10] describe an alternative, minimum χ^2 method for estimating LD50.

A maximum likelihood method can be used to estimate the LC50 instead. Neter et al.[19] explain this method using a hypothetical probability function with one parameter, θ. If the population with the probability function $f(Y, \theta)$ is sampled, the joint probability function for the observations, Y_1 to Y_n (Equation 1.52a in Neter et al.[19]) is

$$g(Y_1 \cdots Y_n) = \prod_{i=1}^{n} f(Y_i, \theta) \tag{13}$$

where θ = a single parameter of the example function. (The Π is similar to the Σ operation except that the product instead of the sum is taken for $f(Y_i, \theta)$.)

The likelihood function for the observations (Equation 1.52b in Neter et al.[19]) is the following.

$$L(\theta) = \prod_{i=1}^{n} f(y_i, \theta) \tag{14}$$

The maximum likelihood method maximizes the likelihood function to yield an estimate of θ. The maximum likelihood equations here have no explicit solutions and, consequently, solutions are reached by iteration. The equations are solved repeatedly using the estimates from the previous iteration as the initial values in the equations. The process is repeated until estimates from consecutive iterations are sufficiently similar to be acceptable, i.e., "close enough."

Estimates from maximum likelihood methods are biased; but, according to Finney,[3] the bias in the estimation of LC50 is usually small except when very few observations are taken. Also, the bias is often no worse than those of alternate techniques.[3] Further, the maximum likelihood procedure has very good precision relative to other estimation techniques. When there are less than two partial kills, the maximum likelihood methods cannot be used without adjustment of the $P = 0.00$ or $P = 1.00$ data such as described for Berkson's substitutions. In summary, the maximum likelihood method produces a relatively precise but biased estimate of the LC50. The iterative nature of the method means that calculations are usually performed with a computer.

The normal (probit), logistic (logit), or Weibull models can be fit using maximum likelihood methods. Weighted or unweighted maximum likelihood methods can be used.[5,9] Consequently, the question now becomes, "Which model fits these data best using the maximum likelihood methods?" Goodness-of-fit estimates can be used to measure how closely a model fits the data set. The χ^2 statistics for the various models can be compared to select between the various models.

Example 3

Analysis of Newman and Aplin's[16] data in Example 2 is extended to maximum likelihood fitting of normal, logistic, and Gompertz models. The fitting is done with the SAS PROC PROBIT.[20] Fitting the Gompertz model is analogous to fitting a Weibull model.[21] As outlined by Dixon,[21] the Gompertz model is the following,

$$P = 1 - e^{-e^{x'b}} \tag{15}$$

where P = effect metameter; x' = metameter for concentration; and b = an estimated parameter.

Example 3 *Continued*

The Gompertz model as fit by the SAS PROC PROBIT is expressed as

$$P = 1 - e^{-e^{a+bx}} \tag{16}$$

where a and b are maximum likelihood parameter estimates. Or

$$P = 1 - e^{-(e^a(e^x)^b)} \tag{17}$$

Or, expressed as the Weibull model (see Equation (7)),

$$P = 1 - e^{-\alpha Z^\lambda} \tag{18}$$

where $\alpha = e^a$; $\lambda = b$; and $Z = e^x$

The SAS code for fitting these three models follows.

```
DATA SALT;
  INFILE CARDS EOF=EOF;
  INPUT DEAD TOTAL CONC @@;
  PROP = DEAD/TOTAL;
  /* PRODUCING CONCENTRATIONS TO BE PLOTTED          */
  /* LATER AGAINST PREDICTED VALUES.                 */
  OUTPUT;
  RETURN;
EOF: DO CONC = 10 TO 21 BY 0.50;
  OUTPUT;
  END;
  CARDS;
16 76 10.30 22 79 10.80 40 77 11.60 69 76 13.20
78 78 15.80 77 77 20.10
;
/* PROBIT MODEL OF NACL-MOSQUITOFISH DATA           */
PROC PROBIT LOG10 INVERSECL LACKFIT;
  MODEL DEAD/TOTAL = CONC/D=NORMAL ITPRINT;
  OUTPUT OUT=P P=PROB STD=STD XBETA=XBETA;
RUN;
PROC PLOT;
  PLOT PROP*CONC="X" PROB*CONC="*"/OVERLAY;
  TITLE 'PROBIT—OBSERVED AND PREDICTED VALUES';
RUN;
/* LOGIT MODEL OF NACL-MOSQUITOFISH DATA            */
PROC PROBIT LOG10 INVERSECL LACKFIT;
  MODEL DEAD/TOTAL = CONC/D=LOGISTIC ITPRINT;
  OUTPUT OUT=L P=LPROB STD=LSTD XBETA=LXBETA;
RUN;
PROC PLOT;
  PLOT PROP*CONC="X" LPROB*CONC="*"/OVERLAY;
  TITLE 'LOGIT—OBSERVED AND PREDICTED VALUES';
RUN;
/* GOMPERTZ MODEL OF NACL-MOSQUITOFISH DATA         */
/* NOTE THAT THE NATURAL LOGARITHM IS USED IN       */
/* THIS MODEL TO MAKE IT EASY TO RELATE TO THE      */
/* WEIBULL MODEL.                                   */
PROC PROBIT LOG INVERSECL LACKFIT;
```

Example 3 *Continued*

```
MODEL DEAD/TOTAL = CONC/D=GOMPERTZ ITPRINT;
OUTPUT OUT=G P=GPROB STD=GSTD XBETA=GXBETA;
RUN;
PROC PLOT;
  PLOT PROP*CONC="X" GPROB*CONC="*"/OVERLAY;
  TITLE 'GOMPERTZ—OBSERVED AND PREDICTED VALUES';
RUN;
```

The Normal Model

The procedure converged within seven iterations. The Pearson Goodness-of-Fit $\chi^2 = 1.4112$ ($df = 4$) with a probability $= 0.8422$. [The Pearson χ^2 is used here. Explanations of this and other pertinent χ^2 statistics can be obtained from Christensen and Nyholm,[9] SAS,[20] Berkson,[22] and Salsburg[23] (pages 60–61).]

The PROBIT procedure generated the following estimates of the LC50 and its 95% C.I.

$$\text{Estimated LC50} = 11.44 \text{ g NaCl/L}$$
$$95\% \text{ C.I.} = 11.24 - 11.66 \text{ g NaCl/L}$$

Buikema et al.[24] urged that, in reporting acute toxicity data, the slope with its limits should be reported as well as the LC50, its 95% confidence limits, and the associated χ^2 statistic. The slope and the LC50 are required to fully describe the toxicity model. For example, two intersecting dose-response curves may have a common LC50 but very different slopes. Considering only the LC50 would not reveal the signficant differences associated with these two curves. Although not provided in this summary of results, the slopes for these models are given in the analysis output.

The confidence limits presented here would be more accurately termed fiducial limits. [Fiducial limits are "a statement that there is a 19 of 20 chance that the LC50 value falls within the specified limits ($P = 0.05$)."[24]] However, the fiducial limits and confidence limits are so similar that they are often treated as the same. Finney[3] and Buikema[24] provide further explanation of the conceptual difference between fiducial and confidence limits.

The Logistic Model

The procedure converged within seven iterations. The Pearson Goodness-of-Fit $\chi^2 = 2.0581$ ($df = 4$) with a probability $= 0.7251$.

$$\text{Estimated LC50} = 11.43 \text{ g NaCl/L}$$
$$95\% \text{ C.I.} = 11.23 - 11.64 \text{ g NaCl/L}$$

The Gompertz (Weibull) Model

The methods converged within five iterations. The Pearson Goodness-of-Fit χ^2 ($df = 4$) $= 0.2635$ with a probability $= 0.9920$.

$$\text{Estimated LC50} = 11.58 \text{ g NaCl/L}$$
$$95\% \text{ C.I.} = 11.35 - 11.80 \text{ g NaCl/L}$$

Model Comparison Using χ^2 Ratios

A χ^2 value was estimated for each model. It is used to compare the expected proportions of dead test organisms as predicted by the model to the observed proportions dead. The

Example 3 *Continued*

larger the χ^2 value, the more extreme the failure of the method to model the tolerance distribution.[6] The χ^2-associated probability calculated for the three models showed no significant heterogeneity ($\alpha = 0.05$): there is no evidence for rejecting any of the three models.

The ratio of χ^2 values for the three models can be used to assess their relative goodness-of-fits.[11] The χ^2 will decrease as the goodness-of-fit increases. Consequently, a ratio of χ^2 values less than 1 indicates that the model associated with the χ^2 in the numerator fits the data set better than the model associated with the χ^2 in the denominator.

$$\chi^2_{\text{Weibull}}/\chi^2_{\text{probit}} = 0.2635/1.4112 = 0.1867$$
$$\chi^2_{\text{Weibull}}/\chi^2_{\text{logit}} = 0.2635/2.0581 = 0.1280$$
$$\chi^2_{\text{probit}}/\chi^2_{\text{logit}} = 1.4112/2.0581 = 0.6857$$

The ratios with the Weibull χ^2 in the numerator were considerably less than 1, indicating that the Weibull model is the best choice. Similarly, the probit model seemed better than the logit model. This conclusion can be confirmed by plotting the data against the predicted values for each from the models.

Christensen,[11] and Christensen and Nyholm[9] explained the differences resulting from using the Weibull model instead of the normal model to fit log dose-response data. Relative to the normal model, the Weibull model predictions at the concentration extremes bow upward from a straight line. This implied to Christensen[11] that most organisms die above a certain concentration threshold. Generally, the Weibull model predicts higher mortality at lower concentrations than the probit model.

3. Trimmed Spearman-Karber Method

The maximum likelihood methods require the assumption of specific models. Use of the probit model assumes a normal distribution, and use of the logit assumes a logistic model. Both the normal and logistic curves are symmetrical about the mean population tolerance. The use of the Weibull metameter also implies a specific model, but the associated curve is not symmetrical about the mean. Unfortunately, the ability to identify the appropriateness of any model is often quite limited.

Another problem that may be encountered with maximum likelihood methods is associated with the process of convergence over iterations.[25] For some data sets, the methods can fail to reach acceptable convergence. Also, it is possible that they can converge on inappropriate solutions depending on the initial values used to begin the iterative process. For this reason, it is prudent to plot the predictions from a fitted model along with the original data to check for the reasonableness of the model estimates.

A modified Spearman-Karber method[25] can be used instead of these methods if a symmetrical distribution can be assumed. A specific model need not be assumed in this approach. The procedure is described by Hamilton et al.[25] in the following steps.

Step 1

Let x_1 to x_k be the natural logarithms of k exposure concentrations with $x_1 < \ldots < x_k$. Also let n_1 to n_k, and r_1 to r_k be the total number of fish exposed at each concentration and the number that died after exposure to each concentration, respectively. If more than one concentration treatment had no mortalities then the logarithm of the highest concentration giving no mortalities is x_1. There should be no mortalities in any concentrations below the x_1. Similarly, if more than one concentration resulted in 100% mortality then the natural logarithm of the lowest concentration producing 100% mortality is x_k. All concentrations above x_k must have 100% mortality.

Step 2
Let p_i be equal to r_i/n_i, the proportion of the exposed fish dying by the endpoint. To perform the trimming, the p_i values need to be adjusted if the following conditions are not present, $p_1 \leq p_2 \leq \ldots \leq p_k$. This must be done because the assumption is made that the proportions increase monotonically, i.e., the true proportions of exposed fish dying (P_i) increase with exposure concentration. New p_i values are calculated for adjacent p_i values that do not fulfill this requirement. For two inconsistent, adjacent p_i values, their new p value would be $(r_i + r_{i+1}) / (n_i + n_{i+1})$.

For example (from Table III of Hamilton et al.[25]), five tanks of increasing toxicant concentrations have r_i/n_i values of 0/20, 1/20, 0/20, 3/19, and 20/20. There is an inconsistency because the second lowest concentration had one death, but the next highest concentration had no mortality. The p_i values for these two tanks are replaced by a value intermediate between the two. The new p_i for these two tanks would be $(r_2 + r_3) / (n_2 + n_3) = (0 + 1) / (20 + 20) = 0.025$. If the r_i/n_i values were 0/20, 2/20, 1/20, 1/20, 3/19, and 20/20 then the process would be the following.

a. Original p_i values: 0.00, 0.10, 0.05, 0.05, 0.15, 1.00

b. $(r_2 + r_3 + r_4)/(n_2 + n_3 + n_4) = 4/60 = 0.067$

c. Adjusted p_i values: 0.00, 0.067, 0.067, 0.067, 0.15, 1.00

The process is continued for all inconsistent pairs until the p_i values are monotonically nondecreasing with concentration.

Step 3
Next, the ln concentrations (x_i) are plotted against the final p_i values.

Step 4
Now the cumulative distribution curve (Step 3) must be trimmed. A percentage (α) of the lower and upper portions of the curve are trimmed away to leave a graph of the central $100 - 2\alpha$ portion of the curve. The p scale is replaced by $_\alpha p = (p - \alpha/100) / (1 - 2\alpha/100)$ and these data ($_\alpha p_i$ versus x_i) are replotted. Do not use $_\alpha p$ values above 1 or below 0 when replotting. Instead, the respective ln concentrations predicted by the new 0.0 and 1.0 values on the $_\alpha p$ scale are used as the points at either end of the new curve.

Step 5
The mean of the cumulative frequency curve is calculated as follows. Assume that $k = 6$ but, with trimming, only four points remain ($_\alpha p_2$, $_\alpha p_3$, $_\alpha p_4$, $_\alpha p_5$) after Step 4. The corresponding ln concentration values are x_2, x_3, x_4, and x_5. There are now five intervals to be considered. Three intervals exist between the 4 x_i values. Two more exist on each end of the curve between the predicted x_i corresponding to $p = 0$ and x_2, and between x_5 and the predicted x_i corresponding to $p = 1.0$. Let an interval of ln concentration be represented generally by (x_{i-1}, x_i) and an interval of $_\alpha p$ be represented by $(_\alpha p_{i-1}, _\alpha p_i)$. The mean of the distribution is then calculated.

$$\text{Mean} = \sum_{i=1}^{j} (_\alpha p_i - _\alpha p_{i-1})\left(\frac{x_{i-1} + x_i}{2}\right) \tag{19}$$

where $_\alpha p_i - _\alpha p_{i-1} =$ the proportion of the total area under the curve in the interval, (x_{i-1}, x_i); and $(x_{i-1} + x_i)/2 =$ the midpoint of the interval, (x_{i-1}, x_i).

The α-trimmed Speerman-Karber estimate of the LC50 is the antilog of the mean calculated with Equation (19).

Step 6

Obviously, an α must be selected from the range $0 \leq \alpha < 50$. An appropriate α can be estimated using the recommendations of Hamilton et al.[25] The method is always accurate when $\alpha \geq 100p_1$ (p_1 is the adjusted p_1 associated with the lowest x_i) and $\alpha \geq 100 (1 - p_k)$ (p_k is the adjusted p_i associated with the highest x_i). The α could then be set at the maximum of $100p_1$ and $100 (1 - p_k)$. As α increases, the sensitivity of the method to anomalous values decreases. This would argue for selection of a large α. However, an upper limit is placed on the possible values of α, as the standard error increases as the α increases. Hamilton et al.[25] suggested that an α of 10% is adequate when the highest concentration results in 95% mortality or more and the lowest concentration results in 5% mortality or less.

Step 7

The variance (and 95% confidence interval) for the trimmed Spearman-Karber estimate can be estimated as outlined in the appendix of Hamilton et al.[25] Simulations by these authors suggested that the variance and 95% confidence interval estimates are conservative.

Let

$A = \alpha/100$,

$L = $ maximum $\{i: p_i \leq A\}$ or MAX $\{i: p_i \leq A\}$,

$U = $ minimum $\{i: p_i \geq 1 - A\}$ or MIN $\{p_i \geq 1 - A\}$,

$x_L = $ the largest ln concentration value with an adjusted $p \leq A$,

$x_U = $ the smallest ln concentration value with an adjusted $p \geq 1 - A$,

$n_L = $ the number of individuals exposed to X_L, and

$n_U = $ the number of individuals exposed to X_U

Necessary, intermediate calculations using these defined variables follow.

$$V_1 = \left[\frac{(x_{L+1} - x_L)(p_{L+1} - A)^2}{(p_{L+1} - p_L)^2} \right]^2 \left[\frac{p_L(1 - p_L)}{n_L} \right] \tag{20}$$

$$V_2 = \left[(x_L - x_{L+2}) + \frac{(x_{L+1} - x_L)(A - p_L)^2}{(p_{L+1} - p_L)^2} \right]^2 \left[\frac{p_{L+1}(1 - p_{L+1})}{n_{L+1}} \right] \tag{21}$$

$$V_3 = \sum_{i=L+2}^{U-2} (x_{i-1} - x_{i+1})^2 \frac{p_i(1 - p_i)}{n_i} \tag{22}$$

$$V_4 = \left[(x_{U-2} - x_U) + \frac{(x_U - x_{U-1})(p_U - 1 + A)^2}{(p_U - p_{U-1})^2} \right]^2 \frac{[p_{U-1}][1 - p_{U-1}]}{n_{U-1}} \tag{23}$$

$$V_5 = \left[\frac{(x_U - x_{U-1})(1 - A - p_{U-1})^2}{(p_U - p_{U-1})^2} \right]^2 \left[\frac{p_U(1 - p_U)}{n_U} \right] \tag{24}$$

$$V_6 = \left[\left[\frac{(x_U - x_{L+1})(1 - A - p_U)^2}{(p_U - p_{L+1})^2} \right] - \left[\frac{(x_{L+1} - x_L)(A - p_L)^2}{(p_{L+1} - p_L)^2} \right. \right.$$
$$\left. \left. + (x_L - x_U) \right]^2 \frac{p_{L+1}(1 - p_{L+1})}{n_{L+1}} \right. \quad (25)$$

The variance about the mean can be estimated from these equations and the following equations for various values of $U - L$.

If $U - L \geq 4$ then

$$\text{Variance} = \frac{(V_1 + V_2 + V_3 + V_4 + V_5)}{(2 - 4A)^2} \quad (26)$$

If $U - L = 3$ then

$$\text{Variance} = \frac{(V_1 + V_2 + V_4 + V_5)}{(2 - 4A)^2} \quad (27)$$

If $U - L = 2$ then

$$\text{Variance} = \frac{(V_1 + V_5 + V_6)}{(2 - 4A)^2} \quad (28)$$

If $U - L = 1$ then

$$\text{Variance} = (x_U - x_L)^2 \left[\left[\frac{(0.5 - p_U)^2}{(p_U - p_L)^4} \right] p_L \frac{(1 - p_L)}{n_L} \right.$$
$$\left. + \left[\frac{(0.5 - p_L)^2}{(p_U - p_L)^4} \right] p_U \frac{(1 - p_U)}{n_U} \right] \quad (29)$$

The 95% confidence interval is twice the square root of the variance estimated with Equation (26), (27), (28), or (29).

Example 4
Calculations for the trimmed Spearman-Karber method are demonstrated with the NaCl-mosquitofish data used in Examples 2 and 3.

According to Step 1, these data (concentration, k, x_i, r_i, n_i) are tabulated here in the first five rows. No adjustments as described in Step 2 are required. The data for the highest concentration are omitted because the next concentration (15.8 g NaCl/L) also resulted in a 100% mortality response. These data (p_i, x_i) are plotted in Figure 3A.

Concen-tration	20.1	15.8	13.2	11.6	10.8	10.3	0.016
k	—	6	5	4	3	2	1
x_i	—	2.760	2.580	2.451	2.380	2.332	− 4.135
r_i	77	78	69	40	22	16	0

Example 4 *Continued*

n_i	77	78	76	77	79	76	79
p_i	—	1.00	0.91	0.52	0.28	0.21	0.00
$_\alpha p$	—	—	—	0.53	0.13	0.02	—

An α is estimated using Step 6. Because the $p_1 = 0.00$, α should be greater than or equal to $100p_1$, i.e., 0%. Because the $p_k = p_6 = 1.00$, α should be greater than or equal to $100 (1 - p_k)$, i.e., 0%. Both estimates result in the trivial recommendation that trimming of this data set should be greater than or equal to 0%. Hamilton et al.[25] recommended an $\alpha = 10\%$ in such a case (See Step 6). For the sake of illustration, let's use the next p values ($k = 2$ and $k = 5$) to estimate α. The α should be the largest of the two values, $100(1.00 - 0.91)$ or 9% and $100(0.21)$ or 21%. For convenience, an α near the larger of these values (21%) is selected. Twenty percent will be trimmed from each tail of the curve (Figure 3A).

Now $_\alpha p$ values $((p - \alpha/100)/(1 - 2\alpha/100))$ are calculated and entered in the table above. As outlined in Step 4, any $_\alpha p$ values greater than 1 or less than 0 are not plotted. Therefore, the values of 1.18 and -0.33 for $i = 5$ and 1 are discarded. Figure 3B is the polygon resulting from plotting $_\alpha p$ values against in ln concentration. The predicted data points corresponding to $p = 0.20$ and $p = 0.80$ have been scaled to $_\alpha p$ value of 0.00 and 1.00, respectively. The mean of the resulting distribution is estimated with Equation (19). The following table summarizes the associated calculations.

Interval $(j, j - 1)$	$_\alpha p_j$	$_\alpha p_{j-1}$	x_j	x_{j-1}	Product
2,1	0.02	0.00	2.332	2.200	0.04
3,2	0.13	0.02	2.380	2.332	0.26
4,3	0.53	0.13	2.451	2.380	0.97
5,4	1.00	0.53	2.550	2.451	1.18
					Σ 2.45

The antilogarithm of 2.45 is the trimmed Spearman-Karber estimate of the LC50. That is LC50 $= e^{2.45} = 11.58$ g NaCl/L. This close to estimates derived from the graphical (11.40–11.59 g NaCl/L) and maximum likelihood (11.43–11.58 g NaCl/L) methods in Examples 2 and 3.

The 95% confidence interval for the LC50 is estimated using Equations (20) to (24), and (26) as outlined in Step 7.

$$A = 0.20 \qquad U\text{–}L = 5\text{–}1 = 4$$
$$p_L = 0.00 \qquad P_U = 0.91$$
$$x_L = -4.135 \qquad X_U = 2.580$$
$$n_L = 79 \qquad n_U = 76$$

With these data and those in the preceding table, V_1 to V_5 are calculated [Equations (20) to (24)].

$$V_1 = 0 \qquad V_2 = 0.000920118$$
$$V_3 = 0.000036137 \qquad V_4 = 0.000116697$$
$$V_5 = 0.000004764$$

The variance is calculated using these estimates in Equation (26) as $U\text{-}L = 4$. The variance $= 0.00107716$. The 95% confidence limits are then calculated.

Example 4 *Continued*

$$95\% \text{ confidence interval} = 0.065640231$$
$$\text{Lower limit} = e^{(2.45-0.065640231)} = 10.85 \text{ g NaCl/L}$$
$$\text{Upper limit} = e^{(2.45+0.065640231)} = 12.37 \text{ g NaCl/L}$$

4. Binomial Method

Stephan[18] argued that partial kills were not necessary to estimate an LC50 and approximate 95% confidence intervals. He reasoned that the LC50 must fall between the highest concentration giving 0% mortality (A) and the lowest concentration giving 100% mortality (B). (This is strictly true only if sample error is assumed to be insignificant.) The estimate of the LC50 is

$$\text{LC50} = (A*B)^{1/2} \tag{30}$$

when A and B are expressed in concentration units. The interval between A and B can be used as the confidence interval for the LC50. This confidence interval is at least a 95% confidence interval if the number of individuals treated at a concentration interval

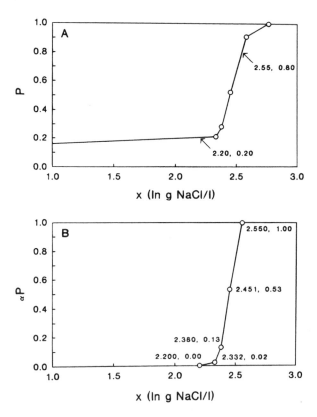

Figure 3 The trimmed Spearman-Karber method for estimating the LC50 and associated 95% confidence interval for the sodium chloride-mosquitofish mortality data. Panel A presents all perinent data and the points to be trimmed. Panel B is the plot resulting from trimming and estimation of $_aP$ calculation (see Example 4 for details).

(n) exceeds 5. For example, the coefficients for the confidence intervals are 93.8, 96.9, 98.4, and 99.2% for $n = 5, 6, 7$, and 8, respectively.[18] If the same number of individuals are exposed at the A and B concentrations, the associated confidence coefficient is[18]

$$\text{Coefficient} = 100[1 - 2(0.5)^n] \qquad (31)$$

If the number of individuals exposed to concentration A (n_A) and the number exposed to concentration B (n_B) are different, Gelber et al.[26] express the confidence coefficient for the interval between A and B in the form,

$$\text{Coefficient} = 100[1 - (0.5)^{n_A} - (0.5)^{n_B}] \qquad (32)$$

5. Moving Average Method

A simple moving average method can be used to estimate the LC50[17,18,26] if the exposure concentrations are set in a geometric pattern, e.g., 2, 4, 8, 16 mg/L, and the same number of individuals are exposed to each concentration. Gelber et al.[26] stated that the restrictions on this method made it less appealing than the others mentioned previously. However, Stephan[18] and Gad and Weil[17] did not believe that the restrictions were so confining as to disfavor its use.

Gelber et al.[26] presented the moving average method with the simple equation,

$$\hat{p}_i = \sum_{j=i-s/2}^{i+s/2} w_j p_j \qquad (33)$$

There are i exposure intervals with associated proportional kills of p_i. Because the concentrations are defined by a geometric series, all the log concentrations have the same span (s) between them. Using the p_j and estimated weights (w_j), a moving rate of mortality is estimated with Equation (33).

The moving average method will mechanically produce results when there are fewer than two partial kills.[18] Initially Stephen[18] considered moving average estimates from data sets with one partial kill as statistically valid but recently pointed out that such estimates should be considered approximations only.[27] With no partial kills and the use of the logarithm of exposure concentrations, the moving average method produces the same approximation of the LC50 as Equation (30).[27]

Because of the restrictions imposed on this technique, no further discussion is given here. Instead the reader is referred to Gad and Weil,[17] who give a detailed explanation with an example and all necessary tables.

6. Coping with Control Mortalities

Frequently, a certain level of mortality in the control group is unavoidable. Various rules of thumb exist for defining acceptable, albeit unwelcome, levels of control mortality. For example, if one can make the optimistic assumption that the underlying cause of mortality associated with the controls does not influence the mortality associated with the toxicant,[18] a range of control mortalities from 5%[28] to 10%[24,28] may be acceptable. Several authors[5,24,28,29] recommend that Abbott's formula[30] be used to adjust the proportions dying in each exposure to account for this acceptable control mortality.

$$p_c = \frac{p - p_0}{1 - p_0} \tag{34}$$

where p_c, p, and p_o = the corrected, original, and control mortality proportions, respectively.

7. Duplicate Treatments

It is often desirable to determine whether duplicate treatments are homogeneous. Stephan[18] suggested that Fisher's exact test could be used for this purpose. Test results are suspect if the assumption of homogeneity is rejected for duplicates.

8. Summary of LC50 Methods

There are several methods for estimating the LC50 (LD50 or EC50) and its associated 95% confidence interval. The simplest graphical techniques are convenient, yet the results are subject to minor inaccuracies of fitting a line by eye. The Litchfield-Wilcoxon method, a semigraphical method, can be used to estimate the LC50 and its 95% confidence interval, but it also is subject to some degree of error associated with fitting a line by eye. The maximum likelihood methods of fitting normal, logistic, or Weibull models assume specific models and require at least two partial kills unless adjustment is made to the proportions responding, e.g., Berkson's substitutions.[22] The ratio of χ^2 values for candidate models can be used to judge the relative appropriateness of each. The trimmed Spearman-Karber method provides a nonparametric alternative requiring only a monotonic increase in proportion responding and a symmetrical distribution. If there are no partial kills then a binomial method can be applied. The moving average method can be applied if the exposure concentrations are set in a geometric pattern and the same number of individuals are exposed to each concentration.

The tedium of calculating these statistics can be minimized by using a variety of software packages. Pertinent procedures in the SAS program have been illustrated. A very convenient program is available to estimate the LC50 and associated statistics using the probit (maximum likelihood method for a normal model), trimmed Spearman-Karber, binomial, and moving average methods. The trimmed Spearman-Karber method was added to software written by C. Stephan of the U.S. Environmental Protection Agency (U.S. EPA) to produce this program (CT-TOX). At this printing, it may be obtained from T. Haze.[31] Another convenient program can be obtained from the U.S. EPA as described in Appendix H of Weber et al.[32] (Mention of specific software throughout this book does not imply endorsement.)

9. Incipient LC50

The incipient (asymptotic, ultimate, or threshold) LC50 is the calculated concentration below which 50% of exposed individuals would live indefinitely relative to the lethal effects of the toxicant. Equally important is the procedural definition of the incipient LC50, the antilogarithm of the asymptotic log concentration of the (log) LC50 versus (log) time curve. (See Figure 4A for the general behavior of toxic response versus time. Although LT50 is used in this figure instead of LC50, the same general behavior is expected. The LC50 decreases with duration of exposure until an asymptotic LC50 is reached.) Mortality during the course of a toxicity test is often recorded at set intervals. The estimated LC50 is calculated for each time. The log of the LC50 for each time (X) is plotted against the log of time (Y) to produce a toxicity curve. The lethal threshold concentration (incipient LC50)[33] is estimated to be the concentration at which the curve begins to run parallel with the X axis.

Sprague[29,33] provides a general discussion of quantitative methods for dealing with the incipient LC50. As pointed out by Chew and Hamilton,[34] there are conceptual

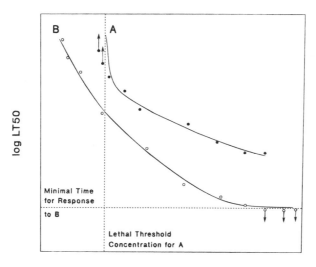

Figure 4 An illustration of the incipient lethal level or incipient threshold concentration (A). As the exposure concentration decreases, a concentration is reached below which 50% or more of the exposed individuals will live indefinitely relative to the effect of toxicant A. Some toxicants (B) may show no apparent lethal threshold concentration. A minimum time for the toxic response may also be apparent (B). Fifty percent of the individuals exposed survive for at least this length of time regardless of toxicant concentration.

problems with this approach. First, LC50 values are often estimated at a series of times during a toxicity test and treated as independent. This is a dubious assumption. Second, concentration is treated as an independent variable when, in practice, it is set in the experimental design.

10. The Significance of the LC50

The LC50 does not indicate environmental safety of a specific concentration of toxicant.[28] Rather, it is a measure of toxicity that is best employed in a relative context, e.g., concentration X of chemical A has a higher toxic effect than concentration X of chemical B after 96 h of exposure, or the toxic effect is less under condition A than under condition B. It is quite dependent on conditions surrounding the toxic response. Furthermore, the time endpoint is often selected for convenience. As discussed earlier, the median value ($p = 0.50$) was selected because it was the most statistically reliable value. Despite the common acknowledgement of this fact, LC50 values are used inappropriately to imply environmental safety. An example might be the misinterpretation of the incipient LC50. Again, the incipient LC50 is the concentration below which 50% of the individuals will live indefinitely (relative to the toxicant effects) under the test conditions. The median value is completely meaningless relative to the survival of an endemic population; it is a statistically convenient value only.

III. TIME-TO-DEATH

A. THE STANDARD APPROACH

1. General

Another approach to analyzing toxicity data is to model time-to-death (or resistance time), the duration of exposure before an individual's death. Sprague[29] described the straightforward analysis of time-to-death data sets. The cumulative proportions of exposed individuals dying is recorded at a series of time intervals. These data are then plotted on logarithm-probability paper (percentage mortality on the probability scale versus the log of time) to yield a line. One line may be generated for each toxicant concentration or level of a stressor. This approach is similar to that already described for dose-response plots, assuming a normal distribution of P at a set endpoint when plotted against the logarithm of dose. From our previous discussion, it follows that the probit (or NED) of the proportion dying at any time plotted against the logarithm of time should also produce a straight line if the assumption of a normal model for these metameters is correct.

The cumulative mortality (the proportion of all individuals dying at exposure duration, t) plotted against the logarithm of duration does not always produce a straight line. If there is a distinct change in slope for the time course of cumulative mortality, the resulting plot is called a split probit. A split probit is often attributed to different mechanisms being dominant at different times or distinct subsets of individuals having different resistances to the stressor.[29] However, inappropriate fitting of the data set to the normal model could be another reason for a split probit, because curve assessment usually ends at visual inspection and alternative models are rarely examined.

Just as the LC50 is useful in endpoint dose-response methods, the median response time is important in the time-to-death approach. The median lethal time (LT50, median period of survival, median resistance time, or median time-to-death) is the duration of exposure corresponding to a cumulative mortality of 50% for the exposed individuals. If a nonlethal or ambiguously lethal response is considered, median effective time (ET50) would be the appropriate term.[29] The median lethal time is analogous to the LC50 and, assuming that the normal model is appropriate for the logarithm of exposure duration-cumulative mortality data, it can be analyzed as described for the probit methods for the LC50. Shepard[35] provided an excellent example of the median response time usage. In addition to discussing split probits and covariates such as acclimation, he also demonstrated an alternate method of estimating the asymptotic lethal concentration using the median response (resistance) time instead of the LC50 as described here. This approach is discussed below in detail.

2. Litchfield Method for Estimating LT50

Litchfield[36] described a rapid graphical method for estimating the LT50 and its 95% confidence limits. It is very similar to the previously described Litchfield and Wilcoxon method published the same year and described previously. With the exception of one exponent, the associated equations are identical with Equations (9) to (12).

Assume that all individuals die during the period of exposure; all individuals have a quantified time-to-death. (If there were survivors, the data would be censored as defined in Chapter 2. The data set would be right-censored because the observations censored are those associated with high values of time, i.e., values to the right of the distribution.) The cumulative percentage of individuals dying (number dead/total number exposed) is plotted against time (duration of exposure) on log-probability paper. The time corresponding to a cumulative percentage of 50% is the LT50.

Calculation of the 95% confidence limit for the LT50 is estimated with Equations (35) to (38). First, the slope function (equivalent to the standard deviation) is estimated by Equation (35).

$$S = \frac{\dfrac{LT84}{LT50} + \dfrac{LT50}{LT16}}{2} \tag{35}$$

where LT16, LT50, LT84 = the times corresponding to the 16%, 50%, and 84% of the cumulative percent mortality.

The 95% confidence interval is estimated using f_{LT50}.

$$f_{LT50} = S^{1.96/\sqrt{N}} \tag{36}$$

where N = the number of individuals exposed. The upper and lower confidence limits are calculated by multiplying or dividing the estimated LT50 by f_{LT50}.

Upper Limit of 95% C.I.:

$$\text{Upper} = LT50 * f_{LT50} \tag{37}$$

Lower Limit of 95% C.I.:

$$\text{Lower} = LT50 / f_{LT50} \tag{38}$$

If there are survivors then the equation used to estimate the f_{LT50} needs to be modified as outlined by Litchfield.[36] Values for E from Bliss's[37] Table VIII are required for this adjustment. Pertinent material from this table is reproduced in the Appendix as Table 8. The E acts to adjust the N to account for the number of survivors. For censored data sets, the following equation becomes pertinent.

$$f_{LT50} = S^{1.96/\sqrt{N_2}} \tag{39}$$

where $N_2 = N/E$. This value is used instead of that from Equation (36) in Equations (37) and (38) to estimate the 95% confidence limits for the LT50 of the censored data set.

Bliss[37] estimated E using an x' value, the point of censoring expressed in terms of the standard deviation. This value is derived with the point of censoring, log LT50, and log standard deviation of the LT50. In this case, the log LT50 and log standard deviation are taken from the graph and Equation (35) as described above. Because these are right-censored data and Bliss's[37] table is produced for left-censored data, the x' is obtained by plotting reaction times ($1000/t$). The point of censoring is defined by inverting each survival time and subtracting 3 from the logarithm of each survival time. The process, as outlined in Litchfield[36] or Bliss,[37] to estimate x' can be simplified if one remembers that the NED expresses a proportion in terms of deviations from the mean. Consequently, the NED (Table 7 in the Appendix) for one minus the proportion dead at the end of the exposure can be used as x' in Table 8.

Example 5

Times-to-death of streptomycin-treated mice with tuberculosis were tabulated by Litchfield[36] for 60 days. Fourteen mice died with the following times-to-death: 26, 29, 32, 32, 37, 39, 39, 42, 43, 43, 44, 47, 52, and 59 days. Six lived beyond 60 days. What is the LT50 and its 95% confidence interval for these mice?

Example 5 *Continued*

Time (days)	Cumulative Number Dead	Number Dead/Total Number
26	1	0.05
29	2	0.10
32	4	0.20
37	5	0.25
39	7	0.35
42	8	0.40
43	10	0.50
44	11	0.55
47	12	0.60
52	13	0.65
59	14	0.70

Using log-probablity paper (Figure 5), an LT50 of 46 days is derived. Extrapolating and using the results in Equation (35),

$$S = [(66/46) + (46/32)]/2 = 1.44 \text{ days}$$

The 95% confidence intervals may be calculated using Equations (39), (37), and (38). But x', E, N_2, and finally, f_{LT50} must be calculated first.

x' = NED for $1.00 - 0.70 = -0.52440$
$E = 1.1390$ (linearly interpolated from Table 8)
$N_2 = N/E = 20/1.1390 = 17.56$
$f_{LT50} = S^{1.96/\sqrt{N_2}} = 1.44^{1.96/4.190} = 1.44^{0.468} = 1.19$
Upper Limit = LT50 (f_{LT50}) = 46 (1.19) = 54.7 days
Lower Limit = LT50/f_{LT50} = 46/1.19 = 38.7 days

The LT50 for these mice is 46 days with a 95% confidence interval of approximately 39 to 55 days.

3. Lethal Threshold Concentration

Sprague[29] illustrated the use of plotting the logarithms of LT50 (Y) and concentration (X) to derive the lethal threshold concentration, the concentration at which 50% of the exposed individuals will survive indefinitely relative to the toxicant effects. When present (Figure 4A), this measure is analogous to the incipient LC50. However, there are data sets that seem to have no apparent threshold concentration (Figure 4B). Further, an asymptotic time may also be present (Figure 4B). The asymptotic time is the minimum duration of exposure necessary for expression of a response.[29] At any time less than the asymptotic time, no response is noted regardless of the toxicant concentration.

Gaddum[5] and Lloyd[38] quantified this curve using two formulae. Ostwald's equation[38] was used when the logarithm of the concentration versus logarithm of LT50 produced a straight line.

$$C^n t = K \qquad (40)$$

where C = the toxicant concentration; t = LT50; and n, K = regression-derived constants. If the plot is curvilinear, the equation of choice[5,38] is the following:

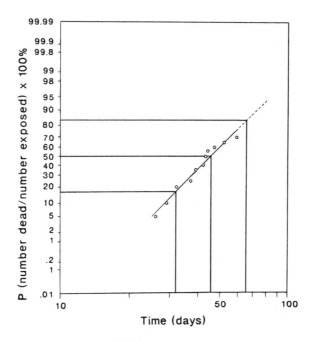

Figure 5 Mortality of streptomycin-treated mice with tuberculosis[36] is plotted using log-probability scales to estimate the LT50 and its 95% confidence interval (Example 5).

$$(C - C_0)^n(t - t_0) = K \tag{41}$$

where C_0 and t_0 = threshold concentration and threshold time, respectively.

The initial estimates for the threshold concentration and time may be extracted by eye during the fitting process.[39] Gaddum[5] suggests plotting C against $1/t$ for large t values to find the value of $1/t$ that equals 0. This is used as the estimate of C_0. Similarly, the plot of t versus $1/C$ for large C values can be used to estimate t_0.

Gaddum[5] also noted that other workers have found the following formulae fit this type of data also.

$$(C - C_0)(1 - e^{-a(t-t_0)}) = k \tag{42}$$

$$(C - C_0)t = K\left(1 + \frac{t_0}{t}\right) \tag{43}$$

$$\left(\frac{C_0}{C} - 1\right)\left(\frac{t}{t_0} - 1\right) = K \tag{44}$$

Gaddum[5] pointed out that Equation (44) is simply another form of Equation (41).

None of these methods for describing the time-response relationship can be clearly linked to any underlying mechanisms, although some linkage was attempted in the associated publications. In contrast, Chew and Hamilton[34] derived a time-concentration-response curve based on pharmacokinetic principles. They assumed that a threshold amount of toxicant existed in some biological compartment such as the gill above which mortality occurs. Then they used a one compartment bioaccumulation model (see

Equation (41) in Chapter 3) to estimate the time to reach that threshold given a series of specified exposure concentrations.

$$C_t = AC_w(1 - e^{-Bt}) \tag{45}$$

where C_t = the concentration in the compartment at time t; C_w = concentration in the source (water); A = a constant; and B = a constant. Their model, as used to fit time-response data, took the form,

$$f(C_w) = \frac{\left[\ln C_w - \ln\left(C_w - \left(\frac{\xi}{A}\right)\right)\right]}{B} \tag{46}$$

where B = "the average rate of entry and elimination from the organism; B determines the curvature of $f(C_w)$"; ξ/A = "the 'asymptotic LC50'; that is, less than half the population of organisms would die due to the toxicant if the environmental concentration were less than ξ/A.", and ξ = the median concentration in the compartment at which half the organisms in the population will die.

Median survival time data for ammonia and cadmium toxicity to fish are fit to this model. (In health sciences, such a model that links pharmacokinetics to toxic or pharmacological action (pharmacodynamics) would be called a pharmacokinetic-pharmacodynamic model.) Nonlinear, least-squares fit of median time-to-death versus concentration is done with a weighting of the reciprocal of the estimated variance of the median time-to-death.

Data from concentration treatments with "too few deaths" are not used in the models. They[34] suggest a rule of thumb for exclusion of treatments. Let k be the smallest integer such that

$$\alpha > (0.5)^N \sum_{j=k}^{N} \binom{N}{j} \tag{47}$$

where α = the level of significance; and N = the number of organisms exposed to the concentration. Exclude any point from the analysis where the number of deaths (d) is less than or equal to k.

The standard notation in Equation (47) (binomial coefficient) can be reexpressed as shown in Equation (48). It denotes the number of possible ways that subsets of size j may be drawn from a population of size N. (See Feller[39], page 34 or Pollard[40], page 4 for a detailed explanation of this notation).

$$\binom{N}{j} = \frac{N!}{j!(N-j)!} = \frac{N(N-1)(N-2)\cdots(N-j+1)}{j!} \tag{48}$$

With the potential for linkage to bioaccumulation models, this approach of Chew and Hamilton[34] has much appeal. It has the potential for elaboration to account for extrinsic and intrinsic factors affecting accumulation and toxicity kinetics. Further evaluations and enrichments of this approach, such as the work of Heming et al.[41] with organochlorine pesticide toxicity, would be valuable contributions to the field.

B. THE SURVIVAL TIME APPROACH

1. General

It is unfortunate that pollution biologists have tended to form a splinter group as far as toxicology is concerned. Standard techniques of analysis, developed by pharmacologists and statisticians for testing drugs, have too often been ignored.[29]

There has been remarkably little attention paid to time-to-death (survival time) methods in the field of aquatic toxicology.[16,42] The reason probably is linked to the tendencies of aquatic toxicologists noted by Sprague in the preceding quote. This is unfortunate because a rich array of pertinent methods is available from the engineering and health sciences literatures. For example, pertinent survival analysis methods have been described in detail in books by Miller[13] and Cox and Oakes.[43]

Such techniques have been used in studies of covariate effects on toxicity.[16,44-46] However, as pointed out in Newman and Aplin[16] and Dixon and Newman,[42] aquatic toxicologists have yet to take full advantage of these techniques. This section's primary objective is to describe these methods with an emphasis on their virtues. However, because of this stress on advantages, it must be emphasized here that the author is not implying that these techniques are superior in all ways to the endpoint methods described earlier. Survival time and endpoint methods both have their place in aquatic toxicology. Sadly, there currently is a gross imbalance in their application.

Several terms and concepts must be presented before discussing specific methods. Chief among these are the survival and hazard functions. Both are used in more in-depth analysis of time-to-death data.

Assume a time course of exposure with individuals dying over period, T. The mortality within the exposed group of individuals can be described as having a probability density function [$f(t)$] and cumulative distribution function [$F(t)$]. An estimate of the $F(t)$ would be the total number of individuals dead at time, t, divided by the total number of exposed individuals.

$$F(t) = \frac{\text{Number dead}_t}{\text{Total number exposed}} \tag{49}$$

The survival function [$S(t)$] is estimated by the number of individuals surviving to time t divided by the total number of individuals exposed to the toxicant. It can be expressed in terms of $F(t)$.

$$S(t) = 1 - F(t) \tag{50}$$

The hazard rate or function [$h(t)$] is the probability of dying during a given interval. It has been interpreted as the force of mortality,[13] instantaneous mortality rate,[12] instantaneous failure rate, or proneness to fail[47] depending on the specific process being studied. Regardless, it can be defined in terms of $f(t)$ and $F(t)$, or $S(t)$.[13,42,47]

$$h(t) = \frac{f(t)}{1 - F(t)} = \frac{-1}{S(t)} \frac{dS(t)}{dt} \tag{51}$$

The cumulative hazard function, $H(t)$, also can be defined in terms of cumulative mortality [$F(t)$].[47,48]

$$H(t) = \int_{-\infty}^{t} h(t)dt = -\ln(1 - F(t)) \tag{52}$$

$F(t)$ can be expressed in terms of $H(t)$

$$F(t) = 1 - e^{-H(t)} \tag{53}$$

With these preliminaries completed, we can now move to a more in-depth time-to-death analysis. It should be noted that, although death (or some surrogate measure of death) is being used in these techniques, other events may also be analyzed with the methods described below. The only restriction is that the event scored for an individual can occur only once during the time course, e.g., time to immobilization, time to first brood, time to recover after intoxication or stupefaction, time to remission, or time to develop cancer.

2. Nonparametric Methods

Two general nonparametric approaches can be used for time-to-death data, the life table and product-limit (Kaplan-Meier) methods.[13,20,43] Neither requires a specific form for the underlying distribution describing survival. The familiar life table or actuarial table places estimates of $S(t)$ in a fixed sequence of intervals, e.g., year age classes of humans. (See Chapter 6 for details.) With the product-limit approach, the intervals may be of any length.[13] Harrell[49] suggests that product-limit methods are preferable to life table methods because they have more resolution and carry fewer assumptions.

The product-limit estimate of $S(t)$ is defined in various sources.[13,20,43,48] The original description can be found in Kaplan and Meier[50] and description of the associated maximum likelihood method is given in Kalbfleisch and Prentice.[51] The notation here is that used by SAS.[20]

$$\hat{S}(t_i) = \prod_{j=1}^{i} \left(1 - \frac{d_j}{n_j}\right) \tag{54}$$

where i = the labels for the failure times, t_i, n_i = the number of individuals alive just before time, t_i, (the number of individuals alive and at risk of dying); and d_i = the number of individuals dying at t_i. This product-limit estimate of $S(t)$ is undefined for times beyond the end of the exposure period (T) if survivors remain.[42]

Greenwood's formula[13,20,42,43,49] can be used to estimate the variance about the product-limit estimate.

$$\hat{\sigma}^2 = \sum_{j=1}^{i} \frac{d_j}{n_j s_j} \tag{55}$$

where $s_j = n_j - d_j$

Dixon and Newman[42] pointed out that Equation (55) reduces to Equation (56) for all times before termination of the experiment if there has been no censoring before termination of the experiment, i.e., all noncensored data.

$$\hat{\sigma}^2(t_i) = \frac{\hat{S}(t_i)[1 - \hat{S}(t_i)]}{N} \tag{56}$$

where N = the total number of individuals exposed. The confidence interval for these estimates can be generated using the square root of the variance estimated in Equation (55) or (56) in Equations (57) and (58).[20]

$$\text{Upper C.I.} = \hat{S}(t_i)(1 + z_{\alpha/2}\hat{\sigma}_i) \qquad (57)$$

$$\text{Lower C.I.} = \hat{S}(t_i)(1 - z_{\alpha/2}\hat{\sigma}_i) \qquad (58)$$

According to Dixon and Newman,[42] Equations (57) and (58) may not be appropriate for small numbers of observations because these equations are based on the assumption that the survival estimate is normally distributed. Dixon and Newman[42] and Harrell[49] referred to other means of estimation for that situation.

The methods just described can be used to estimate the $S(t)$ for an exposed group of individuals. Two or more survival curves can also be examined for equality using these methods. Such methods can be used to test if mortality data generated from replicate tanks can be assumed to be homogeneous. They can also be used to test this same assumption for survival data from different treatments. Both the log-rank and Wilcoxon methods test if "the observed times of death in two (or more) samples come from the same survival distribution."[42] Both give similar results, but the Wilcoxon is more sensitive to deviations in deaths early in the exposure trial that the log-rank test. Both are described generally in Dixon and Newman.[42] The log-rank test is described in detail in Mantel[52] and Cox,[53] and the Wilcoxon test is described in Peto and Peto.[54] (Recently, an alternative weighted Kaplan-Meier statistic has been suggested as described by Pepe and Fleming.[55] Although it will not be used here, the reader is referred to this method if hazard plots cross each other for the two sets of data being compared.)

Example 6

The proportions of mosquitofish dying after 96 hours of exposure to various concentrations of NaCl were used in Examples 2, 3, and 4. Actually, the proportion dying at each salt concentration was derived for pooled numbers from duplicate tanks for convenience of the examples. Further, time-to-death data were collected, not just 96-hour mortality data. Was it appropriate to pool the duplicate tanks in the examples? Let's answer this question (testing for homogeneity) using the null hypothesis that the observed times-to-death in duplicate tanks came from the same survival distribution. The SAS code to implement the nonparametric procedures for the Wilcoxon and log-rank tests is the following:

```
DATA TOXICITY;
    INFILE "B: TOXICITY.DAT";
    INPUT TTD 1–2 TANK $ 4–5 PPT 7–10 WETWT 12–16 STDLGTH 18–20;
    IF TTD>96 THEN FLAG=1;
        ELSE FLAG=2;
    RUN;
PROC SORT;
    BY PPT TANK TTD;
RUN;
PROC LIFETEST;
    TIME TTD*FLAG(1);
    STRATA TANK;
    BY PPT;
RUN;
```

The results are tabulated below for the 10.3, 10.8, 11.6, 13.2, and 15.8 g NaCl/L duplicate tanks. Estimation of statistics for the 20.1 g NaCl/L treatment is impossible because the extremely rapid deaths resulted in insufficient points for the associated survival curve.

Example 6 *Continued*

Treatment	Test	χ^2	df	Associated P
10.3	log-rank	2.7926	1	0.0947
	Wilcoxon	2.8142	1	0.0934
10.8	log-rank	0.6140	1	0.4333
	Wilcoxon	0.7154	1	0.3976
11.6	log-rank	0.2320	1	0.6300
	Wilcoxon	0.1166	1	0.7327
13.2	log-rank	2.8338	1	0.0923
	Wilcoxon	4.4549	1	0.0348
15.8	log-rank	2.8358	1	0.0922
	Wilcoxon	2.8641	1	0.0906

With one exception in the ten tests, the null hypothesis could not be rejected ($\alpha = 0.05$). The χ^2 for the Wilcoxon test using the duplicate 13.2 g NaCl/L treatments had an associated probability less than 0.05 although that for the log-rank had an associated probability greater than 0.05. This is understandable because the Wilcoxon test was sensitive to deviations in deaths early in the exposure trial and there was a slight inequality in salt concentrations as the tanks filled initially. However, beyond the first few points on the survival curves, these curves were close to each other. Given these facts and the results of the log-rank test, the duplicate tanks were deemed sufficiently homogeneous to pool.

3. Parametric and Semiparametric Methods

a. General

Parametric and semiparametric methods are also available for the analysis of time-to-death data. Indeed, survival time methods incorporating specific models are common in health sciences, engineering, and economics.[43] Although their common origins with life table methods can be traced back to the European plague years of the 1660s,[48] the extensive use of survival (or failure) time methods did not begin until World War II.[13] Today, they are used to address diverse topics ranging from mechanical failure[47,56] to coronary disease risk factors[57] to cancer mortality.[58] More recently they have been introduced into ecological disciplines such as population genetics, e.g., Chapter 5 of Manly.[59] Unfortunately, they are only beginning to be used by aquatic toxicologists.[16,44-46]

Survival time models can take several forms. Proportional hazard models use the hazard of a reference group or type as a base hazard, and they scale (make proportional) the hazard of other groups to that baseline hazard. For example, the hazard of contracting lung cancer for smokers may be compared with the baseline hazard for nonsmokers. In contrast, accelerated failure models use functions that describe the change in ln time-to-death resulting from some change in covariates. Continuing the example, the effect of smoking on ln time-to-death may be estimated with an accelerated failure model. Both forms of survival models are described in the following text.

b. Proportional Hazard Models

i. Assuming a Specific Model

The general expression of a proportional hazard model is the following:

$$h(t, x_i) = e^{f(x_i)}h_o(t) \tag{59}$$

where $h(t, x_i)$ = the hazard at time t for a group or class x_i; $h_0(t)$ = the baseline hazard;

and $e^{f(xi)}$ = a function that relates the h (t, x_i) to the baseline hazard. The $f(x_i)$ is some function used to fit the data set. It can be used to fit continuous variables such as fish weight or class variables such as fish sex or type of treatment. A vector of coefficients and a matrix of covariates can be used if more than one covariate is incorporated into the model.

ii. Cox Proportional Hazard Model

The proportional hazard models described here assume that a specific distribution describes the baseline hazard and that hazards between classes are proportional. If the distribution is not apparent or if it is not desirable to select a specific distribution, a semiparametric method is available. The Cox proportional hazard model retains the assumption of proportional hazards but uses a family of Lehmann alternatives (see pages 598–601 in Kotz and Johnson[60]) to fit the baseline hazard.[16] Cox proportional hazard models are common in clinical studies because, in many cases, the underlying distribution is not as important as the relative hazards for the classes of interest.

c. Accelerated Failure Models

Another form of survival model is the accelerated failure time model.

$$\ln t_i = f(x_i) + \epsilon_i \tag{60}$$

where t_i = the time-to-death; $f(x_i)$ = a function that relates $\ln t_i$ to the covariate(s); and ϵ_i = the error term. In this case, the ln time-to-death, not the hazard, is modified by $f(x_i)$.

d. General Form for Survival Time Models

As described by Dixon and Newman,[42] the accelerated failure time model can be converted to the form of a proportional hazard model

$$h(t, x_i) = e^{f(x_i)}h_o(t, e^{f(x_i)}) \tag{61}$$

Dixon and Newman[42] gave examples of the typical types of regression functions that can be used for $f(x_i)$. For example, they included such functions as a linear equation ($a + bX$) for a continuous variable. Linear models for log-transformed variables or polynomial models may be used also. For class variables such as sex or treatment type, the function can simply estimate the mean response for each class. Candidate functions used to model the error distribution include the exponential, Weibull, Gompertz, normal, log normal, log logistic, and gamma distributions. Other, less common functions are tabulated on page 17 of Cox and Oakes.[43]

If the function selected for the error distribution is an exponential, Weibull, or Gompertz function, the model [Equation (61)] can take the form of a proportional hazard model [Equation (59)].[42] The only difference among the three is the way in which time-to-death is incorporated.

Exponential

$$\ln h(t) = a + bX \tag{62}$$

Weibull

$$\ln h(t) = a + bX + c \ln(t) \tag{63}$$

Gompertz

$$\ln h(t) = a + bX + ct \qquad (64)$$

where a, b, c = constants; and X = the independent variable.

The hazard is constant for the exponential model over the duration of exposure (time). For the Weibull distribution, the ln of the hazard either increases ($+c$) or decreases ($-c$) linearly over the duration of exposure. The hazard increases or decreases linearly with duration of exposure for the Gompertz model.[42,61]

e. Selecting the Appropriate Model

Given the variety of candidate functions that can be selected for formulation of survival models, the question becomes "Which function is the best?"

There are several means of addressing this question. Various linearizations are used for the exponential, Weibull, normal, log normal, and log logistic models.

Model	Y	X
Exponential	$\ln S(t)$	t
Weibull	$\ln [-\ln S(t)]$	$\ln t$
Normal	Probit $[F(t)]$	t
Log normal	Probit $[F(t)]$	$\ln t$
Log logistic	$\ln[S(t)/F(t)]$	$\ln t$

The transformed data are plotted for the candidate models. If a model is appropriate, the pertinent plot will produce a straight line. If, for example, there are several classes within the data set, e.g., fish exposed to seven different salt concentrations, curves for each of the seven subsets of fish can be plotted. If a range of individuals differing by a continuous variable is exposed, e.g., fish of different sizes, lines for arbitrary groupings (very small, small, medium, large) can be plotted. This approach can be limited if the number of individuals in the particular group being plotted is small. In that case, the variability in the plot can make it difficult to assess the linearity of the plots.

The log likelihood statistic generated by statistical packages that fit survival models can also be used in model selection. The log likelihood statistic can be used directly if the number of parameters estimated for each candidate model is the same. For example, the Weibull, log normal, and log logistic have the same number of parameters to be estimated; consequently, the associated log likelihood estimates for these three models could be compared directly much as the sums of squares are compared for models fit with least-squares methods. The model with the largest log likelihood value fits the data best. This same approach could also be used in testing various transformations of covariates to select the best transformation.

Harrell[49] and Dixon and Newman[42] described the likelihood ratio test for more formally assessing improvement of fit for nested models. Twice the difference in the log likelihood values for two models approximates a χ^2 distribution for samples with large numbers of observations. (Harrell[49] stated that this χ^2 statistic is roughly equivalent to the Pearson χ^2 statistic.) The χ^2 value for the models being assessed is compared with a critical χ^2 under the null hypothesis of no significant difference. (The associated degrees of freedom are calculated as the difference in the number of estimated parameters between the two models being compared.) Dixon and Newman[42] pointed out that this likelihood ratio test was only valid if one of the models is "nested" in the other. They gave an example of its use for deciding if a linear (oxygen concentration used as a continuous variable)

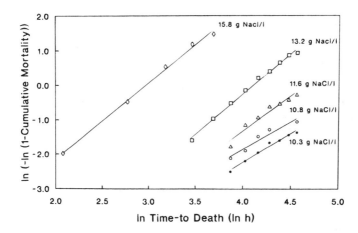

Figure 6 Linear transformation of times-to-death for mosquitofish exposed to five concentrations of sodium chloride (Example 7). The other exposure concentrations (<1 and 20.1 g NaCl/L) produced insufficient numbers of observations to be used in this plot. The generally straight and parallel lines suggest that the Weibull distribution was adequate for describing these data. (Modified from Newman and Aplin[16] with permission.)

or quadratic (oxygen concentration and the square of the oxygen concentration both used as continuous variables) model for $f(x)$ was the most appropriate for describing survival of oxygen-stressed fish.[35]

Akaike's information criterion (AIC) can be used to compare the appropriateness of a series of candidate models varying in complexity, i.e., number of parameters to be fit.[49,62] It adjusts the log likelihood values to account for the differences in the number of parameters available to fit the data set. This criterion, as applied here, has a different form than that given in Chapter 3.

$$\text{AIC} = -2(\text{log likelihood}) + 2P \tag{65}$$

where P = the number of parameters being fit with the model. The AIC values are compared for the various models; the model with the lowest AIC has the best fit.

It must be noted that the methods described above indicate the relative fits of different models to the data set. It is possible that none of the candidate models fit the data. The adequacy of the selected model should always be examined by plotting the predicted and observed data together.

Example 7

The sodium chloride toxicity data generated by Newman and Aplin[16] can now be fully analyzed using survival time modeling methods. Both covariates (fish wet weight and salt exposure concentration) will be incorporated. To illustrate various aspects of the process, the data will be fit with untransformed covariates and then with transformed covariates.

Untransformed Covariates

What distribution should be assumed for the error? The series of linearizations listed earlier in this chapter can be used during the first attempts of addressing this question. $F(t)$ and $S(t)$ are estimated by the cumulative mortality (number dead/total number) and cumulative survival (number surviving/total number) for all times up to the termination of the exposure. The appropriate transformations of these metameters and time (duration

Example 7 *Continued*

of exposure) are then plotted. Figure 6 shows the results for the Weibull model linearization. This linearization seemed to be "best" as judged by visual comparison with those for the other distributions. (Note that no linearization is available for the gamma distribution.) The log likelihood statistic can also be used to compare candidate models if they are estimating the same number of parameters. As the models vary in the number of parameters being estimated, the AIC [Equation (65)] is used. The following SAS code fits these data to survival models assuming exponential, Weibull, log normal, log logistic, and gamma distributions.

```
DATA TOXICITY;
    INFILE "B:TOXICITY.DAT";
    INPUT TTD 1–2 TANK $ 4–5 PPT 6–10 WETWGT 12–16;
    IF PPT>0;
    IF TTD>96 THEN FLAG=1;
        ELSE FLAG=2;
RUN;
PROC SORT;
    BY PPT WETWGT TTD;
RUN;
PROC LIFEREG;
    MODEL TTD*FLAG = PPT WETWGT/DISTRIBUTION=EXPONENTIAL;
    MODEL TTD*FLAG = PPT WETWGT/DISTRIBUTION=WEIBULL;
    MODEL TTD*FLAG = PPT WETWGT/DISTRIBUTION=LNORMAL;
    MODEL TTD*FLAG = PPT WETWGT/DISTRIBUTION=LLOGISTIC;
    MODEL TTD*FLAG = PPT WETWGT/DISTRIBUTION=GAMMA;
RUN;
```

The models generated the log likelihood values tabulated below. The AIC was calculated using the log likelihood statistic and the number of parameters being estimated. (How one determines the number of estimated parameters will be discussed.)

Distribution	Log Likelihood	Number of Parameters	AIC
Exponential	−385	3	776
Weibull	−194	4	396
Log normal	−202	4	412
Log logistic	−198	4	404
Gamma	−193	5	396

The AIC is smallest for the Weibull and gamma distributions, suggesting that they both provide better fits to the data set than the other distributions. Both the Weibull and gamma distributions are generalized exponential functions. However, because the Weibull distribution results in a proportional hazard model, it is used here to illustrate several points. The resulting proportional hazard (Weibull) model follows.

Variable	df	Estimate	(Standard Error)	χ^2	$P > \chi^2$
Intercept, μ	1	7.8579	(0.0853)	8487	< 0.0001
[NaCl] β_s	1	−0.2953	(0.0052)	3258	< 0.0001
Wet Wgt β_w	1	1.0602	(0.2566)	17	0.0001
Scale, σ	1	0.3046	(0.0137)		

Example 7 *Continued*

For this model, four parameters (μ, β_s, β_w, σ) are estimated. The χ^2 estimates for β_s and β_w suggest that the salt concentration and fish wet weight have significant effects on time-to-death ($\alpha = 0.05$). The negative sign associated with β_s indicates that the time-to-death decreases as the salt concentration increases. The positive sign for β_w indicates that time-to-death is shortened as fish wet weight decreases.

The median time-to-death (MTTD) can be estimated with this proportional hazard model (Equation (59)).

$$\text{MTTD} = e^{\mu}e^{\beta_w Wgt + \beta_s [NaCl]}e^{\sigma W} \tag{66}$$

where wgt = wet weight of fish (g); $[NaCl]$ = salt concentration (g NaCl/L); and W = response metameter for Weibull with $P = 0.50$ (Table 7 in the Appendix).

The median time-to-death can be estimated with this model for any salt concentration or fish wet weight within the ranges used to generate the model (Figure 7). The median time-to-death (and the associated standard error) for the average size fish (0.136 g) can be calculated within the range of salt concentrations tested (Figure 8). The resulting curve of median time-to-death versus salt concentration is consistent with the trimmed Spearman-Karber 96-hour LC50 estimate calculated in Example 4. Values for W for other proportions can be used to estimate times-to-death other than that of the median. The 96-hour LC50 may also be estimated directly from the model.

$$96h\ LC50 = \frac{\ln 96 - \mu - \beta_w Wgt - \sigma\ W}{\beta_s} \tag{67}$$

For a fish of average weight (0.136 g), the 96-hour LC50 is estimated to be 11.26 g NaCl/L.

With this proportional hazard model, relative risk can be estimated for fish of various sizes or for fish exposed to various salt concentrations.[42] Newman and Aplin[16] gave the following example (incorrectly). The relative risk of a 0.10-g fish to that of a 1.0-g fish is $e^{-\beta_w(\Delta Wgt)/\sigma}$ where ΔWgt is the difference in weight of the fish. With the Weibull model, the relative risk is $e^{-1.0602(0.9\ g)/0.3046}$ or 22.9. The smaller fish has a risk 22.9 times higher than the larger fish. Similarly, for an increase in salt concentration of 10 g/L, the relative risk for mosquitofish is $e^{-\beta_s(\Delta [NaCL])/\sigma} = 16{,}231.053$. Risk increases approximately 16,231 times with a 10 g/L increase in salt concentration. Further, the risk of a 0.1-g fish at 20 g/L is 22.7*16,231 or 371,690 times greater than a 1.0-g fish at 10 g/L.

Transformed Covariates

The data analysis in Newman and Aplin[16] ended with the above calculations. The covariates, salt concentration, and fish wet weight remained untransformed. However, from the material covered to this point, one could argue that a better concentration metameter to use may have been ln of concentration. Also, as will be discussed soon, one could easily have argued for the use of ln of wet weight, not wet weight. Let's use these arguments to illustrate how one might assess appropriate transformations of covariates using the AIC.

The SAS LIFEREG code given previously was run three additional times using various transformation combinations of wet weight and salt concentration: ln wet weight and g NaCl/L, wet weight and ln g NaCl/L, and ln wet weight and ln g NaCl/L. The first two combinations yielded no improvement in the AIC values relative to the models using

Example 7 *Continued*

wet weight and g NaCl/L. However, the AIC values were improved through the use of logarithms for both covariates (exponential, 758; Weibull, 396; log normal, 380; log logistic, 372; gamma, 380) (Figure 9). The log logistic and log normal models both had AIC values that dropped approximately 30 units when the transformed variables were used. The log logistic model incorporating the transformed variables had the lowest AIC of all models examined. Although the original Weibull model was adequate for the intended purpose, the AIC indicates that other models, e.g., the log logistic model using transformed covariates, are more appropriate for these data. The predicted median times-to-death for a 0.136-g mosquitofish over a range of NaCl concentrations (Figure 8) compares more favorably for the log logistic model with transformed covariates than those for the Weibull model to the widely accepted, trimmed Spearman-Karber 96-hour LC50 value. (Equations (66) and (67) can be used provided the transformed covariates are used and the appropriate "W" parameter for the logistic model is used.)

IV. QUANTIFYING THE EFFECTS OF EXTRINSIC FACTORS

A. OVERVIEW

Quantitative treatment of some of the more common external modifiers of toxic response are presented in this section. Such factors can be incorporated into survival time analyses as demonstrated earlier with salt concentration. Although performed infrequently, they can also be incorporated into endpoint methods such as probit and logit models fit with maximum likelihood methods. For example, the PROBIT procedure in the SAS package[20] incorporates covariates. However, the ability to incorporate covariates effectively is limited with endpoint models relative to survival time models because less information is extracted from the mortality trial. For example, only six data pairs (proportion dead at 96 hours, salt concentration) were used in the probit procedure to estimate the LC50 for the mosquitofish toxicity data. Hundreds of data pairs (time-to-death, salt concentration) were used in analyzing the results of this toxicity test with survival time models.

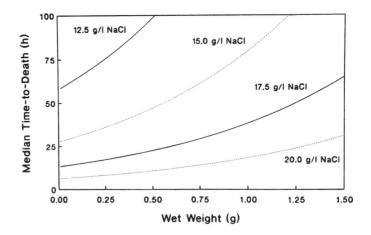

Figure 7 Predicted median times-to-death for mosquitofish exposed to various salt concentrations. Predictions are made across a wide range of fish sizes (wet weight). See Example 7 for further details. (Modified from Newman and Aplin[16] with permission.)

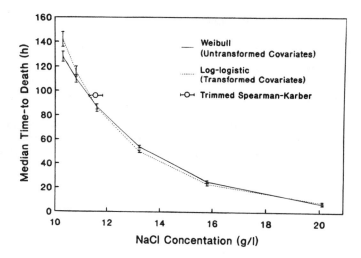

Figure 8 Comparison of the trimmed Spearman-Karber estimate of 96-hour LC50 to concentrations corresponding to predicted median times-to-death of 96 hours using the Weibull model with untransformed variable (salt concentration and wet weight) and log logistic model with transformed variables (In salt concentration, In wet weight). See Example 7 for further details. (Modified from Newman and Aplin[16] with permission).

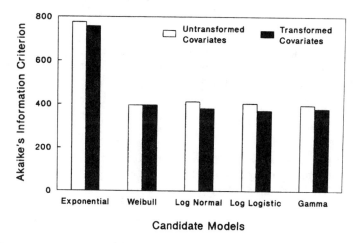

Figure 9 Model selection for sodium chloride-mosquitofish toxicity data using the Akaike's information criterion. See Example 7 for more details.

Many of the factors discussed influence toxic impact by shifting the distribution of the total amount of toxicant among different chemical species and physical phases. Here, they are presented in the context of toxic impact.

B. INORGANIC TOXICANTS
1. Ammonia

The influence of pH and temperature on ammonia toxicity provides a clear illustration of modification of toxic impact by extrinsic factors. Shifts in pH strongly influence ammonia speciation by modifying the equilibrium concentrations of non-ionized (NH_3) and ionized ammonia (NH_4^+).

$$NH_3 + nH_2O \rightleftarrows NH_3 \cdot nH_2O \rightleftarrows NH_4^+ + OH^- + (n - 1)H_2O.$$

Temperature and ionic strength also influence the distribution defined by this equation but to a lesser extent.[63-65]

The non-ionized ammonia seems to be the most toxic form, notionally because it easily diffuses across gills. Ionized ammonia is subject to much slower transport.

An increase in pH shifts the equilibrium to the left in the above equation and, consequently, increases toxic impact. In the range of pH values found in fresh water, an increase of one pH unit is estimated to increase the non-ionized ammonia concentration tenfold. Emerson et al.[66] provided the following convenient method for deriving an estimate of non-ionized ammonia for various pH and temperature conditions at zero salinity.

The ionization constant (K_a) for the abbreviated equation, $NH_4^+ \leftrightharpoons H^+ + NH_3$ is the following,

$$K_a = \frac{[H^+][NH_3]}{[NH_4^+]} \tag{68}$$

Empirical fit of data yielded the following regression line for pK_a.[66] (Remember that $pK_a = -\log K_a$.)

$$pK_a = 0.09018 + \frac{2729.92}{T} \tag{69}$$

where T = temperature in degrees Kelvin (degrees Celsius + 273.15). The percentage of the total ammonia $(NH_3 + NH_4^+)$ present as unionized ammonia is estimated at a specific pH by using the pK_a from Equation (69) in Equation (70).

$$\text{Percent} = \frac{100}{(10^{pK_a - pH} + 1)} \tag{70}$$

Equations (69) and (70) are appropriate for estimating percent of total ammonia present as non-ionized ammonia within the temperature range of 0°C to 30°C, and pH range of 6.0 to 10.0. Because Equation (69) is an empirical relationship, this method cannot be used outside of conditions under which the regression line was generated.

Example 8

What is the change in the percentage of total ammonia present as non-ionized ammonia if pH is increased from 7.0 to 7.9 and temperature is increased from 15°C to 28°C?

a. Percentage at pH = 7.0 and temperature = 15°C.

pK_a estimation using Equation (69):

$$pK_a = 0.09018 + 2729.92/(15 + 273.15) = 9.56413$$

Estimation of percentage non-ionized ammonia from Equation (70):

$$\text{percent} = 100/(10^{9.56413 - 7.0} + 1) = 0.272\%$$

Example 8 *Continued*

b. Percentage at pH = 7.9 and temperature = 28°C.

$$pK_a = 0.09018 + 2729.92/(28 + 273.15) = 9.15516$$
$$percent = 100/(10^{9.15516-7.9} + 1) = 5.264\%$$

c. The percentage of non-ionized ammonia shifted upward from 0.272% to 5.264% or approximately 5% with the change in pH and temperature.

Under the assumption that non-ionized ammonia is the primary toxic species, total ammonia concentrations can be transformed to non-ionized ammonia concentrations before using as a metameter in the methods described above. In doing so, the conditions under which the relationships were fit must be satisfied. Also, Thurston et al.[64] indicate that ionized ammonia can also have a small but significant impact on ammonia toxicity.

2. Metals
a. *Water*
Many factors modify dissolved metal and metalloid toxicity. Examples include the effects of dissolved oxygen,[67] major cations,[68] and phosphate[69] and other ligands. Understanding metal speciation is critical for predicting the influence of many of these extrinsic factors. However, as discussed in the next section, an understanding of speciation is not sufficient. In general, the aquated or "free" metal ion is assumed to be the most toxic form, although other metal-ligand complexes may also be toxic.[70,71] Consequently, speciation is estimated in recent studies of metal toxicity with emphasis on the aquo metal ion. Further, current practice dictates that the chemical composition of waters used for toxicity tests be clearly defined so that speciation can be estimated later. Several programs have been developed to facilitate estimation. Loehle et al.[72] reviewed the general approach and accuracy of several of the more common models. Most are valuable tools if their limitations relative to bioavailability and toxicity are understood. Alternatively, analytical methods may be used to estimate speciation, e.g., specific ion electrodes such as that for the cupric ion. They also provide valuable information regarding the influence of water quality on toxic impact of inorganic toxicants.

Many aspects of metal and metalloid speciation remain impossible to predict accurately. The poorly quantified role of natural organic ligands, poorly defined chemical processes occurring at biological surface microlayers, and the ambiguous importance of chemical kinetics remain problematic. Realizing that a complete knowledge of speciation and associated ramifications is extremely difficult to obtain, empirical methods have been developed to quantify the influence of extrinsic factors on metal toxicity. Today, these methods are used effectively in a variety of applications, including water quality criteria and standard development. Their eventual replacement with models based on first principles remain a major goal in aquatic toxicology.

Perhaps the best example of an empirical relationship of this type is that between water hardness and metal toxicity. Hardness has a strong effect on the toxic impact of dissolved metals such as beryllium,[73] cadmium,[74-76] copper,[77] and zinc.[78] Hardness influences metal toxicity by one or several of the following mechanisms: competition with the toxic metal for biological binding sites, modification of biological processes, or modification of speciation or solubility. Because hardness is often correlated with alkalinity and ionic strength, indirect effects such as shifts in buffering capacity or ligand complexation also have been suggested.

Empirical relationships between metal toxic effect and water hardness commonly involve log transformations of hardness and toxic impact, e.g., see Slonim and Slonim,[73]

Howarth and Sprague,[77] and Nelson et al.[79] This approach is used in water quality criteria for several metals, e.g., see the U.S. EPA criteria documents for copper, cadmium, or zinc.[80-82] Least-squares regression is used to fit the line,

$$\log \text{LC50} = b\log H + \log a \tag{71}$$

where H = hardness; b = the regression slope; and $\log a$ = the regression intercept. Next, the model is backtransformed to the form,

$$\text{LC50} = aH^b \tag{72}$$

However, as discussed in Chapter 3, such a backtransformed model generates biased predictions of mean toxic effect (LC50).[83] This bias arises because the associated error term in the regression model is neglected during the backtransformation. The complete model is

$$\log \text{LC50} = a\log H + \log a + \epsilon \tag{73}$$

where ϵ = the random error term in the model. The error term is incorporated into the complete backtransformed model.

$$\log \text{LC50} = aH^b 10^\epsilon \tag{74}$$

The term 10^ϵ can be estimated using the model error mean square if the residuals are normally distributed.

$$10^\epsilon = 10^{\frac{\text{MSE}}{2}} \tag{75}$$

where MSE = the model mean square error.

If the residuals are not normally distributed, a "smearing estimate of bias"[84] can be used.

$$\text{LC50} = aH^b \left[\frac{1}{N} \sum_{i=1}^{N} 10^{r_i} \right] \tag{76}$$

where N = the number of observation pairs; and r_i = the ith regression residual.

Newman[83] provided examples of this bias using data from U.S. EPA criteria documents. The bias ranged from 2% to 57% for the 12 sets of data examined. (An example of backtransformation bias correction is given in Chapter 3 relative to anthracene elimination kinetics.)

Unless the error term (ϵ) is 0, the LC50 predicted with Equation (72) will not be the mean predicted value of the median lethal concentration (LC50). As discussed in Chapter 3, the median of the median lethal concentration is predicted with Equation (72).[85] If prediction of the median is acceptable or desirable, the above-mentioned bias is not pertinent. However, it should be clear that such a decision has been made because the tendency is to assume that the mean value is predicted. The bias should be estimated if prediction of the mean response is desired. Minimally, the error mean square should be included in any publication reporting such a regression model. This allows future correction if desired.

Before moving our discussion to sediment-associated toxicity, it must be reiterated that many other extrinsic factors influence the toxic impact of dissolved metals and metalloids. Temperature,[86] pH,[87] and salinity[88,89] are three important examples. Unfortunately, much work remains to be done to quantify such effects in terms other than simple chemical speciation.

b. Sediments

Sediment toxicity received much attention during the past decade, e.g., see Cairns et al.,[90] Nebeker et al.,[91] DeWitt et al.,[92] and Burton.[93] Speciation in particles and interstitital water received much deserved attention as sediment toxicity protocols and methods were developed. Despite these efforts, major obstacles impede our understanding of sediment chemistry. Luoma[94] recently acknowledged this lack of essential information regarding sediment speciation and the consequences of this ignorance to prediction of sediment toxicity. Improved methods for analyzing metal distribution among sediment components, improved computational methods for predicting sediment-water exchange, and a better understanding of processes controlling bioaccumulation from solution and food are identified as areas in which the necessary knowledge is clearly lacking.

Although our present knowledge is limited, several quantitative approaches suggest themselves to a researcher interested in quantifying toxic effects of metal-contaminated sediments. Several procedurally defined sediment fractions modify bioavailability of sediment-bound metals in surficial or oxic sediments. Easily extracted iron (1 N HCl or an equivalent extractant) or hydrous iron oxides tend to decrease bioavailability of silver, arsenic, cobalt, lead, and zinc.[95-101] Manganese oxides can also lessen metal bioavailability.[101] Although the organic content of sediment can decrease bioavailability,[102] the extent to which bioavailability is influenced varies widely between the benthic species-metal combinations. Finally, when sequential extractions of sediments are performed, the most readily extracted sediment fractions are generally found to be the most available fractions.[99,101]

As a consequence of these studies, empirical relationships between sediment-bound metals and bioaccumulation in the associated fauna have been developed between accumulated metal and some sediment fraction, e.g., metal in an EDTA extract[103] or exchangeable fraction.[104] Alternatively, because certain sediment qualities influence bioavailability, an empirical relationship can be developed for accumulated metal and normalized sediment metal concentrations, e.g., amount of metal divided by the amount of 1 N HCl extractable iron.[96,98,105] Although these techniques often improve quantitation of metal bioavailability from oxic sediments, it is important to understand that they are empirical relationships, i.e., extrapolation beyond the data set used to generate the relationship is not strictly valid. Also, there is considerable variability in the literature regarding these methods. Although these methods often generate concentration metameters superior to total concentrations of metals, they are still imperfect tools.

Under anoxic conditions, metals in sediments are present primarily as highly insoluble sulfides.[106,107] Consequently, toxicities of metals in anoxic or near anoxic sediments are strongly influenced by S^{2-}.[106,108,109] For this reason, normalization of metal concentrations to sulfide concentration has been used for estimating toxic impact in anoxic sediments.[107,110,111] Solid phase sulfides in sediments are estimated with acid-volatile sulfides (AVS). Procedurally, AVS are the sulfides extractable with cold HCl. They are assumed to be primarily iron and manganese sulfides. The molar concentration of metals extracted in cold HCl [simultaneously extracted metals (SEM)] can be divided by the molar concentration of AVS to yield a normalized estimate of metal available in anoxic sediments to have a toxic impact. For stoichiometric reasons, values of SEM/AVS < 1 imply that all the metal is precipitated with the sulfides and unavailable for toxic action. (The metal in the interstitial water is assumed to be the metal available to have a toxic impact. Those associated with solid phases are not considered. Ingestion and consequent bioaccumulation of solid-phase metals have not yet been considered in this approach.) These methods can be used to enhance quantitation of toxic effect of metals and metalloids in sediments. However, they are empirical relationships using procedurally defined quantities and, as such, they must be used with care. Considerable work remains to be done in this area of aquatic ecotoxicology. Much of the underlying chemistry remains only generally defined.

C. ORGANIC TOXICANTS

Chemical factors such as water hardness and pH modify organic toxicant effects.[112] Physical factors can also modify toxicity. For example, light can modify the toxic impact of photo-labile, organic contaminants that degrade to toxic products, e.g., anthracene.[113] Perhaps the most extensively quantified, extrinsic factors examined to date are the qualities of organic toxicants themselves. The correlation between bioaccumulation (or toxic effect) and K_{ow} is one familiar example (see Chapter 3). Simple empirical models have been generated to predict K_{ow} using known qualities of the parent and substituent structures in an organic chemical.[114,115] These relationships are very useful in predicting the toxic impact of many classes of organic compounds although other factors, e.g., electrophilic qualities of the toxicant[116] or attachment to electron-withdrawing functions on the molecule[115] often must be considered as well. Relationships linking the molecular structure and physical properties of a compound to its biological activity are called structure-activity relationships (SARs). Quantitative structure-activity relationships (QSARs) are SARs that express relationships in quantitative terms. They "are statistical models that are nearly always obtained by regressing values of a common test endpoint for a series of chemicals against one or more quantifiable properties of the chemicals."[116] Properties commonly used are hydrophobic, (e.g., K_{ow}), topological (e.g., molecular connectivity index), electrical (e.g., Hammett constants and ionization potentials), or steric (e.g., total molecular surface area). The QSAR approach, used for many years in drug development, is now producing invaluable empirical relationships in aquatic ecotoxicology. Attempts are at hand to link such relationships to even more basic molecular qualities.[117]

One illustrative approach to quantifying the relationship between toxicity and structure of organic compounds is the additive constitutive approach. This approach estimates the influence of structural components on the lipophilic (hydrophobic) qualities of the toxicant. The lipophilicity is assumed to dictate narcotic activity through its influence on membrane penetration, bioaccumulation in target organs, and binding to proteins and to other cell constituents.[117] The contribution of a structural constituent of the toxicant is estimated using the logarithms of the K_{ow} values for the parent molecule and the parent molecule with the structural component added. A parameter (π_x) is estimated.[118]

$$\pi_X = \log K_{owX} - \log K_{owH} \qquad (77)$$

where π_X = the logarithm of the "partition coefficient" for the structure added to the parent molecule[115]; K_{owX} = the K_{ow} for the derivative compound; and K_{owH} = the K_{ow} of the parent compound.

Tabulations of these π_X values can be used to estimate the K_{ow} of a specific compound. Hansch and Leo[118] provide tabulations of π_X values. A K_{owH} is found from these or other tables for the parent chemical. The π_X values are found for each substituent added to that parent. According to additivity principles, the logarithm of K_{ow} for toxicant can then be estimated as the logarithm of the K_{ow} for the parent molecule plus the sum of the pertinent substituent π_X values.

$$\log K_{ow} = \log K_{owH} + \sum_{i=1}^{n} \pi_i \tag{78}$$

The substituent addition to the parent compound favors the water phase if the Hansch π_X parameter has a negative sign. A positive parameter shifts the distribution in favor of the octanol phase.

This estimate of K_{ow} can be correlated with toxic activity[119] or bioavailability.[120] Toxicity for untested compounds within the class used to establish the QSAR can be predicted with such relationships.

Similarly, molecular topology, electronic, or steric parameters may be used to establish empirical structure-activity relationships.[116,117] Several characteristics may be used together in more complex QSARs.[116] For example, a combined approach incorporating measures of hydrophobicity and degree of disassociation of an active group may be used for transport and toxic action of a weak electrolyte, a substance in equilibrium between non-ionized and ionized forms.[121] Hansch and Leo[118] provided general explanations of the basic types of QSARs. Suter[116] provided a table of QSARs used by the U.S. EPA Office of Toxic Substances. It is strikingly apparent from Suter's table that hydrophobicity is used as the sole or primary quality of interest in most of these QSARs.

Laughlin et al.[119] provided an illustration of QSAR application for a class of similar compounds with a notionally identical mode of action. They produced sound regression models of constituent π_X sums versus toxicity to mud crab zoeae for di- and triorganotin compounds. The sum of the constituent π_X values explain 94% to 95% of the variance in toxicity among organotins. Computer estimations of total molecular surface area for these compounds were used to produce QSARs also. Differences in the total surface area of the various organotins explained 93% to 95% of the variance in toxic impact. These models using total surface areas and Hansch π parameters suggested to Laughlin et al.[119] that partitioning was the best predictor of toxic impact for the organotin antifouling agents. This confirmed earlier work by these authors suggesting that electronic factors had little influence on toxicity. (Hansch and Leo[118] provided a general introduction to electronic parameters such as the σ constants used by Laughlin et al.[119])

Example 9

A study of K_{ow} effects on chlorinated phenol toxicity to freshwater fish produced the relationship, log (1/LC50) = 0.58log K_{ow} − 3.20 (Table 7.1 in Suter[116]). Predict the 96-hour LC50 for 2,4-dichlorophenol using this relationship. First, calculate the predicted K_{ow} for 2,4-dichlorophenol assuming constituent additivity and a phenol parent structure to which the two Cl atoms are bound. Then use the above QSAR to estimate the 96-hour LC50.

The median K_{ow} for phenol calculated from values in Appendix II of Hansch and Leo[118] is 1.485. A π_{Cl} (0.71) is taken from Table VI-1 in Hansch and Leo.[118] These values are used in Equation (78) to estimate the log K_{ow} for this compound. The sum is 1.485 + 0.71 + 0.71 or 2.905. The K_{ow} for this compound is predicted to be the antilogarithm of 2.905. (Aside: The median of the three values given in Appendix II of Hansch and Leo[118] for log K_{ow} of 2,4-dichlorophenol is 3.08.)

The LC50 can now be estimated using the calculated K_{ow}.

log (1/LC50) = 0.58*log K_{ow} − 3.20
log (1/LC50) = 0.58*2.905 − 3.20
log (1/LC50) = −1.5151
1/LC50 = $10^{-1.5151}$ or 0.03054
LC50 = 32.74 or approximately 33 μmol/L

Assessment of the toxic impact of organic compounds in sediments may also take advantage of QSARs. Di Toro et al.[122] used K_{ow} as a measure of hydrophobicity of nonionic organic chemicals in their treatment of sediment toxicity. They assumed that interstitial water concentrations determine the availability and toxicity of associated chemicals. The partitioning of an organic compound among the interstitial water, organic carbon in the sediment solid phases, and the organism was assumed to determine its toxic impact. The toxicity of the organic compound in the interstitial water (as determined by partitioning) was shown to be similar to that of the chemical dissolved in the overlying water. Consequently, sediment toxicity can be predicted from routine toxicity data for the dissolved compound if partitioning between the sediment organic carbon phase and the interstitial waters is quantified [Equation (79)].

$$K_p = \frac{C_s}{C_d} = f_{oc} K_{oc} \qquad (79)$$

where K_p = partition coefficient; C_s = concentration in the sediment; C_d = concentration in the interstitial water; K_{oc} = partition coefficient for the organic carbon phase of the sediments (assumed equal to K_{ow}); and f_{oc} = the mass fraction of organic carbon in the sediments.

The toxic impact of the nonionic organic compound can be estimated with knowledge of the sediment organic carbon content, toxicity of the organic compound when dissolved in water, and its K_{ow}. Implied by Equation (79) is the potential for toxic impact normalization based on the organic carbon content of sediments. Relative to toxic impact, sediment concentrations of nonionic organic compounds in sediments can be expressed more accurately in terms of amount/unit organic carbon than in terms of amount/unit sediment mass. Such a concentration metameter or a transformation of this metameter could be used in any of the above-described endpoint and survival time methods.

V. QUANTIFYING EFFECTS OF INTRINSIC FACTORS

A. OVERVIEW
Only two examples of common intrinsic factors will be discussed for the sake of brevity. Other important factors such as stress hormone response to photoperiod or circadian rhythms[123] will not be discussed in balance with their importance as they are normally controlled rather than quantified. Although the effects of an individual's sex have been quantified[42,44,46,124] and can be striking,[125] they will not be discussed per se as they were discussed briefly in Example 7.

B. ACCLIMATION
The term acclimation, as used by aquatic toxicologists, may have several different meanings. Physiological acclimation refers to an adaptive change in the individual in response to a change in environmental conditions. More specifically, physiological acclimation often refers to shifts taking place under controlled laboratory conditions and acclimatization refers to those taking place under natural conditions. Acclimation may also be used to refer to the time that an organism spends in an exposure system before addition of toxicant, i.e., acclimation to test conditions. Herein, acclimation is used to "refer to a nonlethal exposure and the physiological responses thereto regardless of the effect of these factors on tolerance to subsequent exposure."[126]

Several approaches have been taken to quantify acclimation. For example, Shepard[35] quantified physiological acclimation of trout to low-oxygen conditions. Acclimation involved changes in the oxygen binding capacity of the blood. The incipient lethal level

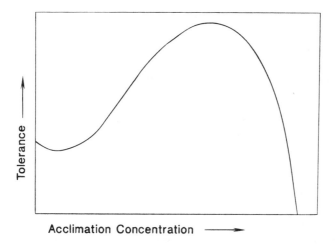

Figure 10 The influence of toxicant acclimation concentration on toxicant tolerance. Toxicant tolerance is expressed here as the difference between the copper incipient lethal level of nonacclimated and acclimated individuals. (Reprinted from Chapman[126] with permission. Copyright ASTM.)

(ILL, mg O_2/L) was linearly related to the acclimation level of oxygen (A, mg O_2/L): ILL $= 0.086A + 0.87$. He estimated the ultimate incipient lethal level (the lowest concentration to which fish may be acclimated and still experience 50% or less mortality) by solving this relationship for ILL $= A$.

Chapman[126] described the general toxic response model incorporating acclimation to metals (Figure 10). This figure was based primarily on LC50 data. There is a general zone similar to that described above for oxygen-acclimated trout within which increasing acclimation concentrations increases the LC50. There is a maximum acclimation concentration above which tolerance decreases. There is also a lower limit below which tolerance decreases. This curve depicts the net results of decreasing tolerance due to damage and response to the metal that enhances tolerance. Induction of changes that lessen toxic impact can have a threshold concentration below which no response occurs, although damage still occurs. This produces a decrease in tolerance at low acclimation concentrations. The upper limit of tolerance is associated with the increasingly significant damage incurred during exposure to acclimation concentrations.[126] Dixon and Sprague[127] observed such a curve when examining copper acclimation by rainbow trout (*Oncorhynchus mykiss*). They used the change in the incipient lethal level for acclimated and nonacclimated trout as the response variable.

Dixon and Sprague[127] also included acclimation time in their treatment of copper toxicity to rainbow trout. By varying acclimation time (0–21 days) as well as acclimation concentration, they derived a response surface for change in tolerance. They described a response surface resembling a ski jump. (Figure 10 with the added dimension of time with tolerance increasing to a plateau at higher acclimation times.) They described the surface quantitatively with a multiple regression model. It incorporated log acclimation concentration (A), a transformation of acclimation time (T), A^2, A^3, T^2, T^3, AT, A^2T, and AT^2.

As suggested by Figure 10, not all preexposures produce enhanced tolerance. Although acclimation to arsenic enhanced tolerance to that metalloid, Dixon and Sprague[128] found a decreasing cyanide tolerance after 7 to 14 days of cyanide preexposure. They attributed this to significant kidney damage at all sublethal, preexposure concentrations.

C. SIZE

Early efforts in drug administration saw drug dosage (the amount of a substance given per unit weight or some other normalizing unit) calculated as directly proportional to an individual's mass, e.g., 3 mg drug/kg of body weight. Moore, as cited in Bliss,[129] argued that dosage should be adjusted to the amount of surface available for drug adsorption. He suggested use of the 2/3 power of body weight to facilitate scaling to absorptive surface. Bliss[129] argued against using a constant factor and developed means of determining empirical constants for each situation. His work is a foundation paper for quantification of size effects on toxicant or drug action.

Bliss used Campbell's time-to-death data for arsenic given orally to silkworm larvae[130] to formulate his approach. First, to combine data from various dosages, he minimized the variance between dosages by converting times-to-death to rates of toxic action (1000/minutes of survival). Next, he used analysis of covariance to demonstrate that toxic action was influenced by animal size (weight) as well as the dosage. The smaller animals were more sensitive to the toxic action of arsenic. He then proceeded to examine various functions of dose (the amount of a substance given to an individual) and body weight that provided convenient linear relationships.

$$f(t) = a + b_1 f(m) - b_2 f(w) \tag{80}$$

where $f(t)$ = some function of survival time t; $f(m)$ = some function of the mass of arsenic in the dose; $f(w)$ = some function of the individual weight; and a, b_1, b_2 = regression constants.

To linearize the effect of dose, the logarithm of dose was plotted against the rate of toxic action (Figure 11A.) He judged that the relationship was linear despite significant variability. The residuals were judged to be normally distributed after visual inspection. The residuals (deviations in rate) from the regression model of log dose versus rate of toxic action were then plotted against log body weight. A linear function was visually determined as adequate to describe the effect of body weight on toxic action (Figure 11B). The following model for size effects was generated.

$$y = a + b_1 \log m - b_2 \log w \tag{81}$$

where y = 1000/time-to-death. This model can be rearranged to estimate the size-specific dosage.

$$y = a + b_1 \left[\log m - \frac{b_2}{b_1} \log w \right] \tag{82}$$

$$y = a + b_1 \log \frac{m}{w^h} \tag{83}$$

where $h = b_2/b_1$. The size factor (h) was relatively constant over several silkworm instars.

Bliss' work remains unquestionably a major contribution. However, several points should be made regarding its general application. Several toxicologists[125,131-133] have used examples to point out that other types of relationships occur. They advised that caution be used before assuming any general law during data analysis. Misapplication can increase variability and confound data interpretation.[125,134] Although Bliss's relationship can be used effectively,[124,134] other relationships should be considered.

Often, smaller individuals are more sensitive than larger individuals,[46,135] but there are toxicant-species pairs for which larger animals are most sensitive[136] or for which

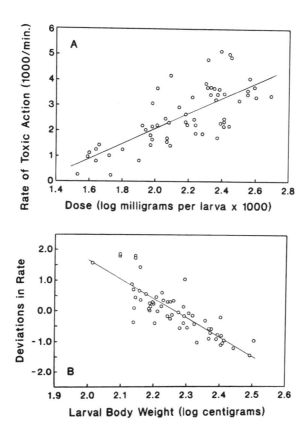

Figure 11 Bliss developed an approach to incorporating animal size into expressions of toxic effects using arsenic toxicity to silkworm larvae. Panel A is a plot and regression line derived from dose rate of toxic action data. Panel B is a plot and associated regression line for the residuals from the first regression line (dose versus rate of toxic action) versus body weight. See the text for more details. (Modified from Figures 1 and 2 of Bliss[129] with permission from the Company of Biologists, Ltd.)

there is no size dependence.[137] Further, size can have an effect on one measure of toxic effect but not on another, e.g., size-dependent rate of mortality but size-independent incipient LC50.[138]

Today, the imperative to transform data to linearize relationships does not exist. In attempting to linearize the relationships, some of Bliss's judgments regarding residual distributions may have been optimistic, especially when the variability about the predicted values is considered. Several of these points can be addressed with computational software such as that used in Example 7. Even if found true for arsenic toxicity to silkworm larvae, the assumption of a normal distribution for the distribution of values for toxic action for all size-toxic impact relationships remains untested. To be of general use, the methods advocated by Bliss must be modified to accommodate censoring (survivors or individuals removed during the experiment). Considering the magnitude of size effects and the variation in size among and within populations, it is surprising that so much refinement of methods remains to be done.

Anderson and Weber[124] extended Bliss's approach to prediction of time endpoints for fish exposed to toxicants dissolved in water. Toxic response of guppies (*Poecilia reticulata*) to dieldrin, potassium pentachlorophenate, potassium cyanide, copper chloride, zinc chloride, and nickel chloride were used. Beginning with Equation (83), they used endpoint data (probits of proportion dead) as the dependent variable. The concentration of toxicant dissolved in the water was used as m. Wet weight (w) was used as a measure of animal size. Multiple regression was used as described previously to estimate h.

They also connected this approach with the common allometric model of LC50 for animal size. The allometric equation begins with the transformed regression model.

$$\log C = \log a + b\log w \qquad (84)$$

where C = LC50 or LD50; w = animal weight; and a,b = regression constants. As mentioned previously, relationships such as described by Equation (84) are often backtransformed to their original units.

$$C = aw^b \qquad (85)$$

Equation (85) can be rearranged.

$$\log(C/w^b) = \log a \qquad (86)$$

In terms of Equation (83), the intercept ($\log a$) in Equation (84) is that value of log (m/w^h) for which y is the metameter for 50% mortality. The b in Equation (84) is equal to the h in Equation (83). In this manner, Anderson and Weber[124] linked Bliss's methods to power relationships commonly used to describe animal size effects on LC50 or LD50. They also recommended that their multivariate approach employing log-transformed values of toxicant concentration and animal size be generally used to scale toxic effects.

Example 10

As suggested earlier in this chapter and by Example 7, survival analysis methods can model allometric effects on toxicity. This suggestion is examined more thoroughly here. Mosquitofish were exposed again to sodium chloride for 96 hours. Three tests were conducted with fish of various sizes. In Test 1, wet weights ranged from 0.005 to 0.047 g (mean, 0.016; standard deviation, 0.006 g). All 280 fish in this test were juveniles. In Test 2, wet weights ranged from 0.033 to 0.223 g (mean, 0.091 g; standard deviation, 0.039 g). The 278 fish in this test were a mixture of juveniles and sexually mature females. In the final Test 3, wet weights ranged from 0.108 to 1.985 g (mean, 0.516 g; standard deviation, 0.330). All 208 fish in the third test were adult females. The overall mean and standard deviation for the 766 fish used in these tests were 0.179 g and 0.271 g, respectively. Through the process described in Example 7, log wet weight and log salt concentration were selected in the various models examined. This is consistent with Bliss's model [Equation (81)]. Using the SAS procedure LIFEREG (see Example 7), exponential, Weibull, log normal, normal, log logistic, and gamma distributions were selected for description of the error function. With the transformations described above, the normal model is similar to Bliss' model [Equation (81)], except time-to-death is used instead of 1000/time-to-death. The other options were not assessed by Bliss and remain unexplored in most subsequent studies.

The following SAS code performs these analyses of pooled data from all three tests and added test ("Run") as a class variable. With minor modification, the code can be used to analyze each test separately.

```
DATA TOXICITY;
INFILE "B:MORT.DAT";
   INPUT RUN 1 UTTD 3-4 LTTD 6-7 WT 9-13 LENGTH 15-16 SALT 18-22;
```

Example 10 *Continued*

```
    LWGT = LOG10 (WT);
    LSALT = LOG10 (SALT);
RUN;
/* UNLIKE PREVIOUS SAS PROGRAMS IN EXAMPLES USING LIFEREG,   */
/* INTERVAL CENSORING IS USED IN THIS EXAMPLE. THE INTER-    */
/* VALS USED IN THIS EXPERIMENT WERE SUFFICIENTLY LARGE      */
/* RELATIVE TO THE TOTAL LENGTH OF EXPOSURE THAT UPPER (UTTD) */
/* AND LOWER (LTTD) LIMITS OF EACH SAMPLING INTERVAL WERE    */
/* USED. IN PREVIOUS EXAMPLES, IT WAS ASSUMED THAT THE SAMP - */
/* ING INTERVALS WERE SUFFICIENTLY SMALL THAT THE TIME OF    */
/* SAMPING WAS AN ADEQUATE MEASURE OF TIME-TO-DEATH. HERE,   */
/* THE SAMPLING TIME IS THE UPPER PORTION OF THE INTERVAL IN */
/* WHICH THE FISH DIED AND THE PREVIOUS TIME OF EXAMINING FOR */
/* DEATH WAS USED AS THE LOWER LIMIT. FOR THE FIRST INTERVAL */
/* WITH DEATHS, THE LTTD IS MISSING. SIMILARLY, FOR THE FINAL */
/* SAMPLING, THE UTTD IS MISSING.                           */
PROC SORT;
    BY RUN LSALT LWT;
RUN;
PROC LIFEREG;
    CLASS RUN;
    MODEL (LTTD, UTTD) = RUN LSALT LWT/D=EXPONENTIAL;
    MODEL (LTTD, UTTD) = RUN LSALT LWT/D=WEIBULL;
    MODEL (LTTD, UTTD) = RUN LSALT LWT/D=LNORMAL;
    MODEL (LTTD, UTTD) = RUN LSALT LWT/D=NORMAL;
    MODEL (LTTD, UTTD) = RUN LSALT LWT/D=LLOGISTIC;
    MODEL (LTTD, UTTD) = RUN LSALT LWT/D=GAMMA;
RUN;
```

The AIC values for each model were the following.

Model	Run 1	Run 2	Run 3	Overall
Exponential	1546	1044	544	3124
Weibull	1270	798	452	2580
Log normal	1236	824	456	2576
Normal	1324	816	464	2658
Log logistic	1240	798	452	2558
Gamma	1236	798	452	2562

In this example, the normal model never provided the best fit (lowest AIC). This observation is consistent with our previous discussions of survival time metameters. The above analysis and the studies cited indicate that several candidate distributions should be examined before modeling size effects on time-to-death. This conclusion can also be extended to toxic endpoint methods because, as outlined above, the endpoint methods of Anderson and Weber[124] are linked to Bliss's approach. Other metameters should be explored in addition to probits suggested by Anderson and Weber.[124]

VI. TOXICANT MIXTURES

Prediction of the combined effects of toxicants has been a goal of aquatic ecotoxicologists because several toxicants are commonly present together in contaminated systems.

Toxicants may interact and influence adsorption, binding to plasma components, tissue distributions, elimination, action at receptor sites, and toxicant metabolism.[139] Sprague[33] used the following example to define the possible joint effects of toxicants. He used the toxic unit scale. In his example, the lethal threshold concentration and dissolved metal concentration are used but other metameters may also be used.

$$\text{Toxic unit} = \frac{[\text{Toxicant}]}{[\text{Lethal threshold}]} \tag{87}$$

where [Toxicant] = the concentration of the toxicant in the exposure solution; and [Lethal threshold] = lethal threshold concentration of the toxicant.

Assume that one half of a toxic unit of toxicant A and one half of a toxic unit of B are administered together. If the combination of the two half-toxic units produces exactly one unit of toxic response then the combination is additive. An additive response is one in which the response is the summation of the responses for each toxicant administered separately. A combined effect may also be less than additive or more than additive. Brown,[140] in noting that additivity was often observed, suggested that additivity can be justified initially with the concept of Selyean stress (see Chapter 1). Each toxicant acts as a Selyean stressor and contributes equally to the degree of experienced shock. (It is ambiguous to this author whether Selyean stress or damage is being discussed in Brown's explanation.) In contrast, Sprague[33] stated that no theoretical basis can be claimed for additivity, and it is assumed for empirical reasons only. Enserink et al.[141] suggested that additivity may be a consequence of similar modes of action for the toxicants in the mixture.

If the toxicants act additively, any combination of toxic units of A and B summing to 1 will have a measured toxic effect of 1. Joint effects of arsenic, cadmium, chromium, copper, mercury, lead, nickel, and zinc to *Daphnia magna* were found to be approximately additive by Enserink et al.[141] Survival during acute exposure, and changes in reproduction and body weight during chronic exposure were used as effects during this study. Toxicant antagonism is present if more than 1 toxic unit is needed with A in the presence of B to have the effect. For example, copper was found to be antagonistic to the toxic action of methylmercury to blue gourami.[142] Zinc is also antagonistic to cygon (*O, O*-dimethyl-*S*-(*N*-methylcarbamoylmethyl) phosphorodithioate) toxicity to zebrafish.[143] If the presence of B enhances the toxic impact of A, the interaction is called potentiation (or synergism). Babich and Stotzky[144] reported potentiation between nickel and copper for microbial growth.

Calmari and Alabaster[139] defined a similar set of terms in their treatment of toxicant mixtures. However, they added the distinction between toxicant interactions at similar or dissimilar sites of action. For example, two toxicants can have a similar, synergistic effect. Interpretation and prediction are greatly enhanced if information is available for making this distinction.

Marking and Dawson[145] developed a scale for assessing toxicant interactions.

$$S = \frac{\text{LC50}_{AC}}{\text{LC50}_{AI}} + \frac{\text{LC50}_{BC}}{\text{LC50}_{BI}} \tag{88}$$

where LC50_{AC} = LC50 for A in the presence of B; LC50_{AI} = LC50 for A in the absence of B; LC50_{BC} = LC50 for B in the presence of A; and LC50_{BI} = LC50 for B in the absence of A.

Values of $S < 1$ indicate a greater than additive response, and values of $S > 1$ suggest a less than additive response. However, the degree of deviation from additivity is not similar for values of the same magnitude in the negative and positive direction from $S = 1$. Marking and Dawson[145] used a modification of S to linearize the scale and used 0 as the reference point (additivity) with this modified index. The modified index was called the additive index or Marking's additive index. For $S \leq 1$, the additive index is $1/S - 1$. It is $-S + 1$ for $S > 1$. The additive index scale is linear with strict additivity indicated by a value of 0. Significant deviation from an additive index of 0 was tested by use of the LC50 95% C.I. values to estimate the acceptable range of additive indices.[146] The lower 95% confidence limits for $LC50_{AI}$ and $LC50_{BI}$ and upper limits for $LC50_{AC}$ and $LC50_{BC}$ were substituted into Equation (88) to estimate the lowest value for the additive index. Upper limits of $LC50_{AI}$ and $LC50_{BI}$ and lower limits of $LC50_{AC}$ and $LC50_{BC}$ are used in Equation (88) to estimate the highest value of the additive index. Inclusion of 0 in the resulting interval suggests a lack of any statistically significant deviation from additivity. Unfortunately, inclusion of 0 has also been used to conclude that toxicant mixtures are additive. For example, Thompson et al.[147] stated: "The value of the [additive index] calculated from the results of this study was +0.218 with a range from -0.64 to $+1.30$. Since 0 is contained within these limits, the toxicity of Zn-Cu mixtures to bluegills was additive under the conditions of these tests." Although the reader can quickly ascertain the true conclusion that the authors were making, the conclusion as stated is inaccurate. This problem stems from some of Marking's own statements such as "Mixtures that resulted in ranges for the additive index that overlapped zero were judged to be only additive in toxicity; ranges that did not overlap zero were either greater or less than additive in toxicity."[146] His meaning is clear, but the conclusion of additivity when ranges include 0 is statistically invalid and must be avoided.

Marking[145,146] also expressed additive indices in terms of magnification factors. For example, additive indices of 9, 1, 0, and -9 have magnification factors of 10, 2, 1, and 0.10 times, respectively. An index of 9 suggests that the combined toxic effect of A and B is 10 times greater than that of A or B alone.

This approach is critical in much work in regulatory ecotoxicology. Di Toro et al.[107] incorporated additivity into methods for assessing sediment quality for metals. Under the assumption of addition, Di Toro et al.[107] suggested that the sum of the metal concentrations in the SEM fraction divided by the AVS should not exceed 1 when several toxic metals are present at high concentrations in the sediment. Spehar and Fiandt[148] recently examined metal interactions in the context of water quality criteria.

The above-mentioned approach can be enriched using factorial experimental designs[139] and response surface methods, e.g., Parker's[149] analysis of combined metal toxicity to protozoa or the examination by Voyer et al.[150] of metal interactions on flounder larvae. Voyer and Heltshe[151] reanalyzed models developed by Parker[149] and Voyer et al.[150] and used the results to distinguish between various types of additivity. For the data of Parker[149] and Voyer et al.,[150] the following models were defined and tested for goodness-of-fit.

$$\hat{y} = a_x + b_y \qquad (89)$$

where $\hat{y}$ = the response variable; a_x = the measured response to the administered level of toxicant x when the concentration of toxicant $y = 0$; and b_y = the measured response to the administered level of toxicant y when the concentration of toxicant $x = 0$.

$$\hat{y} = a + b_1x_1 + b_2x_2 \qquad (90)$$

where x_1 = concentration of toxicant 1; x_2 = concentration of toxicant 2; b_1 = estimated coefficient for toxicant 1; b_2 = estimated coefficient for toxicant 2; and a = estimated parameter.

$$\hat{y} = a + b_1x_1 + b_2x_2 + b_{11}x_1^2 + b_{22}x_2^2 + b_{12}x_1x_2 \qquad (91)$$

where b_{11} = estimated coefficient for quadratic effect associated with toxicant 1; b_{22} = estimated coefficient for quadratic effect associated with toxicant 2; and b_{12} = estimated coefficient for interactive effect.

Equations (89), (90), and (91) describe simple additive, linear additive, and quadratic models for toxicant interactions, respectively. Voyer and Heltshe[151] suggested from their analyses that assessment of alternate models such as Equations (90) and (91) can enhance prediction of effect for toxicant mixtures. Protocol is outlined for conducting factorial experimentation for developing appropriate models. (See Box and Draper,[152] especially Chapter 7, and Chapter 9 in Neter et al.[19] for general discussions of this approach.)

VII. SUMMARY

Two goals were set during the development of this chapter. The first was to present the significant progress made in regulatory ecotoxicology to quantify lethal stress. Valuable methods were described for use in quantifying lethal and other quantal responses to stress. Most grew out of practical or regulatory applications. *En masse,* they were based on a firm, statistical foundation. However, they were developed to their present state to meet regulatory goals, not to elucidate explanatory principles.

The second goal of the chapter was to make apparent our contrastingly slow progress in the context of scientific ecotoxicology. Explanatory principles underlying methods selection, e.g., use of probit versus Weibull metameters, or toxicological definitions, e.g., acute versus chronic toxicity, have often been ignored or discussed superficially. Relatively little effort has been spent in linkage to underlying principles. Very notable exceptions have been the development and application of metal speciation models and QSARs in ecotoxicology and the movement toward their linkage to PBPK and pharmacodynamic models.

In the context of scientific ecotoxicology, too much effort is spent on the "description of the diversity of phenomena"[153] instead of on the determination of underlying themes or theory reduction. This is surprising as regulatory ecotoxicologists have provided a richness of methodologies and information to be bent to these goals. Further, alternative metameters, toxicity endpoints, and methodologies are ignored in favor of methods with no apparent conceptual superiority. Instead, methods are selected for their regulatory significance, e.g., use of a probit estimate of a 96-hour LC50 instead of a survival time model.

Traditional methodologies are not built on or modified as efficiently as possible because of an overzealous adherence to standard methods and a general lack of training in strong inference. A good example is the delay in modifying Bliss's methods for incorporating size into models of toxic impact. Remember that these important techniques were developed initially with an emphasis on linear transformation. Such an approach is understandable if one considers the general inconvenience of executing more involved methods in the 1930s. Regardless, why should these historical constraints still have such a strong influence on the present-day exploration of appropriate models?

REFERENCES

1. Casarett, L. J. "Origin and Scope of Toxicology." In *Toxicology. The Basic Science of Poisons,* ed. L. J. Cassarett and J. Doull. (New York: MacMillan Publishing Co., 1975).

2. Dagani, R. "Aquatic Toxicology Matures, Gains Importance." *Chem. Eng. News* June 30:18–23 (1980).

3. Finney, D. J. *Probit Analysis,* 3rd ed. (London: Cambridge at the University Press, 1971).

4. Bacharach, A. L., M. E. Coates, and T. R. Middleton. "A Biological Test for Vitamin P Activity." *Biochem. J.* 36:407–412 (1942).

5. Gaddum, J. H. "Bioassays and Mathematics." *Pharmacol. Rev.* 5:87–134 (1953).

6. Bliss, C. I. "The Calculation of the Dosage-Mortality Curve." *Ann. Appl. Biol.* 22:134–307 (1935).

7. Finney, D. J. *Statistical Method in Biological Assay,* 3rd ed. (London: Charles Griffin and Company, 1978).

8. Berkson, J. "Why I Prefer Logits to Probits." *Biometrics* 7(4):327–339 (1951).

9. Christensen, E. R., and N. Nyholm. "Ecotoxicological Assays with Algae: Weibull Dose-Response Curves." *Environ. Sci. Technol.* 18(9):713–718 (1984).

10. Armitage, P., and I. Allen. "Methods of Estimating the LD50 in Quantal Response Data." *J. Hyg.* 48:298–322 (1950).

11. Christensen, E. R. "Dose-Response Functions in Aquatic Toxicity Testing and the Weibull Model." *Water Res.* 18(2):213–221 (1984).

12. Pinder, J. E., III, J. G. Wiener, and M. H. Smith. "The Weibill Distribution: A New Method of Summarizing Survivorship Data." *Ecology* 59(1):175–179 (1978).

13. Miller, R. G., Jr. *Survival Analysis.* (New York: John Wiley and Sons, 1981).

14. Trevan, J. W. "The Error of Determination of Toxicity." *Proc. R. Soc. Lond. B. Biol. Sci.* 101:483–514 (1927).

15. Litchfield, J. T., Jr., and F. Wilcoxon. "A Simplified Method of Evaluating Dose-Effect Experiments." *J. Pharmacol. Exp. Ther.* 96:99–113 (1949).

16. Newman, M. C., and M. Aplin. "Enhancing Toxicity Data Interpretation and Prediction of Ecological Risk with Survival Time Modeling: An Illustration Using Sodium Chloride Toxicity to Mosquitofish (*Gambusia holbrooki*)." *Aquat. Toxicol.* 23:85–96 (1992).

17. Gad, S. C., and C. S. Weil. *Statistics and Experimental Design for Toxicologists.* (Caldwell, NJ: Telford Press, 1988).

18. Stephan, C. E. "Methods for Calculating an LC_{50}." In *Aquatic Toxicology and Hazard Evaluation,* ASTM STP 634, ed. F. L. Mayer and J. L. Hamelink. (Philadelphia, PA: American Society for Testing and Materials, 1977).

19. Neter, J., W. Wasserman, and M. H. Kutner. *Applied Linear Statistical Models, Regression, Analysis of Variance and Experimental Designs.* (Boston, MA: Irwin, 1990).

20. SAS Institute Inc. *SAS Technical Report P-179, Additional SAS/STAT Procedures,* Release 6.03. (Cary, NC: SAS Institute Inc., 1988).

21. Dixon, P. M. Personal communication, 1992.

22. Berkson, J. "Maximum Likelihood and Minimum χ^2 Estimates of the Logistic Function." *J. Am. Stat. Assoc.* 50:130–162 (1955).

23. Salsburg, D. S. *Statistics for Toxicologists.* (New York: Marcel Dekker, Inc., 1986).

24. Buikema, A. L., Jr., B. R. Niederlehner, and J. Cairns, Jr. "Biological Monitoring: Part IV—Toxicity Testing." *Water Res.* 16:239–262 (1982).

25. Hamilton, M. A., R. C. Russo, and R. V. Thurston. "Trimmed Spearman-Karber Method for Estimating Median Lethal Concentrations in Toxicity Bioassays." *Environ. Sci. Technol.* 11(7):714–719 (1977).

26. Gelber, R. D., P. T. Lavin, C. R. Mehta, and D. A. Schoenfeld "Statistical Analysis." In *Fundamentals of Aquatic Toxicology,* ed. G. M. Rand and S. R. Petrocelli. (New York: Hemisphere Publishing Corp., 1985).

27. Stephen, C. E. Personal communication, 1992.

28. American Public Health Association. *Standard Methods for the Examination of Water and Wastewater,* 15th ed. (Washington, DC.: American Public Health Association, 1981).

29. Sprague, J. B. "Measurement of Pollutant Toxicity to Fish. I. Bioassay Methods for Acute Toxicity." *Water Res.* 3:793–821 (1969).

30. Tattersfield, F., and H. M. Morris. "An Apparatus for Testing the Toxic Values of Contact Insecticides under Controlled Conditions." *Bull. Entomol. Res.* 14:223–233 (1924).

31. Haze, T. *CT-Tox Multi-Method Program.* Bureau of Water Management, Connecticut Department of Environmental Protection, Water Toxics Laboratory, 122 Washington St., Hartford, CT 06106 (1990).

32. Weber, C. I., W. H. Peltier, T. J. Norberg-King, W. B. Horning, II, F. A. Kessler, J. R. Menkedick, T. W. Neiheisel, P. A. Lewis, D. J. Klemm, Q. H. Pickering, E. L. Robinson, J. M. Lazorchak, L. J. Wymer, and R. W. Freyberg. *Short-term Methods for Estimating Chronic Toxicity of Effluents and Receiving Waters to Freshwater Organisms,* 2nd ed. EPA/600/4–89/001. (Cincinnati, OH: 1989).

33. Sprague, J. B. "Measurement of Pollutant Toxicity to Fish. II. Utilizing and Applying Bioassay Results." *Water Res.* 4:3–32 (1970).

34. Chew, R. D., and M. A. Hamilton. "Toxicity Curve Estimation: Fitting a Compartment Model to Median Survival Times." *Trans. Am. Fish. Soc.* 114:403–412 (1985).

35. Shepard, M. P. "Resistance and Tolerance of Young Speckled Trout (*Salvelinus fontinalis*) to Oxygen Lack, with Special Reference to Low Oxygen Acclimation." *J. Fish. Res. Board Can.* 12(3):387–446 (1955).

36. Litchfield, J. T., Jr. "A Method for Rapid Graphic Solution of Time-Per Cent Effects Curves." *J. Pharmacol. Exp. Ther.* 97:399–408 (1949).

37. Bliss, C. I. "The Calculation of the Time-Mortality Curve." *Ann. Appl. Biol.* 24:815–852 (1937).

38. Lloyd, R. "The Toxicity of Zinc Sulphate to Rainbow Trout." *Ann. Appl. Biol.* 48(1):84–94 (1960).

39. Feller, W. *An Introduction to Probability Theory and Its Application.* (New York: John Wiley and Sons, 1968).

40. Pollard, J. H. *A Handbook of Numerical and Statistical Techniques with Examples Mainly from the Life Sciences.* (New York: Cambridge University Press, 1979).

41. Heming, T. A., A. Sharma, and Y. Kumar. "Time-Toxicity Relationships in Fish Exposed to the Organochlorine Pesticide Methoxychlor." *Environ. Toxicol. Chem.* 8:923–932 (1989).

42. Dixon, P. M., and M. C. Newman. "Analyzing Toxicity Data Using Statistical Models for Time-to-Death: An Introduction." In *Metal Ecotoxicology, Concepts and Applications,* ed. M. C. Newman and A. W. McIntosh. (Chelsea, MI: Lewis Publishers, Inc., 1991).

43. Cox, D. R., and D. Oakes. *Analysis of Survival Data.* (New York: Chapman and Hall, 1984).

44. Diamond, S. A., M. C. Newman, M. Mulvey, P. M. Dixon, and D. Martinson. "Allozyme Genotype and Time to Death of Mosquitofish, *Gambusia affinis* (Baird and Girard), during Acute Exposure to Inorganic Mercury." *Environ. Toxicol. Chem.* 8:613–622 (1989).

45. Diamond, S. A., M. C. Newman, M. Mulvey, and S. I. Guttman. "Allozyme Genotype and Time-to-Death of Mosquitofish, *Gambusia holbrooki,* during Acute Inorganic Mercury Exposure: A Comparison of Populations." *Aqua. Toxicol.* 21:119–134 (1991).

46. Newman, M. C., S. A. Diamond, M. Mulvey, and P. Dixon. "Allozyme Genotype and Time to Death of Mosquitofish, *Gambusia affinis* (Baird and Girard) during

Acute Toxicant Exposure: A Comparison of Arsenate and Inorganic Mercury." *Aquat. Toxicol.* 15:141–156 (1989).

47. Nelson, W. "Theory and Applications of Hazard Plotting for Censored Failure Data." *Technometrics* 14(4):945–966 (1972).

48. Blackstone, E. H. "Analysis of Death (Survival Analysis) and Other Time-related Events." In *Current Status of Clinical Cardiology,* ed. F. J. Macartney. (Boston, MA: MTP Press Limited, 1986).

49. Harrell, F. E., Jr. *Survival and Risk Analysis.* (Durham, NC: Duke University Medical Center, 1988).

50. Kaplan, E. L., and P. Meier. "Nonparametric Estimation from Incomplete Observations." *J. Am. Stat. Assoc.* 53:457–481 (1958).

51. Kalbfleisch, J. D., and R. L. Prentice. *The Statistical Analysis of Failure Time Data.* (New York: John Wiley and Sons, 1980).

52. Mantel, N. "Evaluation of Survival Data and Two New Rank Statistics Arising in Its Consideration. *Cancer Chemother. Rep.* 50:163–170 (1966).

53. Cox, D. R. "Regression Models and Lifetables (with Discussion)." *J. R. Stat. Soc. Ser. B* 34:187–200 (1972).

54. Peto, R., and J. Peto. "Asymptotically Efficient Rank-Invariant Test Procedures." *J. R. Stat. Soc. Ser. A* 135:185–207 (1972).

55. Pepe, M. S., and T. R. Fleming. "Weighted Kaplan-Meier Statistics: A Class of Distance Tests for Censored Survival Data." *Biometrics* 45:497–507 (1989).

56. Nelson, W. "Hazard Plotting for Incomplete Failure Data." *Qual. Tech.* 1(1):27–52 (1969).

57. Pryor, D. B., F. E. Harrell, Jr., K. L. Lee, R. M. Califf, and R. A. Rosati. "Estimating the Likelihood of Significant Coronary Artery Disease." *Am. J. Med* 75:771–780 (1983).

58. Lew, A. A., C. L. Day, T. J. Harrist, W. C. Wood, and M. C. Mihm, Jr. "Multivariate Analysis. Some Guidelines for Physicians." *J. Am. Med. Assoc.* 249(5):641–643 (1983).

59. Manly, B. F. J. *The Statistics of Natural Selection on Animal Populations.* (New York: Chapman and Hall, 1985).

60. Kotz, S., and N. L. Johnson. *Icing the Tails to Limit Theorems.* Encyclopedia of Statistical Sciences, Vol. 4. (New York: John Wiley and Sons, 1983).

61. Brent, E. E., Jr., E. J. Mirielli, Jr., E. Detring, and F. Ramos. *Statistical Navigator Professional User's Guide and Reference Manual Version 1.0.* (Columbia, MO: The Ideas Works, Inc., 1991).

62. Atkinson, A. C. "A Note on the Generalized Information Criterion for Choice of a Model." *Biometrika* 67:413–418 (1980).

63. Hillaby, B. A., and D. J. Randall. "Acute Ammonia Toxicity and Ammonia Excretion in Rainbow Trout (*Salmo gairdneri*)." *J. Fish. Res. Board Can.* 36:621–629 (1979).

64. Thurston, R. V., R. C. Russo, and G. A. Vinogradov. "Ammonia Toxicity to Fishes. Effect of pH on the Toxicity of the Unionized Ammonia Species." *Environ. Sci. Technol.* 15(7):837–840 (1981).

65. Gersich, F. M., and D. L. Hopkins. "Site-specific Acute and Chronic Toxicity of Ammonia to *Daphnia magna* Straus." *Environ. Toxicol. Chem.* 5:443–447 (1986).

66. Emerson, K., R. C. Russo, R. E. Lund, and R. V. Thurston. "Aqueous Ammonia Equilibrium Calculations: Effect of pH and Temperature." *J. Fish. Res. Board Can.* 32:2379–2383 (1975).

67. Clubb, R. W., A. R. Gaufin, and J. L. Lords. "Synergism between Dissolved Oxygen and Cadmium Toxicity in Five Species of Aquatic Insects." *Environ. Res.* 9:285–289 (1975).

68. Muller, H.-G. "Acute Toxicity of Potassium Dichromate to *Daphnia magna* as a Function of the Water Quality." *Bull. Environ. Contam. Toxicol.* 25:113–117 (1980).

172

69. Freedman, M. L., P. M. Cunningham, J. E. Schindler, and M. J. Zimmerman. "Effect of Lead Speciation on Toxicity." *Bull. Environ. Contam. Toxicol.* 25:389–393 (1980).

70. Andrew, R. W., K. E. Biesinger, and G. E. Glass. "Effects of Inorganic Complexing on the Toxicity of Copper to *Daphnia magna.*" *Water Res.* 11:309–315 (1977).

71. Borgmann, U. "Metal Speciation and Toxicity of Free Metal Ions to Aquatic Biota." In *Aquatic Toxicology,* ed. J. O. Nriagu. (New York: John Wiley & Sons, 1983).

72. Loehle, C., P. Bertsch, and G. Mills. "An Evaluation of Chemical Speciation in the MEXAMS Metal Transport Model." *Environ. Software* 1(2):106–112 (1986).

73. Slonim, C. B., and A. R. Slonim. "Effect of Water Hardness on the Tolerance of the Guppy to Beryllium Sulfate." *Bull. Environ. Contam. Toxicol.* 10(5):295–301 (1973).

74. Carroll, J. J., S. J. Ellis, and W. S. Oliver. "Influences of Hardness on the Acute Toxicity of Cadmium to Brook Trout (*Salvelinus fontinalis*)." *Bull. Environ. Contam. Toxicol.* 22:575–581 (1979).

75. Wright, D. A., and J. W. Frain. "The Effect of Calcium on Cadmium Toxicity in the Freshwater Amphipod, *Gammarus pulex* (L.)." *Arch. Environ. Contam. Toxicol.* 10:321–328 (1981).

76. Pascoe, D., S. A. Evans, and J. Woodworth. "Heavy Metal Toxicity to Fish and the Influence of Water Hardness." *Arch. Environ. Contam. Toxicol.* 15:481–487 (1986).

77. Howarth, R. S., and J. B. Sprague. "Copper Lethality to Rainbow Trout in Waters of Various Hardness and pH." *Water Res.* 12:455–462 (1978).

78. Zitko, V., and W. G. Carson. "A Mechanism of the Effect of Water Hardness on the Lethality of Heavy Metals to Fish." *Chemosphere* 5:299–303 (1976).

79. Nelson, H., D. Benoit, R. Erickson, V. Mattson, and J. Lindberg. *The Effects of Variable Hardness, pH, Alkalinity, Suspended Clay, and Humic on the Chemical Speciation and Aquatic Toxicity of Copper,* EPA/600/3–86/023. (Springfield, VA: National Technical Information Service, 1986).

80. U.S. Environmental Protection Agency. *Ambient Water Quality Criteria for Copper—1984,* EPA 440/5-84-031. (Springfield, VA: National Technical Information Service, 1985).

81. U.S. Environmental Protection Agency. *Ambient Water Quality Criteria for Cadmium—1984,* EPA 440/5-84-032. (Springfield, VA: National Technical Information Service, 1985).

82. U.S. Environmental Protection Agency. *Ambient Aquatic Life Water Quality Criteria for Zinc,* EPA 440/5-87-003. (Springfield, VA: National Technical Information Service, 1987).

83. Newman, M. C. "A Statistical Bias in the Derivation of Hardness-Dependent Metals Criteria." *Environ. Toxicol. Chem.* 10:1295–1297 (1991).

84. Koch, R. W., and G. M. Smillie. "Bias in Hydrologic Prediction Using Log-Transformed Regression Models." *Water Res.* 22:717–723 (1986).

85. Miller, D. M. "Reducing Transformation Bias in Curve Fitting." *Am. Stat.* 38(2):124–126 (1984).

86. Cairns, J., Jr., A. G. Heath, and B. C. Parker. "Temperature Influence on Chemical Toxicity to Aquatic Organisms." *J. Water Pollut. Control Fed.* 47(2):267–280 (1975).

87. Newman, M. C., and C. H. Jagoe. "Ligands and the Bioavailability of Metals in Aquatic Environments." In *A Mechanistic Understanding of Bioavailability: Physical-Chemical Interactions,* ed. J. Hamelink and W. Benson. (Chelsea, MI: Lewis Publishers, 1993).

88. Fales, R. R. "The Influence of Temperature and Salinity on the Toxicity of Hexavalent Chromium to the Grass Shrimp *Palaemonetes pugio* (Holthuis)." *Bull. Environ. Contam. Toxicol.* 20:447–450 (1978).

89. MacInnes, J. R., and A. Calabrese. "Combined Effects of Salinity, Temperature, and Copper on Embryos and Early Larvae of the American Oyster, *Crassostrea virginica.*" *Arch. Environ. Contam. Toxicol.* 8:553–562 (1979).

90. Cairns, M. A., A. V. Nebeker, J. H. Gakstatter, and W. L. Griffis. "Toxicity of Copper-Spiked Sediments to Freshwater Invertebrates." *Environ. Toxicol. Chem.* 3:435–445 (1984).

91. Nebeker, A. V., M. A. Cairns, J. H. Gakstatter, K. W. Malueg, G. S. Schuytema, and D. F. Krawczyk. "Biological Methods for Determining Toxicity of Contaminated Freshwater Sediments to Invertebrates." *Environ. Toxicol. Chem.* 3:617–630 (1984).

92. DeWitt, T. H., R. C. Swartz, and J. O. Lamberson. "Measuring the Acute Toxicity of Estuarine Sediments." *Environ. Toxicol. Chem.* 8:1035–1048 (1989).

93. Burton, Jr., G. A. "Assessing the Toxicity of Freshwater Sediments." *Environ Toxicol. Chem.* 10:1585–1627 (1991).

94. Luoma, S. N. "Can We Determine the Biological Availability of Sediment-bound Trace Elements?" *Hydrobiologia* 176/177:379–396 (1989).

95. Luoma, S. N., and E. A. Jenne. "The Availability of Sediment-Bound Cobalt, Silver, and Zinc to a Deposit-Feeding Clam." In *Biological Implications of Metals in the Environment,* CONF-750929, ed. R. E. Wilding and H. Dricker. (Springfield, VA: National Technical Information Service, 1977) 213–230.

96. Luoma, S. N., and G. W. Bryan. "Factors Controlling the Availability of Sediment-bound Lead to the Estuarine Cockle, *Scrobicularia plana.*" *J. Mar. Biol. Assoc. U.K.* 58:793–802 (1978).

97. Cook, M., G. Nickless, R. E. Lawn, and D. J. Roberts. "Biological Availability of Sediment-bound Cadmium to the Edible Cockle, *Cerastoderma edule.*" *Bull. Environ. Contam. Toxicol.* 23:381–386 (1979).

98. Langston, W. J. "Arsenic in U.K. Estuarine Sediments and Its Availability to Benthic Organisms." *J. Mar. Biol. Assoc. U.K.* 60:869–881 (1980).

99. Tessier, A., P. G. C. Campbell, J. C. Auclair, and M. Bisson. "Relationships Between the Partitioning of Trace Metals in Sediments and Their Accumulation in the Tissues of the Freshwater Mollusc *Elliptio complanata* in a Mining Area." *Can. J. Fish. Aquat. Sci.* 41:1463–1472 (1984).

100. Campbell, P. G. C., A. G. Lewis, P. M. Chapman, A. A. Crowder, W. K. Fletcher, B. Imber, S. N. Luoma, P. M. Stokes, and M. Winfrey. "Biologically Available Metals in Sediments." No. 27694 (Ottawa, ONT, Canada: National Research Council Canada, 1988).

101. Young, L. B., and H. H. Harvey. "Metal Concentrations in Chironomids in Relation to the Geochemical Characteristics of Surficial Sediments." *Arch. Environ. Contam. Toxicol.* 21:202–211 (1991).

102. Crecelius, E. A., J. T. Hardy, C. I. Bobson, R. L. Schmidt, C. W. Apts, J. M. Gurtisen, and S. P. Joyce. "Copper Bioavailability to Marine Bivalves and Shrimp: Relationship to Cupric Ion Activity." *Mar. Environ. Res.* 6:13–26 (1982).

103. Ray, S., D. W. McLeese, and M. R. Peterson. "Accumulation of Copper, Zinc, Cadmium and Lead from Two Contaminated Sediments by Three Marine Invertebrates—A Laboratory Study." *Bull. Environ. Contam. Toxicol.* 26:315–322 (1981).

104. Rule, J. H., and R. W. Alden III. "Cadmium Bioavailability to Three Estuarine Animals in Relationship to Geochemical Fractions in Sediments." *Arch. Environ. Contam. Toxicol.* 19:878–885 (1990).

105. Jenne, E. A., and S. N. Luoma. "Forms of Trace Elements in Soils, Sediments, and Associated Waters: An Overview of Their Determination and Biological Availability." In *Biological Implications of Metals in the Environment,* CONF-750929, ed. R. E. Wildung and H. Drucker. (Springfield, VA: National Technical Information Service, 1977) 110–143.

106. Patrick, W. H., Jr., R. P. Gambrell, and R. A. Khalid. "Physiochemical Factors Regulating Solubility and Bioavailability of Toxic Heavy Metals in Contaminated Dredged Sediment." *J. Environ. Sci. Health* A12(9):475–492 (1977).

107. Di Toro, D. M., J. D. Mahony, D. J. Hansen, K. J. Scott, M. B. Hicks, S. M. Mayr, and M. S. Redmond. "Toxicity of Cadmium in Sediments: The Role of Acid Volatile Sulfide." *Environ. Toxicol. Chem.* 9:1487–1502 (1990).

108. Bryan, G. W. "Bioavailability and Effects of Heavy Metals in Marine Species." In *Wastes in the Sea,* Vol. 6: *Near Shore Waste Disposal,* ed. B. Ketchum, J. Capuzzo, W. Burt, I. Duedall, P. Park, and D. Kester. (New York: John Wiley and Sons, 1985).

109. Bjornberg, A., L. Hakanson, and K. Lundbergh. "A Theory of the Mechanisms Regulating Bioavailability of Mercury in Natural Waters." *Environ. Pollut.* 49:53–61 (1988).

110. Carlson, A. R., G. L. Phipps, V. R. Mattson, P. A. Kosian, and A. M. Cotter. "The Role of Acid-Volatile Sulfide in Determining Cadmium Bioavailability and Toxicity in Freshwater Sediments." *Environ. Toxicol. Chem.* 10:1309–1319 (1991).

111. Ankley, G. T., G. L. Phipps, E. N. Leonard, D. A. Benoit, V. R. Mattson, P. A. Kosian, A. M. Cotter, J. R. Dierkes, D. J. Hansen, and J. D. Mahony. "Acid-Volatile Sulfide as a Factor Mediating Cadmium and Nickel Bioavailability in Contaminated Sediments." *Environ. Toxicol. Chem.* 10:1299–1307 (1991).

112. Nimmo, D. R. "Pesticides." In *Fundamentals of Aquatic Toxicology: Methods and Applications,* ed. G. M. Rand, and S. R. Petrocelli, (New York: Hemisphere Publishing Corp., 1985) 335–373.

113. Bowling, J. W., G. J. Leversee, P. F. Landrum, and J. P. Giesy. "Acute Mortality of Anthracene-Contaminated Fish Exposed to Sunlight." *Aquat. Toxicol.* 3:79–90 (1983).

114. Hansch, C., J. E. Quinlan, and G. L. Lawrence. "The Linear Free-Energy Relationship Between Partition Coefficients and the Aqueous Solubility of Organic Liquids." *J. Org. Chem.* 33(1):347–350 (1968).

115. Hansch, C., A. Leo, and D. Nikaitani. "On the Additive-Constitutive Character of Partition Coefficients." *J. Org. Chem.* 37(20):3090–3092 (1972).

116. Suter, G. W. III. *Ecological Risk Assessment.* (Chelsea, MI: Lewis Publishers, 1993).

117. Borman, S. "New QSAR Techniques Eyed for Environmental Assessments." *Chem. Eng. News* February 19:20–23 (1990).

118. Hansch, C., and A. Leo. *Substituent Constants for Correlation Analysis in Chemistry and Biology.* (New York: John Wiley and Sons, 1979).

119. Laughlin, R. B. Jr., R. B. Johannesen, W. French, H. Guard, and F. E. Brinckman. "Structure-Activity Relationships for Organotin Compounds." *Environ. Toxicol. Chem.* 4:343–351 (1985).

120. Neely, W. B., D. R. Branson, and G. E. Blau. "Partition Coefficients to Measure Bioconcentration Potential of Organic Chemicals in Fish." *Environ. Sci. Technol.* 8(13):1113–1115 (1974).

121. Lipnick, R. L. "A Perspective on Quantitative Structure-Activity Relationships in Ecotoxicology." *Environ. Toxicol. Chem.* 4:255–257 (1985).

122. Di Toro, D. M., C. S. Zarba, D. J. Hansen, W. J. Berry, R. C. Swartz, C. E. Cowan, S. P. Pavlou, H. E. Allen, N. A. Thomas, and P. R. Paquin. "Technical Basis for Establishing Sediment Quality Criteria for Nonionic Chemicals Using Equilibrium Partitioning." *Environ. Toxicol. Chem.* 10:1541–1583 (1991).

123. McLeay, D. J., and Munro, J. R. "Photoperiodic Acclimation and Circadian Variations in Tolerance of Juvenile Rainbow Trout (*Salmo gairdneri*) to Zinc." *Bull. Environ. Contam. Toxicol.* 23:552–557 (1979).

124. Anderson, P. D., and L. J. Weber. "Toxic Response as a Quantitative Function of Body Size." *Toxicol. Appl. Pharmacol.* 33:471–483 (1975).

125. Pallotta, A. J., M. G. Kelly, D. P. Rall, and J. W. Ward. "Toxicology of Acetoxycyclo-

heximide as a Function of Sex and Body Weight." *J. Pharmacol. Exp. Ther.* 136:400–405 (1962).

126. Chapman, G. A. "Acclimation as a Factor Influencing Metal Criteria." In *Aquatic Toxicology and Hazard Assessment: Eighth Symposium,* ASTM STP 891, ed. R. C. Bahner and D. J. Hansen. (Philadelphia, PA: American Society for Testing and Materials, 1985) 119–136.

127. Dixon, D. G., and J. B. Sprague. "Acclimation to Copper by Rainbow Trout (*Salmo gairdneri*)—A Modifying Factor in Toxicity." *Can. J. Fish. Aquat. Sci.* 38:880–888 (1981).

128. Dixon, D. G., and J. B. Sprague. "Acclimation-Induced Changes in Toxicity of Arsenic and Cyanide in Rainbow Trout, *Salmo gairdneri* Richardson." *J. Fish Biol.* 18:579–589 (1981).

129. Bliss, C. I. "The Size Factor in the Action of Arsenic upon Silkworm Larvae." *J. Exp. Biol.* 13:95–110 (1936).

130. Campbell, F. L. "Relative Susceptability of Arsenic in Successive Instars of the Silkworm." *J. Gen. Physiol.* 9(6):727–733 (1926).

131. Rall, D. P., and W. C. North. "Consideration of Dose-Weight Relationships." *Proc. Soc. Exp. Biol. Med.* 83:825–827 (1953).

132. Lemanna, C., W. I. Jensen, and I. D. J. Bross. "Body Weight as a Factor in the Response of Mice to Botulinal Toxins." *Am. J. Hyg.* 62:21–28 (1955).

133. Lamanna, C., and E. R. Hart. "Relationship of Lethal Toxic Dose to Body Weight of the Mouse." *Toxicol. Appl. Pharmacol.* 13:307–315 (1968).

134. Hedtke, J. L., E. Robinson-Wilson, and L. J. Weber. "Influence of Body Size and Developmental Stage of Coho Salmon (*Oncorhynchus kisutch*) on Lethality of Several Toxicants." *Fundam. Appl. Toxicol.* 2:67–72 (1982).

135. Heit, M., and M. Fingerman. "The Influences of Size, Sex and Temperature on the Toxicity of Mercury to Two Species of Crayfishes." *Bull. Environ. Contam. Toxicol.* 18(5):572–580 (1977).

136. Hogan, G. R., B. S. Cole, and J. M. Lovelace. "Sex and Age Mortality Responses in Zinc Acetate-Treated Mice." *Bull. Environ. Contam. Toxicol.* 39:156–161 (1987).

137. Angelakos, E. T. "Lack of Relationship Between Body Weight and Pharmacological Effect Exemplified by Histamine Toxicity in Mice." *Proc. Soc. Exp. Biol. Med.* 103:296–298 (1960).

138. Adelman, I. R., L. L. Smith, Jr., and G. D. Siesennop. "Effect of Size or Age of Goldfish and Fathead Minnows on Use of Pentachlorophenol as a Reference Toxicant." *Water Res.* 10:685–687 (1976).

139. Calamari, D., and J. S. Alabaster. "An Approach to Theoretical Models in Evaluating the Effects of Mixtures of Toxicants in the Aquatic Environment." *Chemosphere* 9:533–538 (1980).

140. Brown, V. M. "The Calculation of the Acute Toxicity of Mixtures of Poisons to Rainbow Trout." *Water Res.* 2:723–733 (1968).

141. Enserink, E. L., J. L. Maas-Diepeveen, and C. J. Van Leeuwen. "Combined Effects of Metals; An Ecotoxicological Evaluation." *Water Res.* 25(6):679–687 (1991).

142. Roales, R. R., and A. Perlmutter. "Toxicity of Methylmercury and Copper, Applied Singly and Jointly, to the Blue Gourami, *Trichogaster trichopterus*." *Bull. Environ. Contam. Toxicol.* 12(5):633–639 (1974).

143. Roales, R. R., and A. Perlmutter. "Toxicity of Zinc and Cygon, Applied Singly and Jointly, to Zebrafish Embryos." *Bull. Environ. Contam. Toxicol.* 12(4):475–480 (1974).

144. Babich, H., and G. Stotzky. "Synergism Between Nickel and Copper in Their Toxicity to Microbes: Mediation by pH." *Ecotoxicol. Environ. Saf.* 7:576–587 (1983).

145. Marking, L. L., and V. K. Dawson. "Method for Assessment of Toxicity or Efficacy

of Mixtures of Chemicals." *U.S. Fish Wildl. Serv. Invest. Fish Control* 67:1–8 (1975).

146. Marking, L. L. "Toxicity of Chemical Mixtures." In *Fundamentals of Aquatic Toxicology,* ed. G. M. Rand and S. R. Petrocelli. (New York: Hemisphere Publishing Corp., 1985).

147. Thompson, K. W., A. C. Hendricks, and J. Cairns, Jr. "Acute Toxicity of Zinc and Copper Singly and in Combination to the Bluegill (*Lepomis macrochirus*)." *Bull. Environ. Contam. Toxicol.* 25:122–129 (1980).

148. Spehar, R. L., and J. T. Fiandt. "Acute and Chronic Effects of Water Quality Criteria-Based Metal Mixtures on Three Aquatic Species." *Environ. Toxicol. Chem.* 5:917–931 (1986).

149. Parker, J. G. "Toxic Effects of Heavy Metals upon Cultures of *Uronema marinum* (Ciliophora: Uronematidae)." *Mar. Biol.* 54:17–24 (1979).

150. Voyer, R. A., J. A. Cardin, J. F. Heltshe, and G. L. Hoffman. "Viability of Embryos of the Winter Flounder *Pseudopleuronectes americanus* Exposed to Mixtures of Cadmium and Silver in Combination with Selected Fixed Salinities." *Aquat. Toxicol.* 2:223–233, (1982).

151. Voyer, R. A., and J. F. Heltshe. "Factor Interactions and Aquatic Toxicity Testing." *Water Res.* 18(4):441–447 (1984).

152. Box, G. E. P., and N. R. Draper. *Empirical Model-Building and Response Surfaces.* (New York: John Wiley and Sons, 1987).

153. Rapport, D. J., H. A. Regier, and T. C. Hutchinson. "Ecosystem Behavior under Stress." *Am. Nat.* 125(5):617–640 (1985).

Chapter 5

Hypothesis Tests for Detection of Chronic Lethal and Sublethal Stress

Scientific research is a process of guided learning. The objective of statistical methods is to make that process as efficient as possible.[1]

I. GENERAL

Predictive models such as those described in Chapter 4 can be readily applied to acute toxicity and, in many cases, to chronic lethal or sublethal stress (e.g., Suter et al.[2]). However, as effects become more subtle and difficult to model, predictive models are often replaced by hypothesis tests that simply test for the presence of a significant effect. This chapter examines hypothesis testing as applied to chronic toxicity and sublethal indicators of stress.

Regrettably, application of particular chronic lethal or sublethal responses are not discussed, although such a discussion is obviously desirable. Only general statistical methods are treated as there is insufficient space to describe specific techniques in adequate detail. Fortunately, excellent reviews of such techniques already exist.[3-5]

II. METHOD SELECTION

Weber et al.[6] provided a thorough overview of statistical methods applicable to regulatory assessment of chronic toxicity and sublethal effects. Their approach, as diagrammed for fathead minnow larval survival (Figure 1), is used as an outline for this chapter with supplemental material added as required. Much of the supplemental material is taken from a recent review of post-analysis of variance methods.[7] Techniques associated with the leftmost branch in Figure 1 ("Probit Analysis") are described in detail in Chapter 4. The hypothesis-testing techniques comprising the large, central portion of Figure 1 are discussed in this chapter.

III. ONE-WAY ANALYSIS OF VARIANCE

An experimental design leading to a one-way analysis of variance (ANOVA) is assumed in much of the following discussion. In such an experiment, a series of treatments are randomly assigned, including a control or reference treatment, with several replications of each treatment level. At the end of the experiment, observations from replicates for each treatment are made for a single variable. Observations are assumed to be independent.

Formal requirements of ANOVA techniques to test for equal means include a common variance for observations from all treatments and normal distributions for the observations. Transformations of quantal (e.g., dead/alive) data, proportions (e.g., percent hatched), or other expressions of response are often necessary to satisfy the requirements of homogeneity of variance and normality. The reciprocal of a variable can be used to this end.[6] Often logarithm or square root transformations[6,8,9] are used although other power functions are also available.[9]

The most common transformation for quantal or proportional data is the arcsine square root transformation. The arcsine square root transformation prescribed for proportional survival models are discussed in Chapter 4. Such transforms for various proportions

STATISTICAL ANALYSIS OF FATHEAD MINNOW
LARVAL SURVIVAL AND GROWTH TEST

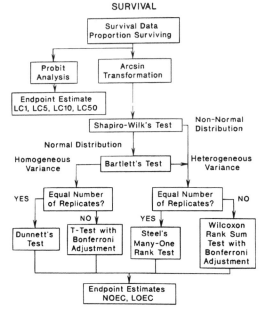

Figure 1 Scheme for analysis of chronic survival and sublethal effects recommended by Weber et al.[6] for regulatory testing.

can be extracted from Table 7 in the Appendix. Weber et al.[6] modified the procedure slightly to improve the homogeneity of variances by recommending that arcsin $\sqrt{1/(4n)}$ and (1.5708 − radians for the proportion $P = 0$) be used when the proportions are 0.0 and 1.0, respectively. (The 1.5708 − radians for $P = 0$ is the arcsine square root transformation of $(n − 1/4)/n$.) Here n is the number of animals exposed in the specific group for which the transformation is being done. Snedecor and Cochran[10] suggest this modification if n is less than 50; however, they also discuss another more accurate alternative for data sets with small n values. The purpose of these transformations is to modify quantal or proportional data to minimize the likelihood of violating the assumption of variance homogeneity required by several of the more powerful, subsequent methods. Also, the distributions for the transformed proportions are more likely to be normal than those for the untransformed proportions.[8]

In an ANOVA, the estimated total sum of squares is broken down into the sum of the among and within treatments sums of squares. The mean sum of squares$_{within}$ is an estimate of the sampling (or "error") variance, and the mean sum of squares$_{among}$ is an estimate of the additional variance associated with the treatments. Both the within and among mean sums of squares estimate a common variance (σ_0^2) if the null hypothesis of no significant difference between treatment means is true. But, if the null hypothesis is not true, the mean sum of squares$_{among}$ will be some value larger than the common variance. Consequently, the mean sum of squares$_{among}$/mean sum of squares$_{within}$, or F statistic, can be used to test the null hypothesis of equal means between treatments.[11] If the calculated value of F is larger than a specified critical value, this indicates a significant deviation, and the null hypothesis of equal means is rejected. However, the particular treatment(s) differing significantly from the others must be determined with additional, post-ANOVA testing.

Example 1

Weber et al.[6] provided the following fathead minnow (*Pimephales promelas*) larval survival data for sodium pentachlorophenol (NaPCP) exposure. Ten fish were exposed in each tank. Proportions of the total number alive at the end of the exposure were reported. We analyzed the arcsine transformed data using a one-way ANOVA.

Toxicant Concentration (μg/L)	Replicate − Proportion Surviving			
	A	B	C	D
0	1.0	1.0	0.9	0.9
32	0.8	0.8	1.0	0.8
64	0.9	1.0	1.0	1.0
128	0.9	0.9	0.8	1.0
256	0.7	0.9	1.0	0.5
512	0.4	0.3	0.4	0.2

These quantal data are transformed using values from Table 7 in the Appendix. For 1.0 (all 10 fish survived the exposure), the transformation used is that recommended by Weber et al.[6] or $1.5708 - \arcsin\sqrt{1/(4n)} = 1.5708 - 0.1588 = 1.412$. Substituting these values into the table.

Toxicant Concentration (μg/L)	Replicate			
	A	B	C	D
0 (Treatment 1)	1.412	1.412	1.249	1.249
32 (Treatment 2)	1.107	1.107	1.412	1.107
64 (Treatment 3)	1.249	1.412	1.412	1.412
128 (Treatment 4)	1.249	1.249	1.107	1.412
256 (Treatment 5)	0.991	1.249	1.412	0.785
512 (Treatment 6)	0.685	0.580	0.685	0.464

According to the general notation of Dixon and Massey,[12] there are six treatments ($q = 6$), including a control or reference treatment (0 μg/L). The six treatments or rows have means, $\mu_1, \mu_2, \ldots \mu_6$. The variance is assumed to be the same (σ_0^2) for all six means. (Normally, this assumption would be tested before an ANOVA.) The treatments have random samples of size, $n_1, n_2, \ldots n_6$ drawn from them. In the preceding table, there are four replicate tanks for each treatment. To develop a working table for an ANOVA, let x_{ij} be the jth observation (replicate) out of $z = 4$ from treatment i.

i	1	2	3	4	Total (T_i)	Mean ($\bar{x}_i$)
			j			
1	1.412	1.412	1.249	1.249	5.322	1.330
2	1.107	1.107	1.412	1.107	4.733	1.183
3	1.249	1.412	1.412	1.412	5.485	1.371
4	1.249	1.249	1.107	1.412	5.017	1.254
5	0.991	1.249	1.412	0.785	4.437	1.109
6	0.685	0.580	0.685	0.464	2.414	0.604
			Grand total (T_G)		27.408	
			Grand mean ($\bar{X}_G$)			1.142

Example 1 *Continued*

The total or overall variance can be estimated by s_T^2, the mean total sum of squares.

$$s_T^2 = \frac{\displaystyle\sum_{i=1}^{q} \sum_{j=1}^{z} (X_{ij} - \overline{X}_G)^2}{N - 1} \tag{1}$$

where N = the total number of observations over all treatments ($N = n_1 + n_2 + \cdots + n_6$). The numerator is the overall or total sum of squares and the denominator is the total degrees of freedom.

The variance for the among treatment means is estimated by the mean sum of squares, s_A^2.

$$s_A^2 = \frac{\displaystyle\sum_{i=1}^{q} \frac{T_i^2}{n_i} - \frac{T_G^2}{N}}{q - 1} \tag{2}$$

The numerator and denominator are the sum of squares among the treatment means and the treatment degrees of freedom, respectively.

The within treatment (or "error" or among replicate) variance is estimated by the mean error sum of squares, s_w^2.

$$s_w^2 = \frac{\displaystyle\sum_{i=1}^{q} \sum_{j=1}^{z} (X_{ij} - \overline{X}_i)^2}{N - q} \tag{3}$$

The numerator and denominator are the within treatment (or "error") sum of squares and the error degrees of freedom, respectively.

An ANOVA table of variance is then constructed.

Variance	Sum of Squares	df	Mean Sum of Squares
Among	1.574	5	0.315 (s_A^2)
Within	0.426	18	0.024 (s_w^2)
Total	2.000	23	0.087 (s_T^2)

The null hypothesis that the treatment means are the same can then be tested with an F statistic. The F calculated for comparison to tabulated values is the ratio of the mean sums of squares (MSS). In this case, $MSS_{among}/MSS_{within} = 0.315/0.024 = 13.125$. An F statistic is extracted from a table given α, $q - 1$, and $N - q$ (e.g., Table 16 in Rohlf and Sokal[13]). In this case, $F_{0.05}(5,18)$ is 2.77. There is a probability of 0.05 that the calculated F would be larger than the tabulated value of 2.77 by chance alone if all μ values were equal. In the present example, the null hypothesis is rejected because 13.125 is greater than 2.77, the critical F value. However, as the analysis stands, it remains ambiguous which particular treatment means are not equal. Tukey's or Scheffe's tests could be used to resolve this ambiguity. They would test for signficant differences among all pairs of means. The significance probability (α) would be based on all possible comparisons of means, i.e., $[q(q - 1)]/2$ or 15 comparisons.

Example 1 *Continued*

The following SAS code performs the calculations described above including Tukey's and Scheffe's tests.

```
DATA FATHEAD;
  INPUT ARC CONC @@;
  CARDS;
  1.4120    0 1.4120    0 1.2490    0 1.2490    0
  1.1071   32 1.1071   32 1.4120   32 1.1071   32
  1.2490   64 1.4120   64 1.4120   64 1.4120   64
  1.2490  128 1.2490  128 1.1071  128 1.4120  128
  0.9912  256 1.2490  256 1.4120  256 0.7854  256
  0.6847  512 0.5796  512 0.6847  512 0.4636  512
  ;
PROC GLM;
  CLASS CONC;
  MODEL ARC=CONC;
  MEANS/TUKEY SCHEFFE;
RUN;
```

The wide range of methods to perform treatment comparisons after an ANOVA was reviewed recently by Day and Quinn[7] and Tukey.[14] Some are concerned only with comparison of a specific pair of means but others involve comparisons among all means. Still others compare one mean such as the control mean to several means. Depending on the specific method, one of several kinds of errors may be pertinent. Consequently, a brief review of statistical errors pertinent to these post-ANOVA hypothesis tests follows.

A Type I error occurs if the null hypothesis is rejected when it is true. The Type I error rate (number of incorrect statements/total number of statements if the experiment was repeated many times) can be expressed either as a significance probability (α) or a confidence probability or coefficient $(1 - \alpha)$.[15] In contrast, a Type II error occurs if the null hypothesis is accepted when it is not true. Its associated probability is designated β. The power of a test is the probability of correctly rejecting the null hypothesis when it is not true, i.e., $1 - \beta$. The Type I error rate for post-ANOVA comparisons can be estimated for a specific comparison or for a set of comparisons. The definition for error rate given here is sufficient without elaboration for the case of a single comparison. An experimentwise (or familywise) Type I error rate is used in the case involving many comparisons. The experimentwise error rate is "the proportion of experiments in which at least one false rejection of the null hypothesis is made."[16] A clear understanding of the distinction between the experimentwise Type I error rate and the Type I error rate for a specific comparison of paired means (pairwise Type I error rate) is critical to understanding the procedures in the rest of this chapter.

IV. TEST OF NORMALITY: SHAPIRO-WILK'S TEST

The assumptions of normality must be assessed before using an ANOVA and post-ANOVA tests such as Dunnett's test, a t-test with Bonferroni's adjustment, Dunn-Šidák's t-test, or Williams's test. These assumptions may be assessed using one of a variety of tests.

Graphical methods including probability or quantile-quantile plotting can be used to visually assess the normality of data. Miller[9] and Sokal and Rohlf[18] provided excellent overviews with examples for several of these techniques. Miller[9] recommended visual inspection of probit or NED plots (untransformed or transformed data) as outlined in

Chapter 4 for detecting nonnormality of data. The n values are ordered from smallest ($i = 1$) to largest ($i = n$). Next, the probit or NED for $i/(n + 1)$ is plotted against the observation value (or its transform if appropriate). If the points produce a straight line, the data are judged to be sufficiently normal. "If a deviation from normality cannot be spotted by eye on probit paper, it is not worth worrying about."[9] Miller[9] suggested that particular attention be paid to deviations at the two ends of the line because they were more important than values toward the center of the line for tests comparing means, e.g., an ANOVA and related tests. The reader interested in fuller discussion, including the specific effects of kurtosis and skewness on ANOVA, is referred to pages 5 through 16 in Miller.[9]

The assumption of normality can be assessed more formally with Shapiro-Wilk's,[17] χ^2, or Kolmogorov-Smirnov tests.[18] Miller[9] argued against the use of the latter two tests because they were less sensitive to deviations at extremes than to the less crucial deviations at the distribution center. The recommended Shapiro-Wilk's test is used here to test the null hypothesis that the distribution from which observations were taken is normal. If the null hypothesis is rejected, the distribution is judged to be nonnormal at a specified α. It is important to note that, in the strictest sense, acceptance of the null hypothesis does not prove that a distribution is normal. Indeed, with a sufficiently large sample, the null hypothesis would be rejected for almost any data set. Fortunately, several of these tests such as ANOVA[9,10,12] produce acceptable results despite moderate violations of normality or homogeneity of variance. In fact, Monte Carlo studies indicate that probabilities derived from an ANOVA are close to the real probabilities if the underlying distribution is at least symmetrical and the variances for the treatments are within three-fold of each other.[19]

Shapiro-Wilk's test compares ordered observations to expected order statistics for a normal distribution. The set of n observations are denoted as $x_1, x_2, \ldots x_n$. The observations are ordered and relabeled such that $y_1 \leq y_2 \leq \cdots \leq y_n$. Then, using these n ordered observations and coefficients from Table 9 in the Appendix, the following calculations are performed.

$$s^2 = \sum_{i=1}^{n} (y_i - \bar{y})^2 = \sum_{i=1}^{n} (x_i - \bar{x})^2 \tag{4}$$

If n is an even number, define $k = n/2$ and estimate b with Equation (5).

$$b = \sum_{i=1}^{k} a_{n-i+1}(y_{n-i+1} - y_i) \tag{5}$$

where a_{n-i+1} = values taken from Table 9 ($n \leq 50$). If n is an odd number, define $k = (n - 1)/2$ and estimate b with Equation (6).

$$b = a_n(y_n - y_1) + \cdots + a_{k+2}(y_{k+2} - y_k) \tag{6}$$

The s^2 from Equation (4) and b^2 as estimated with either Equation (5) or (6) are then used to calculate W.

$$W = \frac{b^2}{s^2} \tag{7}$$

Table 10 in the Appendix gives critical values of W for $n \leq 50$. An estimated W value smaller than the critical W value from Table 10 indicates that the H_0 of a normal distribution is rejected.

Example 2

Was the assumption of normality for the data used in Example 1 appropriate? The observations in the 128 μg/L treatment are used to illustrate each detail of this method; then the assumption of normality for all 24 observations is tested. (Note: These calculations are quite tedious. Fortunately, the SAS procedure UNIVARIATE can perform them if the number of observations is less than 2000.[20] The statement PROC UNIVARIATE NORMAL PLOT also produces a quantile-quantile probability plot.)

The 128 μg/L treatment:

$$n = 4$$
$$x_1, \ldots, x_4 = 1.412, 1.107, 1.249, 1.249$$
$$y_1, \ldots, y_4 = 1.107, 1.249, 1.249, 1.412$$
$$\bar{x} = 1.254$$
$$s^2 = (1.107 - 1.254)^2 + (1.249 - 1.254)^2 + (1.249 - 1.254)^2$$
$$+ (1.412 - 1.254)^2$$
$$= 0.0466$$

Because n is even, b is estimated with Equation (5) and $k = 2$.

$$b = a_4(y_4 - y_1) + a_3(y_3 - y_2)$$
$$= 0.6872(1.412 - 1.107) + 0.1677(1.249 - 1.249)$$
$$= 0.2096$$
$$b^2 = 0.2096^2 = 0.0439$$
$$W = 0.0439/0.0466 = 0.9421$$

The critical value from Table 10 for W at $\alpha = 0.05$ and $n = 4$ is 0.748. Consequently, there is insufficient evidence for a significant deviation from a normal distribution.

All observations: The pertinent treatment mean must be subtracted from each value within the treatment before W can be estimated. The resulting values, adjusted for differences in treatment means, are then arranged from smallest to largest and used according to the procedures described above.

i	Adjusted Value	$(y_i - \bar{y})^2$	a_{n-i+1}	$a_{n-i+1}(y_{n-i+1} - y_i)$
1	−0.324	0.102	0.4493	0.282
2	−0.147	0.020	0.3098	0.116
3	−0.140	0.018	0.2554	0.076
4	−0.122	0.014	0.2145	0.044
5	−0.118	0.013	0.1807	0.036
6	−0.081	0.006	0.1512	0.024
7	−0.081	0.006	0.1245	0.020
8	−0.076	0.005	0.0997	0.012
9	−0.076	0.005	0.0764	0.009
10	−0.076	0.005	0.0539	0.006
11	−0.024	0.000	0.0321	0.001
12	−0.005	0.000	0.0107	0.000
13	−0.005	0.000		$b = 0.626$
14	0.014	0.000		$b^2 = 0.392$

Example 2 *Continued*

i	Adjusted Value	$(y_i - \bar{y})^2$	a_{n-i+1}	$a_{n-i+1}(y_{n-i+1} - y_i)$
15	0.041	0.001		
16	0.041	0.001		
17	0.041	0.001		
18	0.081	0.006		
19	0.081	0.006		
20	0.082	0.006		
21	0.082	0.006		
22	0.158	0.023		
23	0.229	0.050		
24	0.303	0.088		
		$s^2 = 0.382$		

The $\bar{y}$ for the adjusted values is -0.005. The s^2 is the summation of the 24 $(y_i - \bar{y})^2$ values or 0.382. The products, $a_{n-i+1}(y_{n-i+1} - y_i)$ are summed, and that sum is squared to derive b^2. The W is estimated as $b^2/s^2 = 0.392/0.382 = 1.026$. The critical value from Table 10 for $W (\alpha = 0.05, n = 24)$ is 0.916. There is no indication of significant deviation from normality for these adjusted values.

V. TEST FOR HOMOGENEITY OF VARIANCES: BARTLETT'S TEST

The homogeneity of variances must also be examined because it is a formal requirement for ANOVA. Bartlett's test can be used under the assumption that the observations are normally distributed. The null hypothesis is equal variances of the treatments. The resulting statistic is compared with a critical χ^2 with an associated degrees of freedom $(q - 1)$ and α. The calculations are the following.

$$M = \left(\sum_{i=1}^{q} df_i\right)\ln \bar{s}^2 - \sum_{i=1}^{q} df_i \ln s_i^2 \tag{8}$$

where $\bar{s}^2 = $ the mean of all of the individual treatment variances; and $df_i = $ the degrees of freedom for treatment i (e.g., $n_i - 1$).

$$C = 1 + \left[\frac{1}{3(q - 1)}\right]\left[\sum_{i=1}^{q} \frac{1}{df_i} - \frac{1}{\sum_{i=1}^{q} df_i}\right] \tag{9}$$

The M/C is compared to a critical χ^2.

Example 3
Let's apply Bartlett's method to test for variance homogeneity in the data used in Examples 1 and 2.

i	Treatment (μg/L)	s_i^2	$\ln s_i^2$	df_i $(n_i - 1)$
1	0	0.008856	-4.727	3
2	32	0.023256	-3.761	3
3	64	0.006642	-5.014	3

Example 3 *Continued*

4	128	0.015541	−4.164	3
5	256	0.076770	−2.567	3
6	512	0.011099	−4.501	3
			$\Sigma -24.734$	18

$\bar{s}^2 = 0.023694$
$q = 6$
$M = (18)(\ln 0.023694) - (* \ 3 \ -24.734) = 6.837$
$C = 1 + [1/15][6/3 - 1/18] = 1.130$
$M/C = 6.05$

A critical value of 11.07 is obtained from a table of χ^2 values ($\alpha = 0.05$, $df = 5$), e.g., Table 14 in Rohlf and Sokal.[13] As M/C is less than this critical χ^2, the null hypothesis of equal variances is not rejected. Strictly speaking, this indicates no statistically significant deviation from homogeneity of variances. It does not prove that the variances are equal.

Both tests for normality and homogeneity of variance are conducted before an ANOVA to assess underlying assumptions. Further, they are useful in evaluating transformations. The tests are run before and after transformations such as those described above to determine the effectiveness of the trasnformations.

VI. TREATMENT MEANS COMPARED WITH THE CONTROL MEAN

A. DUNNETT'S TEST

1. Equal Number of Observations
Often the null hypothesis that treatment means are the same as a control or reference group mean is tested instead of the null hypothesis of equivalent means for all treatments. In contrast to Example 1 in which Tukey's or Scheffe's tests were mentioned as tests for significant differences among all means, the means for treatments exposing fish to various concentrations of NaPCP might be compared only with that for unexposed fish. Methods such as Dunnett's test[21,22] accommodate multiple comparisons of this nature and provide a more powerful method of testing this hypothesis than application of Tukey's or Scheffe's test. (The increase in power is a consequence of only $q - 1$ comparisons being made instead of $[q(q - 1)]/2$ comparisons.) Dunnett's test is based on the assumptions of normality and equal variances (Figure 1). In the example used above, the differences between the mean of the 0 µg/L and the five other treatments would be tested jointly with an experimentwise α as follows.

Let p be the number of treatments to be compared with the control ($p = q - 1$). The estimated means of the treatments are designated $\bar{x}_1, \ldots, \bar{x}_p$, and the estimated mean of the control is designated $\bar{x}_0$. The standard deviation for the q sets of observations is s. The number of observations in each treatment is designated as n_i.

$$s^2 = \frac{\sum_{i=0}^{p} \sum_{j=1}^{n_i} (x_{ij} - \bar{x}_i)^2}{\left(\sum_{i=0}^{p} n_i\right) - (p + 1)} \tag{10}$$

Notice that Equation (10) is the same as Equation (3) for the within treatment variance, except minor changes in notation were made to conform to Dunnett's original description of the technique.[21] Often, an ANOVA is performed before implementation of Dunnett's test to facilitate estimation of this variance. An ANOVA can also be used with more complicated experimental designs to assure that interaction terms are insignificant. Dunnett[22] discusses estimation if interactions are important.

If the number of observations is the same for all treatments, including the control, an "allowance" is estimated.

$$A = ts\sqrt{2/n} \tag{11}$$

This allowance is used to establish a confidence limit for the differences between the two means beyond which the difference is significant. Table 11 ($\alpha = 0.01$) or 12 ($\alpha = 0.05$) may be used for one-sided tests. Table 13 ($\alpha = 0.01$) or 14 ($\alpha = 0.05$) may be used for two-sided tests.

Example 4
The fathead minnow toxicity data used in the preceding examples are analyzed with Dunnett's procedures. A one-sided test with a lower limit is appropriate because the toxicant is expected to lower, rather than raise the chance of survival. A two-sided test is discussed briefly only to illustrate the associated calculations. Note that the "i" used here is equivalent to $i - 1$ in previous examples.

i	$\bar{x}_i$	$\Sigma(x_{ij} - \bar{x}_i)^2$
0	1.330	0.02657
1	1.183	0.06977
2	1.371	0.01993
3	1.254	0.04662
4	1.109	0.23031
5	0.604	0.03330
		Σ 0.4265

$\Sigma n_i = 24$
$s^2 = 0.4265/(24 - 6) = 0.02369$ (within treatment mean sum of squares in Example 1)
$s = 0.1539$

One-sided Test (Lower Limit): The allowance (A) is the allowable difference in treatment means. In other words, A estimates the lower confidence limit for the differences $\bar{x}_i - \bar{x}_0$.

From Table 12 in the Appendix, t for a one-sided test ($\alpha = 0.05$, $df = 18$, $p = 5$) is 2.41.

$A = 2.41 * 0.1539 * \sqrt{2/4} = 0.2623$

i	$\bar{x}_i - \bar{x}_0$
1	$1.183 - 1.330 = -0.147$
2	$1.371 - 1.330 = 0.041$
3	$1.254 - 1.330 = -0.076$
4	$1.109 - 1.330 = -0.221$
5	$0.604 - 1.330 = -0.726$

Example 4 *Continued*

Only $\bar{x}_5$ is significantly different ($\alpha = 0.05$) from $\bar{x}_0$ because $|-0.726|$ is a larger difference for $\bar{x}_5 - \bar{x}_0$ than 0.2623. No other differences are significant at the chosen α.

Two-sided test: Although not pertinent to this specific example, a two-sided test for these data might involve estimation of both upper and lower confidence limits. The A would be estimated with a t taken from Table 14 instead of Table 12 in the Appendix.

$$t = 2.76$$
$$A = 2.76 * 0.1539 * \sqrt{2/4} = 0.3004$$

The A would be compared with both negative and positive differences between means.

The substitution of the following SAS code for the last five lines of code in Example 1 allows calculation of two-sided and one-sided (lower limit) Dunnett's tests.

```
PROC GLM;
    CLASS CONC;
    MODEL ARC=CONC;
    MEANS/DUNNETT("0") DUNNETTL("0");
RUN;
```

2. Unequal Number of Observations

By design or mishap, the numbers of observations in the various treatments may not be equal. In fact, there are sound reasons for having more observations in the control than in the exposure treatments.[1,21,23] For an α of 0.05 or less, the optimum allocation of observations between the control and treatments can be estimated.[21,23]

$$\frac{n_0}{n_i} = \sqrt{q - 1} \tag{12}$$

In our example, the optimum number of observations (replicate tanks) in the control would be 2.24 times that in an exposure treatment, e.g., nine control tanks and four replicate tanks for each exposure treatment. Dunnett's test may be used with unequal n values for the control and treatments by substituting the A estimated in Equation (13) for that estimated by Equation (11).

$$A = ts\sqrt{\frac{1}{n_0} + \frac{1}{n_i}} \tag{13}$$

However, t values in the associated tables were developed under the condition of equal observation numbers or, more precisely, a correlation of $1/2$.[21,22] An estimate of this correlation if the number of observations differ is given in Equation (14).

$$\rho_{ij} = \frac{1}{\dfrac{n_0}{n_i} + 1} \tag{14}$$

Increasing the number of control observations (n_0) relative to the number in the exposure treatments (n_i) will lower the value of ρ_{ij} to less than 1/2. This will decrease the P associated with the confidence limits below the assumed 0.95 or 0.99. Consequently, use of the approach described above for a one-sided test will produce an approximate (conservative) test when observation numbers are unequal. However, the values of P do not differ much from the tabulated values for a wide range of ρ_{ij} values about 1/2. Day and Quinn[7] recommended Dunnett's test using Kramer's modification[24] [Equation (13)] as the best method for control-treatment comparisons with unequal observation numbers.

For two-sided tests, Dunnett[22] provided a correction factor for adjusting the tabulated t values for unequal observation numbers between the control and the other treatments. First, $1 - (n_i/n_0)$ is calculated. The result is multiplied by the superscript in the appropriate table of t values (Tables 13 and 14 in the Appendix). This factor is the percentage by which the tabulated t value should be increased to allow for the greater number of control observations. If there were nine control replicates in the above-mentioned example, the tabulated t value (2.76) would be adjusted by (3.6(1-4/9)) or 2%. The adjusted t value would be 2.76*1.02 or 2.815. More precise estimates for one- and two-sided tests when the correlation is not 1/2 can be generated easily by comprehensive statistical programs such as SAS. For example, the SAS algorithm used in Example 4 will perform these calculations when observation numbers are unequal between the control and exposure treatments, and among exposure treatments. When the numbers of observations are different among experimental treatments, the correlation is estimated by the SAS program using the harmonic mean of the observation numbers.

B. t-TEST WITH BONFERRONI'S ADJUSTMENT

Weber et al.[6] recommended that Bonferroni's adjusted t-test be used instead of Dunnett's test if the numbers of observations are unequal because the confidence limit for Dunnett's test is approximate with unequal numbers of observations. However, for the two-sided test, adjustment is easily accomplished for Dunnett's test as discussed. Further, this is not a general restriction as some computer applications adjust ρ_{ij} accordingly. The SAS procedure GLM[20] adjusts the ρ_{ij} and can be used as a more powerful tool than the recommended t-test with Bonferroni's adjustment. (The difference in power results from the fact that the t-test with Bonferroni's adjustment sets an upper limit (α) on the experimentwise error rate, but Dunnett's test fixes α for the experimentwise error rate.)

Generally, Bonferroni's adjustment of the t-test modifies the α associated with the hypothesis test for each pair of means to accommodate the multiple comparisons. An upper limit for the experimentwise α (the probability of making at least one Type I error over a series of comparisons) is defined, e.g., 0.05, and an adjusted α' is estimated for use in testing each individual pair in the experiment to maintain this upper limit for the experimentwise error rate. With Bonferroni's adjustment, α' is estimated by α/p, the experimentwise α divided by the number of pairs being tested in the entire experiment. For example, the α' used for testing the individual pairs in Example 1 would be 0.05/5 or 0.01. The α' is used to extract pertinent t values used for the p comparisons. Obviously, such a procedure would require t values for adjusted α values. Such tables of t values are provided in the Appendix (Tables 15 to 18). The experimentwise α, df, and p are used to extract a critical t from these tables. The values in the columns for $p = 1$ (mean of the control versus the mean of one treatment only) are simply Student's t statistics because $\alpha/p = \alpha/1 = \alpha$.

Under the assumptions of normality and homogeneity of variance, t_i is estimated with Equation (15) for each pair of means. The s is estimated with the square root of the within treatment mean sum of squares [Equation (3)].

$$t_i = \frac{\bar{X}_0 - \bar{X}_i}{s\sqrt{\dfrac{1}{n_0} + \dfrac{1}{n_i}}} \tag{15}$$

Each value is compared with tabulated Bonferroni's t values. In the Appendix, Tables 15 ($\alpha = 0.01$) and 16 ($\alpha = 0.05$) are used for one-sided tests, and Tables 17 ($\alpha = 0.01$) and 18 ($\alpha = 0.05$) for two-sided tests. Values in Tables 15 and 16 were generated with the SAS function TINV using as arguments: $1 - \alpha/p$ as the quantile, df, and 0 as the distribution center.[26]

C. DUNN-ŠIDÁK *t*-TEST

A preferred alternative to the t-test with Bonferroni's adjustment is the Dunn-Šidák t-test.[27] Like Bonferroni's adjusted t-test, this test sets an upper limit (α) on the experimentwise error rate and, as a consequence, it also has less power than Dunnett's test. However, the Dunn-Šidák t-test generally has slightly more power than the t-test with Bonferroni's adjustment.[7] (Less conservative estimates of P based on the Bonferroni adjustment are also available, e.g., Wright[28] and provide more power than the commonly used Bonferroni adjustment described above.)

Although the Dunn-Šidák t-test also adjusts the α to get an upper limit for the experimentwise error rate, the adjustment is different from Bonferroni's adjustment. The use of an adjusted α (α') as defined below, results in an experimentwise error rate $\leq \alpha$ for the p comparisons.[18]

$$\alpha' = 1 - (1 - \alpha)^{1/p} \tag{16}$$

The t values associated with the resulting unconventional values of α' cannot be taken directly from most Student's t tables. Instead tables such as Table 19 (one-sided test) or 20 (two-sided test) in the Appendix are required. (Values in Table 19 were generated with the SAS function TINV using as arguments: $(1 - \alpha)^{1/p}$ as the quantile, df, and 0 as the distribution center).[26] Use of two-tailed tables of Dunn-Šidák's t values for one-sided tests by halving the experimentwise α has been recommended by Rohlf and Sokal.[13] However, a minor inaccuracy occurs during such use[30] and tables such as Table 19 in the Appendix should be used instead. Note that there is no column for $p = 1$ in these tables. As described with the Bonferroni's t tables, the Dunn-Šidák t values are simply Student's t values when $p = 1$. The adjustment on the t values as defined by Equation (16) becomes simply $\alpha' = \alpha$ when $p = 1$. Consequently, the t values provided for $p = 1$ in the Bonferroni's t tables (Student's t statistic) can be used as the Dunn-Šidák t for $p = 1$.

Example 5

Both a t-test with Bonferroni's adjustment and a Dunn-Šidák test will be used to analyze the fathead minnow survival data.

i	$\bar{X}_0 - \bar{X}_i$	t_i
1	0.147	1.352
2	−0.041	−0.377
3	0.076	0.699
4	0.221	2.032
5	0.726	6.676

Example 5 *Continued*

t-test with Bonferroni's Adjustment: The t value for $\alpha = 0.05$, $q - 1 = p = 5$, $df_{within} = 18$ is 2.552 (Table 16 in the Appendix). The t_i for treatment 5 (512 μg/L) was larger than the critical t value indicating that the mean for this treatment was significantly different from that of the control (0 μg/L). No other treatment means were significantly different from that of the control.

t-test by Dunn-Šidák's Method: The t value for an experimentwise $\alpha = 0.05$, $p = 5$, and $df_{within} = 18$ is 2.543 (Table 19 in Appendix). The t_i calculated above for treatment 5 is greater than 2.543 leading to rejection of the null hypothesis of equal means. Again, none of the other means differed significantly from the control mean.

VII. MONOTONIC TREND: WILLIAMS'S TEST

Williams[23,31] argued that the methods described above do not make use of all the information available to researchers. If one assumes that response changes monotonically with increasing dose, a more powerful approach (isotonic regression) may be developed. William's approach has a null hypothesis that mean responses are equal among treatment doses and an alternate hypothesis that mean responses monotonically increase or decrease with treatment dose. The test is executed in two stages. In the first stage, the presence or absence of a significant deviation from the null hypothesis is tested. Next, the lowest dose producing a significant mean response is identified.

Williams's procedure with equal numbers of observations (n) among treatments can be summarized as follows. In the design, there is a control ($i = 0$) and p treatments (dose levels, $i = 1$ to p) with dose increasing with i. The mean responses ($\bar{X}_i$ values) are assumed to be independent and normal. The observations have a common variance (σ^2) estimated by s^2, the mean sum of squares$_{within}$.

Next the $\bar{X}_i$ values are used to produce a series of maximum likelihood estimates of expected mean responses assuming the alternate hypothesis of monotonic ordering with dose. The estimates are $M_0 \leq M_1 \leq M_2 \leq M_3 \leq \cdots \leq M_p$ if the mean responses increase with dose. If mean responses are expected to decrease, the signs are changed on the means ($\bar{X}_i$) before use. One of the inequalities between M_i values must be strict, i.e., all cannot be equal. Adjustment must also be made if the $\bar{X}_i$ values (or $-\bar{X}_i$ values) do not satisfy the series, $\bar{X}_0 \leq \bar{X}_1 \leq \bar{X}_2 \leq \bar{X}_3 \leq \cdots \leq \bar{X}_p$. The adjustment process involves the following steps.

Assume that $\bar{X}_0 \leq \bar{X}_i > \bar{X}_{i+1} \leq \bar{X}_{i+2} \leq \cdots \leq \bar{X}_p$. The $\bar{X}_i > \bar{X}_{i+1}$ must be adjusted to satisfy the inequality series for the M_i values. The two means are replaced by a single estimated mean.

$$\bar{X}_{i,i+1} = \frac{w_i \bar{X}_i + w_{i+1} \bar{X}_{i+1}}{w_i + w_{i+1}} \tag{17}$$

The weights (w_i, w_{i+1}) are proportional to the number of observations used to produce each mean. Equation (17) reduces to Equation (18) because n is the same for all treatments.

$$\bar{X}_{i,i+1} = \frac{\bar{X}_i + \bar{X}_{i+1}}{2} \tag{18}$$

This estimated mean ($\bar{X}_{i,i+1}$) replaces the two means ($\bar{X}_i$, $\bar{X}_{i+1}$) in the series. This process is repeated until the series of inequalities is satisfied. For every such replacement, the number of unique means in the series is reduced by one.

The largest M in the final series (M_p) is compared with the $\bar{X}_0$ to test the null hypothesis. The test statistic $(\bar{t}_p)$ is estimated with Equation (19).

$$\bar{t}_p = \frac{M_p - \bar{X}_0}{s\sqrt{2/n}} \qquad (19)$$

If the estimated $\bar{t}_p$ is greater than a critical $\bar{t}_{p,\alpha}$ (Table 21 for $\alpha = 0.01$ or Table 22 for $\alpha = 0.05$ in the Appendix), the null hypothesis is rejected. However, no further conclusion, except that a significant response is present at the highest M, may be made at this point. More information is extracted during a second step in which the lowest dose with a significant response is determined. A $\bar{t}_{p-1}$ is calculated with Equation (19) for M_{p-1} instead of M_p and compared with a critical value of $\bar{t}_{p-1,\alpha}$. The null hypothesis that the M_{p-1} is the same as the M_0 is rejected if the estimated $\bar{t}_{p-1}$ is greater than the critical $\bar{t}_{p-1,\alpha}$. The process is repeated for $M_{p-2}, M_{p-3}, \ldots M_1$ to determine the lowest dose at which a significant response was detected.

Example 6

The fathead minnow survival data are be used to illustrate Williams's test as applied to treatments with equal numbers of observations ($n = 4$). From previous examples, the following information is available: $s^2 = 0.024$, $df = 18$, $p = 5$, $q = 6$.

$\bar{X}_i$	$-\bar{X}_i$	M_i	$df, "p"$	$\bar{t}_p$	$\bar{t}_{p,\alpha}$
1.330	−1.330	−1.330			
1.183	−1.183	−1.277	18,1	0.484	1.734
1.327	−1.371	−1.277	18,2	0.484	1.818
1.254	−1.254	−1.254	18,3	0.694	1.845
1.109	−1.109	−1.109	18,4	2.017	1.859
0.604	−0.604	−0.604	18,5	6.627	1.867

Is there a significant response? Because the assumed response is a decrease in survival with increasing dose, the sign of the mean responses must be changed before proceeding (see column 2, $-\bar{X}_i$). Notice that $-\bar{X}_1$ is greater than $-\bar{X}_2$. This pair must be modified as per Equation (18) to generate the associated M_i

$$\bar{X}_{1,2} = (-1.183 - 1.371)/2 = -1.277$$

The M_i values for all treatment means are provided in column 3 above. Using Equation (19), $\bar{t}_p$ is estimated from the largest M_i.

$$\bar{t}_p = (-0.604 + 1.330)/\sqrt{2*(0.024)/4} = 6.627$$

From Table 22 in the Appendix, a critical value of $\bar{t}_{5,\alpha=0.05}$ (one-sided test) is found to be 1.867. Because 6.627 is greater than 1.867, the null hypothesis is rejected. (The value for "p" = 1 in the above table is taken from a Student's t distribution table for $\alpha = 0.05$, $df = 18$ for a one-sided test. Alternatively, the same value can be obtained from a table for a two-sided test by using $2*\alpha$ or 0.10 as the α for that table. Bonferroni adjusted t values for $p = 1$ within Tables 15 to 18 in the Appendix are Student's t values that may be used for this purpose.)

Example 6 *Continued*

What is the lowest concentration with a significant response? The process is repeated for all M_i values with critical $\bar{t}_{p,\alpha}$ values taken from Table 22 using the *df* and "*p*" listed in column 4 of the table above. The results are provided in columns 5 and 6. The lowest concentration with a significant response was 256 μg/L.

Williams[23] also provides a modification of this approach if the number of observations is not equal between the control and experimental treatments. If the number of control observations (n_0) is equal to or greater than the number of observations in each experimental treatment (n_i), $\bar{t}_i$ can be estimated. Again, there are sound reasons for n_0 to be greater than n_i. Like Dunnett's calculations,[21] Williams's power calculations suggested that n_0/n_i $(= w)$ should be equal to or slightly higher than $\sqrt{q} - 1$.

$$\bar{t}_i = \frac{M_i - \bar{X}_0}{s\sqrt{\dfrac{1}{n_i} + \dfrac{1}{n_0}}} \tag{20}$$

The critical $\bar{t}_{i,\alpha}$ must now be adjusted for n_0/n_i. Recognizing that the $\bar{t}_{i,\alpha}$ decreases linearly with $1/w$, Williams estimated a series of β_t values for extrapolating within Tables 21 and 22 to account for this effect. They are placed as exponents in the tables. The value of $\bar{t}_{i,\alpha}$ with a w of 1 (i.e., $\bar{t}_{i,\alpha(1)}$) is modified with w and β_t using Equation (21).

$$\bar{t}_{i,\alpha(w)} = \bar{t}_{i,\alpha(1)} - 10^{-2}\beta_t\left(1 - \frac{1}{w}\right) \tag{21}$$

The modified $\bar{t}_{i,\alpha(w)}$ is the critical value compared to the $\bar{t}_p$ values estimated with Equation (20).

Example 7

Let's assume that there were eight control observations and four experimental (noncontrol) observations in a data set similar to the fathead minnow data used previously ($w = 8/4 = 2.00$). Assume that the s^2 is 0.024. The *df* would be $28 - 6$ or 22. The $\bar{X}_0$ and M_p are -1.330 and -0.604, respectively. Equation (20) is used to estimate a $\bar{t}_i$ of 7.653.

From Table 22 ($\alpha = 0.05$, *df* = 22, $p = 5$), $\bar{t}_{i,\alpha(1)}$ and β_t are found to be 1.846 and 5, respectively.

$$\bar{t}_{i,\alpha(2)} = \bar{t}_{i,\alpha(1)} - 10^{-2}*\beta_t*(1 - (1/w))$$
$$= 1.846 - 10^{-2}*5*(1 - 0.5) = 1.821$$

The null hypothesis is rejected because $\bar{t}_5$ is greater than $\bar{t}_{i,\alpha(2)}$.

Williams's test can still be used if unequal numbers of observations occur among noncontrol treatments. Williams[23] used the maximum likelihood estimators previously described [Equation (17)] with the number of observations for each treatment used as weights. Equation (22) is a formal expression of the procedure.

$$M_i = \text{MAX}_{1 \leq u \leq i} \text{MIN}_{i \leq v \leq p} \frac{\displaystyle\sum_{j=u}^{v} n_i \overline{X}_i}{\displaystyle\sum_{j=u}^{v} n_i} \tag{22}$$

Equation (22) is used to generate the maximum likelihood estimates of the means. Tables 21 and 22 can still be used provided there are only moderate differences in n_i values. Williams's[23] reported that these tables may be applied with confidence if $0.80 \leq n_i/n_p \leq 1.25$ for all $1 \leq i \leq p - 1$.

Example 8

Williams's test is performed on a modified, fathead minnow survival data set. One additional observation is added to the 0 (transform = 1.412), 32 (transform = 1.107), 64 (transform = 1.249), and 256 (transform = 0.991) μg/L treatments to produce this modified data set. Thus, these four treatments now have five observations each and the remaining two treatments (128 and 512 μg/L) have four observations each. (The new s is estimated to be 0.145. The data passed tests for variance homogeneity and normality at $\alpha = 0.05$. As all n_i/n_p are less than or equal to 1.25, Table 22 can be used.) Only values from Table 22 for the two treatments with observation numbers different from that of the control (128 and 512 μg/L) need to be adjusted with Equation (21) before tabulation below.

$\overline{X}_i$	$-\overline{X}_i$	M_i	df, "p"	$\bar{t}_p$	$\bar{t}_{p,\alpha}$
1.347	−1.347	−1.347			
1.168	−1.168	−1.257	22,1	0.982	1.717
1.347	−1.347	−1.257	22,2	0.982	1.798
1.254	−1.254	−1.254	22,3	0.957	1.825
1.086	−1.086	−1.086	22,4	2.848	1.838
0.604	−0.604	−0.604	22,5	7.644	1.846

The null hypothesis is rejected for the last two treatments. The lowest concentration displaying a mean survival significantly less than that of the control is the 256 μg/L treatment.

If it is ambiguous whether the monotonic trend involves an increase or decrease, a two-sided Williams's test is appropriate. Tables such as Tables 23 ($\alpha = 0.01$) and 24 ($\alpha = 0.05$) in the Appendix can be used for this purpose. The testing process is analogous to that described above (e.g., Examples 7 and 8) for the one-sided tests.

VIII. STEEL'S MULTIPLE TREATMENT-CONTROL RANK SUM TEST

If the assumption of normality is rejected for the data, the parametric methods described should not be used (Figure 1). Less powerful nonparametric tests such as Steel's rank sum tests[16,32,33] must be used instead. Although equal variance is a formal requirement for these tests, they are believed to be relatively robust to variance heterogeneity.[32] They assume a continuous distribution for the measured variable. Steel's test as described initially requires equal numbers of observations for all treatments and the control. Later, a modification for unequal numbers of observations in the control versus the experimental treatments is discussed.

The formal null hypothesis is that all observations come from the same population regardless of treatment. This population is described by the cumulative distribution function, F. The null hypothesis can be written $F_0 = F_1 = \cdots = F_p$ where F_0 is the distribution of the control observations and $F_1, \cdots, F_p$ (or F_i values) are those for the experimental treatments 1 to p. It is tested using a specified experimentwise error rate. The alternative hypothesis is that one or more F_i is "stochastically larger"[32] than F_0 i.e., $F_0 < F_i$, $F_0 > F_i$, or $F_0 \neq F_i$. If $F_0 < F_i$, the median and other percentiles of the experimental treatment are larger than those of the control. The opposite is true if $F_0 > F_i$. Steel[32] describes the locations of the control and treatment distributions as "different" in the case $F_0 \neq F_i$.

To test the null hypothesis, the observations for each control-experimental treatment pair are pooled and ranked from smallest to largest. The average rank is used in the case of ties for each of the tied values. Next, the ranks are summed for the experimental treatment observations. This rank sum for the experimental treatment is designated T_i. Next, the rank sum is calculated for the observations from the control (T_i') using the convenient relationship,

$$T_i' = (2n + 1)n - T_i \tag{23}$$

where n = number of replicates in each treatment.

The ranking and summing process is repeated for each of the p control-experimental treatment pairs. For the results from each pair, the minimum of T_i and T_i' [MIN(T_i, T_i')] is compared with a critical value from Table 26 (two-sided test) in the Appendix. With a one-sided test, whether T_i or T_i' is used as the minimum value for a particular pair for comparison with the critical value (Table 25) will determine whether the treatment is significantly smaller or greater than the control.

Example 9

The transformed fathead minnow survival data tabulated in Example 1 will be used to illustrate Steel's multiple treatment-control rank sum test. The ranking process will be illustrated with the 0 µg/L-32 µg/L pair.

Transformed Survival	Treatment (µg/L)	Rank
1.107	32	2
1.107	32	2
1.107	32	2
1.249	0	4.5
1.249	0	4.5
1.412	0	7
1.412	0	7
1.412	32	7

$T_1 = 2 + 2 + 2 + 7 = 13$
$T_1' = (2*4 + 1)*4 - 13 = 23$

The results of T_i and T_i' calculations for the five treatment pairs as illustrated for the 0 µg/L-32 µg/L pair are summarized.

i	Pair	T_i	T_i'	MIN(T_i, T_i')	Critical T Value
1	0-32	13	23	13	10

Example 9 *Continued*

2	0-64	20	16	16	10
3	0-128	15	21	15	10
4	0-256	14	22	14	10
5	0-512	10	26	10	10

For the one-sided test, the treatment decreases survival so T_i is used. The T_i values are compared with the tabulated critical Steel's rank sums T values. Because T_5 is equal to the critical T value from Table 25 in the Appendix (one-sided test, $\alpha = 0.05$, $n = 4$, $p = 5$), the null hypothesis is rejected. The null hypothesis is not rejected for any of the other pairs as their associated T_i values are greater than this critical T value. Only the 512 μg/L concentration had significantly elevated mortality. This is the same conclusion as that reached with Dunnett's test.

For a two-sided test, the MIN(T_i, T_i') would be compared with a critical value from Table 26. However, for this low number of observations and number of comparisons, no critical value can be estimated for this test statistic.

Tables 25 and 26 are limited to 2 to 9 treatment comparisons and 4 to 20 observations per comparison. To extend these tables, Steel[32] suggested using Dunnett's t statistic (Tables 11 to 14 in the Appendix). The critical T is estimated using three approximations.

$$\mu_T = \frac{n(2n + 1)}{2} \tag{24}$$

$$\sigma_T^2 = \frac{n^2(2n + 1)}{12} \tag{25}$$

$$T = \text{Integer portion of } (\mu_T - t\sigma_T) \tag{26}$$

The t in Equation (26) is obtained from the appropriate Dunnett's t table for $df = \infty$. In using Dunnett's tables for one-sided tests, the difference in the assumed correlation ($\rho_{ij} = 1/2$) as discussed earlier for Dunnett's test and the true correlation (approximately $n/(2n + 1)$) is ignored.

Miller[15] described a procedure for using Steel's test if the number of observations in the control (n_0) is different from the number in the experimental treatments but $n_1 = n_2 = n_3 = \cdots n_p$. All observation numbers for treatments (n_i) are equal to n but different from n_0. Let R_{ij} be the rank of treatment i values' jth observation and $\rho = n/(n + n_0 + 1)$.

$$R_i = \sum_{j=1}^{n} R_{ij} \tag{27}$$

For a one-sided test, the null hypothesis is rejected if MAX($R_1 \cdots R_p$) $\geq r^\alpha$. The critical r^α is estimated using Equation (28).

$$r^\alpha \approx \frac{n(n + n_0 + 1)}{2} + \frac{1}{2} + m_{p(\rho)}^\alpha \sqrt{\frac{nn_0(n + n_0 + 1)}{12}} \tag{28}$$

where $m_{p(\rho)}^\alpha$ = value obtained from tables of Gupta.[34] For a two-sided test, the null hypothesis is rejected if $R^* = MAX(R_1^* \cdots R_p^*) \geq r_*^\alpha$. The R_i^* are the $MAX(R_i, n(n + n_0 + 1) - R_i)$. The r_*^α is estimated with Equation (29).

$$r_*^\alpha \approx \frac{n(n + n_0 + 1)}{2} + \frac{1}{2} + |m|_{p(p)}^\alpha \sqrt{\frac{nn_0(n + n_0 + 1)}{12}} \tag{29}$$

where $|m|_{p(p)}^\alpha$ = value from Dunnett's tables (see Appendix) for $df = \infty$.

Day and Quinn[7] recommended that, unless the correlation was large, Fligner's[35] modification of Steel's test could be used when observation numbers are unequal between the control and experimental treatments, or between experimental treatments. The sum of ranks associated with the ith treatment (R_i) is estimated.

$$R_i = \sum_{b=1}^{n_0} \sum_{c=1}^{n_i} \psi(X_{ic} - X_{0b}) + \frac{n_i(n_i + 1)}{2} \tag{30}$$

where $\psi(a) = 1$ if $a > 0$, 0.5 if $a = 0$, and 0 if $a < 0$. An r_i is estimated as the larger of R_i and $n_i(n_i + n_0 + 1) - R_i$.

$$r_i = MAX(R_i, n_i(n_i + n_0 + 1) - R_i). \tag{31}$$

The r_i values are compared with critical values estimated using Bonferroni's adjustment to generate an experimentwise error rate. Remember that, because the Bonferroni adjustment results in an upper bound for the experimentwise error rate, this modified test will have less power than the unmodified test.

IX. WILCOXON RANK SUM TEST WITH BONFERRONI'S ADJUSTMENT

Weber et al.[6] recommended using the Wilcoxon rank sum test with Bonferroni's adjustment for the experimentwise error rate when the number of observations varied. For each experimental treatment-control pair, observations are combined and ranked. In the one-sided procedure, the values are ranked from smallest to largest if the treatment effect is thought to decrease the value of the variable relative to that of the control. If the treatment effect is thought to increase the values of the variable, the signs of the values are changed before ranking. (Each value is multiplied by -1.) For the two-sided test, the minimum of the treatment rank sum and control rank sum is compared with the test statistic. In the case of ties, each of the tied values is replaced by the average rank. Next, the ranks are summed for the experimental treatment observations (R_T) and for the control observations (R_C). This process is repeated for all treatment-control pairs. The rank sums $(R_T$ and/or $R_C)$ resulting from the p pairs are compared with critical values from Table 27 (one-sided test) or 28 (two-sided test) in the Appendix.

The critical values in Tables 27 and 28 were generated from probability tables for the Wilcoxon statistic[36,37] using experimentwise α values and the Bonferroni adjustment to estimate α'. Values for $m = 9$ or 10 in Table 27 were taken directly from Table 18 of Kokoska and Nevison.[37] The Mann-Whitney U values were estimated from Table 18 of Kokoska and Nevison[37] for $m = 9$ and 10 and used to calculate values of R for $p = 9$ and 10 [Equation (32)]. For other combinations of experimental and control replicate numbers (p and m, respectively), Mann-Whitney U values were extracted from Beyer's[36] Table X.3 using the adjusted α values (α' in Tables 27 and 28) and the number of observations in the treatment (p) and control (m). These U values were then used to produce critical rank sums for the experimental treatment with the relationship,[11]

$$R = U + \frac{p(p + 1)}{2} \qquad (32)$$

If the calculated rank sum is less than or equal to the critical rank sum value in the one-sided test table, the null hypothesis of no difference between the treatment and control would be rejected. For a two-sided test, the MIN(R_T, R_C) is compared with the critical rank sum value in the two-sided test table. (The R_c is the calculated rank sum for the control.)

Example 10

The transformed fathead minnow toxicity data as modified in Example 8 is used to perform the Wilcoxon rank sum test with Bonferroni's adjustment for a one-sided test.

Toxicant Concentration (μg/L)	Replicate				
	A	B	C	D	E
0	1.412	1.412	1.249	1.249	1.412
32	1.107	1.107	1.412	1.107	1.107
64	1.249	1.412	1.412	1.412	1.249
128	1.249	1.249	1.107	1.412	
256	0.991	1.249	1.412	0.785	0.991
512	0.685	0.580	0.685	0.464	

The rank sum values are calculated and compared with critical values in Table 27 (one-sided test with $\alpha = 0.05$).

Pair	R_T	R_C	m	p	Critical Values One-Sided Test
0-32	18.5	36.5	5	5	16
0-64	27.5	27.5	5	5	16
0-128	15.5	29.5	5	4	10
0-256	19.5	35.5	5	5	16
0-512	10.0	35.0	5	4	10

Because the R_T for the control-512 μg/L treatment comparison is equal to the critical value, the null hypothesis is rejected for this pair. None of the other control-treatment pairs were significantly different because the associated R_T values were not less than or equal to the critical rank sum value from Table 27. It is instructive to note that the more powerful Williams's test indicated significant effects at both the 256 and 512 μg/L treatments in this data set (Example 8).

X. INFERRING BIOLOGICAL SIGNIFICANCE FROM STATISTICAL SIGNIFICANCE

The methods described in above test for statistical significance, not biological significance. As pointed out by Salsburg,[19] the Neyman-Pearson sense of the term "statistically significant" is that a test simply "signified that something has happened different from the proposed [null] hypothesis." It indicates plausibility, not importance.[1] Nonetheless, judgments must be made regarding biological importance and methods have been proposed to estimate biological significance from results of hypothesis tests. Conclusions arising from these methods have shortcomings that should be recognized during any such inferential process. The following definitions as used in this process highlight some of the more important limitations.

- No Observed Effect Concentration or Level (NOEC or NOEL):
  ". . . the highest dose for which the difference with the control group is not statistically significant."[38]

  "The highest concentration of a material in a toxicity test that has no statistically significant adverse effect on the exposed population of test organisms as compared with the controls. When derived from a life cycle or partial life cycle test, it is numerically the same as the lower limit of the MATC."[39] This definition is also used by Weber et al.[6]

- Lowest Observed Effect Concentration or Level (LOEC or LOEL):
  "The lowest test dose at which the response is significantly different from the control group."[38]

  "The lowest concentration of a material used in a toxicity test that has a statistically significant adverse effect on the exposed population of test organisms as compared to the controls. When derived from a life cycle or partial life cycle test, it is numerically the same as the upper limit of the MATC [Maximum Acceptable Toxicant Concentration]."[39] This definition is also used by Weber et al.[6]

- Maximum Acceptable Toxicant Concentration (MATC):
  "An undetermined concentration within the interval bounded by the NOEC and LOEC that is presumed safe by virtue of the fact that no statistically significant adverse effect was observed."[6]

- Safe Concentration:
  "The highest concentration of toxicant that will permit normal propagation of fish and other aquatic life in receiving waters. The concept of a 'safe concentration' is a biological concept, whereas the 'no observed effect concentration' is a statistically defined concentration."[6]

Consideration of the definitions for LOEL and NOEL suggest an immediate difficulty. Although the effect often is assumed to be an adverse effect, this is not always demonstrated to be the case. Indeed, Hoekstra and Van Ewijk's definitions[38] do not indicate that the effect must be adverse. Judging whether an effect is adverse or benign may be extremely difficult in the context of the consequent application to estimate the "safe concentration," e.g., the concentration permitting normal propagation of aquatic biota. Even assuming an adverse effect, sound decisions regarding the consequences of toxicant release to an aquatic system require more than these statistical methodologies. A profound lack of any ecological or temporal context for these rudimentary effects (survival, weight gain, maturation, and reproduction) detected during highly structured and temporally deficient experiments often precludes sound decision making.

These definitions of acceptable or unacceptable contaminant levels are based on hypothesis testing methods such as those described above. However, if all assumptions are met, confidence limits for predictive models as described in the previous chapter could be used to define such levels more effectively.[2] Other methods including those based on trend analysis[40] or threshold models[41] are applicable also. For example, Tukey et al.[40] define the "no statistically significant trend dose" or NOSTATSOT dose as "the highest dose for which a trend incorporating responses from all dose levels (including the control) up to and including the given dose, is not statistically significant." Even in the context of hypothesis testing, use of the traditional hypothesis test to imply no difference between treatments if there is no significant evidence to the contrary may not be the best method for assessing adverse effects. Dixon and Garrett[42] argued that bioequivalence tests for insignificant difference in responses of exposure treatments and the control treatment may be more appropriate.

Another concern is the tendency to interpret α values too strictly or improperly. For example, a biologically critical effect with an associated $p = 0.06$ should not be ignored. On the other hand, a highly (statistically) significant effect may be biologically trivial if the variable measured has no effect on the organism. The traditional α values of 0.05 or 0.01 used in hypothesis testing are arbitrary; this must be understood if they are to be used effectively in inferring biological importance. Although few introductory statistics texts fail to discuss this concept, it is too often forgotten in daily practice. Readers made uneasy by these comments may want to review Box et al.[1] (page 109), Snedecor and Cochran[10] (page 27), Noether[11] (page 64), Salsburg[19] (page 84), or Green[43] (page 9). Further, in a strict sense, acceptance of the null hypothesis does not lead one directly to the conclusion of no effect. It simply indicates no significant difference from the null hypothesis at some level of probability.

Most important, the ability to detect a difference when there is one (statistical power) depends on the experimental design and applied statistical methods, not simply the magnitude of the effects. In the methods described above, suboptimal experimental design or high variability due to poor technique favors failure to reject the null hypothesis. Each decision made in Figure 1 that moves the researcher away from the techniques on the left side and toward techniques on the right side lowers statistical power. The general consequence is higher LOEL and NOEL values: suboptimal design and technique are rewarded.

Hoekstra and Van Ewijk[38] and others validly argue that procedures should be developed that reward superior experimental design and technique. Crump,[44] Chen and Kodell,[45] and Hoekstra and Van Ewijk[38] have formulated techniques that move us closer to this goal. The approach of Hoekstra and Van Ewijk[38] will be discussed here.

Hoekstra and Van Ewijk[38] describe a two-step approach in which a dose is identified (step 1) from which linear extrapolation (step 2) is performed to estimate an acceptably small effect. The authors used 1% as their "acceptably small effect." The dose from which extrapolation is performed is called the bounded-effect dose. Specifically, "the bounded-effect dose is simply the highest dose at which the confidence limits for the excess risk [toxic effect relative to the control] do not exceed 25%." The null hypothesis for step 1 is that the toxic effect is 25% or more and the alternative hypothesis is the toxic effect is less than 25%. As weak design or suboptimal technique tend to increase confidence intervals, using the confidence intervals to set the bounded-effect dose rewards superior methods. Linear extrapolation results in a conservative estimate of the acceptably small effect under the assumption that the bounded-effect dose is at or below the point of inflection for the dose-response curve. Based on the commonly used logistic model and its fit to 72 data sets, linear extrapolation from 25% was determined to be generally acceptable as the bounded-effect dose.

Although Hoekstra and Van Ewijk[38] provide examples using t-tests, several of the techniques described above can be used to estimate the confidence intervals for each dose as needed to establish the bounded-effect dose. For example, the confidence limits for the differences for means can be estimated with Dunnett's test and expressed as a percentage of the mean for the control. That value with the limit of approximately 25% (25% or higher) is then used as the bounded-effect dose for extrapolation.

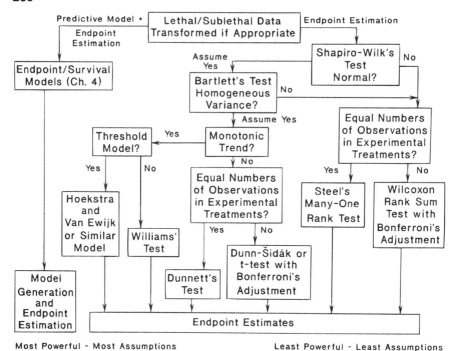

Figure 2 Scheme for analysis of chronic lethal or sublethal effects as recommended in this chapter.

XI. SUMMARY

A wide range of sound statistical methods exist for testing or quantifying chronic lethal and sublethal effects on biota (Figure 2). Unfortunately, the strength of inference derived from these methods is often questionable. Indeed, Platt's[46] comment, "The mathematical box is a beautiful way of wrapping up a problem, but it will not hold the phenomena unless they have been caught in a logical box to begin with" seems very pertinent here.

There is a variety of means to strengthen inference. First, favor statistical methods and experimental designs with the highest power as can be reasonably obtained. (The most powerful test is that with the lowest β given an α)[19] Statistical power may be enhanced by increasing sample size or selecting another test. The methods toward the left-hand side of Figure 2 tend to have the highest power. In contrast to the tendency suggested in Figure 1, the number of control observations could be increased with a consequent increase in statistical power. Certainly, the strength of inference associated with "no significant effect" is clarified if an estimate of power is presented also. (For further discussion of the importance of power in ecological studies, the reader is referred to Toft and Shea,[47] Rotenberry and Weins,[48] Gerrodette,[49] Peterman and Bradford,[50] and Peterman[51]). Second, clearly classify the nature of the measured effect, e.g., Selyean stress, effect with hormesis, preadaptive stress, damage, ambiguous effect, or neutral effect. The type of effect will strongly affect method selection (e.g., a monotonic trend would not be assumed if hormesis was expected) and subsequent inference regarding its biological relevance. Failure to do so weakens inference and muddles conclusions regarding biological significance. Third, clearly state the limitations of inferences derived from the test results. For example, conclusions from tests using time endpoints, e.g., 96 hour LC50 should not be used to imply biological consequences in a field situation involving longer durations of exposure. Fourth, these methods should be used in tandem with those addressing higher and lower order effects on other ecological components to strengthen inferences about biological consequences.

REFERENCES

1. Box, G. E. P., W. G. Hunter, and J. S. Hunter. *Statistics for Experimenters. An Introduction to Design, Data Analysis, and Model Building.* (New York, NY: John Wiley & Sons, 1978).
2. Suter, G. W., II, A. E. Rosen, E. Linder, and D. F. Parkhurst. "Endpoints for Responses of Fish to Chronic Toxic Exposures." *Environ. Toxicol. Chem.* 6: 793–809 (1987).
3. Adams, S. M., ed. "Biological Indicators of Stress in Fish." In *American Fisheries Society Symposium 8.* (Bethesda, MD: American Fisheries Society, 1990).
4. McCarthy, J. F., and L. R. Shugart, eds. *Biomarkers of Environmental Contamination.* (Chelsea, MI: Lewis Publishers, 1990).
5. Huggett, R. J., R. A. Kimerle, P. M. Mehrle, Jr., and H. L. Bergman. *Biomarkers. Biochemical, Physiological, and Histological Markers of Anthropogenic Stress.* (Chelsea, MI: Lewis Publishers, 1992).
6. Weber, C. I., W. H. Peltier, T. J. Norberg-King, W. B. Horning, II., F. A. Kessler, J. R. Menkedick, T. W. Neiheisel, P. A. Lewis, D. J. Klemm, Q. H. Pickering, E. L. Robinson, J. M. Lazorchak, L. J. Wymer, and R. W. Freyberg. *Short-Term Methods for Estimating the Chronic Toxicity of Effluents and Receiving Waters to Freshwater Organisms,* EPA/600/4–89/001. (Cincinnati, OH: Environmental Monitoring Systems Laboratory, Environmental Protection Agency, 1989).
7. Day, R. W., and G. P. Quinn. "Comparisons of Treatments After an Analysis of Variance in Ecology." *Ecol. Monogr.* 59(4): 433–463 (1989).
8. Gelber, R. D., P. T. Lavin, C. R. Mehta, and D. A. Schoenfeld. "Statistical Analysis." In *Fundamentals of Aquatic Toxicology. Methods and Applications,* ed. G. M. Rand and S. R. Petrocelli. (Washington, DC: Hemisphere Publishing Corporation, 1985).
9. Miller, R. G., Jr. *Beyond ANOVA, Basics of Applied Statistics.* (New York, NY: John Wiley & Sons, 1986).
10. Snedecor, G. W., and W. G. Cochran. *Statistical Methods,* 6th ed. (Ames, IA: Iowa State University Press, 1973).
11. Noether, G. E. *Introduction to Statistics. A Fresh Approach.* (New York, NY: Houghton Mifflin Company, 1971).
12. Dixon, W. J., and F. J. Massey, Jr. *Introduction to Statistical Analysis.* (New York, NY: McGraw-Hill Book Co., 1969).
13. Rohlf, F. J., and R. R. Sokal. *Statistical Tables,* 2nd ed. (New York, NY: W.H. Freeman and Company, 1981).
14. Tukey, J. W. "The Philosophy of Multiple Comparisons." *Stat. Sci.* 6(1): 100–116 (1991).
15. Miller, R. G., Jr. *Simultaneous Statistical Inference.* (New York, NY: McGraw-Hill Book Company, 1966).
16. Steel, R. G. D. "A Rank Sum Test for Comparing All Pairs of Treatments." *Technometrics* 2(2): 197–207 (1960).
17. Shapiro, S. S., and M. B. Wilk. "An Analysis of Variance Test for Normality (Complete Samples). *Biometika* 52: 591–611 (1965).
18. Sokal, R. R., and F. J. Rohlf. *Biometry. The Principles and Practice of Statistics in Biological Research,* 2nd ed. (New York, NY: W.H. Freeman and Company, 1981).
19. Salsburg, D. S. *Statistics for Toxicologists.* (New York: Marcel Dekker, Inc., 1986).
20. SAS Institute Inc. *SAS Procedures Guide, Release 6.03.* (Cary, NC: SAS Institute, 1988).

21. Dunnett, C. W. "A Multiple Comparison Procedure for Comparing Several Treatments with a Control." *J. Am. Stat. Assoc.* 50: 1096–1121 (1955).

22. Dunnett, C. W. "New Tables for Multiple Comparisons with a Control." *Biometrics* 20: 482–491 (1964).

23. Williams, D. A. "The Comparison of Several Dose Levels with a Zero Dose Control." *Biometrics* 28: 519–531 (1972).

24. Kramer, C. Y. "Extension of Multiple Range Tests to Group Means with Unequal Numbers of Replications." *Biometrics* 12(3): 307–310 (1956).

25. Bailey, B. J. R. "Tables of the Bonferroni *t* Statistic." *J. Am. Stat. Assoc.* 72: 469–478 (1977).

26. SAS Institute Inc. *SAS Language Guide for Personal Computers, Release 6.03.* (Cary, NC: SAS Institute, 1988).

27. Ury, H. K. "A Comparison of Four Procedures for Multiple Comparisons among Means (Pairwise Contrasts) for Arbitrary Sample Sizes." *Technometrics* 18(1): 89–97 (1976).

28. Wright, S. P. "Adjusted P-Values for Simultaneous Inference." *Biometrics* 48: 1005–1013 (1992).

29. Games, P. A. "An Improved *t* Table for Simultaneous Control on *g* Contrasts." *J. Am. Stat. Assoc.* 72: 531–534 (1977).

30. Dixon, P. M., personal communication, 1993.

31. Williams, D. A. "A Test for Differences Between Treatment Means When Several Dose Levels Are Compared with a Zero Dose Control." *Biometrics* 27: 103–117 (1971).

32. Steel, R. G. D. "A Multiple Comparison Rank Sum Test: Treatments versus Control." *Biometrics* 15: 560–572 (1959).

33. Steel, R. G. D. "Some Rank Sum Multiple Comparison Tests." *Biometrics* 17: 539–552 (1961).

34. Gupta, S. S. "Probability Integrals of Multivariate Normal and Multivariate *t*." *Ann. Math. Stat.* 34: 792–828 (1963).

35. Fligner, M. A. "A Note on Two-Sided Distribution-Free Treatment versus Control Multiple Comparisons." *J. Am. Stat. Assoc.* 79: 208–211 (1984).

36. Beyer, W. H. *Handbook of Tables for Probability and Statistics,* 2nd ed. (Cleveland, OH: CRC Press, 1968).

37. Kokoska, S., and C. Nevison. *Statistical Tables and Formulae.* (New York, NY: Springer-Verlag, 1989).

38. Hoekstra, J. A., and P. H. Van Ewijk. "Alternatives for the No-Observed-Effect Level." *Environ. Toxicol. Chem.* 12: 187–194 (1993).

39. Rand, G. M., and S. R. Petrocelli. *Fundamentals of Aquatic Toxicology. Methods and Applications.* (Washington, DC: Hemisphere Publishing Corp., 1985).

40. Tukey, J. W., J. L. Ciminera, and J. F. Heyse. "Testing the Statistical Certainty of a Response to Increasing Doses of a Drug." *Biometrics* 41: 295–301 (1985).

41. Cox, C. "Threshold Dose-Response Models in Toxicology." *Biometrics* 43: 511–523 (1987).

42. Dixon, P. M., and K. A. Garrett. "Statistical Issues for Field Experimenters." In *Wildlife Toxicology and Population Modeling Integrated Studies of Agroecosystems,* ed. R. J. Kendall and T. Lacher. (Chelsea, MI: Lewis Publishers, 1993).

43. Green, R. H. *Sampling Design and Statistical Methods for Environmental Biologists.* (New York, NY: John Wiley & Sons, 1978).

44. Crump, K. S. "A New Method for Determining Allowable Daily Intakes." *Fundam. Appl. Toxicol.* 4: 854–871 (1984).

45. Chen, J. J., and R. L. Kodell. "Quantitative Risk Assessment for Teratological Effects." *J. Am. Stat. Assoc.* 84: 966–971 (1989).

46. Platt, J. R. "Strong Inference." *Science* 146: 347–353 (1964).

47. Toft, C. A., and P. J. Shea. "Detecting Community-wide Patterns: Estimating Power Strengthens Statistical Inference." *Am. Nat.* 122: 618–625 (1983).

48. Rotenberry, J. T., and J. A. Wiens. "Statistical Power Analysis and Community-wide Patterns." *Am. Nat.* 125: 164–168 (1985).

49. Gerrodette, T. "A Power Analysis for Detecting Trends." *Ecology* 68(5): 1364–1372 (1987).

50. Peterman, R. M., and M. J. Bradford. "Statistical Power of Trends in Fish Abundance." *Can. J. Fish. Aquat. Sci.* 44: 1879–1889 (1987).

51. Peterman, R. M. "The Importance of Reporting Statistical Power: The Forest Decline and Acidic Deposition Example." *Ecology* 71(5): 2024–2027 (1990).

Chapter 6

Effects at the Population Level

Pollutants matter because of their effects on populations, and so, indirectly, on communities too, but pollutants act by their effects on individual organisms.[1]

The populational nature of ecological toxicology is of fundamental importance. Investigations into the patterns underlying the population structure and its dynamics, the population dynamics of different animals and the stability of their populations have long been the focus of attention of ecologists of different specialties. In fact, population is the basic form of existence of organisms and an elementary unit of the evolutionary process.[2]

I. GENERAL

A. POPULATION LEVEL RESEARCH

The critical importance of populations as pointed out by these quotes dictates that more research addressing questions at the population level is needed for sound environmental stewardship. The wealth of information regarding effects on individuals also suggests a need for more linkage of effects measured at the individual level to effects at the population level.[3]

The fields of population biology and genetics are rich in conceptual and mathematical models yet remain relatively wanting for data sets with which to test them. Intelligent blending of these models and theories with the data-rich field of ecotoxicology holds certain promise of exciting contributions in the near future. (Equally certain during this infusion of ideas and methods will be the occasional misapplication of concepts and methods.) It is the goal of this chapter to foster appropriate application of these useful concepts and methods. Because it is impossible to provide sufficient detail on all pertinent materials in one chapter, the reader is strongly encouraged to explore the cited sources.

B. DEFINITION OF POPULATION

A population may be defined as a group of individuals of a species occupying a defined space at a particular time[4,5] or "a collective group of organisms of the same species (or some other groups within which individuals may exchange genetic information) occupying a particular space."[6] Although the concept of population is easily understood, indistinct temporal and spatial boundaries often make operational definition of a study population annoyingly arbitrary. Further, the concept of population is given slightly different emphasis in population genetics than in demography. For example, Cavalli-Sforza and Bodmer[7] defined a Mendelian population as a population of interbreeding individuals sharing a common gene pool. Populations may be divided into demes ("semi-isolated subpopulations with various patterns of genetic exchange or migration between them.")[8] Definition of consequent genetic clines (a steady change in allele frequency across a defined region of study) and related phenomena then become germane to sound genetic studies of populations. Obviously, sound conclusions regarding effects at the population level require a clear definition of the study population.

II. POPULATION SIZE

A. GENERAL

Determining the effects of environmental factors on population size is an integral part of ecology. Liebig's law of the minimum (a population size is limited by some essential factor in its environment that is scarce relative to the amounts of other essential factors)[9] is one excellent example that linked nutritional requirements of individuals to limitations on population size. Shelford's law of tolerance (the tolerance of individuals of a species over an environmental gradient or series of environmental gradients can determine the species' geographical distribution or sizes of local populations)[10,11] spawned innumerable comparisons of laboratory-derived tolerance limits for species to population size or distribution in the field. Refinement of such concepts included the development of tolerance polygons that incorporated acclimation into the pertinent environmental factors.

Concepts linking physiological or nutritional limitations of individuals to population distributions and sizes remain central to ecotoxicological studies at the population level. Laboratory studies provide tolerance limits for various toxicants within a pertinent milieu, and these limits are used with field estimates of bioavailable toxicant to interpret differences in population size among contaminated and uncontaminated sites.

B. MEASUREMENT OF POPULATION SIZE

1. General

Perhaps the most commonly measured population attribute is size. Size is often expressed as relative abundance or density, although size estimates can also be complete or absolute counts of all individuals in a population. Relative abundances are expressed in terms such as catch per unit effort, annual counts on a typical wintering ground, or sightings per Christmas bird count. They have arbitrary units of relative population size. Densities are expressed as number of individuals or biomass per unit of space (area or volume). Densities may be absolute (estimates of the population size in a given area) or relative (estimates of relative densities among populations). The number of individuals of a particular species per square centimeter of Hester-Dendy sampler surface would be used as a relative density for example. Further, densities may be crude or ecological densities. Crude densities are numbers per unit of space, and ecological densities are numbers per unit of suitable habitat space.

Techniques used to determine the size of field populations are categorized as either quadrat, mark-recapture, or removal-based estimates. Quadrat methods use density estimates for sample plots (quadrats) within the population area and the overall area that the entire population occupies to predict the population size. Mark-recapture (capture-recapture, tag-recapture, or Peterson) methods use a precensus marking and release of individuals followed by recapture. The proportion of the total number captured that were marked individuals is used to estimate the population size. Removal estimates involve multiple resamplings without replacement and use the decrease in capture rate to estimate population size. Removal estimates may involve few or many resamplings. Multiple resampling methods fit catch-per-unit effort versus total catch-to-date data to a line and extrapolate the total catch-to-date downward to a catch-per-unit effort of 0.

2. Quadrat Estimates

The total area (or volume) of interest is divided into subunits or "quadrats." The subunits are often arbitrary, e.g., 1 m $\times$ 1 m square plots along a transect. Quadrat methods are most conveniently applied to small, sedentary organisms that are easily counted and unlikely to avoid "capture" in the quadrat during counting.[12] To generate reliable counts within quadrats, three conditions should be met.[5] First, the population within each quadrat should be known precisely. Second, the quadrat area and total area occupied by the population must be known. Third, the quadrat(s) must be representative of the whole area of interest. The third condition is discussed in more detail in Section IV. Weigert's[13] and Hendrick's[14] methods for estimating optimal quadrat size are described with examples and FORTAN code provided by Krebs.[15]

Using the formulations outlined by Pielou,[12] the total number of individuals in the population is N and the number of quadrats (sampling units) within which the entire population is contained is Y. The mean number of individuals per quadrat ($\bar{N}$) is N/Y. The $\bar{N}$ is estimated by the sum of the number of individuals counted in each quadrat (Σn) divided by the number of quadrants counted (y) or $\bar{n} = \Sigma n/y$. The predicted, total population size ($\hat{N}$) is then $Y * \bar{n}$. The precision of this estimate depends on the number of quadrats counted and the evenness of the distribution of individuals among the quadrats. The variance associated with this estimate is calculated as the following,[12]

$$\text{Variance of } \hat{N} = \frac{Y^2}{y(y-1)} \left[1 - \frac{y}{Y} \right] \left[\sum_{i=1}^{y} n_i^2 - \frac{\left(\sum_{i=1}^{y} n_i \right)^2}{y} \right] \tag{1}$$

Equation (1) is appropriate if, during random selection of quadrats, a quadrat could only be selected once (sampling without replacement). Pielou[12] replaced $1 - y/Y$ with 1 in Equation (1) if a quadrat could be picked for sampling more than once (sampling with replacement).

The standard error (square root of the variance or $s_{\hat{N}}$) of the total population size, associated degrees of freedom ($df \leq y - 1$; $df = y - 1$ if the y is specified before the census), and a table of t statistics (e.g., Table 18 ($\alpha = 0.05$) in the Appendix using the column for $p = 1$) can be used to generate a 95% confidence limit (C.L.) for the population size estimate.

$$95\% \text{ C.L.: } \hat{N} \pm ts_{\hat{N}} \tag{2}$$

Pielou[12] also presented estimates of population density. As mentioned, the estimate of the average population density ($\hat{N}_a$) is $\bar{N}$. The average population density has a variance defined by Equation (3).

$$\text{Variance of } \hat{N}_a = \frac{1}{Y^2} [\text{Variance of } \hat{N}] \tag{3}$$

The 95% confidence limits of the mean population density ($\hat{N}_a$) are defined by Equation (4).

$$95\% \text{ C.L.: } \hat{N}_a \pm ts_{\hat{N}_a} \tag{4}$$

Example 1

A population of clams on a beach was decimated after an oil spill. To monitor the reestablishment of the population, a baseline estimate of the population size must be obtained for the 500 m^2 area of beach. The number of clams in each of 50, randomly selected quadrats (1 m $\times$ 1 m squares) is counted. Each quadrat is sampled without replacement, i.e, sampled only once.

Clams/Quadrat	Frequency of Quadrat Count
0	12
1	24

Example 1 *Continued*

2	6
3	3
4	2
5	1
6	1
7	1
≥ 8	0

a. Estimation of $\hat{N}$:

$$\bar{n} = \Sigma n/y = (24 + 12 + 9 + 8 + 5 + 6 + 7)/50 = 1.42 \text{ clams/m}^2 \text{ quadrat}$$
$$\hat{N} = \bar{n}*Y = 1.42 \text{ clams/m}^2*500 \text{ m}^2 = 710 \text{ clams}$$

b. Estimation of associated variance and 95% confidence interval (C.I.)

$$\begin{aligned}
\text{Variance} &= [500^2/(50*49)]*[1 - (50/500)]*[217 - (5041/50)] \\
&= [102.0408]*[0.9]*[116.1800] \\
&= 10669.59
\end{aligned}$$

Standard error = 103.29

$$\begin{aligned}
95\% \text{ C.L.} &= 710 \pm t*103.29 \\
&= 710 \pm 2.0086*103.29 \\
&= 710 \pm 207 \\
95\% \text{ C.I.} &= 503 \text{ to } 917 \text{ clams}
\end{aligned}$$

c. Estimation of $\hat{N}_a$:

$$\hat{N}_a = \bar{N} \approx \bar{n} = 1.42 \text{ clams/m}^2 \text{ quadrat}$$

d. Estimation of associated variance and 95% C.I.

$$\begin{aligned}
\text{Variance} &= [1/Y^2]*[\text{variance of } \hat{N}] \\
&= [1/500^2]*[10669.59] \\
&= 0.04268
\end{aligned}$$

$$\begin{aligned}
95\% \text{ C.L.} &= 1.42 \pm 2.0086*0.04268 \\
&= 1.42 \pm 0.09 \\
95\% \text{ C.I.} &= 1.33 \text{ to } 1.51 \text{ clams/m}^2 \text{ quadrat}
\end{aligned}$$

3. Mark-Recapture Estimates

The characteristics of a species or its habitat often make quadrat methods inappropriate or impractical. For example, placing open quadrats over an area of a pond and counting the number of sunfish in each quadrat generates meaningless results because of the species mobility and avoidance behavior. Mark-recapture methods can be often used in such situations. In a precensus trapping, individuals are marked and then released. The marked animals enter back into the population. A census is taken by retrapping. The numbers of marked and unmarked individuals are used to estimate the total population size.

Such estimates assume several conditions.[12,16] First, marking or tagging of an individual must not alter its behavior or risk of mortality. Second, marked individuals must be scored accurately. This condition is violated if tags are lost or marks become indistinguishable. Third, marked animals must mix randomly within the population and their likelihoods of recapture must be the same as those of unmarked individuals. Estimates can be compromised if there is insufficient time between marking and recapture for the marked individuals to mix thoroughly within the population. Learned behavior associated with being trapped initially (trap addiction or avoidance) may also invalidate this condition. Finally, it is assumed that there are insignificant numbers of births, deaths, and migrants between the initial marking and recapture dates. Such a population which remains constant during sampling is referred to as a closed population. (Methods for open populations do exist[17] but are not discussed here.)

Again, using the formulations and notation of Pielou,[12] the intuitive Lincoln's estimator of population size is calculated from mark-recapture data according to Equation (5). (The Lincoln's estimator is also called the Petersen-Lincoln estimator.)

$$N = \frac{nM}{m} \tag{5}$$

where N = total population size (number of individuals); M = the total number of marked individuals; n = the number of individuals caught at censusing; and m = the number of marked individuals caught at censusing.

Equation (5) overestimates N because m is not independent of n. Various modifications of Lincoln's estimator have been generated to minimize this bias. For example, Emmel[16] recommends Bailey's modification [Equation (6)] as a less biased estimator of N. (Bailey[18] provides more detail on this method.) Equation (8) may be used to calculate the variance for this estimate.[19]

$$\hat{N} = \frac{M(n + 1)}{m + 1} \tag{6}$$

Seber,[19] Pielou,[12] and White et al.[17] recommend Chapman's[20] modification which produces an estimate with negligible bias if $m \geq 7$.

$$\hat{N} = \frac{(M + 1)(n + 1)}{m + 1} - 1 \tag{7}$$

$$\text{Variance of } \hat{N} = \frac{(M + 1)(n + 1)(M - m)(n - m)}{(m + 1)^2(m + 2)} \tag{8}$$

The 95% confidence limits are the following:

$$95\% \text{ C.L.} = \hat{N} \pm 2\sqrt{\text{Variance of } \hat{N}} \tag{9}$$

Example 2

A question arose during a mesocosm study of trophic transfer regarding how much of a pollutant was present in a prey species (sunfish) for possible transfer to largemouth bass. An essential piece of information for the associated calculation was the estimated sunfish population size. In a precensus survey, 42 sunfish were tagged and released back into the pond. During a follow-up censusing, 17 tagged and 21 untagged sunfish were caught (M = 42, n = 17 + 21 = 38, m = 17).

Example 2 *Continued*

   a. N is estimated using Chapman's modification [Equation (7)].

$$\hat{N} = [(42 + 1)(38 + 1)]/(17 + 1) - 1$$
$$= 92.17 \text{ or } 92 \text{ sunfish}$$

   b.   Variance in $\hat{N}$ [Equation (8)] and 95% confidence interval [Equation (9)]

$$\text{Variance} = [(42 + 1)(38 + 1)(42 - 17)(38 - 17)]/[(17 + 1)^2(17 + 2)]$$
$$= [43*39*25*21]/[324*19]$$
$$= 143.02$$

Standard error $= 11.96 \approx 12$

$$95\% \text{ C.L.} = 92 \pm 2*12$$
$$= 92 \pm 24$$

95% C.I.: 68 to 116 sunfish

4. Removal-Based Estimates

Numbers of individuals collected during consecutive catches may also be used to estimate population size. Individuals from previous catches are removed from the population and the decrease in catch is used to estimate size. Removal can be by physical removal or by placing some distinguishing mark on captured individuals before release back into the environment. For example, seines may be placed as obstructing curtains upstream and downstream of a stream segment in which one wishes to estimate the population size of a fish species. After capture, the individuals may be removed by returning them to the stream below the location of the downstream seine. They are physically removed from the population being estimated. Alternatively, fish from a pond may be captured, counted, tagged, and released back into the pond. They are not counted if caught again.

The specific equations used to provide an estimate of population size and the associated variance depend on the number of times the population is sampled. Those outlined by Seber and Le Cren[21] and Pielou[12] for two- and three-catch samplings are presented in detail. Next, methods based on many samplings are discussed. Regardless of the number of catches, these methods are based on the assumption that all individuals in the closed population have the same probability of capture, and this probability does not change significantly between periods of capture. The number of individuals captured each time is assumed to substantially decrease the size of the remaining population. The error structure is assumed to be binomial.[21]

For methods based on two periods of capture, the maximum likelihood estimator of population size and its variance are defined by Equations (10) and (11).

$$E(\hat{N}) = N + \frac{q(1 + q)}{p^3} = N + b \tag{10}$$

$$\text{Variance of } \hat{N} = \frac{Nq^2(1 + q)}{p^3} + \frac{2q(1 - p^2 - q^3)}{p^5} - b^2 \tag{11}$$

where $E(\hat{N}) = $ the expected value of the population size estimate; $N = $ population size; $p = $ probability of capture; $q = $ probability of escaping capture; and $b = $ bias.

Equations (12) and (13) are approximations based on Equations (10) and (11). (Note that the expected values of $C_1 = Np$ and $C_2 = Np(1 - p)$ or Npq.)

$$\hat{N} = \frac{C_1^2}{(C_1 - C_2)} \tag{12}$$

$$\text{Variance of } \hat{N} = \frac{(C_1 C_2)^2 (C_1 + C_2)}{(C_1 - C_2)^4} \tag{13}$$

where $\hat{N}$ = predicted population size (moment estimate); C_1 = the number caught during the first trapping; and C_2 = the number caught during the second trapping.

The probability of capture (p) is estimated using Equation (14). The probability of escaping capture (q) is $1 - p$.

$$\hat{p} = \frac{C_1 - C_2}{C_1} \tag{14}$$

where $\hat{p}$ = predicted probability of capture.

These approximations [Equations (12) and (13)] are adequate if $Np^3 > 16q^2(1 + q)$. Otherwise, the bias (b) in these approximations should be assessed.[21] The bias is estimated by Equation (15).

$$b = \frac{q(1 + q)}{p^3} \tag{15}$$

Estimates of population size and its associated variance can be adjusted for this bias.

$$\hat{N}_{ub} = \hat{N} - b \tag{16}$$

where $\hat{N}_{ub}$ = unbiased, predicted population size.

$$\text{Var}_{ub} = \text{variance of } \hat{N} - \frac{2q(1 - p^2 - q^3)}{p^5} + b^2 \tag{17}$$

where var_{ub} = unbiased variance of $\hat{N}$. If the bias divided by the standard error [calculated by taking the square root of the variance from Equation (13)] is greater than 0.1, Seber and Le Cren[21] suggested that the bias not be corrected in the variance estimate.

Pielou[12] also provided general equations for estimating N from three samplings. Obviously, the addition of more catches should improve the precision of the estimate. Her estimates of N from three catches (C_1, C_2, C_3) and the associated variance are generated with the following equations. (Note that the expected values of $C_1 = Np$, $C_2 = Npq$, and $C_3 = Npq^2$. The q and p are as defined previously.) The population size is estimated by Equation (18).

$$\hat{N} = \frac{C_1 + C_2 + C_3}{1 - q^3} \tag{18}$$

First, an estimate ($\hat{q}$) of q must be derived. Equations (19) and (20) are used to calculate Q for this purpose.

$$T = C_1 + C_2 + C_3 \tag{19}$$

$$Q = \frac{C_2 + 2C_3}{T} \tag{20}$$

The Q from Equation (20) is placed into quadratic Equation (21) to get $\hat{q}$.

$$(2 - Q)q^2 + (1 - Q)q - Q = 0 \tag{21}$$

This is done by using the following solution to the simple quadratic equation of the form given in Equation (22).

$$ax^2 + bx + c = 0 \tag{22}$$

Equation (22) can be solved for x.

$$x = \frac{-b \pm \sqrt{b^2 - 4ac}}{2a} \tag{23}$$

Letting $2 - Q = a$, $1 - Q = b$, and $-Q = c$, q can be calculated

$$q = \frac{-(1 - Q) \pm \sqrt{(1 - Q)^2 - 4(2 - Q)(-Q)}}{2(2 - Q)} \tag{24}$$

Equation (24) can be simplified to Equation (25).

$$q = \frac{(Q - 1) \pm \sqrt{1 + 6Q - 3Q^2}}{2(2 - Q)} \tag{25}$$

Although two roots are possible from Equation (25), only the probability within the range of 0 to 1 is pertinent. The $\hat{q}$ is placed into Equation (18) instead of q to provide an estimate ($\hat{N}$) of N.

The variance and 95% confidence interval for $\hat{N}$ are calculated with Equations (26) and (27).

$$\text{Variance of } \hat{N} = \frac{\hat{N}T(\hat{N} - T)}{T^2 - \dfrac{9\hat{N}(\hat{N} - T)(1 - \hat{q})^2}{\hat{q}}} \tag{26}$$

$$95\% \text{ C.L.: } \hat{N} \pm 2\sqrt{\text{variance of } \hat{N}} \tag{27}$$

Example 3

Population size of trout in a small, glacial lake experiencing acidification is estimated. Adult trout are caught during consecutive samplings, marked, and returned to the lake. The lake is then resampled. The brief time between samplings and the relative isolation of the lake are judged sufficient to satisfy the assumption of a closed population. (Data used here are those simulated for $N = 625$ and $p = 0.40$ by White et al.,[17] Table 4.2 on page 104.)

Example 3 *Continued*

Two samplings (C_1 and C_2) yielded 260 and 141 unmarked trout:

a. $\hat{N}$ is estimated with Equation (12)

$\hat{N} = 260^2/(260 - 141)$
$\qquad = 568.07$ or 568 adult trout

b. Variance of $\hat{N}$ is estimated with Equation (13)

Variance $= [(260*141)^2*(260 + 141)]/(260 - 141)^4$
$\qquad\quad = [1,343,955,600*401]/200,533,921$
$\qquad\quad = 2687.46$

Standard error $= 51.84$ or 52 trout

c. p and q are estimated with Equation (14)

$\hat{p} = [260 - 141]/260$
$\quad = 0.46$
$\hat{q} = 1 - 0.46$
$\quad = 0.54$

d. Bias is estimated with Equation (15)

$b = [0.54*(1 + 0.54)]/0.46^3$
$\quad = 0.8316/0.0973$
$\quad = 8.54$

e. Unbiased estimate of N [Equation (16)]

If $Np^3 > 16q^2(1 + q)$, then bias correction would be made on this estimate of N. The calculated $\hat{N}$, $\hat{p}$, and $\hat{q}$ are used to test this inequality.

$Np^3 \approx 568*0.46^3$
$\qquad \approx 55.287$
$16q^2 \approx 4.666$

Because the estimate of Np^3 is greater than the estimate of $16q^2$, the bias correction should be made.

$\hat{N}_{ub} = 568 - 9$
$\qquad = 559$ adult trout

f. The unbiased estimate of the variance

If b/standard error is greater than 0.1, Seber and Le Cren[21] suggested that no bias correction be made for the variance estimate. Using the above estimates of b and standard error,

$b/\text{SE} \approx 8.54/51.84$
$\qquad \approx 0.16$

No bias correction is recommended. However, for the sake of illustration in this example, Equation (17) and estimates of p, q, and b are used to estimate the unbiased variance.

Example 3 *Continued*

Var_{ub} = Variance of $\hat{N} - [2*0.54*(1 - 0.46^2 - 0.54^3)/0.46^5] + 8.54^2$
   = 2687.46 − 33.08 + 72.93
   = 2727.31

(Unbiased standard error = 52.22)

Three samplings (C_1, C_2 and C_3):
Assume that three samplings were made with counts of 260, 141, and 97 unmarked trout. The methods described here can be used to provide another estimate.

Equation (19) is used to calculate T:

$T = 260 + 141 + 97 = 498$ adult trout

Equation (20) is used to estimate Q:

$Q = [141 + 2*97]/498$
   = 0.6727

Equation (25) is used to estimate q:

$$q = \frac{(0.6727 - 1) \pm \sqrt{1 + 6*0.6727 - 3*0.6727^2}}{2(2 - 0.6727)}$$

$$q = 0.5992$$

$$\approx 0.60$$

a. $\hat{N}$ Estimation:
    These estimates of q and T are used in Equation (18) to calculate the population size.

    $\hat{N} = (260 + 141 + 97)/(1 - 0.60^3)$
       = 635 adult trout

b. Variance of $\hat{N}$ is estimated with Equation (26):

    $$\text{Variance} = \cfrac{635*498*(635 - 498)}{498^2 - \cfrac{9*635*(635 - 498)*(1 - 0.60)^2}{0.60}}$$

    = 43,323,510/39,216
    = 1104.74

    Standard error = 33.24 or 33 trout

With multiple catch methods, the catch per unit effort (e.g., fish caught per day) may be used in linear regression against the total number of individuals caught before that sampling. Extrapolation of total number caught downward to a catch per unit effort of 0 allows estimation of the total population size (X intercept) (Figure 1). This method (Leslie-De Lury Method) is detailed by Seber and Le Cren[21] as follows.

Let the consecutive samplings (C_1, C_2, ... C_k or C_i) be characterized by the same probability of capture (p). Assume constant effort between removals, e.g., catch per unit effort. Plot or fit to a regression line of C_i versus the sum of all catches up to, but not including, C_i.

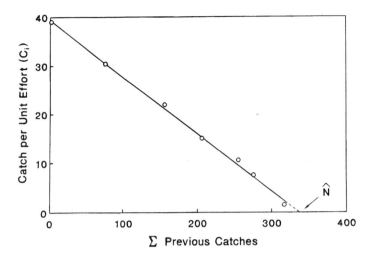

Figure 1 Leslie-De Lury regression method for estimation of $\hat{N}$.

$$C_i = \text{slope}\left(\sum_{j=1}^{i-1} C_j\right) + \text{intercept} \qquad (28)$$

To this point, the assumption of equal p values between sampling periods could not be tested. This is one of the disadvantages of using only one or two samplings to estimate p. However, with the regression method, the line would be straight if p is constant between samplings ($p_1 = p_2 \cdots = p_k$). More formal methods are available for determining if p values are equal.[17,21,22]

Example 4
Assume that four consecutive catches of sizes 260, 141, 97, and 50 were made from the trout population described in Example 3. The population size and associated probability of capture can be estimated.

Linear regression can be performed manually or with the following SAS code to fit the catch per unit effort (C_i, dependent variable) versus sum of all previous catches (independent variable). In the SAS code, CATCH is the number of trout caught during a particular sampling and TOTAL is the total number of trout caught up to that sampling. Four pairs of data are used (CATCH/TOTAL: 260/0, 141/260, 97/401, and 50/498).

```
DATA CATCH;
  INPUT CATCH TOTAL @@;
  CARDS;
  260 0 141 260 97 401 50 498
  ;
  RUN;
PROC GLM;
  MODEL CATCH=TOTAL;
  OUTPUT OUT=PREDICT P=CATHAT R=RESID;
RUN;
```

The r^2 associated with the regression results was 0.995. The slope was -0.4149. The probability of capture can be estimated to be the slope multiplied by -1, e.g., approxi-

Example 4 *Continued*

mately 0.41 (standard error of estimate = 0.02). This is very close to the *p* (0.40) set by White et al.[17] during generation of these simulated data. The regression model can be solved for CATCH = 0 to estimate the population size.

CATCH = −0.4149*TOTAL + 257.2119
    0 = −0.4149*TOTAL + 257.2119
TOTAL = −257.2119/−0.4149
TOTAL = 620 trout

The estimate of the total number of trout ($\hat{N}$ = 620) is slightly less than the actual total number of 625 trout set during generation of these simulated data.

White et al.[17] recommended a maximum likelihood method instead of the regression method described here. More detail regarding this method and other mark-recapture or removal methods for closed populations can be obtained from Otis et al.,[23] White et al.,[17] Krebs,[15] and Skalski and Robson.[24] As of 1993, several programs, including CAPTURE and TRANSECT,[25] implementing many of these and related methods can be obtained from the Utah Cooperative Research Unit, Utah State University, Logan, UT 84322. Krebs[15] also described and provided FORTRAN programs for many of these methods.

C. SIMPLE POPULATION GROWTH

1. The Exponential Growth Model

Descriptions of population dynamics traditionally begin with density-independent exponential (also called geometric or Malthusian) growth, e.g., May.[26] The unrestrained change in population size (*dN/dt*) is expressed as a function of the population size (*N*) and *r*. The constant *r* is referred to as the per capita growth rate, per capita average growth rate, Malthusian constant, intrinsic rate of increase, or instantaneous rate of increase. Often, r_m is used to designate the maximum rate at which a population can grow when completely unrestrained, i.e., it is a physiological quality of the population. The symbol *r* is then reserved for description of population growth that may not be completely unrestrained, e.g., measured growth in a field population experiencing logarithmic growth.

$$\frac{dN}{dt} = rN \tag{29}$$

This equation is transformed to Equation (30) to predict population size at any time.

$$N_t = N_0 e^{rt} \tag{30}$$

where N_t = population size at time *t*; and N_0 = population size at time 0. Note that these equations are similar to those presented in Chapter 3 for elimination and uptake. Indeed, Equation (30) is the negative exponential model for elimination if *r* is negative, i.e., the population is decreasing in size. Consequently, several pertinent aspects of curve fitting are not discussed here. A review of Chapter 3 and associated examples may be helpful if more detail is desired.

Equation (30) may be made linear during data fitting.

$$\ln N_t = rt + \ln N_0 \tag{31}$$

A semilogarithmic-plot ($\ln N_t$, t) will produce a straight line with a slope of r.

Estimates of r from Equation (31) may be used to calculate doubling times (the time, expressed in the same unit as the growth rate, in which the number of individuals doubles in the population, i.e., t for which $N_t/N_0 = 2$).

$$t_d = \frac{\ln 2}{r} \tag{32}$$

If necessary, two points in the growth curve may be used to estimate growth rate. The growth during a specified period (λ) is estimated under the assumption of exponential growth. The λ is referred to as the finite rate of increase; however, May[26] referred to λ as the multiplicative growth factor per generation and considered the term "finite rate of increase" a misnomer. Stated concisely, $N_{t+1} = \lambda N_t$. The λ calculated from the population sizes at two times is used to estimate r. (Recommendations given for estimation of half-life from two points in Chapter 3 are suggested here for estimation of λ with only two points.)

$$\lambda = \frac{N_t}{N_0} = e^r \tag{33}$$

Slightly different constants may be used with the same basic mathematics for microbial growth dynamics.[27] Microbial growth by division is also expressed as an exponential growth rate constant, k. Units are generations or divisions per unit of time.

$$N_t = N_0 2^{kt} \tag{34}$$

$$\ln \frac{N_t}{N_0} = kt \tag{35}$$

The k may be generated by determining numbers of cells at two times during exponential growth. The instantaneous or specific growth rate constant (μ in units of day^{-1} or h^{-1}) may be calculated from k using the relationship $\mu = \ln 2 * k$. The specific growth rate expressed as μ is calculated with Equation (36).

$$\mu = \frac{\ln \dfrac{N_{t2}}{N_{t1}}}{t_2 - t_1} \tag{36}$$

Similar to Equation (30),

$$N_t = N_0 e^{\mu t} \tag{37}$$

A series of specific growth rates may be determined for some microbial population over time. The change in the maximum specific growth rates may then be used as a measure of pollutant effect, e.g., algal growth.[28]

Example 5

Cladoceran reproduction was used by Hamilton[29] to assess the effects of a toxicant. The cumulative number of young produced by a control female over 7 days was monitored during a renewal toxicity test. (Data extracted from Figure 1 of Hamilton.)[29]

Day	Cumulative Number of Young	ln N_t (Including Female)
0	0	0
1	0	0
2	0	0
3	0	0
4	4	1.609
5	14	2.639
6	14	2.639
7	28	3.332

Either linear regression [Equation (31)] or nonlinear regression [Equation (30)] can be used to estimate r from these data. Here, an r is estimated by linear regression and used as an initial r in an iterative, nonlinear regression method.

```
DATA GROWTH;
  INPUT N T @@;
  LNN = LOG(N);
  CARDS;
  1 0 1 1 1 2 1 3 5 4 15 5 15 6 29 7
  ;
RUN;
PROC GLM;
  MODEL LNN=T;
  OUTPUT OUT=PREDICT P=LNNHAT R=RESID;
RUN;
PROC PLOT;
  PLOT LNN*T="*" LNNHAT*T="P"/OVERLAY;
RUN;
/* NONLINEAR FIT USING THE ESTIMATE GENERATED WITH    */
/* THE ABOVE LINEAR METHODS, E.G., 0.56 AS THE        */
/* INITIAL ESTIMATE OF r.                             */
PROC NLIN DATA=GROWTH;
  PARMS R=0.56;
  BOUNDS 0<R;
  DER.R=T*EXP(R*T);
  MODEL N=1*EXP(R*T);
  OUTPUT OUT=CURVE P=NHAT R=CRESID;
RUN;
PROC PLOT;
  PLOT N*T="*" NHAT*T="P"/OVERLAY;
  PLOT CRESID*T="*"/VREF=0;
RUN;
```

The initial estimate of r from the linear regression was 0.56 ± 0.09. This estimate was used in the nonlinear regression to produce a final estimate of 0.48 ± 0.01 (asymptotic 95% confidence interval: 0.45 to 0.50). The doubling time for this clone would be approximately ln 2/0.48 or 1.44 days.

2. The Logistic Growth Model

Populations do not continue unrestrained growth indefinitely. The preceding equations are modified to steadily decrease the growth rate by some factor as the population density increases. This factor K is linked to a constant carrying capacity, the maximum population size (density) that a particular environment is capable of supporting indefinitely.

$$r_{dd} = r\left[\frac{K - N}{N}\right]$$ (38)

where r_{dd} = the population density-dependent r. The growth rate decreases to 0 as N approaches K. The change in N is given for such logistic growth by the differential Equation (39).

$$\frac{dN}{dt} = rN\left[\frac{K - N}{K}\right] = rN\left[1 - \frac{N}{K}\right]$$ (39)

The N at any time is determined for populations with overlapping generations with Equation (40) which describes the classic logistic curve.

$$N_t = \frac{K}{1 + \left[\dfrac{K - N_0}{N_0}\right]e^{-rt}}$$ (40)

The logistic equations may be modified to make them more flexible. For example, Gilpin and Ayala[30] add a parameter (θ) that allows different growth curve symmetries.

$$\frac{dN}{dt} = rN\left[1 - \left(\frac{N}{K}\right)^{\theta}\right]$$ (41)

The differential Equation (39) and integrated Equation (40) are often used for populations undergoing continuous growth with overlapping generations.[26,31] For fitting data, an alternate form of Equation (40) is convenient.

$$N_t = \frac{K}{1 + e^{a - rT}}$$ (42)

The a is an integration constant defining the position of the curve relative to the origin.[5,32] Odum[6] shows that a is equal to $\ln((K - N)/N)$ at time = 0. [At time = 0, $N = N_0$ so a becomes $\ln((K - N_0)/N_0)$. Substitution of this for a in Equation (42) yields Equation (40).] Equation (42) may be converted to the linear form,

$$\ln\left[\frac{K - N}{N}\right] = a - rt$$ (43)

The K is estimated by eye from a graph of N versus time and then used to estimate

$\ln((K - N)/N)$ for each t. A line is then fit to $\ln((K - N)/N)$ versus t. The slope is $-r$ and the intercept is a.

Example 6

The population growth of a marine amphipod on a clean sediment and sediment spiked with crude oil is monitored for 160 days. The r and K for the amphipods growing in the clean and spiked sediments are estimated. (The data below are fabricated from those of Gause.)[33]

	Population Size (Number/Aquarium)	
Time (days)	Clean Sediment	Spiked Sediment
0	2	2
10	50	50
30	200	240
40	300	230
60	1100	500
85	1250	480
100	1800	600
120	1600	580
140	1850	600
160	1500	580

These data are graphed (N versus t), and estimates of K for each sediment are extracted from the graph. These estimates are then used to calculate $\ln((K - N)/N)$. Linear regression models of $\ln((K - N)/N)$ versus t were used to generate estimates of parameters a (Clean, 3.00; Spiked, 1.60) and r (Clean, 0.04; Spiked, 0.03). The estimates of K and r were then used in a nonlinear regression procedure to fit these data to Equation (40). The estimates and their associated asymptotic standard errors and asymptotic 95% confidence intervals are given below.

	Sediment	
Parameter	Clean	Spiked
r		
Estimate (SE)	0.12 ± 0.01	0.14 ± 0.01
95% C.I.	$0.11 - 0.14$	$0.12 - 0.16$
K		
Estimate (SE)	1611 ± 78	561 ± 27
95% C.I.	$1430 - 1791$	$500 - 623$

The crude oil spike had a very clear effect of lowering the carrying capacity of the sediments, with K dropping from approximately 1430 to 561 after spiking. The effect on population r was not as clear.

The SAS code used to perform these calculations follows. The program listed here was used in two phases. First, the linear regression procedures (PROC GLM) were performed to estimate r. Next, the nonlinear regression procedures (PROC NLIN) were added. Estimates of K and r from the PROC GLM procedures were used as initial values in the iterative fitting procedure to produce the final parameter values.

Example 6 *Continued*

```
DATA LOGISTIC;
  INPUT SEDIMENT $ N T @@;
    IF SEDIMENT="CLEAN" THEN K=1900;
    IF SEDIMENT="CLEAN" THEN NINIT=2;
    IF SEDIMENT="SPIKED" THEN K=650;
    IF SEDIMENT="SPIKED" THEN NINIT=2;
    Y=(K-N)/N;
    LNY=LOG(Y);
    CARDS;
    CLEAN 2 0 CLEAN 50 10 CLEAN 200 30 CLEAN 300 40 CLEAN 1100 60
    CLEAN 1250 85 CLEAN 1800 100 CLEAN 1600 120 CLEAN 1850 140
    CLEAN 1500 160 SPIKE 2 0 SPIKE 50 10 SPIKE 240 30 SPIKE 230 40
    SPIKE 500 60 SPIKE 480 85 SPIKE 600 100 SPIKE 580 120 SPIKE 600
    140 SPIKE 580 160
    ;
RUN;
PROC SORT;
  BY SEDIMENT;
RUN;
PROC GLM;
  MODEL LNY=T;
  BY SEDIMENT;
RUN;
DATA CLEAN;
  SET LOGISTIC;
  IF SEDIMENT="CLEAN";
RUN;
DATA SPIKE;
  SET LOGISTIC;
  IF SEDIMENT="SPIKE";
RUN;
PROC NLIN DATA=CLEAN;
  PARMS R=0.04 KAY=1900;
  BOUNDS 0<R;
  BOUNDS 0<KAY<2500;
  MODEL N=KAY/(1+((KAY-NINIT)/NINIT)*EXP(-R*T));
RUN;
PROC NLIN DATA=SPIKE;
  PARMS R=0.04 KAY=1900;
  BOUNDS 0<R;
  BOUNDS 0<KAY<2500;
  MODEL N=KAY/(1+((KAY-NINIT)/NINIT)*EXP(-R*T));
RUN;
```

Difference equations are used for populations with nonoverlapping generations or discrete growth such as those of many insects and annual plants. The following difference equations are commonly used for discrete populations.[26,31]

Growth unrestrained by population density:

$$N_{t+1} = \lambda N_t \tag{44}$$

Density-dependent (restrained) growth:

$$N_{t+1} = N_t e^{r(1-\frac{N_t}{K})} \tag{45}$$

or

$$N_{t+1} = \lambda N_t [1 + aN_t]^{-\beta} \tag{46}$$

or

$$N_{t+1} = \frac{\lambda N_t}{1 + [aN_t]^b} \tag{47}$$

or

$$N_{t+1} = N_t e^{r(1-\frac{N_t^\theta}{K})} \tag{48}$$

Equation (45) is the classic Ricker model[34] with an equilibrium N of K. Equation (48) is a modified Ricker or θ-Ricker model[35] that allows for a more general growth symmetry to the curve, i.e., allows a more flexible shape. It grew out of Gilpin and Ayala's[30] modification of the logistic model. It also has an equilibrium N of K. Equation (46) from Hassell et al.[36] includes a and β, constants that define the density dependence of the population. The slope (β) is a measure of density effects and a determines curve shape. The equilibrium N for Equation (46) is $[\lambda^{1/\beta} - 1]/a$. The similar Equation (47) from Watkinson[37] has an equilibrium population size of $[\lambda - 1]^{1/b}/a$.

Data may be fit to the model described by Equation (45) by linear regression with following equation.[5] The estimated values of A and B are used to calculate r and K.

$$\frac{N_{t+1} - N_t}{N_t} = A - BN_{t+1} \tag{49}$$

where N_t, N_{t+1} = population densities at times t and $t + 1$; A = a constant equal to $e^r - 1$; and $B = A/K$. Hassell[38] describes similar procedures for fitting data to the model described by Equation (46).

These equations represent only a few of the most common models used to describe simple, density-dependent growth. The interested reader is referred to May and Oster[31] for a tabulation of additional models. None of these equations has any clear superiority to another except for the purely historical dominance of the Ricker model. Indeed, even the general value of the logistic model for description of population growth remains in active debate, e.g., Watkinson,[37] Szathmary,[39] and Ginzburg.[40,41]

In the equations pertinent to overlapping generations (continuous growth), the population responds instantaneously to density effects. This is not realistic in many instances. Hutchinson[42] introduced a "delay differential equation" that incorporates a lag in response (decrease in population growth rate) to density effects. The response is to the population density at some time prior to N_t.

$$\frac{dN_t}{dt} = rN_t \left[1 - \frac{N_{t-T}}{K} \right] \tag{50}$$

where T is the time lag in response. More complicated models incorporating lags are common. Krebs[5] and Odum[6] both presented models with two lags. The first (T or reaction time lag) is that incorporated into Equation (50). It is the time lag before a negative effect of crowding is realized. The second (g or reproductive lag) is the time to respond (begin increasing) to a favorable change in the environment.

$$\frac{dN_t}{dt} = rN_{t-g}\left[1 - \frac{N_{t-T}}{K}\right] \tag{51}$$

For the equations pertinent to discrete populations (nonoverlapping generations), a time delay or lag is implicit in their structure.[26] For example, the increase in N from t to $t + 1$ may be the change in population density per generation. Consequently, incorporation of lag terms into these simplistic models analogous to the lag terms in the continuous growth models is not required. Regardless of their form, time delays in both the continuous and discrete logistic models strongly influence the equilibrium stability as discussed in Section C.3.

3. Population Stability

In Chapter 1, Popper[43] was quoted as follows: "The empirical basis of objective science has thus nothing 'absolute' about it. Science does not rest upon solid bedrock. The bold structure of its theory rises, as it were, above a swamp, but not down to any natural or 'given' base; and if we stop driving the piles deeper, it is not because we have reached firm ground. We simply stop when we are satisfied that the piles are firm enough to carry structure, at least for the time being." This quote is extremely pertinent to the topic of population stability. Presently, the piles are being driven deeper into the swamp as the scientific structure above the foundation becomes more and more massive.

The concept that population density approaches a single equilibrium point is an unrealistic depiction based on expediency and history. It emerged from the simplistic "balance in nature" concept that has a long history in Western society[44,45] and it remains at the foundation of many present-day arguments or decisions in our field. For example, the concept (expectation) of an equilibrium population density permeates Sarokin and Schulkin's[46] excellent discussion of pollution-related population disturbances. In their review, a "surprisingly meager" amount of information forced upon the authors the assumption that any large deviation in population size implies dysfunction. This assumption is probably not true for many populations.

For populations displaying continuous growth, Equation (50) may be used to incorporate a lag in response (change in the rate of growth). Although treated as such to this point, this model does not necessarily describe a gradual growth to K. The dynamics of this model depend on the relative sizes of T and the characteristic return time (T_R).

The r may be used to estimate a characteristic return time, a parameter that indicates the rate at which the population will return toward K (or any analogous equilibrium density). The T_R is the inverse of r, i.e., $T_R = 1/r$. As T increases relative to T_R, population density becomes increasingly unstable.[26] The population density tends to overshoot and then to undershoot K as T increases relative to T_R. Figure 2 illustrates three general regions of population stability described by May.[26]

$0 < rT < e^{-1}$	Damped stable point
$e^{-1} < rT < 0.5\pi$	Oscillatory, damped, stable point
$0.5\pi < rT$	Stable-limit cycles

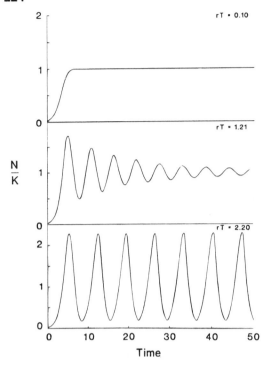

$\dfrac{N}{K}$

Figure 2 Population dynamics for a continuous growth model [Equation (50)] with increasing values of r.

Damped stable-point dynamics for population density involve convergence at a single point such as K. Oscillatory, damped, stable-point dynamics include swings back and forth with eventual convergence to K. A population displaying stable limit cycles will oscillate with a fixed period and amplitude about K. The cycles may involve oscillation between two or more points. (See Schaffer and Kot.[47])

Discrete growth models also are capable of surprisingly complex dynamics. The dynamics of Equation (45) become increasingly complex as r increases.[26,48] May[48] provided stability qualities for various ranges of r in Equation (45).

$0 < r < 2.000$	Stable point
$2.000 < r < 2.526$	Stable, oscillating cycle between 2 points
$2.526 < r < 2.656$	Stable, oscillating cycle among 4 points
$2.656 < r < 2.685$	Stable, oscillating cycle among 8 points
$2.685 < r < 2.692$	Stable, oscillating cycle among 16 to 64 or more points
$2.692 < r$	Chaotic dynamics

Figure 3 illustrates the associated dynamics using N/K to scale population densities relative to the carrying capacity. Stable point (convergence to one population density) and stable cycle (stable oscillations about a point) dynamics have been previously described. Chaotic behavior is characterized as apparently random dynamics arising from a deterministic basis such as Equation (45) with specific parameter values. With chaotic dynamics, the population density cannot be predicted at any time in the future.

The remarkable work of May[48] suggested that, under some conditions, it is impossible to predict N even for simplistic models such as Equation (45). This has stimulated almost

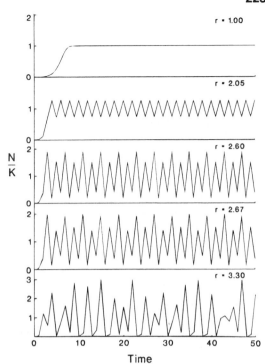

Figure 3 Population dynamics for a discrete growth model [Equation (45)] with increasing values of r.

two decades of reexamination of population dynamics. Using the model described by Equation (46) to fit insect population data, it was quickly discovered that r values begetting chaotic dynamics were unrealistically high.[36] Only one laboratory population had a sufficiently high r to display chaotic dynamics. This fostered scientific opinions that field populations likely do not display chaotic dynamics. Thomas et al.[35] and Philippi et al.[49] found no indication of chaotic behavior in laboratory populations of *Drosophila* fit to Equation (48) and also concluded that populations are unlikely to exhibit chaotic behavior. May[50,51] suggested that the addition of more details to these models fosters chaotic behavior. Other criticisms[52] focused on the loss of many qualities of chaos in population models when stochastic components were incorporated. The suggestion was that natural populations experiencing random perturbations would be unlikely to display any detectable chaotic dynamics. The substance of this criticism was strengthened by May's own conclusions[53] regarding the extreme difficulty of discerning chaotic behavior in the presence of stochastic components.

This aspect of population behavior remains an exciting area of research. Unfortunately, it has not been applied to population ecotoxicology yet. This is surprising as Allen[54] suggested that periodic forcing of a population such as that associated with weather can increase the likelihood of chaos and instability. Many contaminant effects on populations share qualities similar to those described by Allen for periodic forcing by weather. Further, the logic leading to Pool's suggestion[45] that insect populations sprayed with pesticides may be a fertile area for detecting chaos can be applied equally well to nontarget species that are of interest to ecotoxicologists. (Note that, as mentioned earlier, many of these equations are similar to those used for bioaccumulation kinetics. There is no reason to assume that such complex dynamics are not present under certain situations involving bioaccumulation. This also would be an exciting area to address in the near future.) In addition to the references provided in the text, the interested reader is directed toward Schaffer and Kot,[47] May et al.,[55] May and Anderson,[56] Holden,[57] Glass and Mackey,[58] Kot et al.,[59] Nychka et al.,[60] and Olsen and Schaffer[61] for further discussions of pertinent concepts and methods.

The stochastic and deterministic complexities of laboratory and field populations assure that the debate regarding whether or not chaos occurs in populations will continue for many years. Regardless of the outcome, Robert May made a major contribution to population biology by shifting attention away from stable points, e.g., K, and toward the variation that should be expected for populations.

III. DEMOGRAPHY

A. GENERAL

In our discussion to this point, a population is described simply as a group of homogeneous individuals. This is obviously an oversimplification. A population is characterized by individuals differing in age and sex. Consequently, there are important differences between individuals relative to their contribution of offspring to the population or likelihood of dying at any given time. These differences greatly influence the dynamics and qualities of the population through time. Indeed, r and K are really summary statistics arising from the dynamics of age- and sex-structured populations. To understand populations more completely, the consequences of differences in age- and sex-specific recruitment and loss should be examined. A brief outline of demography (the quantitative study of birth, death, migration, age, and sex within populations) will be provided here for that purpose.

B. LIFE TABLES

1. General

Demographers developed life tables to describe age-specific birth and death within populations. Some life tables include only mortality data (l_x schedules). Sex-specific mortality is frequently calculated. More comprehensive tables may include age-specific natality ($l_x m_x$ schedules). Often, males are not tabulated as their contributions to natality are considerably more difficult to quantify than those of females.

Life tables may be developed with different types of data. Cohort (horizontal or dynamic) life tables follow a group of individuals born at the same time. The age-specific mortality and natality of all or a specific number of individuals, such as 1000, are followed in the cohort. A dynamic-composite life table may be constructed if cohort results from a population observed during different years are composited into a single table. A time-specific (vertical, static, stationary, or current) life table examines the individuals in a population at a specific moment. The population may include many cohorts. Cohort and time-specific tables are equivalent only if birth and mortality rates are constant and independent of which cohort is contributing to the estimate,[5,62] e.g., the population is at equilibrium and the environment does not change between samplings.

Time intervals used to construct life tables differ among studies. Several years may be used for long-lived organisms, whereas months or weeks may be more appropriate for short-lived species. Regardless, repeated samples such as those used in cohort table construction must be taken at consistent intervals and times, e.g., annually during spawning season.

Life table analyses have remarkably consistent terminology. An age interval is usually designated as x. The x then becomes the subscript for a series of age-specific variables, e.g., the number of survivors that are present at x is n_x. The proportion or number of the original group of individuals surviving to the beginning of interval $x + 1$ is l_x. (The interval x to $x + 1$ is often designated X in life tables.) The number or proportion dying

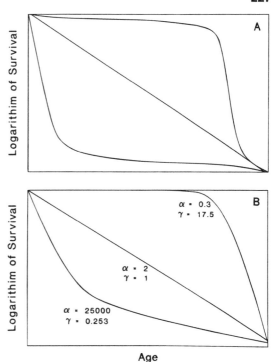

Figure 4 Deevey[63] described Type I, II, and III survival curves as shown in panel A. Pinder et al.[64] demonstrated that a Weibull distribution can be used to describe these and intergrading survival curves (panel B).

in the interval x to $x + 1$ is d_x. The probability of dying in interval x to $x + 1$ is q_x. The estimated average number of offspring produced by a female in the interval x to $x + 1$ is m_x.

With these quantities defined for each interval, several statistics can be calculated to aid in understanding the dynamics of a population. With survivorship data, the mean expected life span for an individual alive at age x (e_x) can be estimated. Rate of increase (r) and mean generation time (T_c) may be estimated with survival and natality data.

2. Death

Deevey[63] describes three general survivorship curves (Figure 4A). To a certain age, the probability of survival is high with species conforming to the Type I curve. Survival rates drop drastically after that age. Human populations in developed countries conform to the Type I curve. For Type II curves, there is a steady decrease in survival probability through time with the rate of survival remaining constant. Many adult bird populations display such survivorship curves. With Type III curves, there is high initial mortality followed by a slow decrease in survival probability. Oyster populations can display Type III curves with very high mortality at early life stages. These survivorship curves and many intergrading curves may be better described by the Weibull model[64] or other generalized exponential models such as the gamma model. (See Chapter 4 for more details.) Figure 4B illustrates the use of the Weibull model [Equation (52)] to produce curves similar to the qualitative curves of Deevey.[63]

$$\text{Survival Probability} = e^{-(\alpha t)^\gamma} \tag{52}$$

The Weibull model may be fit to survivorship data and used to estimate l_x, d_x, or q_x. Pinder et al.,[64] Rago and Dorazio,[65] Fox,[66] and Lebreton et al.[67] demonstrated the specific application of the Weibull and related models to life tables. Associated statistical inference methodologies are discussed in Chapter 4.

Count data are used directly in many life table analyses. Age-specific mortalities and life expectancies are calculated from these counts of individuals.

Example 7

The mortality rates and life expectancies for individuals within a fish population from an area receiving heated effluent are required for an environmental assessment. One thousand individuals were marked and their numbers monitored until there were no survivors (6 years). The survival data (l_x) were used to estimate the annual mortality both in numbers of individuals (d_x) and the proportion of the individuals dying who were present at the beginning of the interval (q_x). The q_x was easily estimated by d_x/l_x. From these data, the average years lived (L_x) was estimated as $(l_x + l_{x+1})/2$. The total years lived (T_x) was estimated by summing the L_x values from the bottom up to the pertinent age class. The mean expected life span (e_x in years) was then calculated to be T_x/l_x for each age interval.

Age (yrs)	l_x	d_x	q_x	L_x	T_x	e_x
0–1	1000	800	0.800	600	810	0.81
1–2	200	122	0.610	139	210	1.05
2–3	78	50	0.641	53	71	0.91
3–4	28	24	0.857	16	18	0.64
4–5	4	4	1.000	2	2	0.50
5–6	0					

Fish in the first year had a mean life expectancy of 0.81 years. The life expectancy increased slightly for the next 2 years but then declined until the last year of life when it was only 0.50 years.

These data could also have been analyzed using techniques described in Chapter 4. Significant differences between survival curves also could have been tested.

---

3. Birth and Death

Natality can now be added to survival tables to estimate several population characteristics. Normally, only females are included in these tables because of the ease of determining female contributions to productivity relative to those of males. The mean number of females born per female in age class x (m_x or age-specific birth rate) is used as a measure of age-specific productivity. Once the age-specific birth rate is estimated, the net reproductive rate (R_0) can be calculated. [The net reproductive rate is the expected number of females produced during the lifetime of a newborn female, i.e., (number of daughters born in $x + 1$)/(number of daughters born in x)]. The products $l_x m_x$ are summed over all the x values to estimate R_0. The mean generation time (T_c) is then estimated by dividing the sum of the $x l_x m_x$ products by R_0. (Note that the midpoint of each interval is used for x in this calculation, e.g., 0.5 for the interval 0 to 1 year old.) The instantaneous rate of increase (r) for the population is $\ln R_0/T_c$.

Example 8

A cohort of female crayfish was followed for 6 years in a contaminated mesocosm. The table below summarizes the calculations necessary to estimate the R_0 and T_c for these decapods. (Note that l_x in this table is expressed as a proportion, not the number of individuals.)

Example 8 *Continued*

Age (x)	l_x	m_x	$l_x m_x$	$x l_x m_x$
0–1 or 0.5	1.000	0.000	0.000	0.000
1–2 or 1.5	0.312	2.238	0.698	1.047
2–3 or 2.5	0.095	1.390	0.132	0.330
3–4 or 3.5	0.037	0.410	0.015	0.053
4–5 or 4.5	0.003	0.400	0.001	0.005
5–6 or 5.5	0.000		$\Sigma = 0.846$	$\Sigma = 1.435$

The R_0 is 0.846 females left during the lifetime of a female. The R_0 is below a rate of simple replacement ($R_0 = 1$). Assuming that the results were generated from a population with a stable structure, this low R_0 indicates that the population size will slowly decline with time. The generation time is 1.435/0.846 or 1.70 years. The estimated r is approximately ln 0.846/1.70 or -0.098.

These and more involved calculations are performed by a variety of PC-based software packages, e.g., RAMAS/age[68] and TIME-ZERO.[69] Perusal of the classic matrix approach to life tables[70,71] is also recommended to the reader contemplating a demographic study of populations exposed to pollutants.

The Euler-Lotka equation may be used to determine r if the population has a stable structure. Stearns[62] argues that moderate violations of this assumption will have little effect on the results. The population is assumed to be changing exponentially, and to have constant birth and death rates.

$$\sum_{x=0}^{\infty} l_x m_x e^{-rx} = 1 \tag{53}$$

The Euler-Lotka equation is applied by repeated substitution of estimates of r until the left side of Equation (53) is "close enough" to 1. The r estimated using the life table in Example 8 could be used to provide the first approximation of r in Equation (53). The value of the left side of the equation will be larger or smaller than 1. The r is adjusted slightly so that the value of the left side is closer to 1. For example, if insertion of r results in the left side of Equation (53) being 1.20; the r is increased slightly and used again in Equation (53). The process is repeated until the final r results in the solution to the left side of the equation that is "close enough" to 1. (Stearns,[62] page 38 provides a convenient PASCAL program that performs these calculations.)

In the preceding paragraph, comment without explanation was made regarding the stability of population structure. Given a specific r or λ, a population will eventually establish a constant or stable distribution of individuals between the age classes. This structure is referred to as the stable age structure of the population. This distribution of individuals between age classes can be predicted if age-specific births and death rates are relatively constant and the population is changing exponentially. (See Krebs,[5] Stearns,[62] and Smith[72] for more details.)

$$C_x = \frac{\lambda^{-x} l_x}{\displaystyle\sum_{i=0}^{\infty} \lambda^{-i} l_i} \tag{54}$$

where C_x = the proportion of all individuals in age class x. If the population size is constant, the stable age distribution is then termed the stationary or life-table age distribution.[5,62]

Example 9

Estimate the stable structure for a population of fish with an $r = 0.113$ ($\lambda = 1.12$) and the l_x values listed below.

x (yrs)	l_x	λ^{-x}	$\lambda^{-x}l_x$	c_x
0	1.000	1.000	1.000	0.574
1	0.513	0.893	0.458	0.263
2	0.291	0.797	0.232	0.133
3	0.074	0.712	0.053	0.030
4	0.000	0.636	0.000	0.000
			$\Sigma = 1.743$	$\Sigma = 1.000$

Respectively, 57.4, 26.3, 13.3, and 3.0% of the fish will be in the 0, 1-, 2-, and 3-year classes.

An approach such as that described above to estimate r for an age-structured population may also be used to assess the effect of contaminants on populations. For example, Daniels and Allan[73] calculated r for zooplankton populations exposed to pesticides. A distinct drop in the population parameter (r) was noted for both a copepod and cladoceran when dieldrin concentrations exceeded specific levels. They argued that, with the same amount of effort, a more ecologically realistic measure of population health (r) could be generated than by standard toxicity testing techniques. Demographic analyses would be valuable additions to the methods recommended by Mount and Norberg[74] and Hamilton[29] for analysis of zooplankton toxicity test data. Munzinger and Guarducci[75] found that standard methods were poor indicators of effects on populations exposed to metals and recommended demographic methods as being superior. Similarly, Pesch et al.[76] examined the effects of sediment contamination on λ for laboratory populations of the polychaete, *Neanthes arenaceodentata*. In using this approach to compare populations under various contamination regimes, the precautions of Rago and Dorazio[65] regarding the distributions to be expected for λ should be remembered. They pointed out that the distributions of λ are often skewed and normal statistics can be inappropriate for their analysis.

For a stationary population, reproductive value (V_A) or the expected contribution of offspring for the lifetime of an individual[8] for any age class of females can be estimated.[62]

$$V_A = \sum_{x=A}^{\infty} \frac{l_x}{l_A} m_x \tag{55}$$

The V_0 (V_A for age class 0) is R_0. By definiton, $R_0 = 1$ for a stationary population. Therefore, the reproductive values are defined relative to a newborn female with $V_A = V_0 = 1$.

Example 10

Calculate V_A for the year classes of the fish population with the demographic qualities tabulated below.

Example 10 *Continued*

x (yrs)	l_x	m_x	V_A
0	1.000	0.000	1.000
1	0.513	3.005	4.160
2	0.291	2.018	2.035
3	0.074	0.068	0.068
4	0.000	0.000	0.000

V_A: $V_0 = 1.000$ by definition
$V_1 = ((0.513/0.513)*3.005) + ((0.291/0.513)*2.018) + ((0.074/0.513)*0.068) = 4.160$
$V_2 = ((0.291/0.291)*2.018) + ((0.074/0.291)*0.068) = 2.035$
$V_3 = ((0.074/0.074)*0.068) = 0.068$

To the author's knowledge, V_A has not been used in predicting potential, age-dependent effects on populations from contaminated sites, despite its obvious value. (Barnthouse et al.[77] used a very similar approach.) For example, a toxic event in which most of the 3-year-old fish described in Example 10, were left alive, but none of the younger fish, would have a devastating impact on the population because of the low V_A for 3-year-old fish. Based on this type of analysis, Petersen and Petersen[78] questioned the common assumption that, because young organisms are more sensitive than older organisms to toxicants, there will be a large effect on a population by loss of young. Clearly, the V_A values must be considered when making such judgments.

4. Other Considerations

In this discussion of demography, several important factors were ignored out of necessity. Obviously, the sex ratio is important. With age-specific sex ratios, males can be incorporated more fully into life table analyses. Emmel[16] and Smith[72] described the general effect of age-specific sex ratios on demography. Migration can also contribute to demographic qualities of populations. More complex models including those involving genetic components incorporate migration.

IV. SPATIAL DISTRIBUTION OF INDIVIDUALS

A. GENERAL

To this point, it has been assumed that individuals were randomly distributed within the area or volume containing the population. However, deviations from randomness lead to very different behaviors at the population level. This is particularly true for sessile or sedentary species existing within relatively large areas.[32] Consequently, the distribution of individuals within a population or study area can be a critical quality to assess. This section presents general methods that may be applied to this task. The associated descriptions and terminology are extracted from Pielou[12,32] and Ludwig and Reynolds.[79] The reader is encouraged to review these references for more detail. Further, several of the methods described in this section may be implemented with program code provided in Ludwig and Reynolds,[79] and Krebs.[15] Manly[80] also provided code for performing nearest neighbor and other methods using Monte Carlo and randomization techniques.

232

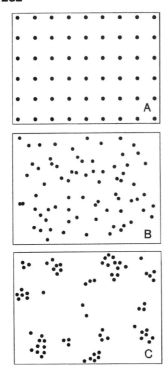

Figure 5 Three types of spatial patterns for individuals. A, uniform; B, random; and C, clumped.

The three general distribution patterns are uniform, random, and clumped (aggregated, clustered, or patchy) (Figure 5). A uniform distribution of individuals within an area is the least common pattern. More likely, individuals will be randomly distributed or clumped. Further, clumps themselves may be distributed in any of these three patterns. Patterns may also differ in intensity and grain (Figure 6). A coarse-grained pattern has large gaps between large clusters, e.g., the pattern in Figure 6A has a coarser grain than that of Figure 6B. A high-intensity pattern has large differences in densities, e.g., the pattern in Figure 6A has a higher intensity than that of Figure 6C. In Figure 6, the distribution shown in panel A has both coarser grain and higher intensity than that in panel D.

Pielou[32] distinguished between patterns associated with arbitrary sampling units and discrete habitat sites or sampling units (SU). Examples of patterns for discrete sampling units are the pattern of crayfish numbers under each of 1000 rocks; that for number of individuals for an insect species per stream snag along a length of stream; or that for the distribution of individuals for a particular parasitic species amongst 200 hosts. The sampling units (rocks, stream snags, and hosts) in which the individuals are distributed are discrete units. In contrast, units of space may be arbitrary if the individuals are contained within a continuous habitat, e.g., the number of clams per square meter of sediment surface or number of a species of zooplankton per cubic meter of water. Approaches used for testing and describing spatial patterns are not equally applicable to discrete and arbitrary sampling units.

B. INDICES FOR DISCRETE SAMPLING UNITS

If individuals are randomly distributed among the sampling units, their frequency distribution can be described by a Poisson distribution. Because the variance and mean of the Poisson distribution are equal, estimates of σ^2/μ ($s^2/\bar{x}$) may be used as a measure of

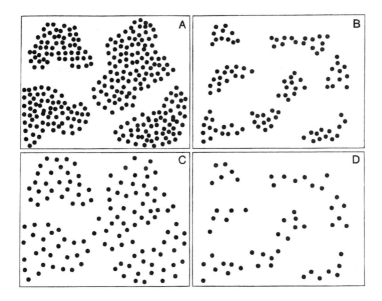

Figure 6 An example of grain and intensity (modified from Pielou[12]).

general conformity to the Poisson distribution. Indeed, the variance divided by the mean is referred to as the index of dispersion. A Poisson distribution is indicated if σ^2 is equal to μ and implies a random pattern. Similarly, a σ^2 less than μ indicates a negative binomial distribution and implies a clumped pattern. A σ^2 greater than μ suggests a positive binomial distribution and implies a uniform distribution. Pielou[32] provided a χ^2 test of fit to the Poisson distribution based on this ratio of variance to mean. Ludwig and Reynolds[79] gave similar tests for small ($N < 30$) and large ($N \geq 30$) numbers of SU values. Krebs[15] provided FORTRAN code as well as an excellent explanation of these methods. (The reader should note that Hurlbert[81] argued that this ratio was not a good measure of departure from randomness.) As will be shown, this ratio is used as a test statistic and as a descriptive statistic for spatial distribution.

Pielou,[32] Ludwig and Reynolds,[79] and Krebs[15] provided straightforward χ^2 tests for deviations from the Poisson or negative binomial distributions. (Ludwig and Reynolds[79] and Krebs[15] also provided computer programs for implementing these tests.) Tests for uniform distributions were not provided by these authors as this distribution was not likely to be pertinent in most cases. Rejection with these tests suggests inadequacy of the distributions for description of the observed spatial patterns, e.g., the pattern is not random (rejection of Poisson model) or the pattern is not clumped (rejection of the negative binomial model). Again, in a strict sense, failure to reject the null hypothesis does not prove that the distribution is appropriate. Failure to reject the null hypothesis can arise from testing inadequacies, e.g., lack of statistical power, as well as the qualities of the distribution.

The null hypothesis (H_0) that the Poisson distribution describes the data (implying a random pattern) is tested with a minimum of 30 SU values.[79] The probabilities of counts of x of 1, 2, 3, ... r individuals per SU is described by Equation (56) for a Poisson distribution.

$$P_i = \frac{\lambda^i e^{-\lambda}}{i!} \approx \frac{\bar{x}^i e^{-\bar{x}}}{i!} \tag{56}$$

where P_i = probability of i individuals per SU ($i = 0, 1, 2 \ldots r$); λ = mean of the number of individuals per SU; and $\bar{x}$ = estimated mean number of individuals per SU. The observed frequencies (F_i) of counts of individuals per SU are tabulated first. The expected frequencies (E_i) corresponding to each of the $r + 1$ observed frequencies are estimated with the total number of individuals observed on all SU values (N) and the estimated P_i from Equation (56).

$$E_i = \frac{\bar{x}^i e^{-\bar{x}}}{i!} N \tag{57}$$

A χ^2 statistic can then be calculated using the observed and expected frequencies for the data.

$$\chi^2 = \sum_{i=0}^{r} \frac{(F_i - E_i)^2}{E_i} \tag{58}$$

The H_0 is rejected if this χ^2 value is greater than a tabulated value (e.g., Table 14 in Rohlf and Sokal[82]) with an associated $\alpha = 0.05$ and $df = (r + 1) - 1 = r$. (The $df = r$ is used because this is the number of frequency classes − the number of estimated parameters.)

Ludwig and Reynolds[79] describe a very similar method for testing the H_0 that the spatial pattern can be described by a negative binomial distribution. Acceptance of the H_0 implies, but does not prove, that these data have a clumped pattern. Let μ be the mean number of individuals per SU, and k a measure of the degree of clumping. Again, a table of observed frequencies of numbers of individuals per SU (F_i) is constructed. The expected probabilities are then calculated under the assumption of a negative binomial distribution.

$$P_i = \left[\frac{\mu}{\mu + k} \right]^i \left[\frac{(k + i - 1)!}{[i!(k - 1)!]} \right] \left[1 + \frac{\mu}{k} \right]^{-k} \tag{59}$$

Ludwig and Reynolds[79] provide the following forms of Equation (59) to facilitate calculations.

$$P_0 = \left[1 + \frac{\bar{x}}{\hat{k}} \right]^{-k} \tag{60}$$

$$P_1 = \left[\frac{\bar{x}}{(\bar{x} + \hat{k})} \right] \left[\frac{\hat{k}}{1} \right] [P_0] \tag{61}$$

$$\bullet$$
$$\bullet$$
$$\bullet$$

$$P_r = \left[\frac{\bar{x}}{\bar{x} + \hat{k}} \right] \left[\frac{\hat{k} + r - 1}{r} \right] [P_{r-1}] \tag{62}$$

The μ in Equation (59) is estimated by $\bar{x}$, and the k is estimated by the following procedure. First, Equation (63) is used for an initial estimate of k.

$$\hat{k} = \frac{\bar{x}^2}{s^2 - \bar{x}} \tag{63}$$

where $s^2 =$ the variance estimated from all of the SU values.

If the original $\bar{x} > 4$ and the estimated $\hat{k} > 4$, this estimate of k is used in Equations (60) to (62). Otherwise, the following iterative process is used to estimate k in Equations (60) to (62). The $\hat{k}$ from Equation (63) is inserted into Equation (64) and both sides of Equation (64) calculated.

$$\log \frac{N}{N_0} = \hat{k} \log\left(1 + \frac{\bar{x}}{\hat{k}}\right) \tag{64}$$

where $N =$ the total number of SU values; and $N_0 =$ the number of SU values with 0 individuals.

If the value on the right side of Equation (64) is smaller than the value on the left side of this equation, the $\hat{k}$ is increased slightly and the right side of the equation solved again. The $\hat{k}$ is decreased if the value of the right side is larger than that of the left side. This process is repeated until both sides of Equation (64) are "sufficiently similar." The final $\hat{k}$ from this process is used in Equations (60) to (62) to estimate the P_i values.

As described for the Poisson distribution, a χ^2 statistic is generated with a resulting table of expected and observed frequencies [Equation (58)]. The estimated χ^2 is compared with a table of χ^2 with an associated α. However, the $df = r + 1 - 2 = r - 1$ because the number of parameters being estimated is now 2.

Example 11

The gregarious behavior of a particular benthic species results in a clumped distribution of individuals under intertidal rocks. This behavior is maintained under normal laboratory conditions as well. The species behavior is examined in mesocosms spiked with petroleum waste. The distribution of individuals among 109 rocks was measured. The H_0 is posed that the distribution can be described by a negative binomial function—the associated implication being that a negative binomial is a good model for the usual clumped distribution of individuals. The observed frequencies (F_i) for i values from 0 to 11 individuals per rock are tabulated in columns 1 and 2 in the following table.

i (number/rock)	F_i (Observed P)	$E_i(P_iN)$ (Expected P)	$\dfrac{(F_i - E_i)^2}{E_i}$
0	5	1.53	7.87
1	6	5.89	<0.00
2	5	11.99	4.08
3	15	16.79	0.19
4	21	18.31	0.39
5	30	16.46	11.14
6	10	12.75	0.59
7	5	8.72	1.59
8	4	5.34	0.34
9	4	3.05	0.30
10	2	1.64	0.08
11	2	0.76	2.02
			Sum = 28.59

Example 11 *Continued*

The mean number of individuals per rock ($\bar{x}$) is 4.59 with a variance of 5.37 ($N = 109$). The initial estimate of $\hat{k}$ is found with Equation (63) to be $4.59^2/(5.37 - 4.59) = 27.01$. Because $\bar{x} > 4$ and $\hat{k} > 4$, this initial estimate is adequate for use in further calculations. Using $\bar{x}$ and $\hat{k}$ in Equations (60) to (62), the expected probabilities for the $r + 1$ counts are calculated.

$$P_0 = \left[1 + \frac{4.59}{27.01} \right]^{-27.01} = 0.014$$

$$P_1 = \left[\frac{4.59}{4.59 + 27.01} \right] [27.01] \, [0.014] = 0.054$$

Similarly, P_2 to P_{11} are calculated to be 0.110, 0.154, 0.168, 0.151, 0.117, 0.080, 0.049, 0.028, 0.015, and 0.007, respectively. The P_i values are multiplied by the total number of sampling units (109 rocks) to get the predicted frequencies (column 3 in the preceding table). The differences between the 11 pairs of observed and expected frequencies are squared (column 4), divided by expected frequencies and summed to get the χ^2 value of 28.59. This statistic is compared with a χ^2 with $\alpha = 0.05$ and $df = r - 1 = 10$. A value of 18.307 is found in Table 14 of Rohlf and Sokal.[82] The negative binomial model is rejected for description of the species pattern under these contaminated conditions.

The index of dispersion is also used as a descriptive statistic with values ranging from 0 (extreme uniformity) to 1 (random) to N (extreme clumping). Often this index is modified by subtraction of 1 to produce the index of clumping.[83] The index allows comparison between populations with different means but identical total numbers of individuals sampled.[32]

$$\text{Index of clumping} = \frac{s^2}{\bar{x}} - 1 \tag{65}$$

The w statistic is used for this purpose.[32] The index of clumping is $s_j^2/\bar{x}_j - 1$ with $j = 1$ and 2 for the two sets of data to be compared.[32] If w is beyond $\pm 2.5/\sqrt{(n - 1)}$ then the indices from the two data sets differ at $\alpha = 0.05$.

$$w = -0.5 \ln \frac{(s_1^2/\bar{x}_1)}{(s_2^2/\bar{x}_2)} \tag{66}$$

Pielou[32] made a pertinent observation relative to using this index. She suggested that, if deaths occurred for some reason, and the original and surviving population were sampled, this comparison could be used to test if deaths were random. Obviously, the mean would drop with mortality. Perhaps not so obvious is the fact that the index of clumping would also decrease if deaths were random relative to the differing densities within the sample area. The index (I) will decrease by $\theta*I$ where θ is the proportion of the original population surviving. Deviations from this predicted decrease would suggest nonrandom (density-dependent) mortality.

The values for both the indices of dispersion and clumping are dependent on the total number of individuals sampled (n). Green's Index is a modification of these indices that corrects for this dependence on n. Consequently, Green's Index allows comparisons between samples varying in total number of individuals sampled (n), mean number of individuals per SU ($\bar{x}$), and number of SU values (N). Ludwig and Reynolds[79] gave the extreme range of Green's Index to be $-1/(n - 1)$ (completely uniform) to 0 (random) to 1 (maximum clumping). Continuing the argument of Pielou, any significant change in Green's Index with loss of a portion of the population suggests nonuniform loss of individuals relative to spatial density.

$$\text{Green's Index} = \frac{\dfrac{s^2}{\bar{x}} - 1}{n - 1} \tag{67}$$

Many other indices for discrete sampling units have been developed. The reader is directed to Pielou[12,32] and Ludwig and Reynolds[79] for more discussion of these indices.

C. INDICES FOR ARBITRARY SAMPLING UNITS

1. Indices Based on Quadrats

Often arbitrary sampling units are quadrats such as those used earlier to estimate population densities. The variance to mean ratio for quadrats differing in size can be used to assess clumping of individuals. This can be seen readily using Figure 5. If randomly placed quadrats of increasing size are placed over a uniform pattern, the variance to mean ratio will remain relatively constant. With a random pattern, the ratio will fluctuate in a random fashion. Clear peaks (or trends) in the ratio will emerge with clumped distributions.

Alternatively, the variance to mean ratio for contiguous pairs of quadrats of increasing size may be used to assess spatial patterning.[12] (Often a row of quadrats is sampled, and adjacent quadrats are sequentially pooled to obtain blocks of quadrats of increasing size.) The likelihood of smaller pairs of blocks being within the same clump or gap between clumps is higher than that for larger pairs. Consequently, the variance to mean ratio from a clumped pattern will increase as block size increases. This approach may be generalized to an ANOVA if desired. A peak in the variance to mean ratio (or mean square for number of quadrats per block) will occur at roughly the mean clump size.[12] Blocks of 2^n (e.g., 1, 2, 4, 8, 16, 32) contiguous quadrats are often plotted against the logarithm of mean squares in such exercises.

Example 12

Eight 1×1 m contiguous quadrats are sampled in a row across a beach at the mean low tide line. The numbers of species A found in the eight quadrats are 5, 4, 8, 12, 14, 4, 2, and 4 across the beach. The mean for the entire sample of 53 individuals in the eight quadrats is 6.62 individuals per quadrat. Estimate the variance to mean ratio for quadrat blocks of sizes 1 through 4.

Block = 1 quadrat:
Four block (= quadrat) pairs are available for estimating the ratio. The block pair means are the total number of individuals in all pertinent quadrats (number in quadrat 1 + number in quadrat 2 of the pair in this case) divided by the total number of quadrats in the pair (2 in this case). The pair variances are $[\Sigma(x - \bar{x})^2]/2$.

$$\text{Pair 1(5,4):} \quad \bar{x} = (5 + 4)/2 = 4.5$$
$$s^2 = [(5 - 4.5)^2 + (4 - 4.5)^2]/2 = 0.25$$
$$\text{Pair 2(8,12):} \quad \bar{x} = 10.00 \quad s^2 = 4.00$$

Example 12 *Continued*

$$\text{Pair } 3(14,4): \quad \bar{x} = 9.00 \qquad s^2 = 25.00$$
$$\text{Pair } 4(2,4): \quad \bar{x} = 3.00 \qquad s^2 = 1.00$$

The average s^2 for blocks of one quadrat is $(0.25 + 4.00 + 25.00 + 1.00)/4$ or 7.56. The block average $(\bar{x}_b)$ is 53/8 blocks of 2, 1×1 m quadrats or 6.62. The $s^2/\bar{x}_b$ is 7.56/6.62 or 1.14.

Block = 2 quadrats:
Two pairs of contiguous blocks can be made by combining adjacent quadrats, $(5 + 4, 8 + 12)$ and $(14 + 4, 2 + 4)$.

Pair 1(9,20):
$$\bar{x} = (9 + 20)/4 = 7.25$$
$$s^2 = [(9 - 7.25)^2 + (20 - 7.25)^2]/2 = 82.81$$
Pair 2(18,8):
$$\bar{x} = (18 + 6)/4 = 6.00$$
$$s^2 = [(18 - 6)^2 + (8 - 6)^2]/2 = 74.00$$

The average s^2 for blocks of 2 quadrats is $(82.81 + 70.56)/2$ or 76.69. The block average $(\bar{x}_b)$ is 53 individuals/4 blocks or 13.25. The $s^2/\bar{x}_b$ is 74.00/13.25 or 5.58.

Block = 4 quadrats:
One pair of blocks $(5 + 4 + 8 + 12, 14 + 4 + 2 + 4)$ can be used.

Pair 1(29,24):
$$\bar{x} = (29 + 24)/8 = 6.62$$
$$s^2 = [(29 - 6.62)^2 + (24 - 6.62)^2]/2 = 401.46$$

The average s^2 for blocks of 4 quadrats is 401.46. The block average $(\bar{x}_b)$ is 53 individuals/2 blocks or 26.50. The $s^2/\bar{x}_b$ is 401.46/26.50 or 15.15.

These $s^2/\bar{x}_b$ are plotted against the block sizes to get an estimate of clumping.

As mentioned, an ANOVA using 1, 2, and 4 quadrat blocks can be performed for this purpose also. The mean squares for the block classes will be in the same proportion as the relative $s^2/\bar{x}_b$ estimated above, i.e., approximately 1.14 to 5.79 to 15.15 (1:5.08:13.29). Clearly, the mean squares could be plotted against block size instead of the $s^2/\bar{x}_b$ to reveal the clumping in the data.

Ludwig and Reynolds[79] presented additional quadrat-based methods including a refinement of the method described above. This paired-quadrat variance method estimates the variance between pairs of fixed-size quadrats at specific distances between the (paired) quadrats. The variance for a spacing of 1 is calculated with Equation (68).

$$s^2 = \left(\frac{1}{N - 1} \right) \left[\left(\frac{x_1 - x_2}{2} \right)^2 + \left(\frac{x_2 - x_3}{2} \right)^2 + \cdots + \left(\frac{x_{N-1} - x_N}{2} \right)^2 \right] \qquad (68)$$

where N = the total number of quadrats from which the counts were obtained.

The s^2 values for subsequent spacings are obtained as illustrated here for spacings of 2.

$$s^2 = \left(\frac{1}{N-2}\right)\left[\left(\frac{x_1 - x_3}{2}\right)^2 + \left(\frac{x_2 - x_4}{2}\right)^2 + \cdots + \left(\frac{x_{N-2} - x_N}{2}\right)^2\right] \quad (69)$$

For larger spacings, the denominator in the first term in Equation (69) and the intervals between quadrats used in each variance estimate are adjusted accordingly. The variance values are then plotted against spacings between quadrats. Ludwig and Reynolds[79] recommended that a conservative estimate of the maximum number of spacings to be calculated is 10% of N. However, they also state that 20% of N may be acceptable.

Example 13

The quadrat data from Example 12 are reanalyzed using the paired-quadrat variance method just described. The 1×1 m contiguous quadrats (x_1, x_2, x_3, x_4, x_5, x_6, x_7, and x_8) contains 5, 4, 8, 12, 14, 4, 2, and 4 individuals, respectively. For illustrative purposes, we are ignoring the recommendation of Ludwig and Reynolds[79] calculating variances for spacings up to 10% of N, e.g., 0.1*8 = 0.8 or 1 m spacing. Instead, an arbitrary number of four spacings are used here for illustrative purposes.

s^2 for spacing = 1:

$$\left(\frac{1}{8-1}\right)\left[\left(\frac{5-4}{2}\right)^2 + \left(\frac{4-8}{2}\right)^2 + \left(\frac{8-12}{2}\right)^2\right.$$
$$\left. + \left(\frac{12-14}{2}\right)^2 + \left(\frac{14-4}{2}\right)^2 + \left(\frac{4-2}{2}\right)^2 + \left(\frac{2-4}{2}\right)^2\right] = 5.18$$

s^2 for spacing = 2:

$$\left(\frac{1}{8-2}\right)\left[\left(\frac{5-8}{2}\right)^2 + \left(\frac{4-12}{2}\right)^2 + \left(\frac{8-14}{2}\right)^2\right.$$
$$\left. + \left(\frac{12-4}{2}\right)^2 + \left(\frac{14-2}{2}\right)^2 + \left(\frac{4-4}{2}\right)^2\right] = 13.21$$

s^2 for spacing = 3:

$$\left(\frac{1}{8-3}\right)\left[\left(\frac{5-12}{2}\right)^2 + \left(\frac{4-14}{2}\right)^2 + \left(\frac{8-4}{2}\right)^2\right.$$
$$\left. + \left(\frac{12-2}{2}\right)^2 + \left(\frac{14-4}{2}\right)^2\right] = 18.25$$

s^2 for spacing = 4:

$$\left(\frac{1}{8-4}\right)\left[\left(\frac{5-14}{2}\right)^2 + \left(\frac{4-4}{2}\right)^2 + \left(\frac{8-2}{2}\right)^2 + \left(\frac{12-4}{2}\right)^2\right] = 11.31$$

Plotting of the variances with distances would show a distinct clumping at 3-m spacings.

Many of these methods are sensitive to quadrat size. A quadrat too small will result in many 0 counts, but a quadrat too big will obscure spatial detail. In their review[84] of the literature, Upton and Fingleton suggested quadrat sizes resulting in an average count of 1.0 to 4.0 individuals per quadrat. Consequently, quadrat techniques become inefficient if individuals are widely dispersed. The distance-based methods described next are more appropriate in such situations.

2. Indices Based on Distance

Numerous distance techniques have been developed. Distances can be between individuals and their nearest neighbors, e.g., the classic work of Clark and Evans.[85] Distances to the second, third, fourth, or further nearest neighbors can also be determined and used to calculate spatial patterning. Pielou[12,32] and Cressie[86] provided detailed discussion and illustration of these techniques. Krebs[15] provided FORTRAN code for nearest neighbor analysis.

Alternatively, the distance from a randomly selected point to the nearest individual can be compared with the distance between that individual and its nearest neighbor. The T-square index is one such technique recommended by Ludwig and Reynolds[79] and Cressie.[86] Ludwig and Reynolds's[79] treatment of the technique is provided here as an example. (Ludwig and Reynolds,[79] and Krebs[15] also listed BASIC and FORTRAN code for computing the associated statistics including tests of significance.)

First, a point O is selected randomly from within the area containing the individuals. The distance x between point O and the nearest individual (P) is measured. A perpendicular line intersecting line OP at P is then drawn. This perpendicular line defines two half-planes, one of which contains the original randomly selected point O. The nearest neighbor (Q) to point P is identified on the other side of the perpendicular line, i.e., on the half-plane opposite that containing point O. The distance (y) between P and Q is measured. This process is repeated for N sample points. Cressie[86] indicated that the number of random sampling points should be less than or equal to the number of individuals in the area divided by 10. Normal approximations are appropriate for testing if N is approximately 10 or more.

The C statistic is calculated with Equation (70).

$$C = \frac{\sum_{i=1}^{N} \frac{x_i^2}{(x_i^2 + 0.5\, y_i^2)}}{N} \tag{70}$$

C is 0.5 for a random pattern. It decreases as the pattern tends toward uniformity and increases as the pattern tends toward clumping. Significant deviations from 0.5 can be tested with z.

$$z = \frac{C - 0.5}{\sqrt{\dfrac{1}{12N}}} \tag{71}$$

This calculated z is compared with a tabulated z for a standard normal distribution, e.g., 1.96 for $\alpha = 0.05$.

Example 14

Normally, individuals of clam species X will distribute themselves randomly in a homogeneous substrate. Random patches of fly ash-contaminated sediments are formed within the otherwise clean sediment of a mesocosm. One hundred clams are then placed randomly into the mesocosm and allowed 1 month to move about in the sediments. Did the spatial pattern of clams remain random? (Control mesocosms show that clams maintain a random pattern in the absence of contaminated patches. One month is adequate for extensive movement about the mesocosm by the clams.) Ten random points ($N = 10$) are used to generate x_i, y_i data pairs. The C is then estimated with Equation (70).

x_i (cm)	x_i^2	y_i (cm)	y_i^2	$x_i^2/(x_i^2 + 0.5\, y_i^2)$
63	3969	15	225	0.9724
37	1369	5	25	0.9910
56	3136	30	900	0.8745
71	5041	6	36	0.9964
58	3364	34	1156	0.8534
63	3969	23	529	0.9375
72	5184	31	961	0.9152
75	5625	21	441	0.9623
59	3481	23	529	0.9294
64	4096	29	841	0.9069
				$\Sigma = 9.3390$

From Equation (70), $C = 9.3390/10 = 0.9339$.

From Equation (71),

$$z = \frac{0.9339 - 0.50}{\sqrt{\dfrac{1}{(12)(10)}}} = 4.7530$$

Because z (4.7530) is greater than 1.96, the H_0 that the clams are randomly distributed is rejected. The large value of C suggests significant clumping in the heterogeneous habitat.

D. SOURCE AND SINK CONSEQUENCES

Spatial heterogeneity was ignored in the discussions of population density, demographics, and dynamics. However, uneven distributions of individuals within a mosaic of habitats of varying quality has a strong effect on the overall population dynamics.[87] Those groups of individuals in excellent habitats may have birth rates exceeding death rates, and those in poor habitats may have higher death rates than birth rates. The consequent differential fitness of individuals within different areas produces sources and sinks, i.e., some areas produce surplus offspring that move into the poorer habitats.[88,89] Pulliam and Danielson[90] demonstrated the importance of these processes that have been downplayed in most ecological models.

Two important qualities of source-sink populations are directly pertinent to ecotoxicology. First, a habitat may be rendered unsuitable for sustaining a population by contamination, but it still may have a steady number of individuals in it if there is a suitable source nearby. This important concept is useful in reconciling laboratory results with field data. Further, it would be important in ecological assessment. Second, all habitat losses of similar size may not have equivalent effects on a population's ability to survive. A small loss of source habitat may be devastating; a large loss of sink habitat less so. Source-sink aspects of population dynamics must be kept clearly in mind despite our traditional preoccupation with homogeneous populations "that [is] increasingly conditioning our perception of reality."[87]

V. POPULATION GENETICS

A. BASIC CONCEPTS

1. Evolution by Natural Selection

The theory of evolution by natural selection is based on several assumptions. First, it is assumed that populations are capable of producing a surplus of offspring. Second, individuals within a population occupying a specific environment will differ in their abilities to survive and reproduce, i.e., their fitness will differ. Third, differences in fitness have a hereditary basis. Mendel provided a mechanism for maintaining this variation in fitness without blending or dilution between generations. As a consequence of the surplus of offspring and heritable differences in fitness, the genes of more fit individuals are overrepresented in each successive generation. This "process by which genotypes with greater fitness leave, on the average more offspring than do less fit genotypes" is natural selection.[8] It is the mechanism suggested by Darwin for evolutionary change.

2. Hardy-Weinberg Equilibrium

The frequency of genotypes within populations remain stable through time if certain conditions exist: (1) the population in equilibrium is assumed to be a large population of randomly mating, diploid species with overlapping generations, (2) no natural selection is occurring, and (3) mutation and migration rates are assumed to be negligible. The Hardy-Weinberg (or Castle-Hardy-Weinberg) principle defines the expected frequencies of genotypes for such a population. Let two alleles (100, 165) be present for a gene at frequencies of p and q, respectively. Predicted genotype frequencies for a two allele locus (e.g., 100 and 165) at Hardy-Weinberg equilibrium are: 100/100, p^2; 100/165, $2pq$; and 165/165, q^2.

Conformity of genotype frequencies to the Hardy-Weinberg model can be determined with a χ^2 test. The observed number of individuals of each genotype (e.g., $O_{100/100}$, $O_{100/165}$, $O_{165/165}$) are compared with the expected number of individuals of that genotype under the Hardy-Weinberg model (e.g., $E_{100/100}$, $E_{100/165}$, $E_{165/165}$). For a polymorphic locus with two alleles, the χ^2 value is estimated as

$$\chi^2 = \sum_{i=1}^{3} \frac{(O_i - E_i)^2}{E_i} \qquad (72)$$

where O_i = the number of individuals observed for the three ($i = 3$) possible genotypes; and E_i = the expected number of individuals for the three possible genotypes. This χ^2 is compared with a tabulated critical χ^2 value with a specified α and $df = 1$. (The df is the number of genotypes minus the two parameters in the data (p and q).) A calculated χ^2 greater than the critical χ^2 would lead to rejection of the H_0 and indicate that significant deviation from Hardy-Weinberg expectations has occurred.

Example 15

A population in a contaminated area experienced a decimating exposure to a toxicant. The population was examined the next year before breeding began. Among the parameters measured is the frequency of genotypes for an electrophoretic trait in 1250 individuals (N). The genotype frequencies have met Hardy-Weinberg expectations during several preexposure samplings. Is the population still in Hardy-Weinberg equilibrium relative to this trait?

Genotype	Observed Numbers	Expected Numbers
100/100	201	167.445
100/165	513	580.110
165/165	536	502.445

The expected frequencies for the 100/100, 100/165, and 165/165 genotypes are predicted from the observed allele frequencies. Each diploid individual has two alleles, i.e., there is a total of 2*1250 or 2500 alleles in the 1250 individuals. Of these alleles, 2*201 + 1*513 or 915 of the alleles present are the 100 allele. The frequency for the 100 allele (p) is 915/2500 or 0.366. The frequency of the 165 allele (q) is $1 - 0.366$ or 0.634. The expected frequency of the 100/100 genotype is $p^2*N = 0.366^2*1250 = 167.445$ Similarly, the expected frequencies of the 100/165 (2*0.366*0.634*1250) and 165/165 (0.634^2*1250) genotypes are 580.110 and 502.445, respectively.

$$\chi^2 = \frac{(201 - 167.445)^2}{167.445} + \frac{(513 - 580.110)^2}{580.110} + \frac{(536 - 502.445)^2}{502.445} \approx 16.73$$

The critical χ^2 ($\alpha = 0.05$, $df = 1$) is 3.84; therefore, the H_0 is rejected. The genotype frequencies are significantly different from Hardy-Weinberg expectations. This deviation may be a consequence of one or several violations of the assumptions underlying the Hardy-Weinberg model. There may have been selection against a sensitive genotype (the heterozygote, 100/165). The removal of individuals from the habitat could have resulted in a rapid movement of migrants into the area. The general deficit of heterozygotes favors, but, certainly does not prove, the possibility that significant migration produced the disequilibrium. (See Wahlund effect below.)

The assessment of Hardy-Weinberg equilibrium for two alleles can be extended to three alleles at a locus. Predicted frequencies of the six associated genotypes can be made if frequencies for the three alleles are known. For example, frequencies of the 66 (g), 100 (p), and 165 (q) alleles would be used to predict the following six genotype frequencies.

Genotype	Frequency
66/66	g^2
66/100	$2gp$
66/165	$2gq$
100/100	p^2
100/165	$2pq$
165/165	q^2

The six pairs of observed and expected genotype frequencies would be used in Equation (72) with $i = 1$ to 6. The degrees of freedom associated with the χ^2 would be the number of genotypes minus the number of alleles, i.e., 6-3 or 3 df.

3. Genetic Drift

a. General

Hardy-Weinberg expectations of no change in genotype frequencies through time are based on specific assumptions. The violation of the assumption of a large (infinite) population has many ramifications pertinent to ecotoxicology. With a reduction in population size, random genetic drift (random change in allele frequencies) can be accelerated. Thus, drift may be sufficiently large to result in fixation or loss of an allele at a locus. (The probability of fixation for an allele is equal to its frequency.)

b. Effective Population Size

The rate of genetic drift is related to the number of individuals contributing genes to the next generation, the effective population size (N_e). Obviously, because not all individuals contribute genes to the next generation, the N_e will often be some number less than the total number of individuals in the population. (If females store sperm, the N_e could be larger than the total number of individuals because absent males may be contributing via stored sperm.) Further, N_e changes through generations as the demographic composition and total number of individuals fluctuates in the population.

The effective population size, N_e, for a population with nonoverlapping generations may be estimated over many generations of changing population size as follows:[8]

$$\frac{1}{N_e} = \left[\frac{1}{t}\right]\left[\frac{1}{N_1} + \frac{1}{N_2} + \cdots + \frac{1}{N_t}\right] \tag{73}$$

where t = time in generations; and N_i = population size at time i. This harmonic mean estimate of N_e has the advantage of weighting the small N values more heavily than the large N values. This weighting accommodates the greatly increased probability of change in allele frequency with decreasing N. (Indeed, a very low N such as that associated with a drop to a low population size after a toxic event can produce a genetic bottleneck. The bottleneck reflects the decreased number of individuals available to carry an allele into the next generation.)

If the number of males and females contributing genes to the next generation is unequal, the effective population size (population with nonoverlapping generations) may be estimated with Equation (74).[8]

$$N_e = \frac{4N_m N_f}{N_m + N_f} \tag{74}$$

In populations with overlapping generations, estimation of N_e becomes more difficult. Hartl and Clark[8] provided a general ($N_e \approx N/2$ where N = number of individuals in the population) and a more detailed estimate [Equation (75)] for this purpose.

$$N_e = \frac{4N_a L}{\sigma_n^2 + 2} \tag{75}$$

where N_a = the natality within a unit of time; L = the mean generation time; and σ_n^2 = the variance in brood size. Crow and Kimura[91] estimated the influence of N_e and the initial frequency (p) of a neutral allele (an allele not experiencing selection) on the rate at which it will be lost ($p \rightarrow 0.0$) or driven to fixation ($p \rightarrow 1.0$). Excluding cases for which the allele is lost, the average time expressed in generations to fixation is estimated by Equation (76).

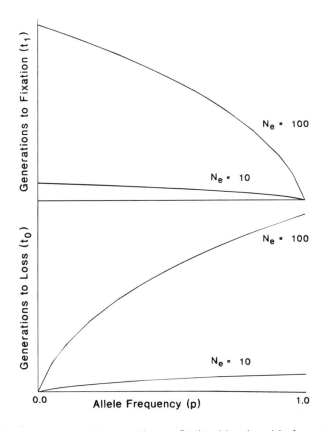

Figure 7 The time expressed in generations to fixation (t_1) or loss (t_0) of a neutral allele as functions of N_e and initial allele frequency (p). Note the accelerating influence of a drop of N_e from 100 to 10 individuals.

$$\bar{t}_1 = -\frac{1}{p} [4N_e(1 - p) \ln (1 - p)] \tag{76}$$

In cases for which alleles do not go to fixation, the average time to loss for a neutral allele as a function of N_e is defined by Equation (77).

$$\bar{t}_0 = -4N_e \left[\frac{p}{1 - p}\right] \ln p \tag{77}$$

Figure 7 summarizes the importance of p and N_e on the time to fixation or loss for a neutral allele.

Several important points can be made from Equations (73) to (77). First, although it is a common assumption in ecotoxicology,[92-94] change in allele frequency associated with pollution does not necessarily indicate selection. It may arise from accelerated genetic drift as a consequence of a generally reduced N_e or an abrupt shift associated with a genetic bottleneck. Differences in survival between sexes may exacerbate this condition [Equation (74)]. Demographic shifts also modify N_e [Equation (75)] and, under certain situations, influence genetic drift. As the N_e is lowered by a toxic event, the rates at which a rare allele ($p < 0.5$) is lost or a common allele ($p > 0.5$) becomes fixed are greatly accelerated. Additionally, the probability of a loss of total genetic diversity would be increased if contamination reduces N_e.

4. Wahlund Effect

The Hardy-Weinberg model also assumes that there is no significant migration into a population. If this is not true, the observed genotype frequencies measured in the "population" are composites of frequencies of the initial population and the population of migrants. The Wahlund effect can occur if these separate populations are themselves in Hardy-Weinberg equilibrium. The Wahlund effect is simply that, immediately after combination but before breeding, the composite population such as that just described will have a deficit of heterozygotes.[8] (This may have been the reason for the decrease in heterozygotes noted in Example 15.) Endler[95] noted that the Wahlund effect can befuddle detection of selection in natural populations if the population dynamics are inadequately understood. He further suggested that, if the sampling area changes between sampling periods, the Wahlund effect can occur due to undetected differences in genetic structure over the various areas sampled.

One generation after mixing and mating, the resulting population will come to Hardy-Weinberg equilibrium and the frequencies of homozygotes in the resulting population will be lower than those in the separate populations. This reduction in homozygotes after mixing and breeding is called the Wahlund principle. (For richer detail, see Hartl and Clark.[8])

5. Natural Selection

Deviation from Hardy-Weinberg expectations could also indicate selection. As discussed, selection acts on differences in fitness between individuals. Specifically, fitness differences exist in mating or fertilization ability, fertility, fecundity, or survival.[95] Selection may be directional. For a quantitative trait (a "continuous" trait affected by several loci), this may involve a directional shift in the distribution of individual qualities, e.g., toward individuals able to produce the most stress protein. Directional selection can occur for a "Mendelian" trait involving one locus if one homozygote has higher fitness than the other genotypes at the pertinent locus. Alternatively, selection may be normalizing. Fitness may be highest for individuals toward the middle of a distribution for a quantitative character or for the heterozygote for a simple, "Mendelian" character. In contrast, with disruptive selection, fitness is lowest for the individuals toward the center of a distribution or for the heterozygote.

Selection associated with mortality (viability) can be measured by examining the genotype frequencies of adults and comparing these frequencies with Hardy-Weinberg expectations. For example, let a two allele locus have expected genotype frequencies of $100/100 = p^2$, $100/165 = 2pq$, and $165/165 = q^2$. With selection, the measured frequencies are w_1p^2, $w_2 2pq$, and w_3q^2, respectively. The values of w_1, w_2, and w_3 are the relative fitnesses of the three genotypes.

Example 16

Mosquitofish populations from areas experiencing chronic exposure to agricultural pesticides are found to possess high frequencies of a tolerance allele, T. Controlled matings of laboratory-reared fish produce 200 TT, 200 TS, and 200 SS fish. Both the T (p) and S (q) allele frequencies are 0.50. After exposure to pesticide, the surviving numbers of the three genotypes were TT = 195, TS = 123, and SS = 14. The relative fitnesses for the TT (w_1), TS (w_2), and SS (w_3) genotypes were estimated.

Example 16 *Continued*

Genotype	Initial Number	Surviving Number	Proportion Alive	Normalized to TT (w)
TT	200	195	0.975	1.000
TS	200	123	0.615	0.631
SS	200	14	0.070	0.072

The relative fitnesses of the SS homozygote ($w = 0.072$) and TS heterozygote ($w = 0.631$) are considerably lower than that of the TT homozygote.

Natural selection is extremely difficult to demonstrate convincingly in natural populations. Estimation of power is critical in such studies, because the absence of statistically significant differences in fitness between genotypes does not prove that there are no biologically significant differences in fitness. Lewontin and Cockerham[96] presented statistical methods of testing for significant natural selection and for estimation of associated power. It is illustrative to note that, with differences in fitnesses as large as $w = 1.0$ for genotype bb and $w = 0.5$ for genotype BB, sample sizes of 378 individuals are required to ensure detection of selection in 90% of all studied cases (Lewontin and Cockerham,[96] Table 1). With $w = 0.8$ for bb and $w = 1.0$ for BB, the required sample size increases to 3405! These are impressive sample sizes if one considers that a selection coefficient in the range of a 1% difference can have a very significant effect on the change in allele frequencies.[8] (The relationship between fitness and the selection coefficient is $s = 1 - w$. It is the relative decrease with selection).[97] Hartl and Clark[8] suggested that studies lacking the power to detect differences of 10% or less are of questionable value.

Inherent in the task of detecting natural selection are many logistical problems as tabulated by Endler[95] (pages 99, 108, and 116). Further, approaches taken in many cases are inferentially weak. Endler[95] cited three flaws in most such studies. First, many studies only focus on one component of fitness and fail to estimate the lifetime fitness of individuals. For example, survival under stress is examined but reproductive fitness is ignored. Second, only one or a few traits are studied with little attention to other traits or trait interactions. Third, the function of the trait being studied is often vague or undefined. The reason or mechanism for differential fitness may remain undetermined.

This last point is particularly damning in ecotoxicology in which understanding of the mechanism is often as important as detecting the presence of selection.[98] Fitness is a term linked to a specific set of conditions. As conditions shift, the relative fitnesses may also change. Without knowledge of the underlying mechanisms, inference about selection under differing conditions is questionable. "The time has passed for 'quick and dirty' studies of natural selection."[95]

There is insufficient space here to describe detailed methods used to test for directional, disruptive, or stabilizing selection. The reader is directed to Chapter 6 of Endler,[95] Chapter 5 and other chapters in Manly,[99] and Arnold and Wade[100] for more information.

6. Quantitative Genetics

Some traits are influenced by many genes and their expression may be strongly influenced by the environment. The time-to-death for an organism is an excellent example of such a trait. Such quantitative traits are not often analyzed as "Mendelian" traits. Methods developed by agricultural scientists and breeders are applied instead.

Often regression analysis is used to examine such traits. A large set of paired measurements for a trait expressed in an offspring and its parent may be used. A significant slope suggests a relationship; however, an environment shared by the parent and offspring may also result in a significant slope. (The phenotype, not the genotype, is being measured.) Mitchell-Olds and Shaw[101] provided a current review of regression techniques in this area.

Wilson and Bossert,[97] Hartl and Clark,[8] and Stearns[62] defined phenotypic variance with Equation (78). This equation is based on the assumption of no significant genetic-environmental interactions.

$$P = \mu + G + E \qquad (78)$$

where P = the phenotype of an individual; μ = the population mean of the trait; G = the deviation from μ resulting from the individual's genotype; and E = the deviation from μ resulting from the individual's microenvironment or consequences of development. The total phenotype variance in a population (σ_p^2) is the sum of the variance due to the genotype (σ_g^2) and variance due to the environment (σ_e^2). The σ_g^2 is composed of the additive effects of all pertinent genes (σ_a^2), epistatic variance due to gene interactions (σ_i^2), and dominance variance (σ_d^2). The heritability of a trait may be described in these terms.

Narrow sense heritability (h^2) is simply the sum of the variance due to additive genetic factors divided by the total phenotype variance, σ_a^2/σ_p^2.[8,62] Narrow sense heritability can be calculated from the slope (b) of the regression model described above.

$$b = h^2/2 \qquad (79)$$

Broad sense heritability includes narrow sense heritability plus variance due to environmental and other genetic effects. It is defined as σ_g^2/σ_p^2.[8,62] An analysis of covariance (ANCOVA) may be used to determine whether significant broad sense heritability is present.

Example 17

Lee et al.[102] exposed mother and offspring mosquitofish to inorganic mercury and measured times-to-death. After birth, individuals from each brood were placed into identical 100-L pools and allowed to mature before toxic exposure to mercury. Their times-to-death were noted during acute exposure. Individuals within each brood shared a common mother and environment during maturation. All broods shared similar rearing conditions. Mothers were also exposed to inorganic mercury and their times-to-death recorded. Narrow and broad sense heritability for mercury tolerance were examined.

Narrow Sense Heritability: Ten mother-offspring pairs were used to build a regression model. Mother times-to-death, offspring median times-to-death, and offspring wet weight were included in this model. The slope was not statistically significant at $\alpha = 0.05$ (H_0: slope = 0, $T = 0.64$, $P = 0.54$) indicating no detectable narrow sense heritability. The addition of covariates such as fish wet weight, log of fish wet weight, or fish standard length failed to improve the model. (Although insignificant narrow sense heritability was noted here, it has been detected by others for mercury[103] or lead[104] exposure in other fish species. Klerks and Levinton[105] and Posthuma et al.[106] also calculate high heritability for metal tolerance in oligochaetes and springtails.)

Broad Sense Heritability: Seventeen broods of offspring were used in an ANCOVA with fish wet weight included to adjust for brood size differences. (See Stearns[62] for more detail.) The SAS code pertinent to the analysis was the following:

Example 17 *Continued*

```
PROC GLM;
   CLASS BROOD;
   MODEL KIDTTD = BROOD KIDWGT;
RUN;
```

Source	df	Sums of Squares	Mean Square	F	Probability > F
Model	17	8,877.07	522.18	2.98	0.0003
Error	115	20,134.32	175.08		
Corrected Total	132	29,011.39			
Brood	16	8,290.88	518.18	2.96	0.0040
Wet Weight	1	586.19	586.19	3.35	0.0699

The model clearly indicated a significant effect of brood on time-to-death for the 133 fish ($p = 0.0040$).

The lack of detectable narrow sense, but detectable broad sense, heritability relative to time-to-death suggests that components of variance involving nonadditive genetic, genetic-environmental, or environmental factors during maturation were important in determining time-to-death.

In ecotoxicology it is also important to understand clearly the relationship (reaction norm) in which the phenotype for a particular genotype varies as a continuous function of an environmental factor.[107] It defines the ability of a particular genotype to express different phenotypes over a gradient of some environmental factor such as temperature or contaminant level. Figure 8 illustrates the reaction norm concept. In panels A and B, for a particular genotype, a range of phenotypes is expressed over a range of intensities for some environmental factor. These phenotypes may be some surrogate measure of fitness, e.g., growth rate, fecundity, or stress protein production. The relationship may be a straight line (panel A) or a curved line (panel B).

Reaction norms for a series of genotypes may be compared with reveal important detail as shown in Figure 8C for fecundity as influenced by contaminant level. Genotypes 1 and 2 have parallel reaction norms suggesting no obvious genetic-environment interactions in the context of fecundity. Their response is similar over the range of environmental conditions although the fecundity of genotype 2 remained consistently lower than that of genotype 1. Nonparallel reaction norms suggest genetic-environment interactions. In the context of genotype 3 versus the other two genotypes, there was a significant genetic-environment interaction. The fecundity of genotype 3 dropped much faster with increasing contaminant level than those of genotypes 1 and 2. It is critical to note that, the point at which genotypes 1 and 3 cross, there would be no detectable difference in fecundity. A study examining genotype effects at this contaminant level would lead one to the incorrect conclusion that genotype had no effect on fecundity under contaminant exposure. In fact, at lower levels of contamination, genotype 3 could have the highest fecundity of the three genotypes, yet it would have the lowest fecundity at higher levels.

Reaction norms are important to remember because genetic-environmental interactions are common.[62] The reaction norm influences the rate of response to a selective agent and the reaction norms themselves can be subject to natural selection.[8] Stearns[62,107]

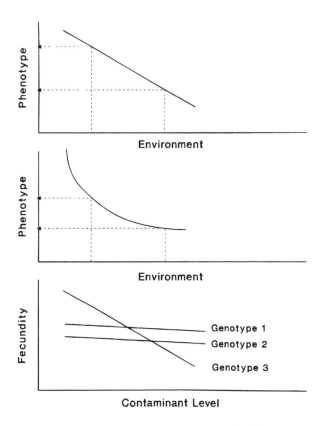

Figure 8 An illustration of reaction norms. Panel A shows the phenotypes expressed by a particular genotype over the range of an environmental factor. Panel B shows a nonlinear reaction norm for a genotype. Panel C illustrates parallel (no significant environment by genotype interactions for genotypes 1 and 2) and nonparallel (significant environment by genotype interactions, genotype 3 versus the other two genotypes) reaction norms.

provided a thorough discussion of associated methods. Westcott[108] discussed other methods of analyzing genetic-environmental interactions.

B. LETHAL STRESS (VIABILITY)

Genetic differences in survival can be assessed using techniques described in Chapters 4 and 5. Manly[99] suggested that the time-to-death approach described in Chapter 4 could be used very effectively for this purpose. Time-to-death methods are also beginning to be used in ecotoxicology[98,109–113] and are illustrated here. Other methods such as those described in Chapters 4 and 5 for endpoint tests are used for these purposes, e.g., Nevo et al.,[93] Lavie et al.,[114] Lavie and Nevo,[94] Hughes et al.,[115] and Kopp et al.,[116] However, statistical power is often lower with such methods and, as previously discussed, power is critically important in selection studies.[96]

An illustration of the time-to-death approach is provided in Example 18 for a single genetic locus. Multiple locus effects on survival[112,113,117] may be similarly assessed by defining classes based on multiple locus combinations for incorporation into the model.

Example 18

Diamond et al.[109] exposed mosquitofish of various genotypes to inorganic mercury. Their data are reanalyzed here with omission of all genetic data except that for the PGI-2 locus. The times-to-death were used to model and test for the potential effects of PGI-2 genotype on time-to-death. Fish were sexed and weighed before electrophoretic determination of their genotypes. The SAS PROC LIFEREG was used to produce the following output based on a Weibull distribution model.

Variable	df	Estimate (SE)	χ^2	Probability of χ^2	Label
Intercept (μ)	1	4.134(0.130)	1004.1	0.0001	
Sex (β)	1		61.5	0.0001	
	1	0.358(0.046)	61.5	0.0001	Female
	0	0	—	—	Male
Wet weight (β)	1	3.157(0.361)	76.3	0.0001	
PGI-2 Geno-type(β)	5		17.33	0.0039	
	1	0.370(0.117)	9.90	0.0017	100/100
	1	0.468(0.117)	16.06	0.0001	100/66
	1	0.362(0.123)	8.66	0.0033	100/38
	1	0.389(0.125)	9.64	0.0019	66/66
	1	0.339(0.141)	5.79	0.0161	66/38
	1	0	—	—	38/38
Scale (σ)	1	0.514(0.091)			

The wet weight and sex of a mosquitofish have significant ($\alpha = 0.05$) effects on time-to-death as indicated by the probability of getting a χ^2 as large as the calculated χ^2 by chance alone. As discussed in Chapter 4, the estimates of β for the two sexes indicate that males were more sensitive than females during exposure to mercury. The positive β for wet weight also suggests that smaller fish were more sensitive than larger fish. The overall χ^2 for PGI-2 genotype indicates a significant genotype effect on time-to-death. The reference genotype (38/38) had a time-to-death significantly shorter than those of all other genotypes as indicated by the associated χ^2 values. For example, $\chi^2 = 9.90$ was calculated for the H_0 that the β for the 100/100 genotype was not significantly different from that of the 38/38 genotype. (Because the 38/38 genotype was the reference genotype, its β was set at 0.)

The relative risks of the various genotypes can be used to estimate relative fitnesses. As described in Chapter 4, relative risk for a class variable such as genotype or sex is $e^{-\beta/\sigma}$.

Genotype	β	Relative Risk	w
100/100	0.370	$e^{-0.370/0.514} = 0.487$	0.82
100/66	0.468	$e^{-0.468/0.514} = 0.402$	1.00
100/38	0.362	$e^{-0.362/0.514} = 0.494$	0.81
66/66	0.389	$e^{-0.389/0.514} = 0.469$	0.86
66/38	0.339	$e^{-0.339/0.514} = 0.517$	0.78
38/38	0	$e^{-0/0.514} = 1.000$	0.40

These relative risks (column 3) are interpreted as described previously. For example, the risk of death for the 38/38 homozygote is roughly twice (1.000/0.487) that of the 100/100 homozygote. These relative risks can be converted to relative fitnesses (w values).

> **Example 18** *Continued*
>
> This is done by making the w for the most fit genotype (100/66) equal to 1 (0.402/0.402) and all other genotypes' relative fitnesses some amount less than 1 (right column above) by dividing 0.402 by the relative risk of the genotype in question. These w values may now be used in conventional population genetics models.

The number of scored loci for which an individual is heterozyous has been correlated with a variety of surrogate measurements of fitness.[118–125] For example, Diamond et al.[109] and Newman et al.[98] examined the number of heterozygous loci over eight scored loci as a measure of heterozygosity and found correlation with mosquitofish time-to-death during acute mercury or arsenate exposure. These correlations between measures of fitness and heterozygosity may be a product of heterosis such as that described by Watt.[126] (Heterosis is the superior performance of heterozygotes.) Trehan and Gill[127] support this mechanism by demonstrating superior acid phosphatase activity for heterozygous *Drosophila*. Inbreeding depression and overdominance may also provide mechanisms for this multiple locus heterozygosity advantage.[128,129] However, in some cases, an apparent heterozygosity effect is simply an artifact arising from the additive effects of individual loci.[98] In still other cases, there is no detectable multiple locus heterozygosity advantage.

A multiple locus heterozygosity effect on survival has been identified tentatively in some studies of pollutant effects on aquatic populations.[98,109,116] However, as just mentioned, Newman et al.[98] demonstrated that multiple locus heterozygosity effects can be artifacts. Furthermore, Kopp et al.[116] found lower levels of heterozygosity at contaminated (low pH/high aluminum) sites than at clean sites, results inconsistent with their laboratory studies that predicted high heterozygosity in survivors of contaminated conditions. Field surveys[116,130] also note declines in multiple locus heterozygosity with increased pollution. The mechanism invoked most often to explain this drop in multiple locus heterozygosity is selection against heterozygotes at several loci.[116,130] However, several of the mechanisms described already (e.g., genetic bottlenecks, accelerated drift associated with low N_e, Wahlund effect) seem probable but untested explanations. Furthermore, interpretation of such findings is confounded by selection involving other unexamined components of fitness. Indeed, Nadeau and Baccus[131] and Clegg et al.[132] suggest that selection for reproductive components is more common than survival-related selection and can be in the opposite direction as survival selection. Thus, there is a high risk of inaccurate conclusions based only on survival differentials. Unfortunately, these studies are too preliminary to provide any strong inferences at this time. However, they do suggest an additional tool with which population responses may be assessed.

In addition to relationships between individual heterozygosity and fitness, averaged over all individuals, heterozygosity describes population genetic diversity. A drop in mean heterozygosity due to any contaminant-related mechanism can indicate a decrease in the overall genetic diversity of the population. Because the long-term viability of the population (evolutionary potential) depends on genetic diversity, these changes are important regardless of any implied selection based on multiple locus heterozygosity.

In addition to heterozygosity, several indices also describe genetic diversity. These include counts of polymorphic loci (e.g., 9 of 12 loci were polymorphic) often expressed as proportions (e.g., $P = 0.66$), or mean number of alleles per scored locus (e.g., 1.75 alleles/scored locus). Leberg[133] suggested that the number of polymorphic loci or average number of alleles per locus provide the best measures of genetic diversity if genetic bottlenecks are suspected.

The average heterozygosity and associated within-population variance for a sample of n individuals could be estimated.[134]

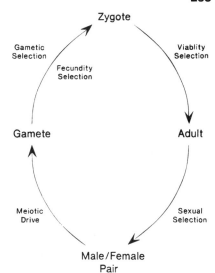

Figure 9 Components of selection (modified from Figure 1 in Chapter 4 of Hartl and Clark[8]).

$$\overline{H}_l = \frac{1}{n} \sum_{j=1}^{n} x_{jl} \tag{80}$$

$$\text{Variance } \overline{H}_l = \frac{1}{n} \overline{H}_l(1 - \overline{H}_l) \tag{81}$$

where $x_{jl} = 1$ if the individual is a heterozygote, or 0 if the individual is a homozygote. Equation (80) estimates average heterozygosity for a locus (l) by simply dividing the sum of x_{jl} values by n. H_l values may be averaged for several (m) loci.

This average heterozygosity fails to consider genetic variance information associated with different homozygotes. Weir[134] suggested that, for inbred populations that have few heterozygotes but several different homozygotes, other indices of genetic diversity such as Equation (82) are better than Equation (80) as a measure of genetic variability. The average genetic diversity [Equation (83)] averaged over m loci can also be used to estimate genetic variation.

$$D_l = 1 - \sum_{u=1}^{f} \overline{P}_{lu}^2 \tag{82}$$

$$\overline{D} = 1 - \frac{1}{m} \sum_{l=1}^{m} \sum_{u=1}^{f} \overline{P}_{lu}^2 \tag{83}$$

where $\overline{P}_{lu}$ = the measured frequency of the uth allele at the lth locus; and f = the total number of alleles at the lth locus.

C. SELECTION COMPONENTS

Differential survival (viability or zygotic selection) is only one of several important components of selection. (Figure 9 shows the major selection components in the context of the life cycle of a sexual organism.) Viability selection begins at zygote formation and continues through the lifetime of the individual. It may be examined at various ages or stages, e.g., Christiansen et al.[135] Viability selection can also affect reproductive success because an individual living longer than another can have more opportunity to mate and produce more offspring. Consequently, it is important to understand the demographic qualities of a species population when interpreting selection components.[131,136] Sexual selection is the differential mating success of individuals with specific genotypes. It may involve mechanisms such as the classic male-male competition for access to females or female mate selection. Meiotic drive, the differential production of gametes by heterozygote parents, may also occur.[8,137] Gametic selection involves the differential success of gametes produced by heterozygotes.[8] In the classic method papers of Christiansen and co-workers,[137–139] these last two selection components are tested together as gametic selection. In their methods, gametic selection is defined as a distorted segregation in heterozygotes. Finally, more offspring (zygotes) may be produced from matings involving different pairs of genotypes (fecundity selection). Fecundity selection is expressed through mating pairs, not individuals. Selection at these various stages in combination with the demographic qualities of a population can produce distinct differences in the rate of allele frequency change.[8] Also, several components can act to balance each other with an epiphenomenal maintenance of polymorphism[8] that would be inexplicable if only one component of fitness were quantified.

Prout[140] sampled individuals in successive generations to examine selection components. Christiansen and Frydenberg[137] developed a selection component analysis scheme that can be applied to populations of sexual, live-bearing species. Genotypes were determined for adults, gravid females, and one offspring per gravid female. In their analysis, expected and observed frequencies were tested with χ^2 statistics for the following series of sequential hypotheses. It is important to note that the hypotheses that follow are nested and, upon rejection of a hypothesis, no further hypotheses can be tested. The sequence of hypotheses from Table IV of Christiansen and Frydenberg[137] is outlined below.

H_1: Half of the offspring from heterozygous females are heterozygous, i.e., there is no selection among female gametes. Rejection of this hypothesis suggests gametic selection in heterozygous females.

H_2: The frequency of transmitted male gametes is independent of female genotype. The male gametes are transmitted identically among all female genotypes. Provided H_1 was not rejected, rejection of H_2 suggests nonrandom mating with female sexual selection.

H_3: The frequency of transmitted male gametes is equal to the frequency in adult males, i.e., all males mate with equal success. Rejection implies differential male mating success and gametic selection in males if H_1 and H_2 have not been rejected before this test.

H_4: The frequencies of genotypes are equal among gravid and nongravid (mature) females. Rejection implies differential female mating success if previous hypotheses have not been rejected.

H_5: Equal frequencies of genotypes between adult males and females. Rejection suggests that zygotic selection is not the same for males and females if H_1 to H_4 have not been rejected previously.

H_6: The adult population frequency is the same as that estimated for the zygotic population. Rejection suggests zygotic (survival or viability) selection.

Acceptance of all six hypotheses implies no detected selection at any component, i.e., no evidence that the gene is not neutral.

The space available here is insufficient for providing additional details of this approach. More information can be derived from Christiansen and Frydenberg,[137] Hartl and Clark,[8] and Weir.[134] Nadeau and Baccus[131] and Nadeau et al.[136] described a modification which uses genotypes of all offspring for each female to remove the interdependence of the sequence of hypotheses outlined. By performing this additional work, most hypotheses can then be tested independently. Williams et al.[141] provide a comprehensive selection component technique using linear modeling methods.

Example 19

Genetic changes were noted in mosquitofish (*Gambusia holbrooki*) populations exposed to inorganic mercury in 7250 L mesocosms (example derived from Mulvey et al.[142]) Fish were harvested from the mesocosms (duplicate mesocosms A and B for each treatment) and their genotypes at eight loci determined by starch gel electrophoresis. One late stage embryo was removed from each gravid female for electrophoresis. The results (probabilities associated with the χ^2 tests) from selection component analysis of data for the PGI-2 locus are shown below. Three allozymes were segrating for this locus. For ease of analysis, the rare PGI-2 alleles (38, 66) were pooled in these analyses yielding three different genotypes, 100/100, 100/−, or −/−. The − indicates either a 38 or 66 allele. (The FORTRAN program for most of these analyses is listed in the Appendix as Table 29. The fecundity selection was tested separately with an ANCOVA using mother size, genotype and fecundity (number of late-stage embryos).

| | Treatment | | | |
| | 0 µg/L | | 18 µg/L | |
Hypothesis Implying	A	B	A	B
Female gametic selection	0.54	0.64	0.07	0.71
Random mating	0.76	0.96	0.51	0.91
Male reproductive selection	0.70	0.88	0.07	0.73
Female sexual selection	0.55	0.52	0.01	0.09
All components listed above	0.76	0.85	0.01	0.54
Zygotic selection equal in sexes	0.54	0.18	<0.01	0.26
Zygotic selection	0.68	0.19	0.42	0.58
All zygotic selection components	0.62	0.16	0.02	0.35
Total fit to random mating and neutrality	0.78	0.61	<0.01	0.51

There was no evidence of selection at the PGI-2 locus in the 0 µg/L mesocosms. In the 18 µg/L mesocosms, there was significant ($\alpha = 0.05$) deviation from the expected equality of genotypes between gravid and nongravid females in duplicate A. Female sexual selection was implied. Note that testing stops at this hypothesis, and no further hypotheses (male sexual selection or zygotic selection) can be tested.

Which genotypes did best relative to the female sexual selection? The percentage of mature females that were gravid are the following:

PGI-2 Genotype	0 µg/L (%)	18 µg/L (%)
100/100	69.6	43.3
100/−	68.8	70.7
−/−	67.8	68.0

Clearly, the PGI-2 100/100 homozygotes had lower reproductive success (female sexual

Example 19 *Continued*

selection component) under mercury exposure than the other genotypes. An ANCOVA including female size and genotype versus fecundity (the number of late-stage embryos) indicated that the 100/100 genotype also produced significantly fewer young than the other genotypes under mercury stress.

D. TOLERANCE

The term tolerance has been used to imply physiological acclimation, and the term resistance has been used to imply genetic adaptation. Conforming to Weis and Weis,[143] the term tolerance is used here to cover both acclimation and genetic adaptation. Distinction is made by using the terms acclimation and genetic adaptation, not tolerance and resistance. The focus in this portion of the chapter is genetic adaptation (to toxicants) as expressed through the phenotype, "a change in a phenotype that occurs in response to a specific environmental signal and has a clear functional relationship to that signal that results in an improvement in growth, survival, or reproduction."[62]

Adaptation to pollutants can involve a wide range of mechanisms including detoxification,[144-147] transport,[148,149] essential element regulation,[150] and behavior.[151] Adaptation can occur rapidly. For example, Klerks and Levinton[105] estimate heritabilities of 0.59 to 1.08 for metal tolerance in an oligochaete species that result in clear adaptation within one to four generations. Once selection pressure is removed, loss of tolerance may be equally rapid due to the costs of maintaining the tolerance mechanism.[152]

The rate of tolerance acquisition is modified by many factors as tabulated by Mulvey and Diamond.[152] Generally, selection is faster for a trait under monogenic control than for a trait under polygenic control. Tolerance associated with a dominant gene is selected much more rapidly in early generations than tolerance associated with a recessive gene. The rate of adaptation is generally faster with shorter generation times and higher population growth rates. The presence of uncontaminated patches acting as refugia or a significant influx of migrants lowers the rate of adaptation.

An extremely important but generally neglected aspect of adaptation is the concept of adaptive constraint. Stearns[62] provides the following explanations of constraint: "any pattern or state that can be attributed to phylogeny, as opposed to recent microevolution within the currently existing population," or "genes act through proteins that change the properties of cells, that cells are the key players in development, and that cells interact in processes constrained by physics in ways that cannot be changed by simple gene substitutions, steered a bit—modified slightly, but not fundamentally changed. Only a small portion of phenotype space is then available for exploration by gene substitutions." Although considerable effort is expended in detecting and quantifying population adaptation to pollutants, the limits of such adaptation often remain vaguely defined. Klerks and Weis[153] and Mulvey and Diamond[152] correctly point out that observed enhanced tolerance in extant populations from contaminated environments likely represents the exceptional outcome of species fate. The general loss of species reflects the other more common fate of species as a consequence of adaptive constraints.

Such adaptive constraints can beget selection constraints: selection can enhance toxicant tolerance only so far. Further, strong selection can lead to reduced genetic diversity and loss of alleles. Selection constraints can become more severe with reductions in population size and genetic variation within the population.[62] Consequently, reduction in population size or genetic variation by natural or pollution-related causes can restrict the population's ability to adapt to future stressors.

VI. SUMMARY

This chapter aims to provide a population biology and genetics framework for measuring pollution effects. Most of the chapter presents basic ecological concepts and methods directly applicable to ecotoxicology. Examples are used to demonstrate the pertinence of each method to ecotoxicology. Basic methods are outlined to assess population size, dynamics, and demographics. Important concepts regarding the stability of population densities are discussed and related to ecotoxicology. Finally, population genetics concepts and methods are illustrated and their relevance to ecotoxicology argued. The importance of examining all selection components is discussed.

REFERENCES

1. Moriarty, F. *Ecotoxicology. The Study of Pollutants in Ecosystems.* (London: Academic Press, Inc., 1983).
2. Bezel, V. S., and V. N. Bolshakov. "Population Ecotoxicology of Mammals." In *Bioindications of Chemical and Radioactive Pollution,* ed. D. A. Krivolutsky. (Boca Raton, FL: CRC Press, 1990).
3. Calow, P., and R. M. Sibly. "A Physiological Basis of Population Processes: Ecotoxicological Implications." *Funct. Ecol.* 4: 283–288 (1990).
4. Reid, G. K. *Ecology of Inland Waters and Estuaries.* (New York: Van Nostrand Reinhold Company, 1961).
5. Krebs, C. J. *Ecology. The Experimental Analysis of Distribution and Abundance.* (New York: Harper and Row Publishers, 1972).
6. Odum, E. P. *Fundamentals of Ecology,* 3rd ed. (Philadelphia: W. B. Saunders Company, 1971).
7. Cavalli-Sforza, L. L., and W. F. Bodmer. *The Genetics of Human Populations.* (San Francisco: W. H. Freeman and Co., 1971).
8. Hartl, D. L., and A. G. Clark. *Principles of Population Genetics,* 2nd ed. (Sunderland, MA: Sinauer Associates, Inc., 1989).
9. Liebig, J. *Chemistry in Its Application to Agriculture and Physiology.* (London: Taylor and Walton, 1840).
10. Shelford, V. E. "Physiological Animal Geography." *J. Morphol.* 22: 551–618 (1911).
11. Shelford, V. E. *Animal Communities in Temperate America.* (Chicago: University of Chicago Press, 1913).
12. Pielou, E. C. *Population and Community Ecology. Principles and Methods.* (New York: Gordon and Breach Science Publishers, 1974).
13. Wiegert, R. G. "The Selection of an Optimum Quadrat Size for Sampling the Standing Crop of Grasses and Forbs." *Ecology* 43: 125–129 (1962).
14. Hendricks, W. A. *The Mathematical Theory of Sampling.* (New Brunswick: Scarecrow Publishers, 1956).
15. Krebs, C. J. *Ecological Methodology.* (New York: Harper Collins Publishers, 1989).
16. Emmel, T. C. *Population Biology.* (New York: Harper and Row Publishers, 1976).
17. White, G. C., D. R. Anderson, K. P. Burnham, and D. L. Otis. *Capture-Recapture and Removal Methods for Sampling Closed Populations, LA-8787-NERP.* (Los Alamos, NM: Los Alamos National Laboratory, 1982).
18. Bailey, N. T. J. "On Estimating the Size of Mobile Populations from Capture-Recapture Data." *Biometrika* 38: 293–306 (1951).
19. Seber, G. A. F. "The Effects of Trap Response on Tag Recapture Estimates." *Biometrics* 26: 13–22 (1970).

20. Chapman, D. G. "Some Properties of the Hypergeometric Distribution with Applications to Zoological Samples Censuses." *Univ. Calif. Publ. Stat.* 1(7): 131–160, (1951).

21. Seber, G. A. F., and E. D. Le Cren. "Estimating Population Parameters from Catches Large Relative to the Population." *J. Anim. Ecol.* 36: 631–643 (1967).

22. Zippin, C. "The Removal Method of Population Estimation." *J. Wildl. Manage.* 22: 82–90 (1958).

23. Otis, D. L., K. P. Burnham, G. C. White, and D. R. Anderson. "Statistical Inference from Capture Data on Closed Animal Populations." *Wildl. Monogr.* 62: 1–135 (1978).

24. Skalski, J. R., and D. S. Robson. *Techniques for Wildlife Investigations. Design and Analysis of Capture Data.* (New York: Academic Press, 1992).

25. Laake, J. L., K. P. Burnham, and D. R. Anderson. *User's Guide for Program TRANSECT.* (Logan UT: Utah State University Press, 1979).

26. May, R. M. *Theoretical Ecology. Principles and Applications.* (Philadelphia: W.B. Saunders Co., 1976).

27. Stanier, R. Y., M. Doudoroff, and E. A. Adelberg. *The Microbial World,* 3rd ed. (Englewood Cliffs, NJ: Prentice-Hall Inc., 1970).

28. Christensen, E. R., and N. Nyholm. "Ecotoxicological Assays with Algae: Weibull Dose-Response Curves." *Environ. Sci. Technol.* 18 (9): 713–718 (1984).

29. Hamilton, M. A. "Statistical Analysis of the Cladoceran Reproductivity Test." *Environ. Toxicol. Chem.* 5: 205–212 (1986).

30. Gilpin, M. E., and F. J. Ayala. "Global Models of Growth and Competition." *Proc. Natl. Acad. Sci. USA* 70: 3590–3593 (1973).

31. May, R. M., and G. F. Oster. "Bifurcations and Dynamic Complexity in Simple Ecological Models." *Am. Nat.* 110 (974): 573–599 (1976).

32. Pielou, E. C. *An Introduction to Mathematical Ecology.* (New York: John Wiley & Sons, 1969).

33. Gause, G. F. "The Influence of Ecological Factors on the Size of Population." *Am. Nat.* 65: 70–76 (1931).

34. Ricker, W. E. "Stock and Recruitment." *J. Fish. Res. Board Can.* 11: 559–623 (1954).

35. Thomas, W. R., M. J. Pomerantz, and M. E. Gilpin. "Chaos, Asymmetric Growth and Group Selection for Dynamical Stability." *Ecology* 61 (6): 1312–1320 (1980).

36. Hassell, M. P., J. H. Lawton, and R. M. May. "Patterns of Dynamical Behavior in Single-Species Populations." *J. Anim. Ecol.* 45: 471–486 (1976).

37. Watkinson, A. "Intuition and the Logistic Equation." *Trends Ecol. Evol.* 7(9): 314–315 (1992).

38. Hassell, M. P. "Density-Dependence in Single-Species Populations." *J. Anim. Ecol.* 44: 283–296 (1974).

39. Szathmary, E. "Simple Growth Laws and Selection Consequences." *Trends Ecol. Evol.* 6: 366–370 (1991).

40. Ginzburg, L. R. "Evolutionary Consequences of Basic Growth Equations." *Trends Ecol. Evol.* 7: 133–134 (1992).

41. Ginzburg, L. R. "Reply from L. Ginzburg." *Trends Ecol. Evol.* 8: 69–70 (1993).

42. Hutchinson, G. E. "Circular Causal Systems in Ecology." *Ann. N.Y. Acad. Sci.* 50: 221–246 (1948).

43. Popper, K. R. *The Logic of Scientific Discovery.* (London: Hutchinson and Company, 1968).

44. Ehrlich, P. R., and L. C. Birch. "The 'Balance of Nature' and Population Control." *Am. Nat.* 101 (918): 97–107 (1967).

45. Pool, R. "Ecologists Flirt with Chaos." *Science* 243: 310–313 (1989).

46. Sarokin, D., and J. Schulkin. "The Role of Pollution in Large-Scale Population Disturbances, Part 1: Aquatic Populations." *Environ. Sci.-Technol.* 26(8): 1476–1484 (1992).

47. Schaffer, W. M., and M. Kot. "Chaos in Ecological Systems: The Coals that Newcastle Forgot." *Trends Ecol. Evol.* 1(3): 58–63 (1986).
48. May, R. M. "Biological Populations with Nonoverlapping Generations: Stable Points, Stable Cycles, and Chaos." *Science* 186: 645–647 (1974).
49. Philippi, T. E., M. P. Carpenter, T. J. Case, and M. E. Gilpin. *"Drosophila* Population Dynamics: Chaos and Extinction." *Ecology* 68(1): 154–159 (1987).
50. May, R. M. "Nonlinear Phenomena in Ecology and Epidemiology." *Ann. N.Y. Acad. Sci.* 357: 267–281 (1980).
51. May, R. M. "Chaos and the Dynamics of Biological Populations." *Proc. R. Soc. Lond.* A413: 27–4 (1987).
52. Renshaw, E. *Modelling Biological Populations in Space and Time.* (Cambridge: Cambridge University Press, 1991).
53. May, R. M. "The Chaotic Rhythms of Life." *New Sci.* 18 November: 37–41 (1989).
54. Allen, J. C. "Factors Contributing to Chaos in Population Feedback Systems." *Ecol. Modell.* 51: 281–298 (1990).
55. May, R. M., G. R. Conway, M. P. Hassell, and T. R. E. Southwood. "Time Delays, Density-Dependence and Single-Species Oscillations." *J. Anim. Ecol.* 43: 747–770 (1974).
56. May, R. M., and R. M. Anderson. "Epidemiology and Genetics in the Coevolution of Parasites and Hosts." *Proc. R. Soc. Lond. B Bio. Sci.* 219: 281–313 (1983).
57. Holden, A. V. *Chaos.* (Princeton NJ: Princeton University Press, 1986).
58. Glass, L., and M. C. Mackey. *From Clocks to Chaos. The Rhythms of Life.* (Princeton: Princeton University Press, 1989).
59. Kot, M., W. M. Schaffer, G. L. Truty, D. J. Graser, and L. F. Olsen. "Changing Criteria for Imposing Order." *Ecol. Modell.* 43: 75–110 (1988).
60. Nychka, D., S. Ellner, D. McCaffrey, and A. R. Gallant. "Statistics for Chaos." *Stat. Comput. Stat. Graphics. Newsletter.* October: 4–11 (1990).
61. Olsen, L. F., and W. M. Schaffer. "Chaos versus Noisy Periodicity: Alternative Hypotheses for Childhood Epidemics." *Science* 249: 499–504 (1990).
62. Stearns, S. C. *The Evolution of Life Histories.* (Oxford: Oxford University Press, 1992).
63. Deevey, E. S. "Life Tables for Natural Populations." *Q. Rev. Biol.* 22: 283–314 (1947).
64. Pinder, J. E., III, J. G. Weiner, and M. H. Smith. "The Weibull Distribution: A New Method of Summarizing Survivorship Data." *Ecology* 59(1): 175–179 (1978).
65. Rago, P. J., and R. M. Dorazio. "Statistical Inference in Life-Table Experiments: The Finite Rate of Increase." *Can. J. Fish. Aquat. Sci.* 41: 1361–1374 (1984).
66. Fox, G. A. "Life Tables and Statistical Inferences." *Bull. Ecol. Soc. Am.* 70: 229–230 (1989).
67. Lebreton, J.-D., R. Pradel, and J. Clobert. "The Statistical Analysis of Survival in Animal Populations." *Trends Ecol. Evol.* 8(3): 91–95 (1993).
68. Ferson, S., and H. R. Akcakaya. *RAMAS/age User Manual, Modeling Fluctuations in Age-Structured Populations.* (Setauket, NY: Exeter Software, 1990).
69. Kirchner, T. B. *TIME-ZERO. The Integrated Modeling Environment.* (Fort Collins, CO: Quaternary Software, Inc., 1990).
70. Leslie, P. H. "On the Use of Matrices in Certain Population Mathematics." *Biometrika* 33: 183–212 (1945).
71. Leslie, P. H. "Some Further Notes on the Use of Matrices in Population Mathematics." *Biometrika* 35: 213–245 (1948).
72. Smith, R. L. *Ecology and Field Biology,* 3rd ed. (New York: Harper and Row Publishers, 1980).
73. Daniels, R. E., and J. D. Allan. "Life Table Evaluation of Chronic Exposure to a Pesticide." *Can. J. Fish. Aquat. Sci.* 38: 485–494 (1981).

74. Mount, D. I., and T. J. Norberg. "A Seven-Day Life-Cycle Cladoceran Toxicity Test." *Environ. Toxicol. Chem.* 3: 425–434 (1984).

75. Munzinger, A., and M.-L. Guarducci. "The Effect of Low Zinc Concentrations on Some Demographic Parameters of *Biomphalaria glabrata* (Say), Mollusca: Gastropoda." *Aquat. Toxicol.* 12: 51–61 (1988).

76. Pesch, C. E., W. R. Munns, and R. Gutjahr-Gobell. "Effects of a Contaminated Sediment on Life History Traits and Population Growth Rate of *Neanthes arenaceodentata* (Polychaeta: Nereidae) in the Laboratory." *Environ. Toxicol. Chem.* 10: 805–815 (1991).

77. Barnthouse, L. W., G. W. Suter II, A. E. Rosen, and J. J. Beauchamp. "Estimating Responses of Fish Populations to Toxic Contaminants." *Environ. Toxicol. Chem.* 6: 811–824 (1987).

78. Petersen, R. C., Jr., and L. B.-M. Petersen. "Compensatory Mortality in Aquatic Populations: Its Importance for Interpretation of Toxicant Effects." *Ambio* 17(6): 381–386 (1988).

79. Ludwig, J. A., and J. F. Reynolds. *Statistical Ecology. A Primer on Methods and Computing.* (New York: John Wiley and Sons, 1988).

80. Manly, B. F. J. *Randomization and Monte Carlo Methods in Biology.* (New York: Chapman and Hall, 1991).

81. Hurlbert, S. H. "Spatial Distribution of the Montane Unicorn." *Oikos* 58: 257–271 (1990).

82. Rohlf, F. J., and R. R. Sokal. *Statistical Tables,* 2nd ed. (New York: W. H. Freeman and Company, 1981).

83. David, F. N., and P. G. Moore. "Notes on Contagious Distributions in Plant Populations." *Ann. Bot. (Lond.)* 18: 47–53 (1954).

84. Upton, G. J., and B. Fingleton. *Spatial Data Analysis by Example,* Vol. 1: *Point Pattern and Quantitative Data.* (New York: John Wiley and Sons, 1985).

85. Clark, P. J., and F. C. Evans. "Distance to Nearest Neighbor as a Measure of Spatial Relationships in Populations." *Ecology* 35: 445–453 (1954).

86. Cressie, N. A. C. *Statistics for Spatial Data.* (New York: John Wiley and Sons, Inc., 1991).

87. Weins, J. A. "Population Responses to Patchy Environments." *Annu. Rev. Ecol. Syst.* 7: 81–120 (1976).

88. Lewin, R. "Supply-side Ecology." *Science* 234: 25–27 (1986).

89. Lewin, R. "Sources and Sinks Complicate Ecology." *Science* 243: 477–478 (1989).

90. Pulliam, H. R., and B. J. Danielson. "Sources, Sinks, and Habitat Selection: A Landscape Perspective on Population Dynamics." *Am. Nat.* 137: S50–S66 (1991).

91. Crow, J. F., and M. Kimura. *An Introduction to Population Genetics Theory.* (Minneapolis: Burgess Publishing Company, 1970).

92. Nevo, E., T. Shimony, and M. Libni. "Pollution Selection of Allozyme Polymorphisms in Barnacles." *Experientia* 34: 1562–1564 (1978).

93. Nevo, E., T. Perl, A. Beiles, and D. Wool. "Mercury Selection of Allozyme Genotypes in Shrimps." *Experientia* 37: 1152–1154 (1981).

94. Lavie, B., and E. Nevo. "Genetic Selection of Homozygote Allozyme Genotypes in Marine Gastropods Exposed to Cadmium Pollution." *Sci. Total Environ.* 57: 91–98 (1986).

95. Endler, J. A. *Natural Selection in the Wild.* (Princeton, NJ: Princeton University Press, 1986).

96. Lewontin, R. C., and Cockerham, C. C. "The Goodness-of-Fit for Detecting Natural Selection in Random Mating Populations." *Evolution* 13: 561–564 (1957).

97. Wilson, E. O., and W. H. Bossert. *A Primer of Population Biology.* (Sunderland, MA: Sinauer Associates, Inc., 1971).

98. Newman, M. C., S. A. Diamond, M. Mulvey, and P. Dixon. "Allozyme Genotype and Time to Death of Mosquitofish, *Gambusia affinis* (Baird and Girard) During Acute Toxicant Exposure: A Comparison of Arsenate and Inorganic Mercury." *Aquat. Toxicol.* 15: 141–156 (1989).

99. Manly, B. F. J. *The Statistics of Natural Selection on Animal Populations.* (New York: Chapman and Hall, 1985).

100. Arnold, S. J., and M. J. Wade. "On Measurement of Natural and Sexual Selection: Theory." *Evolution* 38: 709–719 (1984).

101. Mitchell-Olds, T., and R. G. Shaw. "Regression Analysis of Natural Selection: Statistical Inference and Biological Interpretation." *Evolution* 41: 1149–1161 (1987).

102. Lee, C. J., M. C. Newman, and M. Mulvey. "Time to Death of Mosquitofish (*Gambusia holbrooki*) During Acute Inorganic Mercury Exposure: Population Structure Effects." *Arch. Environ. Contam. Toxicol.* 22: 284–287 (1992).

103. Blanc, J. M. *Genetic Aspects of Resistance to Mercury Poisoning in Steelhead Trout (Salmo gairdneri).* Masters thesis, Oregon State University, 1973.

104. Burger, C. V. *Genetic Aspects of Lead Toxicity in Laboratory Populations of Guppies (Poecilia reticulata).* Masters thesis, Oregon State University, 1974.

105. Klerks, P. L., and J. S. Levinton. "Rapid Evolution of Metal Resistance in a Benthic Oligochaete Inhabiting a Metal-Polluted Site." *Biol. Bull.* 176: 135–141 (1989).

106. Posthuma, L., R. F. Hogervorst, E. N. G. Joose, and N. M. Van Straalen. "Genetic Variation and Covariation for Characteristics Associated with Cadmium Tolerance in Natural Populations of Springtail *Orchesella cincta* (L.)." *Evolution* 47: 619–631 (1993).

107. Stearns, S. C. "The Evolutionary Significance of Phenotypic Plasticity." *Bioscience* 39: 436–445 (1989).

108. Westcott, B. "Some Methods of Analyzing Genotype-Environment Interaction." *Heredity* 56: 243–253 (1986).

109. Diamond, S. A., M. C. Newman, M. Mulvey, and D. Martinson, "Allozyme Genotype and Time to Death of Mosquitofish, *Gambusia affinis* (Baird and Girard), during Acute Exposure to Inorganic Mercury." *Environ. Toxicol. Chem.* 8: 613–622 (1989).

110. Dixon, P. M., and M. C. Newman. "Analyzing Toxicity Data Using Statistical Models for Time-to-Death: An Introduction." In *Metal Ecotoxicology. Concepts and Applications,* ed. M. C. Newman, and A. W. McIntosh. (Chelsea, MI: Lewis Publishers, 1991).

111. Newman, M. C., and M. S. Aplin. "Enhancing Toxicity Data Interpretation and Prediction of Ecological Risk with Survival Time Modeling: An Illustration Using Sodium Chloride Toxicity to Mosquitofish (*Gambusia holbrooki*)." *Aquat. Toxicol.* 23: 85–96 (1992).

112. Benton, M. J., and S. I. Guttman. "Allozyme Genotype and Differential Resistance to Mercury Pollution in the Caddisfly, *Nectopsyche albida,* I: Single-Locus Genotypes." *Can. J. Fish. Aquat. Sci.* 49: 142–146 (1992).

113. Benton, M. J., and S. I. Guttman. "Allozyme Genotype and Differential Resistance to Mercury Pollution in the Caddisfly, *Nectopsyche albida,* II: Multilocus Genotypes." *Can. J. Fish. Aquat. Sci.* 49: 147–149 (1992).

114. Lavie, B., E. Nevo, and U. Zoller. "Differential Viability of Phosphoglucose Isomerase Allozyme Genotypes of Marine Snails in Nonionic Detergent and Crude Oil-Surfactant Mixtures." *Environ. Res.* 35: 270–276 (1984).

115. Hughes, J. M., M. W. Griffths, and D. A. Harrison. "The Effects of an Organophosphate Insecticide on Two Enzyme Loci in the Shrimp *Caradina* sp." *Biochem. Syst. Ecol.* 20: 89–97 (1992).

116. Kopp, R. L., S. I. Guttman, and T. E. Wissing. "Genetic Indicators of Environmental Stress in Central Mudminnow (*Umbra limi*) Populations Exposed to Acid Deposition in the Adirondack Mountains." *Environ. Toxicol. Chem.* 11: 665–676 (1992).

117. Lavie, B., and E. Nevo. "Multilocus Genetic Resistance and Susceptibility to Mercury and Cadmium Pollution in the Marine Gastropod, *Cerithium scabridum.*" *Aquat. Toxicol.* 13: 291–296 (1988).

118. Samallow, P. B., and M. E. Soule. "A Case of Stress Related Heterozygote Superiority in Nature." *Evolution* 37: 646–649 (1983).

119. Garton, D. W., R. K. Koehn, and T. M. Scott. "Multiple Locus Heterozygosity and Physiological Energetics of Growth in the Coot Clam, *Mulinin lateralis,* from a Natural Population." *Genetics* 108: 445–455 (1984).

120. Koehn, R. K., and P. M. Gaffney. "Genetic Heterozygosity and Growth Rate in *Mytilus edulis.*" *Mar Biol.* 82: 1–7 (1984).

121. Danzmann, R. G., M. M. Ferguson, F. W. Allendorf, and K. L. Knudsen. "Heterozygosity and Developmental Rate in a Strain of Rainbow Trout (*Salmo gairdneri*)." *Evolution* 40: 86–93 (1986).

122. Ferguson, M. M. "Developmental Stability in Rainbow Trout Hybrids: Genomic Coadaptation or Heterozygosity?" *Evolution* 40: 323–330 (1986).

123. Mitton, J. B., C. Carey, and T. D. Kocher. "The Relation of Enzyme Heterozygosity to Standard and Active Oxygen Consumption and Body Size of Tiger Salamanders, *Ambystoma triginum.*" *Physiol. Zool.* 59: 574–582 (1986).

124. Leary, R. F., F. W. Allendorf, and K. L. Knudsen. "Differences in Inbreeding Coefficients Do Not Explain the Association Between Heterozygosity at Allozyme Loci and Developmental Stability in Rainbow Trout." *Evolution* 41: 1413–1415 (1987).

125. Pemberton, J. M., S. D. Albon, F. E. Guinness, T. H. Clutton-Brock, and R. J. Berry. "Genetic Variation and Juvenile Survival in Red Deer." *Evolution* 42: 921–93 (1988).

126. Watt, W. B. "Adaptation at Specific Loci, II: Demographic and Biochemical Elements in the Maintenance of the *Colias* PGI Polymorphism." *Genetics* 103: 691–724 (1983).

127. Trehan, K. S., and K. S. Gill. "Subunit Interaction: A Molecular Basis of Heterosis." *Biochem. Genet.* 25: 855–862 (1987).

128. Turelli, M., and L. R. Ginzburg. "Should Individual Fitness Increase with Heterozygosity?" *Genetics* 104: 191–209 (1983).

129. Smouse, P. E. "The Fitness Consequences of Multiple-Locus Heterozygosity under the Multiplicative Overdominance and Inbreeding Depression Models." *Evolution* 40: 946–957 (1986).

130. Battaglia, B., P. M. Bisol, V. U. Possato, and E. Rodino. "Studies on the Genetic Effects of Pollution in the Sea." *Rapp. P. V. Reun. Cons. Int. Explor. Mer.* 179: 267–274 (1980).

131. Nadeau, J. H., and R. Baccus. "Selection Components of Four Allozymes in Natural Populations of *Peromyscus maniculatus.*" *Evolution* 35: 11–20 (1981).

132. Clegg, M. T., A. L. Kahler, and R. W. Allard. "Estimation of Life Cycle Components of Selection in an Experimental Plant Population." *Genetics* 89: 765–792 (1978).

133. Leberg, P. L. "Effects of Population Bottlenecks on Genetic Diversity as Measured by Allozyme Electrophoresis." *Evolution* 46: 477–494 (1992).

134. Weir, B. S. *Genetic Data Analysis. Methods for Discrete Population Genetic Data.* (Sunderland, MA: Sinauer Associates, Inc. Publishers, 1990).

135. Christiansen, F. B., O. Frydenberg, A. O. Gyldenholm, and V. Simonsen. "Genetics of *Zoarces* Populations, VI: Further Evidence, Based on Age Group Samples, of a Heterozygote Deficit in the EstIII Polymorphism." *Hereditas* 77: 225–236 (1974).

136. Nadeau, J. H., K. Dietz, and R. H. Tamarin. "Gametic Selection and the Selection Component Analysis." *Genet. Res.* 37: 275–284 (1981).

137. Christiansen, F. B., and O. Frydenberg. "Selection Component Analysis of Natural Polymorphisms Using Population Samples Including Mother-Offspring Combinations." *Theor. Popul. Biol.* 4: 425–445 (1973).

138. Bungaard, J., and F. B. Christiansen. "Dynamics of Polymorphism, I: Selection

Components in an Experimental Population of *Drosophila melanogaster.*" *Genetics* 71: 439–460 (1972).

139. Siegismund, H. R., and F. B. Christiansen. "Selection Component Analysis of Natural Polymorphisms Using Population Samples Including Mother-Offspring Combinations, III." *Theor. Popul. Biol.* 27: 268–297 (1985).

140. Prout, T. "The Estimation of Fitness from Genotypic Frequencies." *Evolution* 19: 546–551 (1965).

141. Williams, C. J., W. W. Anderson, and J. Arnold. "Generalized Linear Modeling Methods for Selection Component Experiments." *Theor. Popul. Biol.* 37: 389–423.

142. Mulvey, M., M. C. Newman, A. Chazal, M. M. Keklak, M. G. Heagler, and S. Hales. "Genetic and Demographic Changes in Mosquitofish (*Gambusia holbrooki*) Populations Exposed to Mercury." *Environ. Toxicol. Chem.* (submitted).

143. Weis, J. S., and P. Weis. "Tolerance and Stress in a Polluted Environment. The Case of the Mummichog." *Bioscience* 39: 89–95 (1989).

144. Fabacher, D. L., and H. Chambers. "Rotenone Tolerance in Mosquitofish." *Environ. Pollut.* 3: 109–141 (1972).

145. Chambers, J. E., and J. D. Yarbrough. "A Seasonal Study of Microsomal Mixed-Function Oxidase Components in Insecticide-Resistant and Susceptible Mosquitofish, *Gambusia affinis.*" *Toxicol. Appl. Pharmacol.* 48: 497–507 (1979).

146. Angus, R. A. "Phenol Tolerance in Populations of Mosquitofish from Polluted and Nonpolluted Waters." *Trans. Am. Fish. Soc.* 112: 794–799 (1983).

147. Klerks, P. L., and P. R. Bartholomew. "Cadmium Accumulation and Detoxification in a Cd-Resistant Population of the Oligochaete *Limnodrilus hoffmeisteri.*" *Aquat. Toxicol.* 19: 97–112 (1991).

148. Yarbrough, J. D. "Insecticide Resistance in Vertebrates." In *Survival in Toxic Environments,* ed. M.A.Q. Khan and J. P. Bederka, Jr. (New York: Academic Press, Inc., 1974).

149. Wood, J. M., and H.-K. Wang. "Microbial Resistance to Heavy Metals." *Environ. Sci. Technol.* 17: 582A–590A (1983).

150. Beeby, A. "Toxic Metal Uptake and Essential Metal Regulation in Terrestrial Invertebrates: A Review." In *Metal Ecotoxicology. Concepts and Applications,* ed. M. C. Newman and A. W. McIntosh. (Chelsea, MI: Lewis Publishers, 1991).

151. Kynard, B. "Avoidance Behavior of Insecticide Susceptible and Resistant Populations of Mosquitofish to Four Insecticides." *Trans. Am. Fish. Soc.* 3: 557–561 (1994).

152. Mulvey, M., and S. A. Diamond. "Genetic Factors and Tolerance Acquisition in Populations Exposed to Metals and Metalloids." In *Metal Ecotoxicology. Concepts and Applications,* ed. M. C. Newman and A. W. McIntosh. (Chelsea, MI: Lewis Publishers 1991.)

153. Klerks, P. L., and J. S. Weis. "Genetic Adaptation to Heavy Metals in Aquatic Organisms: A Review." *Environ. Pollut.* 45: 173–205 (1987).

Chapter 7

Effects at the Community Level

Ecotoxicology is a three-legged stool that, to date, has balanced on two legs: chemical fate and single-species toxicology. These are necessary but insufficient components of the field. The third leg, that of organism interactions, has been slower to develop as a tool for assessing how an ecosystem is affected by and recovers from chemical stresses.[1]

I. GENERAL

In studies of community ecotoxicology, as in many endeavors believed to involve extreme complexity, there is a tendency to retreat to a descriptive approach to accruing knowledge. Although descriptive studies are unquestionably essential, predominance of such an approach to the detriment of experimental approaches eventually begets a weak inferential structure to the associated knowledge base. (See Chapter 1 for discussion of strong inference.) Fortunately, there has been a noticeable shift to include more experimentation in community ecotoxicology. Microcosm, mesocosm, and whole-lake studies have been especially valuable in this regard.

A brief quantitative overview of species interactions as influenced by toxicants is presented, beginning with two-species interactions and ending with community functions within ecosystems. Initially, the influence of toxicants on a species' niche integrity is emphasized. The intent is to demonstrate that the integrity of a species' role, interactions, and habitat in an ecosystem (herein referred to as niche integrity) is as critical as its physiological or demographic integrity in determining the final outcome of exposure to toxicant in an ecological arena. Various measures of community integrity are then discussed relative to the effects of toxicants. Finally, trophic transfer of toxicants is discussed briefly.

Throughout the chapter, the term niche is used in the Hutchinsonian context.[2] Wetzel clearly presented the Hutchinsonian niche as

. . . a certain biological activity space in which an organism exists in a particular habitat. This space is influenced by the physiological and behavioral limits of a species and by the effects of environmental parameters (physical and biotic, such as temperature and predation) acting on it. Each of these parameters can be ordinated on an axis, and can be thought of as a dimension in space. The fundamental niche, then, can be viewed as an n-dimensional space or hypervolume, with each of its n axes or dimensions corresponding to the range of an environmental parameter over which the organism can exist. Since many physical and biotic factors interact, each species occupies only a portion of its fundamental niche; this portion can be referred to as the species realized niche.[3]

In terms of individual fitness, the niche is the volume within n-dimensional space in which fitness is positive.[4]

II. SIMPLE SPECIES INTERACTIONS

A. PREDATOR-PREY INTERACTIONS

One of the best known examples of natural selection in the wild is industrial melanism, the gradual increase to predominance of melanic forms in industrialized areas. The increase in dark morphs is correlated with a general darkening of habitat with soot and loss of lichens. With this darkening, the advantage due to camouflage shifts to favor dark over light forms with an epiphenomenal increase in the frequency of darker individuals. The mechanism is demonstrated in Kettlewell's[5] classic study of differential bird predation on color morphs of the peppered moth (*Biston betularia*) in which predator-prey interactions are carefully quantified in enclosure (aviary) and mark-release field studies.

Although Kettlewell's[5] enclosure and field studies remain an integral lesson during the training of every ecologist including the ecotoxicologist (see Moriarty,[6] pages 79–84), studies of pollution effects on predator-prey interactions remain biased toward highly structured laboratory assays.[7,8] Fewer studies than warranted[9] use enclosure or field designs for predator-prey studies. This neglect compromises our understanding of toxicant effects because predator-prey interactions can be a crucial component of individual fitness and, in the special case of cannibalism,[10] population demography. Furthermore, if prey with highest body burdens of toxicants are most prone to predation due to their weakened state,[11] such information can also be important in fully understanding trophic transfer.[12] Indeed, Goodyear[12] and Atchison et al.[7] suggest that results of traditional acute or chronic mortality tests are less relevant to assessing the effects of pollutants than those focused on ecological mortality (toxicant-related diminution of fitness within an ecosystem context of a magnitude sufficient to be equivalent to physiological death, e.g., death due to a compromised ability to avoid predation).

The most common approach to quantifying toxicant influence on predator-prey interactions involves laboratory experiments in which the prey alone or both, predator and prey, are exposed to toxicant. Predation rate at intervals throughout the trial or the intensity of predation measured at the end of a trial period is compared between control and exposed animals. For example, Kania and O'Hara[11] placed mercury-exposed and unexposed mosquitofish (*Gambusia holbrooki,* formerly *Gambusia affinis*) into a chamber with a narrow shelf acting as a refuge from a predator (largemouth bass, *Micropterus salmoides*). After 60 h, the numbers of exposed and nonexposed mosquitofish falling prey to the bass were compared with a χ^2 test. For all exposures more than 0.005 µg/L of inorganic mercury, there was a significant increase in predation as a consequence of the compromised ability of the prey to avoid the predator. (With more involved designs, many of the methods described in Chapter 4 or 5 could be used to analyze this type of data.) Goodyear[12] obtained similar results for the effects of γ radiation on mosquitofish predation by bass; however, the percentage of fish surviving predation is noted at intervals over the entire time course. This type of data is amenable to nonparametric, semiparametric, and parametric methods described in Chapter 4 for time-to-death studies. Techniques in Chapter 4 have the advantage of allowing model development as well as hypothesis testing. Other approaches share this advantage. For example, in studying cadmium effects on fish predation, Sullivan et al.[13] also quantify the magnitude of effect in addition to testing for a significant effect.

Example 1

Effects of an insecticide (mirex or dodecachlorooctahydro-1,3,4-metheno-2*H*-cyclobuta[*cd*]pentalene) on predation of grass shrimp (*Palaemonetes vulgaris*) by pinfish (*Lagodon rhomboides*) were quantified by Tagatz.[14] After 13 days of exposing shrimp to sublethal concentrations of mirex, pinfish were placed in each exposure tank. Predation was measured in control and mirex-exposed tanks during the next 3 days. The results for day 3 were the following: Control, 42 of 177 shrimp survived predation; mirex-exposed, 5 of 115 shrimp survived predation. (These test statistics differ slightly from those in the original paper because they are estimated from total counts and survival percentages in the original paper.) A χ^2 test can be used to test for a significant deviation from the expected intensity of predation if mirex has no effect.

Example 1 *Continued*

First, a contingency table is generated. The observed and expected scores are placed in the table along with marginal column and row totals. The expected scores are placed in brackets within the table. The expected scores of dead and living animals in the table below are generated with the marginal column totals [expected = (column total*row total)/grand total]. These expected scores are used assuming that the overall predation results reflected in the marginal column totals are independent of treatment category (mirex exposure or control). Consequently, one would expect to see them reflected also in results for the mirex-exposed and control treatments. These expected scores are then used together with the observed scores to calculate a χ^2 statistic.

	Alive	Dead	Row Total
Control	42 (28.5)	135 (148.5)	177
Mirex	5 (18.5)	110 (96.5)	115
Column total	47	245	292

Expected number of control shrimp alive: (47*177)/292 = 28.5.
Expected number of control shrimp dead: (245*177)/292 = 148.5.
Expected number of mirex-exposed shrimp alive: (47*115)/292 = 18.5.
Expected number of mirex-exposed shrimp dead: (245*115)/292 = 96.5.

A χ^2 statistic is calculated as described in Chapter 6 [e.g, Equation (71)].

$$\chi^2 = \frac{(42 - 28.5)^2}{28.5} + \frac{(135 - 148.5)^2}{148.5} + \frac{(5 - 18.5)^2}{18.5} + \frac{(110 - 96.5)^2}{96.5} = 19.36$$

The calculated χ^2 statistic (19.36) is compared with a tabulated value with the appropriate α and degrees of freedom. The df are (number of rows − 1)(number of columns − 1) = (2 − 1)(2 − 1) = 1. The χ^2 for df = 1 and α = 0.05 (3.841) is taken from a table such as Table 14 of Rohlf and Sokal.[15] The calculated χ^2 is larger than this critical χ^2. Consequently, the null hypothesis of independence (number of deaths from predation was independent of treatment) is rejected: exposure to mirex did enhance predation.

This example is quite straightforward. More often several treatments are assessed as in the original work of Tagatz.[14] Further information on more detailed analyses is available in most statistical textbooks, e.g., Chapter 13 of Dixon and Massey.[16] The difference between pairwise and experimentwise Type I error rates should be considered during such tests. (See Chapter 5 for a discussion of these error rates.) Adjustment of the critical χ^2 values may be necessary for multiple comparisons as discussed for critical t statistics in Chapter 5. Rohlf and Sokal[15] provided a table (Table 15) of critical χ^2 values based on Šidàk's adjustment for multiple comparisons.

Analyses of predator-prey interactions based on formal ecological models have also been applied effectively, but not frequently enough, in ecotoxicology. Analyses involving ecological models have the advantage of direct linkage to underlying principles. However, use of ecological models tends to be overshadowed by designs involving simple statistical hypothesis tests. Basic predator-prey models are described here in an ecotoxicological context to foster interest.

Predator-prey population dynamics have traditionally been explained with the Lotka-Volterra model. This model can be described with the clear notation of Emmel[17] and Smith.[18]

$$\frac{dN_1}{dt} = r_1 N_1 - PN_1 N_2 \tag{1}$$

$$\frac{dN_2}{dt} = P_2 N_1 N_2 - d_2 N_2 \tag{2}$$

where N_1 = prey density; N_2 = predator density; r_1 = instantaneous rate of increase for prey species (predator absent); P = predation coefficient; P_2 = predation effectiveness coefficient; and d_2 = density-dependent death rate of the predator.

The prey population is assumed to increase exponentially in the absence of the predator ($dN_1/dt = r_1 N_1$) and its growth is adversely affected by an increasing probability of predator-prey interaction ($N_1 N_2$). In contrast, the growth of the predator is enhanced by any increase in the probability of predator-prey interaction. The $PN_1 N_2$ in Equation (1) is the functional response term.[19] As reflected in this term, one possible effect of prey densities on the predator is a functional change in predatory behavior such as a shift in number of prey taken per predator, shift in size distribution of prey taken, or reallocation of time spent in various aspects of foraging for prey. The numerical response term ($P_2 N_1 N_2$) in Equation (2) reflects a change in the number of predators in response to change of the prey density. Note that the predator population is not density-dependent. According to this model, the predator and prey populations will oscillate about each other through time.

The simple Lotka-Volterra model [Equations (1) and (2)] generally has failed to describe accurately predator-prey dynamics, yet as done here, it is still widely used to introduce students to the oscillatory nature of predator and prey populations, e.g., Gause[20] and Huffaker.[21] May[19] pointed out mathematical difficulties associated with this model and outlined several modifications of Equations (1) and (2) resulting in more realistic models.

$$\frac{dN_1}{dt} = rN_1\left(1 - \frac{N_1}{K}\right) - N_2 F(N_1, N_2) \tag{3}$$

$$\frac{dN_2}{dt} = N_2 G(N_1, N_2) \tag{4}$$

where K = the prey carrying capacity; $F(\ )$ = a specified function for the functional response; and $G(\ )$ = a specified function for the numerical response.

The prey population is made density-dependent with the term, $1 - N_1/K$. Simple functional and numerical response terms are replaced by more realistic functions of N_1 and N_2 that account for such processes as predator saturation/satiation, search image acquisition, or switching to alternate prey. These functions, $F(\ )$ and $G(\ )$, take a variety of forms. May[19] listed a series of candidate forms including $F(\ ) = PN_1$ and $G(\ ) = P_2 N_1 - d_2$ for the simple Lotka-Volterra model [Equations (1) and (2)] (see May's Table 4.1), and provided details for each of these functions. For example, the predator can become less efficient at taking prey as prey densities increase, e.g., Equation (5) ("Holling Type II, Invertebrate" in May[19]) or Equation (6) ("Holling Type III, Vertebrate" in May[19]). The $G(\ )$ function can be made logistic with N_1 controlling the predator carrying capacity [Equation (7)]. Alternatively, $G(\ )$ can be linearly related to $F(\ )$ (May[19]), e.g., Equation (8).

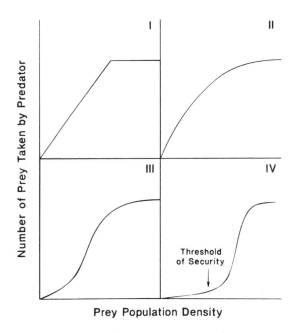

Figure 1 Four types of predator functional responses to prey density. See text for details. (Modified from Figure 8 in Holling.[23])

$$F(\ ) = \frac{kN_1}{N_1 + D} \qquad (5)$$

$$F(\ ) = \frac{kN_1^2}{N_1^2 + D^2} \qquad (6)$$

where D = prey density at which attack capability begins to saturate for the predator[22]; and k = the constant attack (or capture) rate (prey per predator).

$$G(\ ) = r_2\left(1 - \frac{N_2}{\gamma N_1}\right) \qquad (7)$$

where the carrying capacity for the predator (γN_1) is proportional (γ) to the prey population density (N_1).[22] The r_2 is the intrinsic rate of increase for the predator population.

$$G(\ ) = P_2 + d_2 F(\ ) \qquad (8)$$

From empirical data, Holling[23] reasoned that functional and numerical responses can take several general forms. He described four types of functional responses (Figure 1). The number of prey consumed per predator is directly proportional to prey density in the simplest or Type I functional response. It is a positive, linear relationship until saturation. Prey consumption is constant beyond a maximum prey density. Poole[24] suggested that filter-feeding predators can display a Type I response. The most commonly used response (Type II) is often associated with predation by invertebrates.[18,19] Poole[24] suggested that it can also describe responses of some fish species. This functional response also increases as prey density increases but at a gradually decreasing rate. One model describing such a response is provided in Equation (5). It accommodates more aspects of predation than the Type I response model, e.g., prey acquisition and handling times. Such details of prey acquisition and handling are illustrated with Holling's disc model [Equation (9)]. (See Poole[24] for detailed discussion.) Again, the clear notation of Smith[18] is borrowed here.

$$\frac{N_a}{P} = \frac{aNT}{1 + aT_h N} \tag{9}$$

where N_a = number of prey taken or attacked; N = total number of prey; P = number of predators; a = attack rate constant; T = total amount of time available for predator-prey interactions ($T = T_s + T_h N_a$); T_h = handling time spent for a prey item (pursue, capture, consumption, and digestion); and T_s = time spent searching for a prey item.

If learning is involved in the predator's actions, as is the case with many vertebrate predators, the Type III response would be more appropriate[24] than the Type II model. Holling's study[23] of mouse (*Peromyscus maniculatus bairsii*) predation on European pine sawfly (*Neodiprion sertifer*) cocoons is the classic example of a Type III response. Type III functional responses can also involve several prey species with the predator choosing to allocate effort among them. For example, predation or other factors can result in the density of one prey species dropping below a certain level. The predator can then choose to spend an increasingly disproportionate amount of its attention on another, more abundant prey species as the density of the first prey species drops (prey switching). The first prey species population then has an opportunity to reestablish itself. The Type IV functional response mentioned in Holling's classic work[23] has often been discussed in recent texts as a Type III response. With this relationship, there is a critical prey density (threshold of security) below which there is insufficient prey stimulus to elicit any significant predator response.

Numerical responses can involve enhanced predator fecundity or immigration with increasing prey densities. Holling[23] described three numerical response models (Figure 2). Type A responses are direct (positive) responses to increasing prey density, Type B is simply no apparent numerical response to prey density, and Type C is an inverse response to prey density.

Taken together, functional and numerical responses produce a wide range of predator responses to prey density. Figure 3 shows only four of these responses. Underlying each is a complex array of behaviors affecting the outcome of the predator-prey relationship over time.

The change in predator response to prey densities under the influence of toxicants has infrequently drawn the attention of ecotoxicologists. A study of ammonia effects on largemouth bass consuming mosquitofish (*G. affinis*) by Woltering et al.[25] was a refreshing exception. They found that the shape of the functional response curve changed with increasing ammonia concentration. From control to 0.34 mg/L to 0.63 mg/L, there is a lowering of the rate at which consumption of mosquitofish increases with increasing prey densities. Indeed, the mosquitofish, being less sensitive to the effect of ammonia than the bass, harass the bass in treatment tanks with the highest concentrations (0.86 mg/L) and mosquitofish densities! This precipitates a drop in predation rate and loss of weight by the predator.

As should be obvious from Equation (9) and the weight loss noted for bass in the study just described, foraging behavior must be energetically efficient for a predator (or a grazer) to optimize its fitness. Equation (9) indicates that a predator must expend time and energy to find, pursue, capture, consume, and digest prey. Effective selection of

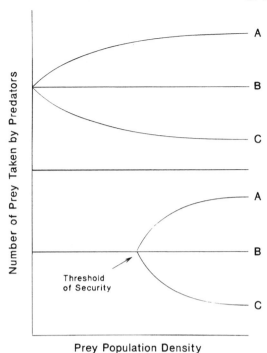

Figure 2 Three types of predator numerical responses to prey density. Shown are positive (A), neutral (B), and negative (C) responses with (bottom panel) or without (top panel) a threshold of security. See text for details. (Modified from Figure 8 of Holling.[23])

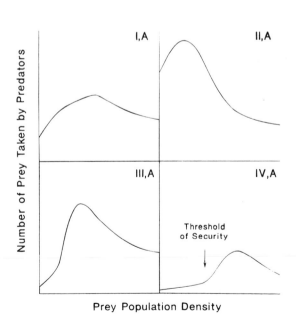

Figure 3 Four examples of the net response to prey density. Only a positive numerical response combined with Types I, II, III, and IV functional responses are shown. (Modified from Figure 8 of Holling.[23])

prey items, e.g., large versus small prey, fast versus slow prey, widely dispersed versus aggregated prey, is crucial to optimal foraging. Decisions regarding time spent within a patch of prey versus time expended searching for new patches to exploit are also important. An enormous, theory-rich literature is available for addressing questions of optimal foraging behavior, e.g. Stephens and Krebs,[26] yet ecotoxicologists have only begun to exploit the associated theory and methodology.

Atchison and co-workers[7,8,27] advocated using the optimal foraging theory to assess toxicant effects. They observed that most studies of predator-prey behavior under toxicant stress drew on empirical data lacking any theoretical foundation. This severely restricted the generality of conclusions and the ability to formulate hypotheses for further elicidation of underlying mechanisms. They used a widely accepted model [Equation (10)] as an example of one optimal foraging model that could be used by ecotoxicologists.

$$\frac{E_n}{T} = \frac{\sum_{i=1}^{z} B_i E_i}{1 + \sum_{i=1}^{z} B_i H_i} - C_s \tag{10}$$

where E_n = net energy (J); T = time (s); $E_i = e_i - C_h H_i$; e_i = net energy obtained from prey size i (J); B_i = prey encounter rate with size i (per s); H_i = prey handling time for size i (s); C_h = energetic cost of handling (J/s); and C_s = energetic cost of searching (J/s).

Given a range of prey sizes (z), the ideal predator selects a subset of prey sizes that maximizes net energy intake (E_n/T). Sandheinrich and Atchison[27] provided numerous references suggesting that key components of Equation (10) are influenced for fish predators exposed to a wide range of toxicants. Application of optimal foraging theory to toxicant effects on predator (and grazer) foraging behavior promises to provide very useful information based on explanatory principles.

B. INTERSPECIES COMPETITION
1. Two Species
Competition between two species was described in the 1920s using the Lotka-Volterra (Volterra, Gause-Volterra) equations.

$$\frac{dN_1}{dt} = r_1 N_1 \left[1 - \frac{N_1}{K_1} - \frac{\alpha_{12} N_2}{K_1} \right] \tag{11}$$

$$\frac{dN_2}{dt} = r_2 N_2 \left[1 - \frac{N_2}{K_2} - \frac{\alpha_{21} N_1}{K_2} \right] \tag{12}$$

These equations can be made more general.

$$\frac{dN_i}{dt} = r_i N_i \left[1 - \frac{N_i}{K_i} - \frac{\alpha_{ij} N_j}{K_i} \right] \tag{13}$$

where N_i, N_j = the densities of species i and j; K_i = the carrying capacity of species i; r_i = intrinsic rate of increase for species i; and α_{ij} = the competition coefficient or the linear reduction of growth rate of species i relative to its carrying capacity K_i as a result of competition with species j.[28] Gause and Witt[29] developed graphical means for

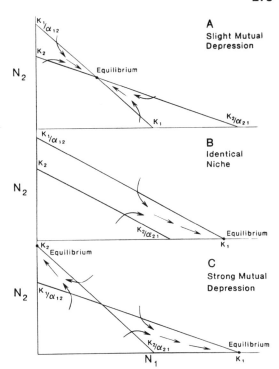

Figure 4 The possible outcomes of interspecies competition based on the linear model of Gause and Witt.[29] Coexistence (A) will result from slight mutual depression (at the noted point of equilibrium). With identical niches, an equilibrium with only one species will occur (competitive exclusion principle). In panel B, species 1 completely excludes species 2 at equilibrium. However, species 2 would prevail if line K_1/α_{12}, K_1 fell below line K_2/α_{21}, K_2. In the region of strong mutual depression, one species or the other remains at final equilibrium depending on the initial densities of species 1 and 2. (Modified from Figures 1, 2, and 3 in Gause and Witt.[29])

determining the outcome of two species competition (Figure 4) based on the intersection of the two straight lines defined by (X intercepts, Y intercepts) K_1/α_{12}, K_1 and K_2, K_2/α_{21} in a plot of species 1 density versus species 2 density (Figure 4). Assuming linear relationships, competitive coexistence can occur in the presence of slight mutual depression by the competitors (Figure 4A). If the two competitors had identical niches (Figure 4B), one would always exclude the other, i.e., the principle of competitive exclusion. If there is a strong mutual depression during competition, one or the other competitor would exclude the other; however, which species is the successful competitor would be determined by the initial N_1 and N_2 (Figure 4C). (See Gause and Witt[29] or Vandermear[30] for a detailed discussion of this topic.)

Gilpin and Ayala[28] added an additional parameter to Equation (13) based on results from competition experiments with *Drosophila* species. This parameter (θ_i) quantifies the nonlinearity of intraspecific growth regulation as described previously for the modified logistic model [Equation (47) in Chapter 6].

$$\frac{dN_i}{dt} = r_i N_i \left[1 - \left[\frac{N_i}{K_i} \right]^{\theta_i} - \frac{\alpha_{ij} N_j}{K_i} \right] \tag{14}$$

Let's examine Gilpin and Ayala's experiments used to select Equation (14) because they contain the means for extending the simple linear approach described for determining equilibrium conditions. Various initial population densities of *Drosophila willistoni* and *Drosophila pseudoobscura* were combined and, after 1 week, the changes in densities for both species (dN_1/dt, dN_2/dt) were measured. The changes for each species in various initial combinations with the other species were fit by regression methods to Equation (14) to generate estimates of r_i, K_i, α_{ij}, and θ_i for each species. Furthermore, 19 vectors

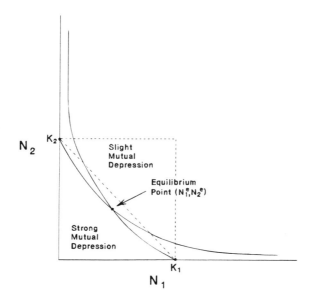

Figure 5 Model for interspecific competition demonstrating species coexistence in the region of strong mutual depression. (Modified from Figure 1 of Gilpin and Ayala[28] and Figure 1 from Gilpin and Justice.[32])

of change in species densities were generated and plotted on a graph of N_1 versus N_2 (Figure 5). The vectors for both species were used to visually fit lines for $dN_1/dt = 0$ and $dN_2/dt = 0$. The intersection of these two lines (N_1^e, N_2^e) is the two-species equilibrium point ($dN_1/dt = dN_2/dt = 0$). Note that the lines derived were not straight as assumed in the work by Gause and Witt[29] (Figure 4) discussed previously.

With traditional discussion of Lotka-Volterra equations, such an equilibrium point would be said to be stable (competitive coexistence) if intraspecific competition was stronger than interspecific competition.[19,31] In Figures 4 or 5, this would be in the region of slight mutual depression above the line connecting the carrying capacities for the two species (K_1, K_2). However, Gilpin and co-workers[28,32] indicated that a stable equilibrium (species coexistence) can also occur in the region of strong mutual depression if the conditions, $K_1 < K_2/\alpha_{21}$ and $K_2 < K_1/\alpha_{12}$, were met. [As clearly seen with Figure 4, these conditions for species coexistence hold for the original Lotka-Volterra model as well as Gilpin's modification described by Equation (14).]

Example 2

Two *Daphnia* species are grown together for 2 weeks beginning with initial densities, N_1 and N_2. Treatments consist of copper-contaminated or uncontaminated diets. Regression analyses of the changes in species densities are then used to generate the following estimates for Equation (14). (This example is fabricated by modifying Table 1 of Gilpin and Ayala[28]).

	Species 1	Species 2
Control		
r	1.496 ± 0.167	4.513 ± 0.259
K	1332 ± 128	791 ± 43
α	0.713 ± 0.077	0.087 ± 0.006
θ	0.35 ± 0.04	0.12 ± 0.02

Example 2 *Continued*

Copper diet

r	1.210 ± 0.215	4.014 ± 0.318
K	450 ± 200	780 ± 53
α	0.903 ± 0.103	0.090 ± 0.006
θ	0.34 ± 0.10	0.10 ± 0.08

Can the two species coexist indefinitely under the test conditions? The above-mentioned models suggest that competitive coexistence is possible if $K_1 < K_2/\alpha_{21}$ and $K_2 < K_1/\alpha_{12}$.

With the control diet: $K_1 = 1332$; $K_2/\alpha_{21} = 791/0.087 = 9092$; $K_2 = 791$; $K_1/\alpha_{12} = 1332/0.713 = 1868$. The inequalities are satisfied for the control populations. With the elevated copper diet: $K_1 = 450$; $K_2/\alpha_{21} = 780/0.090 = 8667$; $K_2 = 780$; $K_1/\alpha_{12} = 450/0.903 = 498$. The second inequality $(K_2 < K_1/\alpha_{12})$ is not satisfied.

The decrease in carrying capacity of species 1 and the reduction in its growth due to competition increases with exposure to copper. Consequently, the necessary inequalities are not met when the diet contains elevated levels of copper, i.e., the ability to compete and coexist is lost for this species pair.

2. Several Species

Vandermeer[33] and Gilpin and Ayala[28] suggested that higher order interactions, e.g., the effect of species 3 on the competition between species 1 and 2, can often be ignored. Because the linear effects model reflected by simple α_{ij} values for species competition is a reasonable caricature of species competition in communities involving several species, the interactions between all competitors can be summarized as a simple matrix of α_{ij} values. Such a competition matrix[30] is a specific type of community matrix in which all α_{ij} values have a positive sign, i.e., the effect of species j on i (α_{ij}) is inhibitory, not stimulatory. (The presence of mutualism and other beneficial species interactions result in negative α_{ij} values if interactions other than competition are incorporated in a more general matrix of interactions or community matrix.)

To generate a competition matrix, a series of competition experiments as described previously is completed for all possible species pairs in the defined community of competitors.[24] The resulting α_{ij} values are then organized into a matrix. The diagonal series of 1 values are the values of α for all species effects on themselves.

$$
\begin{matrix}
1 & \alpha_{12} & \alpha_{13} & \alpha_{14} & \cdot & \cdot & \alpha_{1m} \\
\alpha_{21} & 1 & \alpha_{23} & \alpha_{24} & \cdot & \cdot & \alpha_{2m} \\
\alpha_{31} & \alpha_{32} & 1 & \alpha_{34} & \cdot & \cdot & \alpha_{3m} \\
\cdot & \cdot & \cdot & 1 & \cdot & \cdot & \cdot \\
\cdot & \cdot & \cdot & \cdot & 1 & \cdot & \cdot \\
\cdot & \cdot & \cdot & \cdot & \cdot & 1 & \cdot \\
\alpha_{m1} & \alpha_{m2} & \alpha_{m3} & \alpha_{m4} & \cdot & \cdot & 1
\end{matrix}
$$

In its simplest application, this matrix allows visualization of the distribution of intensities of competitive interactions within a community. More involved analyses are possible. For example, with the vector of carrying capacities generated in the experiments described for all competitors, the matrix can be used to predict equilibrium species densities of each competing species. (See the appendix in Vandermeer[30] and Chapter 15 in Poole[24] for details.) Vandermeer[30] used the competition matrix approach to estimate the number of species that can be maintained in a community at equilibrium. May[22] used this approach to detail conditions for community stability.

Many species interactions can be incorporated into the generalized Lotka-Volterra model [Equation (13)]. Equation (15) is a modification of the multiple species model described in Vandermeer.[30]

$$\frac{dN_i}{dt} = r_i N_i \left[1 - \frac{N_i}{K_i} - \frac{\sum\limits_{j \neq i}^{m} \alpha_{ij} N_j}{K_i} \right] \tag{15}$$

It can be modified to include Gilpin's θ_i [Equation (14)].

$$\frac{dN_i}{dt} = r_i N_i \left[1 - \left(\frac{N_i}{K_i}\right)^{\theta_i} - \frac{\sum\limits_{i \neq j}^{m} \alpha_{ij} N_j}{K_i} \right] \tag{16}$$

Such multiple species competition models can have relatively complex dynamics. Gilpin[34] provided a means of assessing the dynamics within communities containing odd and even numbers of species.

To this point, multiple species competition has focused on contrived experimental designs that, for logistical reasons, involve relatively few species. It could be argued that, as a consequence, results from such studies have little value in environmental decision making because of their lack of realism. However, a strong counterpoint could be made that information from such studies is no less valuable or realistic than that used routinely and pervasively to make regulatory decisions, e.g., LC50 or EC50 based on death, or inhibition of reproduction or growth. Information generated by this means would be a valuable supplement to information collected at other levels of ecological organization and derived from field surveys. The ecological realism of such studies can be enhanced if the design involves field enclosures or mesocosms. Regardless, several important points emerge from these models. First, shifts in competition can eliminate species as surely as outright physiological death or reproductive failure. Second, by their very nature, population densities in systems involving species competition will exhibit cycles.[34] The dynamics of these cycles are complex functions of the vector of the carrying capacities of competing species and the matrix of competition coefficients [Equations (15) and (16)]. Both of these points should be considered during assessment of toxicant impact.

Competition in natural communities is extremely difficult to quantify, and interpretation of the associated results is much more ambiguous than quantification of data from laboratory or enclosure studies. Regardless, this interpretation is possible[35,36] and highly desirable because of the associated enhanced ecological realism. Poole[24] described Levins'[37] approximation of α_{ij} values for field populations sharing a common resource [Equation (17)].

$$\alpha_{12} = \frac{\sum\limits_{j=1}^{r} P_{1j}P_{2j}}{\sum\limits_{j=1}^{r} P_{1j}^2} \qquad (17)$$

where P_{1j} = proportion that resource j is of all resources used by species 1; P_{2j} = proportion that resource j is of all resources used by species 2; and r = number of resources used, e.g., habitat types, foods, times available to feed.

Use of this measure of niche overlap in a community matrix is considered inappropriate by most ecologists.[38] One difficulty has been defining the qualities of natural habitats that influence competition.[38] Also, resource abundance or availability has been ignored.[39] If a resource is not limiting, the overlap of two species cannot be used to imply competition.[19] Hurlbert[39] described several similar indices linked to competition coefficients along with their application. He concluded that such use was valid only under specific, controlled conditions.

C. SYMBIOSIS

Symbiotic relationships including host-parasite, host-pathogen, or mutualistic associations have received scant attention. However, a perusal of the literature suggests that these relationships are not trivial. For example, Sarokin and Schulkin[40] reviewed population-level effects of pollutants and, in most of their examples, suggested that pollutant exposure played a key role in the adverse outcome of infections. In their whole-lake acidification studies, Schindler et al.[41] indicated that the demise of a major species (*Orconectes virilis*) resulted from enhanced microsporozoan and fungal infections after lake acidification. In the few studies of contaminants and infectious diseases, significant effects are noted for diseases caused by bacteria,[42] protozoans,[43] and metazoans.[44,45]

Other symbiotic relationships are important in the fate and effects of toxicants. An intriguing example involves the giant clam-algal zooxanthellae symbiotic relationship.[46] From the phosphorus-deficient waters of the Great Barrier Reef, arsenate is taken up, converted to organic form, and concentrated by zooxanthellae within clam tissues. As a consequence, arsenic is accumulated to extraordinarily high concentrations in giant clams, especially in the kidneys (500 to 1000 μg/g).

Although ecological models exist for examination of symbiotic relationships (e.g., May,[19] Pacala et al.[47]), none have been applied to ecotoxicology. Hopefully, this will change as ecotoxicology matures.

III. COMMUNITY STRUCTURE AND FUNCTION

A. GENERAL

Three general types of experimental units are used to study structural or functional shifts in communities exposed to toxicants: microcosms, mesocosms, and natural ecosystems. The term microcosm as used in ecotoxicology does not have precisely the same meaning as in general use, i.e., a system acting as a representation of a larger system. Instead, microcosms are "laboratory systems that are intended to physically simulate an ecosystem or a major subsystem of an ecosystem."[48] Mesocosms are "outdoor experimental systems that are delimited and to some extent enclosed."[45] Here, the combining form, meso-, is strictly consistent with general usage and means middle, i.e., between the true ecosystem and a contrived laboratory system such as a microcosm. The distinction between microcosms and mesocosms seems to be a matter of location and size. Natural ecosystem

Table 1 **Anticipated changes in ecosystems experiencing stress (from Odum[55])**

Category	Trend
Energetics	1. Increase in community respiration
	2. Unbalanced production to respiration ratio ($P/R < 1$ or $P/R > 1$)
	3. Increase in maintenance to biomass, i.e., production/biomass and respiration/biomass
	4. Increase in importance of auxiliary energy (energy originating from outside the ecosystem)
	5. Increase in exported primary production
Nutrients	6. Increased turnover of nutrients
	7. Decreased cycling of nutrients
	8. Increased loss of nutrient as a consequence of Trends 6 and 7
Community Structure	9. Increased proportion of species that are r-strategists
	10. Decreased size of organisms
	11. Decreased life spans
	12. Shortened food chains
	13. Decreased species diversity and increased species dominance (The reverse may occur if the original diversity was low.)
Ecosystem	14. Internal cycling decreased and input/output from outside ecosystem becomes more important
	15. Regression to earlier successional condition
	16. Decreased efficiency of resource utilization
	17. Decreased positive (e.g., mutualism) and increased negative (e.g., parasitism) interactions
	18. Functional processes such as community metabolism tend to be more robust than species composition or other "structural" properties.

studies involve manipulation of an entire system such as a small lake or watershed, or a large portion of a natural system.

The clear advantage of microcosms is the ability to examine community or ecosystem processes under strictly controlled conditions, to easily manipulate conditions, and to incorporate true replicates.[1,49] Their obvious disadvantage is the loss of ecological realism relative to the other approaches. Mesocosms gain back some realism but become less yielding to control, manipulation, and replication. They have been used very effectively to assess community and ecosystem level processes (e.g., Wilbur,[50] Crosland and La Point,[51] and Liber at al.[52]). Whole or even partial ecosystem studies are expensive, difficult to manipulate, and often depend on pseudoreplication.[41,53,54] Regardless, their associated realism is invaluable. Intelligent melding of all three approaches has produced much valuable information.

Many methods are used to imply toxicant impact on communities or ecosystems. They range from semiquantitative indices to ANOVA methods to inverse regression methods. Three examples are provided below as applied to natural ecosystems, mesocosms, and microcosms.

Odum[55] listed several changes expected in ecosystem qualities as a consequence of stress (Table 1). Most involved changes in biotic (community) rather than abiotic compo-

Table 2 Critical factors and qualitative rankings for ecosystem elasticity[57]

	Qualitative Rank of Importance		
Factor	**1**	**2**	**3**
a. Presence of nearby epicenters	Poor	Moderate	Good
b. Transportability of dissemules	Poor	Moderate	Good
c. Habitat condition	Poor	Moderate	Good
d. Presence of residual toxicants	Much	Intermediate	Low
e. Water quality	Very poor	Partially restored	Normal
f. Management capabilities	None	Some	Strong

nents of the ecosystem. Many (Points 1, 2, 3, 6, 7, 8, 13, and 15 in Table 1) are similar to those outlined by Rapport et al.[56] In addition to a list of consequences similar to Odum's,[55] Rapport et al.[56] provided references to studies documenting such changes.

Cairns[57,58] outlined the qualities contributing to natural ecosystem vulnerability to stress. (Sheenan[59] detailed these and additional ecosystem qualities as influenced by toxicants.) Vulnerability is defined as the susceptibility of the ecosystem to irreversible damage. Implied in the phrase "irreversible damage" is a pragmatic time scale, i.e., irreversible within decades not millennia. Vulnerability is a complex function of ecosystem elasticity (the ability to return to its original, prestress condition) and inertia (the ability to resist change in its function or structure). Resilience, a measure of the number of times that the ecosystem is able to recover to its normal state, is also identified as playing an important role. Cairns[57,58] used factors contributing to elasticity (Table 2) and inertia (Table 3) to develop an intentionally broad, simplistic index for gauging ecosystem vulnerability. Although derived with considerable subjective opinion, these factors focus attention on qualities underlying ecosystem vulnerability and, therefore, warrant discussion.

Six factors contributing to ecosystem elasticity are used to generate a simple recovery index [Equation (18)].

Table 3 Critical factors and qualitative rankings for ecosystem inertia[57]

	Qualitative Rank of Importance		
Factor	**1**	**2**	**3**
a. Biota adapted to significant variability in the environment	Poor	Moderate	Good
b. Much structural and functional redundancy	Poor	Moderate	Good
c. Mixing capability	Poor	Moderate	Good
d. Chemical characteristics	Poor	Moderate	Good
e. Proximity to ecological threshold	Close	Some margin of safety	Large margin
f. Management capabilities	Poor	Moderate	Good

$$\text{Recovery index} = \prod_{i=a}^{f} \text{Score}_i \qquad (18)$$

where a = rank for existence of epicenters; b = rank for transportability of dissemules; c = rank for habitat condition; d = rank for amount of residual toxicant; e = rank for physicochemical water quality; and f = rank for capability for regional management. The first factor (a) is scored on the presence of refugia or other sources of species for reestablishment of the impacted area. The second (b) incorporates the ease with which eggs, young, or other forms of dissemules are transported into the area. The third (c) incorporates any change in the environment such as a change in sediment composition due to siltation. The amount of toxicant remaining in the ecosystem (d) and extent to which the general water quality has been altered (e) are also included. The final factor (f) requires an arbitrary judgment regarding the potential for active remediation or some other action to lessen the impact. A calculated recovery index of 400 or more suggests rapid recovery. A score between 55 and 399 implies a fair to good chance of a rapid rate of recovery. Below 55, chances of a rapid recovery are judged to be poor.

Similarly, the subjective rankings from Table 3 can be used to generate an inertial index.

$$\text{Inertial index} = \prod_{i=a}^{f} \text{Score}_i \qquad (19)$$

where a = rank for the extent to which species are accustomed to wide variability in environmental conditions; b = rank for structural and functional redundancy; c = rank for ability to dilute and dissipate the toxicant; d = rank associated with the influence of water quality on the toxicant effect; e = rank associated with a nearby area of ecological threshold or transition such as an estuary; and f = rank for capability of regional management. Three general ranges of inertial stability (high, fair to good, poor) are defined by scores of 400 or more, 55 to 399, and less than 55.

Kersting[60] developed a formal method of estimating ecosystem strain (distance moved in a state space relative to a reference state). Although it is applicable to many of the measures described in the remainder of this chapter, the approach is illustrated with zooplankton and algal densities (states) associated with a laboratory microcosm. *Daphnia magna* are grown in a microcosm with fluctuating algal densities, and a plot of *Daphnia* densities versus algal densities is constructed. Because there is a lag before zooplankton abundance responds to changes in algal density (numerical response), the zooplankton density 1 week after an observed algal density is paired with that particular algal density. A 95% tolerance ellipse is then drawn about the points on this plot (state plane) from observations made before the introduction of the toxicant (Figure 6). (See Sokal and Rohlf,[61] pages 594–601, for details on construction of a tolerance ellipse.) Data are next taken for the microcosm after toxicant introduction, placed onto the plot, and compared with the ellipse. For example, a 95% tolerance ellipse is constructed as shown in Figure 6. The point (X) associated with the microcosm after application of herbicide (Dichlobenil) is placed on the plot, and the distance from the center of the 95% tolerance ellipse (C) to point X in the state space (e.g., distance A) is divided by the distance from the center of the ellipse along that same line to a point on the 95% tolerance ellipse (e.g., distance B). The index A/B is called the normalized ecosystem strain (S). If $S \leq 1$ the system is within its normal operating range. It is judged outside of its normal operating range if $S > 1$. Kersting[60] suggested that this approach could also be taken

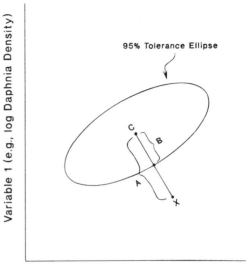

95% Tolerance Ellipse

Figure 6 Estimation of normalized ecosystem strain ($S = A/B$). See text for details. (Modified from Figure 4 of Kersting.[60])

Variable 1 (e.g., log Algal Density)

for more complex systems provided interactions between fluctuating state variables were taken into consideration.

Liber et al.[52] applied ANOVA (see Chapter 5) and inverse regression methods to assess the effect of 2,3,4,6-tetrachlorophenol on zooplankton in mesocosms. Their inverse regression approach incorporated several enclosures spiked with varying amounts of toxicant. Inverse regression involves prediction of an X (not Y) value from the regression line of Y on X. (See Draper and Smith,[62] pages 47–51, Sokal and Rohlf,[61] pages 496–498, and Neter et al.,[63] pages 173–176, for more detail.)

Initially, the log of zooplankton abundance is estimated in seven nonspiked mesocosms. The 95% confidence interval is estimated for the mean of the mesocosms (day 0 in Figure 7) before spiking.

$$\text{Confidence interval for } \mu_0 = \bar{x} \pm \frac{t_{\alpha,\text{2-sided}}\, s}{\sqrt{n}} \qquad (20)$$

where $t_{\alpha,\,\text{2-sided}} = t$ for two-sided test with $\alpha = 0.05$ (Table 14 in Appendix with $p = 1$); and $s =$ standard deviation for the n observations.

The mesocosms are then spiked and, after a certain time (day N in Figure 7), the resulting zooplankton abundances are determined. The observations are used to produce a regression line for the log of 2,3,4,6-tetrachlorophenol concentration versus the log of total zooplankton abundance. The concentration having an observed effect was estimated using the 95% confidence interval for this regression line.

The mean square deviation of the actual data points from the regression [mean square of the error or MSE Equation (21)] is used to generate the 95% confidence interval for the predicted mean values for log zooplankton abundance [Equation (22)].

$$\text{MSE} = \frac{n-1}{n-2}\,(s_y^2 - b^2 s_x^2) \qquad (21)$$

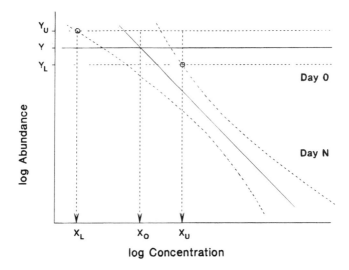

Figure 7 Inverse regression method for estimation of the concentration eliciting a significant effect on zooplankton abundance. See text including Example 3 for details. (Reprinted from Figure 7 in Liber et al.,[52] with kind permission from Pergamon Press Ltd., Headington Hill Hall, Oxford OX3 0BW, UK. Copyright 1992.)

where n = the number of data pairs (x,y); b = the regression slope; s_y^2 = variance of the y values; and s_x^2 = variance of the x values.

$$95\% \text{ C.I.} = Y_p \pm t_{\alpha,2\text{-sided}} \text{MSE} \sqrt{\frac{1}{n} + \frac{(x - \bar{x})^2}{(n - 1)s_x^2}} \qquad (22)$$

where Y_p = the predicted value of y for a given x; $t_{\alpha,2\text{-sided}}$ = t for a given α and $df = n - 2$ (e.g., Table 14 with $p = 1$ in Appendix); $\bar{x}$ = mean of the x values; and x = the specified x value at which the interval is estimated.

Liber et al.[52] used the point at which the upper limit for the day N regression line intersects the lower limit of the day 0 log abundances (point X_U, Y_L in Figure 7) as the no observed effect concentration (NOEC). (See Chapter 5 for discussion of NOEC/NOEL.) The X_U is obtained by inverse regression methods with Y_L (the 95% lower limit for day 0 log abundances). Equation (23) from Draper and Smith[62] predicted this point by making $Y_0 = Y_L$. The $\pm$ sign becomes positive in the numerator in this case. If the toxicant effect were to increase with concentration (i.e., the regression line has a positive slope), the Y_0 would have been set to Y_U to predict the log concentration at which the effect reaches the upper limit (X_L) defined with the day 0 observations.

$$X_U = \bar{X} + \frac{b(Y_0 - \bar{Y}) \pm ts \sqrt{\left(\frac{(Y_0 - \bar{Y})^2}{S_{xx}}\right) + \frac{b^2}{n} - \frac{t^2 s^2}{n S_{xx}}}}{b^2 - \frac{t^2 s^2}{S_{xx}}} \qquad (23)$$

where $S_{xx} = \Sigma(x_i - \bar{x})^2$; t = t for a given α (two-sided) and $df = n - 2$ (e.g., Table 14 with $p = 1$ in Appendix); Y_0 = the lower 95% limit of log abundance; and s = MSE.

Example 3

Data of Liber et al.[52] are used here to illustrate the inverse regression method described previously. With these data, the NOEL (NOEC) is estimated for 2,3,4,6-tetrachlorophenol effect on log zooplankton abundance in the mesocosms. The following SAS code contains the data supplied by K. Liber and generates the necessary summary statistics and regression analysis.

```
OPTIONS PS=58;
DATA LIBER;
  INPUT STATE $ ABUND CONC @@;
  CARDS;
B 1944.0 0    B 2082.0 0    B 1619.0 0    B 1807.0 0    B 1575.0 0
B 1630.0 0    B 2016.0 0    A 2396.0 .25 A 2754.0 .25 A 3411.0 .25
A  692.0 .50 A  614.8 .50 A  793.3 .50 A  220.3 1.0  1  203.8 1.0
A  230.3 1.0 A   69.4 2.0 A   51.8 2.0 A   60.1 2.0 A   27.5 4.0
A   30.3 4.0 A   38.6 4.0 A   20.9 7.3 A   22.6 7.3 A   17.6 7.3
;
RUN;
DATA BEFORE;
  SET LIBER;
  IF STATE="B";
  LABUND=LOG10(ABUND);
RUN;
PROC UNIVARIATE DATA=BEFORE;
  VAR LABUND;
RUN;
DATA EXPOSED;
  SET LIBER;
  IF STATE="A";
  IF CONC<4.0; /* ONLY CONC'S BETWEEN 0.25 & 2.0 MG/L WERE USED /*
  LABUND=LOG10(ABUND);
  LCONC=LOG10(CONC);
RUN;
PROC UNIVARIATE DATA=EXPOSED;
  VAR LABUND LCONC;
RUN;
PROC GLM DATA=EXPOSED;
  MODEL LABUND=LCONC;
RUN;
```

The 95% confidence interval is estimated with Equation (20) for the log of zooplankton abundances in the "prespiked" mesocosms ($n = 7$; $\bar{x} = 3.2553$; $s = 0.0498$). The $t_{2\text{-sided}}$ for $\alpha = 0.05$ with $df = n - 1 = 6$ is 2.45 (Table 14 in the Appendix with $p = 1$).

$$\text{C.I.} = 3.2553 \pm \frac{(2.45)(0.0498)}{\sqrt{7}}$$

where C.I. = 3.2553 ± 0.0461; $3.2092 < \mu < 3.3014$. The lower limit (3.2092) is used as Y_L.

Next, the MSE for the spiked mesocosm data is calculated with Equation (21) or taken from the SAS output resulting from the above code ($n = 12$; $s_y^2 = 0.41887$; $s_x^2 = 0.12357$; $b = -1.83449$).

Example 3 *Continued*

$$MSE = \frac{12 - 1}{12 - 2} [0.41887 - ((-1.83449)^2(0.12357))]$$

$$MSE = 0.00331.$$

The MSE is used as s in Equation (23), where $Y_0 = Y_L = 3.2092$; $S_{xx} = 1.35929$; $b = -1.83449$; t (two-sided, $\alpha = 0.05$, $df = n - 2 = 10$) = 2.2281; $s = MSE = 0.00331$; $\bar{X} = -0.15051$; $\bar{Y} = 2.60247$.

$$X_U = -0.15051 + \frac{-1.11304 + 0.00738\sqrt{0.27082 + 0.28044 - 0.00000}}{3.36534 - \frac{0.00005}{1.35929}}$$

$X_U = -0.47962$. The antilog of X_U is 0.33 mg/L.

The inverse regression estimate of the NOEL for 2,3,4,6-tetrachlorophenol would be 0.33 mg/L. Note that this value is not the same as that calculated by Liber et al.[52] (0.42 mg/L) because all 12 points were used here instead of the average for each concentration ($n = 4$ concentration averages).

The last two techniques discussed here and methods in other chapters (e.g., ANOVA methods in Chapter 5) can be applied to microcosm, mesocosm, or natural ecosystem data to assess toxicant effects on community or ecosystem qualities. In the next sections of this chapter, candidate qualities of study for communities are discussed. These qualities are divided into two broad categories: community structure and community function.

B. COMMUNITY STRUCTURE

1. General
Although they are invaluable for summarizing data and assessing general shifts in community structure, no quantitative index or graph depicting community structure can replace a sound understanding of species natural histories combined with an accurate and thorough survey of species abundances. Often the subtle roles of species such as keystone species (species that influence the community by their activities or roles, not their numerical dominance), or the extreme sensitivity of indicator species, must be understood to adequately predict or document the consequences of toxicant exposure. For example, experimental removal of a starfish[64] or sea urchin[65] from the intertidal zone significantly alters community structure. Removal of keystone species through the action of a toxicant would have a similar effect. Further, valuable clues such as the loss of a toxicant-sensitive indicator species could easily be missed if one relied solely on summary indices or graphs. Combined effects of toxicant sensitivity and species function within the community can produce results only interpretable with a detailed understanding of the associated species. For example, loss of one species of a mutualistic pair (e.g., Osman and Haugsness[66]) due to extreme toxicant sensitivity could result in the loss of the second species regardless of its high tolerance to the toxicant.

2. Species Abundance

a. General
The relative abundances of species within a community are often used to summarize structure and imply processes occurring in the community. In one approach, all species

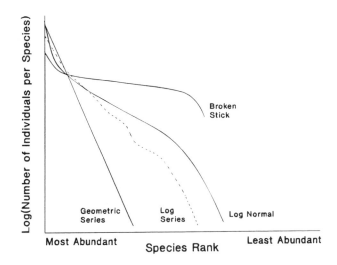

Figure 8 Four common models of species abundance. See text for a detailed description of these models.

are ranked from those with highest number of individuals to those with the lowest number of individuals (Figure 8). Alternatively, relative cover or biomass may be used instead of numbers of individuals. Four basic types of abundance-species rank curves (geometric series, log normal, log series, and broken stick) are used to suggest underlying mechanisms determining community structure. They are based on the assumption of resource competition. Although these explanations provide insight, they may or may not be helpful depending on how closely the operationally defined community ("all the organisms in a chosen area that belong to the taxonomic group the ecologist is studying.")[67] reflects the true ecological community of species competing for resources. For example, interpretation of curve shape based on competition theory is questionable if only one genus of aquatic insects is enumerated but important competing species in other genera are ignored. However, such interpretation can provide very useful insight for a survey of a particular ecological guild ("a group of functionally similar species whose members interact strongly with one another but weakly with the remainder of the community.")[68]

b. Geometric Series Model

The geometric series curve is interpreted using the niche preemption hypothesis. Species abundance (number of individuals, biomass, or percent cover) is thought to reflect the amount of resource that a particular species draws from the ecosystem. Magurran[69] illustrated this hypothesis using the example of a vacant niche volume into which species arrive at regular intervals. When the most dominant species arrives, it takes a proportion (k) of the available resources. When the next most dominant species arrives, it takes the same proportion of the resources that remain. The repetition of this process for all competing species results in a geometric series curve for species abundance. There are a few dominant species with most species being rare. Such curves are uncommon but can be associated with species-poor communities[69] with competition for one or a few crucial resources[70] or with severe environments[18] such as those that are polluted.[19]

Magurran[69] provided the following description of the geometric series. The S species are ranked from the most abundant ($i = 1$) to the least ($i = S$) abundant.

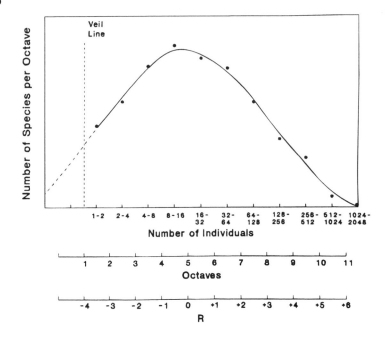

Figure 9 Typical plot used to analyze species abundance data under the assumption of a log normal model. See text including Example 4 for details.

$$n_i = NC_k k(1 - k)^{i-1} \tag{24}$$

where n_i = the number of individuals in species i; N = the total number of individuals in all species; $C_k = 1/[1 - (1 - k)^s]$.

c. Log Normal Model

The log normal is the most common species abundance curve. Indeed, May[19] stated that it rarely provided poorer description than the other abundance curves. The associated hypothesis for this curve is based on the assumption of a large community in which many factors influence species interactions during resource partitioning. These factors are assumed to be relatively independent.[18,69] The log normal distribution results from the complex, multiplicative effects of these factors on the partitioning of resources between species.[9] As a consequence, there are many more species of intermediate abundances in the associated community than in the one just described for the geometric series.

The log normal model is usually analyzed using methods developed by Preston[71,72] and illustrated in Figure 9. The log-transformed abundances (number of individuals counted for a species) are plotted against the number of species in a particular abundance class, e.g., five species are represented by two individuals in the sample. The $\log_2$ is used most often but any log transformation can be used. These $\log_2$ classes or octaves represent doublings in abundance (i.e., 1, 2, 4, 8, 16, 32, ... individuals). In Figure 9, there is a point (veil line) below which the species are so rare that they are not found in the sample. The position of the veil line will vary depending on the size of the sample and the qualities of the community being used as a measure of resource use, e.g., number of individuals, biomass, or percent cover. With counts of individuals, the limit is less than one individual counted per species. The veil line indicates a point of truncation similar to the point of censoring as discussed in Chapter 2 regarding observations below a limit of detection.

Above the veil line, the number of species per octave versus the octave number are plotted. Units of the X axis can be numbers of individuals octaves, or R. The R transformation is $\log_2[N/N_0]$ where N_0 is the number of individuals in the modal octave (R_0). It simply expresses octaves as deviations from the modal octave. See for example, Figure 9 in which $N_0 = 8$. The R will be 0 for R_0 (octave 5) and increasingly positive or negative with each octave to the right or left of R_0.

The number of species in the Rth octave on either side of the modal octave (S_R) can be estimated.

$$S_R = S_0 e^{-a^2 R^2} \tag{25}$$

where S_0 = the number of species in the modal octave; and a = the inverse width of the log normal distribution. The a is approximately 0.2 in most cases when the calculations are performed using $\log_2$. Under the assumption of a known mode, the a may be estimated with the relationship,[70]

$$a \approx \sqrt{\frac{\ln \dfrac{S_0}{S_{R_{MAX}}}}{R_{MAX}^2}} \tag{26}$$

where S_0 = the number of species in the modal octave; R_{MAX} = the octave with the maximum species abundance; and $S_{R_{MAX}}$ = the number of species in R_{MAX}.

Alternatively, Preston[72] and Krebs[38] provided a more involved estimate of a which does not require that the mode be fixed at a known value.

$$a = \frac{1}{\sqrt{2\sigma^2}} \tag{27}$$

where σ^2 = distribution variance estimated by s^2.

The presence of the veil line complicates estimation of s^2 because points below this line are not available for estimation with standard formulae. However, Cohen[73] provided methods for such calculations. Estimates of the mean and standard deviation for the truncated log normal distribution are very similar to estimation of the summary statistics for a censored log normal distribution described in Chapter 2. However, because censoring and truncation are different, the reader should be aware that they often require slightly different methods. Briefly, a data set is censored if a known number of values is missing below a known point for a data set, e.g, the limit of detection. For truncation, the specific point (e.g., veil line) is known, but the number of observations below it is not known. Traditionally, Cohen's[73] methods of maximum likelihood are applied to truncated species abundance curves, although other methods in Chapter 2 are also appropriate. [See Equations (8) and (9) in Chapter 2 and associated text for details on Cohen's[73] method for censored data.] The data are first $\log_2$ transformed and estimates of the mean ($\bar{x}_t$) and variance (s_t^2) for the data above the veil line are determined. Next, these statistics are used in Equations (28) and (29) to derive estimates of the mean and variance for the complete data set including values falling below the veil line.

$$\bar{x} = \bar{x}_t - \theta(\bar{x}_t - x_0) \tag{28}$$

$$s^2 = s_t^2 + \theta(\bar{x}_t - x_0)^2 \tag{29}$$

where x_0 = the $\log_2$ of the veil line midpoint position; and θ = a value from Table 1 in Cohen.[73]

Cohen's table of θ's is provided in Appendix Table 30. To extract a θ from this table, γ must be calculated first.

$$\gamma = \frac{s_t^2}{(\bar{x}_t - x_0)^2} \tag{30}$$

Using linear interpolation, this γ is used to extract a θ for use in estimating the mean and variance for the untruncated data. The variance estimate can then be used in Equation (27) to calculate a.

An additional (γ) parameter (not the same as that just described for Cohen's technique) is often estimated for log normal species abundance data. It relates the species abundance curve discussed to this point with the analogous individuals curve. The individuals curve is a plot of the total number of individuals per octave (Y axis) versus the abundance octave (X axis). As such, it shares a common X axis with and can be plotted with the species abundance curve, e.g., Figure 2.8 in Magurran.[69] (Preston[72] provided a detailed explanation of the individuals curve.) The estimated value of γ is normally in the range of 1.

$$\gamma = \frac{R_0}{R_{MAX}} \tag{31}$$

where R_{MAX} = the octave with the most abundant species from the species abundance curve; and R_N = the modal octave for the individuals curve.

It is important to note from these equations that fitting a log normal model to any data set lacking a clear mode, i.e., a mode assumed to be below the veil line, is not justifiable.[38] Such a curve cannot be distinguished from a log series as discussed under Section III.B.2.d.

With the information generated above, several parameters can be estimated, including a better estimate of S_0 (number of species in the modal octave), the predicted total number of species (S_T), and the predicted number of species in each octave (S_R). Ludwig and Reynolds[70] provided the following estimator of S_0.

$$S_0 = e^{(\text{mean of } \ln S_R\text{'s} + a^2 R^2)} \tag{32}$$

where mean of $\ln S_R$ values = mean of the logarithms of the observed number of species in each octave; and $\bar{R}^2$ = mean of the R^2 values.

Krebs[38] provided the following approach for predicting the total number of species. The $\log_2$ of the midpoint for the veil line (e.g., $\log_2 0.5$) and mean and standard deviation derived with Equations (28) and (29) are used to generate a z (standard normal deviate) for the veil point.

$$z = \frac{x_0 - \bar{x}}{s} \tag{33}$$

This z is used in a table of areas under the unit normal curve to estimate the area to the left of this point of truncation (p_0). This point of truncation expressed as an area under the normal curve (p_0) and the observed number of species (S_{ob}) are used in Equation (34) to estimate the total number of species (S_T).

$$S_T = \frac{S_{ob}}{1 - p_0} \tag{34}$$

A χ^2 goodness-of-fit test can be used to assess data fit to the log normal model. Expected values from Equation (25) and observed values are used to calculate a χ^2 statistic with the associated df dependent on the method used. The χ^2 is used to measure goodness of fit. It can be recalculated after repeated substitutions of a and S_0 values until the pair producing the lowest χ^2 is obtained. Ludwig and Reynolds[70] and Krebs[38] described several ways of performing these calculations and provided programs (LOG-NORM) for fitting species abundance data to a log normal model. Ludwig and Reynolds[70] used Equations (26) and (32) under the assumption of a known mode. Consequently, the df associated with the hypothesis test is the number of octaves minus 2. Krebs[38] described an approach [Equations (27) to (30)] that does not make the assumption of a known mode and uses an estimate of μ [Equation (28)]. Consequently, the df must be reduced to the number of octaves minus 3 for the χ^2 hypothesis test. The approach described by Krebs has the advantage of avoiding the assumption of a known mode.

d. Log Series Model

Magurran[69] explained the log series as she did the geometric series, but the species arrivals at the vacant niche volume were at random, not regular intervals. Like the geometric series curve, the log series curve describes a situation with few dominants and many uncommon species.[69] Along with the log normal curve, it reflects intermediate conditions between the geometric series and broken stick curves. The log series curve is used less often than the log normal curve but it has its advocates, e.g., Magurran.[69]

The total number of species (S) for a log series is defined by Equation (35).

$$S = \alpha[-\ln(1 - x)] = \alpha \ln\left(1 + \frac{N}{\alpha}\right) \tag{35}$$

where S = the total number of species; N = the total number of individuals; x, α = parameters estimated as shown below. The x is usually greater than 0.9 and is estimated by an iterative process. Values of x are substituted into Equation (36) until the equality is "close enough."

$$\frac{S}{N} = \left(\frac{1 - x}{x}\right)(-\ln(1 - x)) \tag{36}$$

The α and its variance are estimated using x.[38,69]

$$\alpha = \frac{N(1 - x)}{x} \tag{37}$$

$$\text{Variance of } \alpha = \frac{\alpha}{-\ln(1 - x)} \tag{38}$$

This α is often used as a heterogeneity ("diversity") index with large values indicating many species and small values indicating few species.

Expected numbers of species in the various abundance classes can be predicted with these estimates of x and α: one species expected, αx; two species expected, $\alpha x^2/2$; three species expected, $\alpha x^3/3$; four species expected, $\alpha x^4/4$; etc. As described for the log normal model, a χ^2 statistic may be used to assess data fit to the log series model (df = number of classes − 1).

e. Broken Stick Model

MacArthur's[74] broken stick curve is usually interpreted with the random niche boundary hypothesis (e.g., Smith,[18] Magurran[69]). The niche volume is randomly divided into S portions for the available species to occupy in much the same way that a stick might be snapped at random intervals along its length. No assumption of preemptive use of resources by a species is made. The species utilization of the resource(s) is assumed to be nonoverlapping[70] and much more equitable than in any of the other models described. There are many species with intermediate abundances in the community. The species are assumed to be influenced only by one shared resource.[69] Despite the historical importance of the broken stick model, Smith[18] and Ludwig and Reynolds[70] suggested that it was seldom used because it described an uncommon situation. Magurran[69] indicated that the broken stick curve could appear during surveys of taxonomically very similar species.

To analyze data using the broken stick model, species are rank ordered by abundance. The number of individuals in the ith most abundant species (N_i) is estimated with Equation (39).[69]

$$N_i = \frac{N}{S} \sum_{z=i}^{S} \frac{1}{z} \tag{39}$$

where N = the total number of individuals in the sample from the community; and S = the total number of species in the sample.

The expected number of species in an abundance class (S_n) is predicted with Equation (40).[69]

$$S_n = \left(\frac{S(S-1)}{N} \right) \left(1 - \frac{n}{N} \right)^{S-2} \tag{40}$$

where n = the number of individuals in the abundance class. The expected values, the observed values, and df (number of abundance classes − 1) can be used in a χ^2 goodness-of-fit test.

f. Pollution Effects on Abundance Curves

Patrick[75] indicated that the shapes of species abundance curves for communities in polluted environments exhibit characteristic changes. With pollution, the species abundance curve moves away from the typical log normal curve with good species richness toward an aberrant log normal or geometric series curve. The aberrant log normal curve has a lowered mode and, perhaps, a broadening to cover more abundance octaves. May[19] hypothesized that these shifts in communities inhabiting contaminated environments represented successional reversions. In early community succession, processes associated with the geometric series/niche preemption hypothesis are thought to be important but, with time, those processes are replaced by processes associated with the log normal hypothesis.

Example 4

Diatom species sampled from a stream above (clean) and below (polluted) an outfall are counted and the number of species per octave tabulated. As done by Krebs,[38] intervals of 2–3, 4–7, etc. were used in calculations to account for species numbers at octave junctures, e.g., 4, 8. Alternatively, if several species have abundances at an octave juncture, half can be assigned to the lower octave and half to the higher octave.[70] $\log_2$ is used as per the original descriptions of the method by Preston,[71,72] although $\log_{10}$ or ln could be used. (Note: $\log_2 x = (\log_{10} x)/(\log_{10} 2)$.)

Octave (Interval)	Interval Midpoint ($\log_2$ Midpoint)	Clean Site	
		Number of Species	R
1 (1)	1 (0)	80	−3
2 (2–3)	2.5 (1.3219)	110	−2
3 (4–7)	5.5 (2.4594)	129	−1
4 (8–15)	11.5 (3.5236)	138	(R_0) 0
5 (16–31)	23.5 (4.5546)	131	+1
6 (32–63)	47.5 (5.5699)	109	+2
7 (64–127)	95.5 (6.5774)	78	+3
8 (128–255)	191.5 (7.5812)	53	+4
9 (256–511)	383.5 (8.5831)	29	+5
10 (512–1023)	767.5 (9.5840)	14	+6
11 (1024–2047)	1,535.5(10.5845)	5	+7
12 (2048–4095)	3,071.5(11.5847)	3	+8
13 (4096–8191)	6,143.5(12.5848)	2	(R_{MAX}) +9

Octave (Interval)	Interval Midpoint ($\log_2$ Midpoint)	Polluted Site	
		Number of Species	R
1 (1)	1 (0)	34	−4
2 (2–3)	2.5 (1.3219)	42	−3
3 (4–7)	5.5 (2.4594)	48	−2
4 (8–15)	1.5 (3.5236)	54	−1
5 (16–31)	23.5 (4.5546)	56	(R_0) 0
6 (32–63)	47.5 (5.5699)	52	+1
7 (64–127)	95.5 (6.5774)	50	+2
8 (128–255)	191.5 (7.5812)	42	+3
9 (256–511)	383.5 (8.5831)	34	+4
10 (512–1,023)	767.5 (9.5840)	23	+5
11 (1,024–2,047)	1,535.5(10.5845)	17	+6
12 (2,048–4,095)	3,071.5(11.5847)	10	+7
13 (4,096–8,191)	6,143.5(12.5848)	7	+8
14 (8,192–16,383)	12,287.5(13.5849)	4	+9
15 (16,384–32,767)	24,575.5(14.5849)	2	+10
16 (32,768–65,535)	49,151.5(15.5849)	1	(R_{MAX}) +11

Estimates of a for these two collections are made with Equation (26): Clean = 0.228; Polluted = 0.182. The other, more involved methods described in Krebs[38] [i.e., Equations (27) to (30)] are also used. The mean and variance for the truncated data (data above the veil line) are estimated using the $\log_2$ (interval midpoint) and the associated number of species in each octave: Clean, $\bar{x}_T = 4.0500$, $s_T^2 = 6.2699$; Polluted, $\bar{x}_T = 5.3656$, $s_T^2 = 10.7383$.

Example 4 *Continued*

Cohen's[73] methods are used to correct for the effect of truncation on the means and variances. First, γ values are calculated with Equation (30). The midpoint between 0 and the smallest abundance value (one individual per species) is used as an estimate of x_0. The $x_0 = \log_2 ((0 + 1)/2) = \log_2(0.5) = -1$. Clean: $\gamma = 6.2699/(4.0500 - (-1))^2 = 0.246$. Polluted: $\gamma = 10.7383/(5.3656 - (-1))^2 = 0.265$.

The θ values are then extracted from Table 30 of the Appendix with these γ values. Clean: $\theta = 0.04845$; Polluted: $\theta = 0.06332$.

Equations (28) and (29) are used to estimate the mean and variance for the two data sets with correction for truncation at the veil line. Clean: $\bar{x} = 4.0500 - 0.04845(4.0500 - (-1)) = 3.8053$. $s^2 = 6.2699 + 0.04845(4.0500 - (-1))^2 = 7.5055$. Polluted: $\bar{x} = 5.3656 - 0.06332(5.3656 - (-1)) = 4.9625$. $s^2 = 10.7383 + 0.06332(5.3656 - (-1))^2 = 13.3041$.

Another set of estimates for a is generated with Equation (27) and the estimates of σ^2. Clean: $a = 0.2581$; Polluted: $a = 0.1939$.

The number of species in the modal octave can be calculated with Equation (32). Clean: mean of ln S_R values = 3.5516; $\bar{R}^2 = 23$; $S_0 = 161.4$. Polluted: mean of ln S_R values = 2.9108; $\bar{R}^2 = 33.5$; $S_0 = 64.7$.

The predicted total number of species is then calculated. First, Equation (33) is used to calculate z. Clean: $z = (-1 - 3.8053)/2.7396 = -1.7540$. Polluted: $z = (-1 - 4.9625)/3.6475 = -1.6347$.

From Table 7 in the Appendix, the NED values corresponding with these z values are used to extract p_0 values for Equation (34). (Alternatively, Table 11 in Rohlf and Sokal[15] or a similar table could be used to estimate the area to the left of the veil line.) Clean: $p_0 \approx 0.040$; Polluted: $p_0 \approx 0.050$.

The predicted total number of diatom species in these two samples was estimated with Equation (34). Clean: $S_T = 881/(1 - 0.040) = 917$; Polluted: $S_T = 476/(1 - 0.050) = 501$.

Next, the values of a and S_0 can be used to predict S_R values. However, because there were two estimates for each of these parameters, the question arises about which estimates are best. Initially, all four possible combinations of a [from Equation (26) or Equation (27)] and S_0 [observed or from Equation (32)] are used to predict S_R values, and χ^2 statistics are calculated from the observed and expected species numbers in each R. The combination with the minimum χ^2 is judged to have the best fit. (The BASIC program LOGNORM written by Ludwig and Reynolds[70] was used to perform these tedious calculations.)

a	S_0	χ^2
	Clean	
0.228	138	10.62
0.228	161.4	47.36
0.258	138	5.27
0.258	161.4	17.59

Example 4 *Continued*

	Polluted	
0.182	56	0.41
0.182	64.7	10.77
0.194	56	3.85
0.194	64.7	7.51

The best combinations of estimates for the samples from the clean and polluted sites are $a = 0.258$ and $S_0 = 138$, and $a = 0.182$ and $S_0 = 56$, respectively. Ludwig and Reynolds[72] suggested that iterative substitution of a and S_0 values may be used at this point with the goal of minimizing the associated χ^2 to improve these estimates. When this is done, the final estimates remain the same for the sample from the polluted site, but the best estimates for the sample from the clean site are $a = 0.247$ and $S_0 = 138$ (final $\chi^2 = 1.87$). Comparing the two samples, the total and modal numbers of species decreases and the R_{MAX} increases in the sample from below the effluent relative to the sample from above the point of discharge.

3. Species Richness

Species richness is the number of species in a community. Although the species richness concept is easily grasped, the inability to clearly delineate the "community" can make it easy to incorrectly interpret associated measures of richness. The community is often some taxonomic group such as fish or insects in a region of interest. Pielou[76] referred to such an operationally defined "community" as a taxocene. It is extremely important that the distinction between the abstract concept of a community and the actual sample from which an index is derived be kept in mind during interpretation of data.

The number of species counted in an area will be influenced by the number of samples taken. It will be some increasing function of the log of the area collected.[77] With increased numbers of samples or quadrats, the cumulative number of species counted increases, but at an ever decreasing rate, toward an asymptotic value. One could laboriously sample more and more quadrats at each site until the curve approaches its asymptotic value for the number of species. Alternatively, species richness can be normalized to a set sample size, e.g., James and Rathbun.[78] Hurlbert[79] provided a rarefaction method for determining species richness.

$$\hat{S}_n = \sum_{i=1}^{s} 1 - \frac{\binom{N - N_i}{n}}{\binom{N}{n}} \tag{41}$$

where $\hat{S}_n$ = the predicted number of species in a sample of size N; N_i = the number of individuals of species i; and n = the sample size (number of individuals) to which normalization is to be done.

For readers unfamiliar with the binomial coefficient notation in this equation, an explanation with solution is provided in Chapter 4 [Equation (48)]. Tables, explanation, and simple properties of the binomial coefficient are also provided in Beyer,[80] pages 362–366.

The large sample variance for $\hat{S}_n$ can be estimated with a rather involved equation.[38,81] It can be made clearer by simplifying some terms first.

$$a = \binom{N}{n}$$

$$b = \binom{N - N_i}{n}$$

$$c = \binom{N - N_j}{n}$$

$$d = \binom{N - N_i - N_j}{n}$$

The variance is estimated with these simplifications.

$$\text{Variance } \hat{S}_n = a^{-1}\left[\sum_{i=1}^{S} b\left(1 - \frac{b}{a}\right) + 2\sum_{i=1}^{S-1}\sum_{j=i+1}^{S}\left(d - \frac{bc}{a}\right)\right] \tag{42}$$

This rarefaction method is preferred over simpler methods when samples (number of individuals) vary in size. Krebs[38] suggested that it should be used only for comparing taxonomically similar groups from similar habitats sampled by identical techniques. The assumption is made that the underlying species-individual relationship is the same between units being compared.[82] The methods are also based on the assumption that individuals are randomly distributed although, with very large samples, the bias associated with any clumping tends to be minimized.

Hurlbert,[79] Ludwig and Reynolds,[70] Magurran,[69] and Krebs[38] provided further details regarding this index. Ludwig and Reynolds[70] and Krebs[38] offered BASIC and FORTRAN code, respectively, for the associated calculations.

Heltshe and Forrester[83] developed a jackknife procedure for estimating species richness from quadrat data. Only the presence or absence of the species is noted for each quadrat. Calculations focus on the number of unique species (those occurring in only one quadrat regardless of their numbers within that quadrat).

$$\hat{S} = S + \left(\frac{n - 1}{n}\right)k \tag{43}$$

where n = the number of quadrats sampled; S = the total number of species in the n quadrats; and k = the number of unique species in the n quadrats.

The variance and approximate 95% confidence interval for this index of species richness can be estimated.

$$\text{Variance } \hat{S} = \frac{n - 1}{n}\left[\sum_{j=0}^{S} j^2 f_j - \frac{k^2}{n}\right] \tag{44}$$

where f_j = the number of quadrats containing j unique species.

$$\text{C.I.} = \hat{S} \pm t_{2\text{-sided},\alpha}\sqrt{\text{Variance } \hat{S}} \tag{45}$$

where $t_{2\text{-sided},\alpha} = t$ from Table 14 in the Appendix for $df = n - 1$ and $\alpha = 0.05$.

This estimator is biased (overestimates the number of species),[83] but the bias is usually smaller than that for the actual observed number of species (underestimates the number of species). Heltshe and Forrester[83] suggested that this jackknife method not be used if there was a large number of unique species or if S was very low. Krebs[38] provided FORTRAN code for implementing the method.

If a large number of quadrats are sampled, Krebs[38] recommended a bootstrap method described by Smith and van Belle[84] instead of the jackknife method described above. (Although neither Smith and van Belle[84] nor Krebs[38] provided code for this approach, it can be implemented with SAS or specialized packages such as RT[85] or RESAMPLING STATS.[86]) Both the bootstrap and jackknife methods described by Smith and van Belle[84] were biased. Smith and van Belle[84] recommended the jackknife method when the number of quadrats sampled is small and the bootstrap method when the number of quadrats is large. Mingoti and Meeden[87] argued that both the jackknife and bootstrap methods were unacceptably biased and suggested an empirical Bayes estimator instead.

Example 5

Species in a particular guild are sampled at eight sites around an outfall, and the number of individuals per species (N_i) is tabulated. Site locations are designated as kilometers upstream ($-$) or downstream ($+$) of the outfall. As the total number of individuals per sample (N) differs, the rarefaction method is used to predict species richness at a sample size of 40 individuals ($n = 40$).

Species	N_i at Sites (Site = Kilometers from the Outfall)							
	-1.0	-0.5	0	$+0.5$	$+1.0$	$+1.7$	$+2.7$	$+5.3$
1	12	12	58	18	11	8	12	10
2	11	12	21	16	11	12	10	11
3	10	10	3	15	11	13	8	8
4	8	8	2	5	8	3	11	6
5	8	8	1	4	7	3	8	6
6	5	4			3	3	4	5
7	2	3			2	1	1	4
8	2	2			1	2	1	2
9	1	2			1		2	1
10		1						1
N	59	62	85	58	55	45	57	54

Using Equations (41) and (42), an estimate of species richnesses and the associated variances for each sample is made at $n = 40$. Krebs'[38] program RAREFACT[38] is used for these calculations.

Statistic	N_i at Sites (Site = Kilometers from the Outfall)							
	-1.0	-0.5	0	$+0.5$	$+1.0$	$+1.7$	$+2.7$	$+5.3$
$\hat{S}_n$	8.476	9.347	4.050	4.991	8.367	7.877	8.312	9.414
Variance	0.386	0.472	0.556	0.009	0.461	0.110	0.485	0.432

Example 5 *Continued*

Above the outfall, the species richness is estimated to range from 8.476 to 9.347 (expected species in a sample of $n = 40$). The species richness drops at the outfall out gradually increases again with distance downstream.

Calculations for the sample taken from the $+0.5$ km site are detailed here. Equation (41) is used to estimate $\hat{S}_n$.

$$\hat{S}_n = \left[1 - \frac{\binom{40}{40}}{\binom{58}{40}} \right] + \left[1 - \frac{\binom{42}{40}}{\binom{58}{40}} \right] + \left[1 - \frac{\binom{43}{40}}{\binom{58}{40}} \right] + \left[1 - \frac{\binom{53}{40}}{\binom{58}{40}} \right] + \left[1 - \frac{\binom{54}{40}}{\binom{58}{40}} \right]$$

$$\hat{S}_n \approx [1 - 0] + [1 - 0] + [1 - 0] + [1 - 0.002] + [1 - 0.007] \approx 4.991$$

The variance calculation [Equation (42)] is tedious and best done with a computer or programmable calculator to minimize error.

A wide range of designs can be used to analyze toxicant effect on species richness. Obviously, ANOVA could be applied to the design described above. Alternative methods outlined in other chapters could be applied with proper care. For example, Niederlehner et al.[88] use an index of percentage of tax affected (Probit [1- (species richness with exposure/species richness without exposure)]*100) and ln of concentration to perform a traditional probit analysis of protozoan community change with exposure to cadmium.

4. Species Heterogeneity ("Species Diversity")

In addition to species richness, the concept of species diversity often includes the evenness or equitability with which individuals in a community are allocated among species. For example, if all species in a community have the same number of individuals, this community's diversity is thought to be higher than that of another community having the same number of species but most of the individuals associated with only one or a few species. Many species heterogeneity indices (also called diversity indices) account simultaneously for both species richness and evenness. Three will be discussed although, under the assumption of a log series species abundance model and sufficient number of species to construct the curve, the α described already may be used as a heterogeneity measure also.[67] Because these indices measure the combined influence of species richness and evenness, effective interpretation of any heterogeneity index necessitates estimating species richness as well as evenness.

The Simpson (Yule) index,[89] a species dominance type of heterogeneity index, is a weighted mean of the species proportional abundances. According to Poole,[24] this index quantifies the dominance of one or a few of the most common species in the community. It is relatively insensitive to species richness because it weights the most abundant species most heavily.[69] Depending on its form, it estimates the probability that two individuals randomly taken from an infinitely large (indefinitely large) population will be of the same (λ) or different ($1 - \lambda$) species.[67,82] (The term "population" refers to the statistical population, i.e., all the individuals in the community.) Hurlbert[79] and Pielou[76] suggest that this index can also be interpreted as the probability of encounter between individuals of the same species.

$$\lambda = \sum_{i=1}^{S} p_i^2 \tag{46}$$

where S = the total number of species; and p_i = the proportion of the total number of individuals in the community that are members of the ith species.

Practical application of Simpson's index entails substitution of n_i (the number of individuals of species i in the total sample) divided by N (the total number of individuals in the sample) for p_i. Further, because the entire population is not sampled, a bias appears that must be corrected by adding the terms $(n_i - 1)$ and $(N - 1)$ to the index. Thus, substitution of n_i/N into Equation (46) provides an estimate of λ for an (effectively) infinite population and Equation (47) provides an estimate for a finite population.[67]

$$\hat{\lambda} = \sum_{i=1}^{S} \frac{n_i(n_i - 1)}{N(N - 1)} \tag{47}$$

Simpson's index can range from 0 to 1. Obviously, species heterogeneity is low when $\hat{\lambda}$ (probability of encounter between two individuals of the same species) is high. A simple modification of this index is often made so that its value increases as heterogeneity increases, i.e., $1 - \hat{\lambda}$. Simpson[89] provided methods for estimating the standard error of this index. Heltshe and Forrester[90] described jackknife methods for estimating the confidence interval for this index.

Two information theory-based indices (Shannon's and Brillouin's) are also used routinely for measuring heterogeneity. Both attempt to quantify the amount of information contained in each individual in the community. However, Shannon's index produces a measure for the effectively infinite population from which the sample was taken, but Brillouin's index provides a measure for the specific collection of individuals in the sample. The Shannon (also known as Shannon-Wiener or Shannon-Weaver) index [Equation (48)], being based on the often dubious assumption that all species are represented in the sample from a population of infinite (or effectively infinite) size, is used ideally with truly random samples from a very large population and a known number of individuals.[38] Unlike Brillouin's index, Shannon's index does not vary with sample size if the number of species and their relative proportions remain constant.[24,69] Brillouin's index [Equation (49)] is preferable if the sample is taken in a nonrandom fashion as is often the case with field collections.[69] Poole[24] recommended it also if all the individuals in a population (community) are counted and assigned to species.

$$H' = \sum_{i=1}^{S} p_i \ln p_i \tag{48}$$

$$H = \left(\frac{1}{N}\right) \ln \left(\frac{N!}{\prod_{i=1}^{S} n_i!}\right) \tag{49}$$

Bases for the logarithms in Equations (48) and (49) vary between applications. Units can be bits/individual ($\log_2$ binary units), decits/individual ($\log_{10}$: bel or decimal units), or nits/individual ($\ln$: natural bel).[24,38,69]

As with Simpson's index, n_i/N values are used instead of p_i values to estimate Shannon's index from a sample. Use of n_i/N values instead of p_i values (i.e., the maximum likelihood estimator of H') introduces a bias that worsens as the sample size decreases. Although Peet[82] and Magurran[69] suggested that this bias rarely had a magnitude that seriously impacted results, Kaesler and Herricks[91] put this bias forward as one reason for avoiding Shannon's index and using Brillouin's index instead. The variance for this estimate of H' can be derived as detailed in Poole[24] and Magurran.[69] (Heltshe and Forrester[90] provided jackknife procedures for estimating a confidence interval about Brillouin's index.)

Unlike Shannon's index, Brillouin's index does not treat the sample as a random sample from an infinitely large population. Instead it treats the sample as a collection of individuals and estimates the heterogeneity of that collection.[69,91] Consequently, Shannon's index provides a heterogeneity measure for a community, but Brillouin's index provides a heterogeneity measure for a sample. Although Kaesler and Herricks[91] recommended avoiding Shannon's index in applied ecology due to its unrealistic assumptions and bias, Krebs[38] argued that using Shannon's or Brillouin's index usually did not produce very different results for large samples. (Brillouin's index will be slightly lower than Shannon's index as seen in Example 6.)

Example 6

These indices are estimated for the fictitious data used in Example 5. Krebs' program DIVERS[38] produces index estimates for all sites and, as shown in this table, indicates a distinct drop in heterogeneity regardless of the index. Only calculations for the +0.5 km site are detailed here. The $\log_2$ is used in the example so $\log_2$ would be substituted for ln in Equations (48) and (49). The units are bits/individual for H' and H.

N_i at Sites (Site = Kilometers from the Outfall)

Index	−1.0	−0.5	0	+0.5	+1.0	+1.7	+2.7	+5.3
$1 - \hat{\lambda}$	0.863	0.871	0.477	0.762	0.853	0.816	0.857	0.578
H'	2.867	2.986	1.248	2.112	2.789	2.572	2.810	3.016
H	2.557	2.658	1.145	1.928	2.471	2.241	2.499	2.653

Estimation of $1 - \hat{\lambda}$ for the +0.5-km site [Equation (47) for a finite population]:

$$\hat{\lambda} = \frac{(18)(17)}{(58)(57)} + \frac{(16)(15)}{(58)(57)} + \frac{(15)(14)}{(58)(57)} + \frac{(5)(4)}{(58)(57)} + \frac{(4)(3)}{(58)(57)}$$

$$= 0.2384$$

$$1 - \hat{\lambda} = 0.7616$$

Estimation of $1 - \hat{\lambda}$ for the +0.5-km site [Equation (46) for an infinite population]:

$$\hat{\lambda} = \left(\frac{18}{58}\right)^2 + \left(\frac{16}{58}\right)^2 + \left(\frac{15}{58}\right)^2 + \left(\frac{5}{58}\right)^2 + \left(\frac{4}{58}\right)^2$$

$$= 0.2515$$

$$1 - \hat{\lambda} = 0.749$$

Example 6 *Continued*

Estimation of H' for the +0.5-km site:

$$H' = -\left[\frac{18}{58}\log_2\left(\frac{18}{58}\right) + \frac{16}{58}\log_2\left(\frac{16}{58}\right) + \cdots + \frac{4}{58}\log_2\left(\frac{4}{58}\right)\right]$$

$$= -(-0.5239 - 0.5125 - 0.5046 - 0.3048 - 0.2661)$$

$$= 2.112 \text{ bits/individual}$$

Estimation of H for the +0.5-km site:

$$H = \frac{1}{58}\log_2\left(\frac{58!}{(18!)(16!)(15!)(5!)(4!)}\right)$$

$$= 1.929 \text{ bits/individual}$$

5. Species Evenness

Species evenness is the equitability of abundances among species in a community.[92] Although only three evenness indices are described here, many have been developed. The first two to be described are the quotients of the observed H' or H over their estimated maxima. As such, the index (J') based on H' estimates the evenness for the (effectively infinite) population from which the sample was taken, and the index (J) based on H estimates the evenness for the collection of individuals in the sample.[67] The J' will overestimate the true evenness of the community, because it is impossible to census all species in the infinite population.[67]

The maximum of Shannon's index is ln S. The associated estimate of evenness (Pielou's J') is calculated with the following equation.

$$J' = \frac{H'}{\ln S} \tag{50}$$

where S = the number of species in the sample. An evenness index (Pielou's J) can also be estimated with the ratio of the Brillouin index (H) to its maximum value.

$$J = \frac{H}{H_{MAX}} \tag{51}$$

The maximum value for H is generated with Equation (52).[24,67,76]

$$H_{MAX} = \frac{1}{N}\ln\left[\frac{N!}{([N/S]!)^{S-r}(([N/S] + 1)!)^r}\right] \tag{52}$$

where N = total number of individuals summed over all species; $[N/S]$ = the integer part of the quotient N/S; and $r = N - S[N/S]$.

Both J' and J are familiar tools to ecologists yet the first is used most widely. Unfortunately, it is dependent on species richness as well as species evenness.[82] Alatalo[92] recommended a modified Hill ratio estimator of evenness as an index insensitive to species richness. Ludwig and Reynolds[70] also pointed out that this modified Hill ratio estimator was not as sensitive to variation in the number of species between samples. It is the ratio of $e^{H'}$ to $1/\hat{\lambda}$ after correction of each for their minima (i.e., subtraction of 1). Simpson's index ($\hat{\lambda}$) is estimated with Equation (47).

$$E = \frac{\frac{1}{\hat{\lambda}} - 1}{e^{H'} - 1} \qquad (53)$$

The $e^{H'}$ (Hill's N_1) reflects the number of "abundant species", and $1/\hat{\lambda}$ (Hill's N_2) reflects the number of "very abundant species."[92] (See Hill[93] for more information.) This index decreases as evenness decreases and approaches 0 as one species approaches complete dominance.[70]

Example 7

Evenness indices are estimated for the fictitious data used in Examples 5 and 6. Calculations are shown only for the sample from the +0.5 site. For illustrative purposes, the natural log is used here instead of the $\log_2$ used in the previous examples. Consequently, units change from bits/individual to nits/individual. Most calculations are done with the BASIC program SPDIVERS[70] although Krebs'[38] FORTRAN program DIVERS can also be used. The program ECOSTAT[94] is used to estimate H and J.

				N_i at Sites (Site = Kilometers from the Outfall)				
Index	−1.0	−0.5	0	+0.5	+1.0	+1.7	+2.7	+5.3
S	9	10	5	5	9	8	9	10
N	59	62	85	58	55	45	57	54
$\hat{\lambda}$ [Eq (47)]	0.137	0.129	0.523	0.238	0.147	0.184	0.143	0.122
$1 - \hat{\lambda}$	0.863	0.871	0.477	0.772	0.853	0.816	0.857	0.878
H'	1.987	2.070	0.865	1.464	1.933	1.783	1.948	2.091
H	1.772	1.842	0.793	1.337	1.712	1.554	1.732	1.839
J'	0.905	0.899	0.537	0.910	0.880	0.857	0.887	0.908
J	0.905	0.898	0.526	0.907	0.879	0.854	0.887	0.909
E	1.002	0.975	0.664	0.962	0.983	0.898	0.993	1.012

All indices (J', J, and E) show that species evenness decreases near the outfall. This is consistent with the species list (Example 5) that indicates species 1 becomes more dominant at the outfall, and the abundances of the other species decrease.

Estimation of J' for the +0.5-km site:
Before J' can be calculated with Equation (50), H' must be calculated with Equation (48).

$$H' = -\left[\frac{18}{58} \ln\left(\frac{18}{58}\right) + \frac{16}{58} \ln\left(\frac{16}{58}\right) + \cdots + \frac{4}{58} \ln\left(\frac{4}{58}\right) \right]$$

$$= -[-0.3631 - 0.3553 - 0.3498 - 0.2113 - 0.1844]$$

$$= 1.464 \text{ nits/individual}$$

$$J' = H'/\ln S = 1.464/\ln(5) = 0.910$$

Example 7 *Continued*

Estimation of J for the +0.5-km site:

Before J can be calculated with Equation (51), H and H_{MAX} must be derived with Equation (49) and (52), respectively.

$$H = \frac{1}{58} \ln\left(\frac{58!}{(18!)(16!)(15!)(5!)(4!)}\right)$$

$$= 1.337 \text{ nits/individual}$$

For Equation (52), [N/S] is estimated from 58/5 to be 11. The r is $58-5(11)$ or 3.

$$H_{MAX} = \frac{1}{58} \ln\left[\frac{58!}{(11!)^2((11+1)!)^3)}\right] = 1.474$$

$$J = \frac{H}{H_{MAX}} = \frac{1.337}{1.474} = 0.907$$

Estimation of E for the + 0.5-km site:

Before E can be calculated with Equation (53), $\hat{\lambda}$ was determined with Equation (47). The $\hat{\lambda}$ was found in Example 6 to be 0.2384.

$$E = \frac{\frac{1}{\hat{\lambda}} - 1}{e^{H'} - 1} = \frac{\frac{1}{0.2384} - 1}{e^{1.4639} - 1} \approx 0.961$$

6. Community Similarity

Indices of species abundance, evenness, and heterogeneity are most effectively used together in assessing changes in community structure. Additionally, an index of similarity between communities can be used, as in the studies of Pontasch et al.[95] and Snoeijs.[96] Indices of similarity compare two communities based on species presence-absence data or species abundance data. Similarity indices based on presence-absence data are very popular, e.g., Snoeijs.[96] Two such indices are described here using the modified notations of Magurran[69] and Krebs.[38]

The most common similarity indices based on presence-absence data are the Jaccard and Sorenson indices. Both range from 0 (the communities are completely dissimilar) to 1 (the communities are identical). Magurran[69] suggested that the Sorenson index performed best of the similarity indices based on presence-absence data.

$$\text{Jaccard Index (JI)} = \frac{j}{a + b - j} \tag{54}$$

$$\text{Sorenson Index (SI)} = \frac{2j}{a + b} \tag{55}$$

where j = the number of species occurring in both samples (sites); a = the number of species occurring in the first sample; and b = the number of species occurring in the second sample.

There are numerous indices based on species abundances data. Only four (Sorenson, Euclidean distance, Bray-Curtis, and Morisita-Horn) are described here. Magurran's review[69] of the literature suggested that the Morisita-Horn index is best. She noted that the Morisita-Horn index is not as sensitive to species richness and sample size as the other abundance-based indices.

The Sorenson index based on species abundances (number of individuals, biomass, percent cover) follows.[69]

$$\text{Sorenson Index (Abundance) (SIA)} = \frac{2N_i}{N_a + N_b} \tag{56}$$

where N_i = the sum of the smallest of the two abundances for the i species collected from the sites being compared; N_a = the abundance (total number of individuals) for the first site (a); and N_b = the abundance for the second site (b).

The Euclidean distance index for samples for sites a and b is defined in Equation (57).[38]

$$\Delta = \sqrt{\sum_{i=1}^{n} (N_{ia} - N_{ib})^2} \tag{57}$$

where N_{ia} = abundance of species i in the sample from site a; N_{ib} = abundance of species i in the sample from site b; and n = the total number of species.

Krebs[38] noted that this index increases as the number of species in the samples increases. Consequently, he suggested that the following average distance measure be used.

$$d = \sqrt{\frac{\Delta^2}{n}} \tag{58}$$

Values of d range from 0 (sites are completely similar) to infinity.[38,70] Another abundance-based index, the Bray-Curtis index, ranges from 0 (complete similarity) to 1. According to Krebs,[38] it weights the most abundant species most heavily and, consequently, is insensitive to rare species.

$$B = \frac{\sum_{i=1}^{n} |N_{ia} - N_{ib}|}{\sum_{i=1}^{n} (N_{ia} + N_{ib})} \tag{59}$$

Magurran[69] favored the final abundance-based index (Morisita-Horn index). Wolda[97] suggested that this index is the best abundance-based index. Krebs[38] referred to this index as the simplified Morisita index and noted that it was nearly independent of sample size.

$$\text{MH} = \frac{2 \sum_{i=1}^{n} (N_{ia}N_{ib})}{(da + db)N_a N_b} \tag{60}$$

where N_{ia}, N_{ib} = number of individuals of species i in the samples from sites a and b, respectively.

As previously defined, N_a and N_b are the number of individuals in the samples from sites a and b. The total number of species is n. The da and db in Equation (60) are calculated as shown.

$$da = \frac{\sum_{i=1}^{n} N_{ia}^2}{N_a^2} \tag{61}$$

$$db = \frac{\sum_{i=1}^{n} N_{ib}^2}{N_b^2} \tag{62}$$

Example 8

The data for the -0.5-km and $+0.5$-km sites in Example 5 are used to illustrate the calculation of these indices. The programs SUDIST,[70] SIMILAR,[38] and ECOSTAT[94] calculate these and many other similarity indices with the exception of the Morisita-Horn index recommended by Magurran.[69]

$$\text{JI} = \frac{5}{10 + 5 - 5} = 0.500$$

$$\text{SI} = \frac{(2)(5)}{10 + 5} = 0.667$$

$$\text{SIA} = \frac{(2)(12 + 12 + 10 + \cdots + 0)}{62 + 58} = 0.7167$$

$$\Delta = \sqrt{(12 - 18)^2 + (12 - 16)^2 + (10 - 15)^2 + \cdots + (1 - 0)^2} = 11.662$$

$$d = \sqrt{\frac{11.662^2}{10}} = 3.688$$

$$B = \frac{|12 - 18| + |12 - 16| + \cdots + |1 - 0|}{(12 + 18) + (12 + 16) + \cdots + (1 + 0)} = 0.283$$

To calculate the Morisita-Horn index, da and db are calculated first.

$$da = \frac{12^2 + 12^2 + 10^2 + \cdots + 1^2}{62^2} = 0.143$$

Example 8 *Continued*

$$db = \frac{18^2 + 16^2 + \cdots + 4^2}{58^2} = 0.251$$

$$MH = \frac{2((12)(18) + (12)(16) + \cdots + (0)(1))}{(0.143 + 0.251)(62)(58)} = 0.889$$

C. COMMUNITY FUNCTION

1. General

Although a wide range of functional responses exists, ecotoxicologists tend to favor qualities of community structure.[98] Matthews et al.[98] broke down functions into two general groups. The first included biological, chemical, and physical rate constants such as photosynthesis, respiration, sedimentation, bioturbation, nitrogen fixation, sediment oxygen consumption, zooplankton filtration rates, and chelation. The second group included taxonomic functions such as rates of recovery after exposure, rates of colonization, or rates of succession. They also noted that enumeration of functional groups, e.g., scrapers, shredders, collectors, and predators can be used to assess functional changes in benthic macroinvertebrate communities. Four examples reflecting community functions will be discussed here. Sheenan[99] provided more detail and specific examples of studies documenting toxicant-induced functional changes in ecosystems.

It is critical to understand the natural variation and trends in these functions over which the effects of a toxicant or effluent might be imposed. Failure to do so can result in falsely attributing an effect to a contaminant. Examples of such natural trends are elucidated here using the river continuum theory. This theory suggests that the community functions associated with different reaches of a river will differ due to physical changes in the river from headwaters to mouth.[100] For example, in the headwaters (stream orders 1 to 3) where there is considerable shading and input of allochthonous material, the productivity to respiration ratio (P/R) can be less than 1. However, the P/R can increase above 1 in the middle reaches of the river (stream orders 4 to 6) where shading is reduced and ample nutrients are present. In the lower reaches, turbidity and depth can limit productivity and the P/R ratio can drop below 1 again. Similarly, the macroinvertebrate functional groups (detritus shredders, collectors, scrapers, and predators) shift with the availability of associated resources. Shredders will be prominent in headwaters where there is an abundance of coarse organic particulate matter such as leaves. They will decrease with distance downriver. Grazers will be most abundant in the middle regions where productivity is high. Collectors will be most abundant in the lower reaches. These general patterns along the gradient from first order to higher order channels are modified in areas of confluence with lower order tributaries. It is critical to consider such natural gradients as these when interpreting community functional shifts at contaminated sites.

2. Productivity and Respiration

A wide range of studies, e.g., Cairns,[58] Kondratieff et al.,[101] Blanck,[102] Goldsborough and Robinson,[103] Amblard et al.,[104] and Day[105] used measures of community photosynthesis and respiration in combination with structural changes to imply toxicant impact. For example, Amblard et al.[104] measured chlorophyll a, ATP/ADP, photosynthetic activity (^{14}C method), and photoheterotrophic incorporation of amino acids in periphyton communities. They found an increase in heterotrophic biomass and activity near a pulp and paper mill effluent. Kondratieff et al.[101] also measured chlorophyll a and ATP. They calculated an autotrophic index with these indicators of community function.

$$AI = \frac{\hat{B}_{ATP}(mg/L)}{chlorophyll\ a(mg/l)} \tag{63}$$

The biomass used in the autotrophic index was estimated with the ATP measurements.

$$\hat{B}_{ATP} = \frac{ATP(ng/L)}{2400\ (mg\ biomass/ng\ ATP)} \tag{64}$$

This index (AI) increases with a proportional increase in heterotrophic activity. The autotrophic index suggests a shift to heterotrophy at sites closest to the outfall (sewage and electroplating waste). Kondratieff et al.[101] also used ANOVA to detect a shift in invertebrate functional groups. Near the effluent, differences in means of invertebrate densities and ash-free dry weights suggested that collector-gatherers and filter feeders dominated.

3. Detritus Processing

A wide range of stressors influence detritus processing.[106] For example, Giesy[107] found that cadmium slows leaf decomposition and leaf colonization by fungi and bacteria. Leaf,[108] macrophyte, and algal[109] detritus processing can also be modified by shifts in pH- and acidification-related phenomena.

Allred and Giesy[110] examined leaf litter processing in streams using open-mesh bags of leaves. The following model is fit to the weight loss in the bags over time (211 days).

$$W_t = W_0 e^{-kt} \tag{65}$$

where W_t = weight (total or ash-free dry weight) at time t; W_0 = initial weight (t = 0); and k = the rate constant for weight loss.

The form of this model is the same as that used in Chapter 3 [Equation (2)] for toxicant elimination kinetics. All elaborations (e.g., half-life or mean lifetime estimates), alternatives [e.g., Equation (11)], and examples (e.g., Examples 1 to 3) in Chapter 3 are potentially pertinent to the analysis of such data. For example, Mulholland et al.[108] found that a linear model (Equation (1) in Chapter 3) can fit leaf decomposition data as well as Equation (65). Multiple compartment formulations can also be appropriate[111] in describing detrital decomposition [e.g., Equation (8) in Chapter 3].

4. Nutrient Spiraling

Cairns[58] argued that it is important to consider nutrient cycling during assessment of biological integrity ("maintenance of the community structure and function characteristic of a particular locale or deemed satisfactory to society"). Later, Cairns and Pratt[112] added that basic functional attributes such as nutrient spiraling could be used for this purpose. Unfortunately, studies of such attributes are uncommon. The basic premise and approach to nutrient spiraling is sketched out here to stimulate more interest in this and similar attributes of nutrient cycling.

Nutrient cycles in lotic (flowing) systems are open. As nutrients move downstream with the hydraulic flow, they are taken up by and become associated with benthic communities and abiotic solid phases. Elwood et al.[113] suggested that, in healthy streams, biotic processes dominate. The nutrients are eventually released to continue their movement downstream. This process of uptake and release during net movement downstream is called nutrient spiraling. Because decomposition rates and algal biomass accumulation in lotic systems can be nutrient limited,[113,114] this attribute of nutrient movement is important in flowing systems. Productivity tends to be high with tight spiraling because nutrients are retained efficiently.[115]

Spiraling length (S) is used as a measure of nutrient retention in lotic systems.[113,115] It is defined as the average distance downstream to complete one cycle for a nutrient. It is estimated with the following equation.[115]

$$S = \frac{F_T}{Uw} \tag{66}$$

where F_T = the total downstream flux of the nutrient (g/s); U = the uptake rate of the nutrient per unit of benthic surface (g/m^2/s); and w = the stream width (m). Although toxicants likely affect spiraling length, this aspect of nutrient cycling in flowing systems has been neglected.

5. Colonization and Succession

The potential impacts of toxicants on successional processes have been touched on very briefly in earlier discussions of species abundance models. Cairns and co-workers' studies of toxicant effects on protozoan colonization (increase to species abundance during primary succession) are examined here. They[116–119] use the MacArthur-Wilson theory of island colonization to model protozoan colonization of polyurethane foam substrates (islands). (See MacArthur and Wilson[120] and Simberloff and Wilson[121] for details.)

The MacArthur-Wilson model is identical in form with the model for bioaccumulation described in Chapter 3. Indeed, substituting C_s in Equation (42) from Chapter 3 into Equation (41) in Chapter 3 will produce the same model formulation, except the process being modeled is toxicant accumulation in an organism rather than species accumulation on an island. Therefore, many of the comments regarding the fit of data to bioaccumulation models in Chapter 3 are pertinent here. The reader is encouraged to review Chapter 3 before applying these methods.

The model considers the arrival (invasion or immigration) and extinction rates of colonizing species from a source community, e.g., a mainland (Figure 10). The arrival rate is a function of time and the distance from the source community to the island. Once on the island, species can remain extant or become extinct. The extinction rate is a function of time and the size of the island. Extinction rates are generally lower for larger islands. For a given island with a fixed size and distance from source community, the rate of species arrivals decreases with time and the rate of species extinctions increases with time until an equilibrium is reached, i.e., the rate of species additions equals that of species loss. Equations (67) and (68) from MacArthur and Wilson[120] describe this approach to an equilibrium number of species.

$$G = \frac{d\mu}{dS} - \frac{d\lambda}{dS} \tag{67}$$

where S = the number of species; μ = the rate of species extinctions when S species are present; and λ = the rate of species arrivals when S species are present.

The constant G is analogous to k_e in Chapter 3. It is a summary statistic incorporating individual species constants and island qualities such as size and distance from the source community.

$$S_t = S_{EQ}(1 - e^{-Gt}) \tag{68}$$

where S_{EQ} = the equilibrium number of species; S_t = the number of species at time t; and t = the time (duration) over which colonization has occurred.

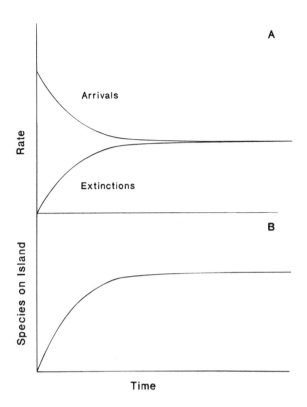

Figure 10 Time course of island colonization based on the MacArthur-Wilson model. (Reproduced from MacArthur and Wilson[120] by permission of Princeton University Press. Copyright 1967 by Princeton University Press.)

MacArthur and Wilson[120] estimated the time to approach 90% of the equilibrium number of species in a manner analogous to that used for bioaccumulation kinetics in Chapter 3 [Equations (3) to (7)].

$$t_{0.90} = \frac{2.303}{G} \tag{69}$$

This is the time to reach all but 10% (0.10) of the equilibrium number of species. Clearly, the time to reach any proportion of the S_{EQ} can be calculated, e.g.,

$$t_{0.90} = \frac{\ln\left(\frac{1}{0.10}\right)}{G} \tag{70}$$

Cairns et al.[119] used this model to assess the effect of cadmium and an organic pesticide used against larval lamprey (TFM or 3-trifluoromethyl-4-nitrophenyl) on protozoan colonization. A naturally colonized polyurethane foam substrate is placed in the center of a microcosm containing noncolonized foam substrates. The noncolonized substrates form a ring of islands equidistant from the naturally colonized epicenter. The microcosms then receive various treatments and the increase in species abundance in the islands is quantified over a period of time. These data are fit to the MacArthur-Wilson model using nonlinear least squares regression. For TFM, S_{EQ} and G decrease abruptly between the 1.0 and 10.0 mg/L TFM treatments. The S_{EQ} decreases linearly with the ln of cadmium concentration in the microcosm water. Inverse regression[116] methods are used to estimate the concentration of cadmium corresponding to an EC20, the concentration at which the S_{EQ} is reduced by 20% of the control S_{EQ}.

Example 9

As described above, Niederlehner et al.[117] measured protozoan colonization of foam substrates in the presence of a series of cadmium concentrations. Data from the control and highest cadmium concentration (9.5 μg/L) treatments were visually extracted from their Figure 2 and used here to illustrate the associated data analysis. These analyses are very similar to those shown in Example 8 in Chapter 3 for bioaccumulation. Although not shown here, the actual numbers of species and those predicted from the regression results should be plotted with time to assess the fit of the nonlinear model. Also, the regression residuals should be plotted against time to detect any trends in the residuals. The codes to perform these plots for assessing regression model fit to these data are given in Example 2 in Chapter 3.

```
DATA COLONIZ;
   INPUT CD DAY SPECIES @@;
   CARDS;
   0.0  0  0 0.0  1  1 0.0  1  5 0.0  3  8
   0.0  7 29 0.0  7 27 0.0 14 38 0.0 14 36
   0.0 21 37 0.0 21 32 0.0 28 32 0.0 28 28
   9.5  0  0 9.5  1  2 9.5  1  3 9.5  3  9
   9.5  3  7 9.5  7 14 9.5  7 16 9.5 14 13
   9.5 14 18 9.5 21  9 9.5 21 12 9.5 28 10
   9.5 28 14
   ;
PROC SORT;
   BY CD DAY;
RUN;
DATA ZERO;
   SET COLONIZ;
   IF CD=0.0;
RUN;
PROC NLIN;
   PARMS SEQ=36 G=0.5;
   BOUNDS 0<SEQ<50 0<G<1;
   MODEL SPECIES = SEQ*(1-EXP(-G*DAY));
   OUTPUT OUT=SPEC1 P=PSPEC1 R=RSPEC1;
RUN;
DATA NINE;
SET COLONIZ;
   IF CD=9.5;
RUN;
PROC NLIN;
   PARMS SEQ=12 G=0.5;
   BOUNDS 0<SEQ<50 0<G<1;
   MODEL SPECIES=SEQ*(1-EXP(-G*DAY));
   OUTPUT OUT=SPEC2 P=PSPEC2 R=RSPEC2;
RUN;
```

Example 9 *Continued*

The control treatment (0 μg/L):
  $S_{EQ} = 34.69$
  Asymptotic standard error for $S_{EQ} = 2.31$
  Asymptotic 95% confidence intervals for S_{EQ} = 29.54 to 39.84
  $G = 0.19$
  Asymptotic standard error for $G = 0.05$
  Asymptotic 95% confidence intervals for G = 0.08 to 0.30
Cadmium Treatment (9.5 μg/L):
  $S_{EQ} = 13.25$
  Asymptotic standard error for $S_{EQ} = 1.08$
  Asymptotic 95% confidence intervals for S_{EQ} = 10.88 to 15.62
  $G = 0.36$
  Asymptotic standard error for $G = 0.12$
  Asymptotic 95% confidence intervals for G = 0.09 to 0.63

The S_{EQ} in the 9.5 μg/L treatment (approximately 13 species) was less than half of that for the control treatment (approximately 35 species). Although the confidence intervals for the G values for the two treatments overlapped, the predicted G values can be used to generate approximate times to 90% equilibrium [Equation (69)]. The $t_{0.90}$ was shortened by cadmium exposure from 12 to 6 days.

IV. COMPOSITE INDICES

Species richness, heterogeneity (diversity), and evenness, and community similarity indices as described above are among the most commonly used indices in studies of community level effects of toxicants. Sheenan[59] tabulated these indices and described specific examples of their successful application. He also tabulated more specialized indices for pollution effects. The purpose of these indices is to include additional information from the community to gain further insight into shifts caused by presence of toxicant. They range in complexity from simple to complex indices requiring considerable expertise for accurate application. For illustrative purposes, several of these composite indices are discussed here.

Karr[122] defined the biological integrity of an ecological system as the ability to support and maintain "a balanced, integrated, adaptive community of organisms having a species composition, diversity, and functional organization comparable to that of natural habitat of the region." Indices for measuring biological integrity or some similar property have been developed by combining various community qualities. Most of these indices are calibrated to qualities of undisturbed similar systems in the same region, e.g., a low-gradient, warm-water stream in the midwestern United States.

Composite indices can be simple as those described by Gammon[123] and Maurer et al.[124] Gammon[123] bases his index for lotic systems on the assumption of high abundance and species heterogeneity (diversity) in healthy fish communities. His index of well-being (I_{wb}) is calculated with fish species and abundance data generated by electrofishing streams. High values of I_{wb} indicate a relatively healthy state.

$$I_{wb} = 0.5 \ln N + 0.5 \ln W + D_N + D_W \tag{71}$$

where N = the number of fish taken/km of stream; W = weight (kg) of fish taken/km of stream; D_N = Shannon's index based on number of individuals; and D_W = Shannon's index based on weight of fish.

Maurer et al.[124] generated a coefficient of pollution (p) applicable to coastal marine systems. First, the number of benthic species and total number of individuals expected in an area are estimated using empirical relationships between these metrics and sediment type and water depth. The expected number of species (g') and individuals (i_0) if the site were undisturbed are then combined with the observed number of species (g) and individuals (i) to generate a coefficient of pollution (p).

$$p = \frac{g'}{g\sqrt{\dfrac{i}{i_0}}} \tag{72}$$

Both of these simple indices draw on measures of total abundance and species heterogeneity (diversity). They are relative indices interpreted by comparison of values from suspect sites to those expected for unimpacted sites.

More complex indices are also common. The index of biological integrity (IBI) is a useful index that, in its original form, assesses impact using 12 attributes (metrics) of fish communities from warm-water, low-gradient, lotic systems.[122,125] These community metrics are grouped into three types: species richness and composition, trophic composition, and abundance and condition metrics. Species richness and composition metrics include the total number of species, number of darter species, number of sunfish species, number of sucker species, number of intolerant species, and proportion of the total number of individuals that are green sunfish (a species indicative of a disturbed system). Trophic composition metrics include the proportion of all individuals that are omnivores, insectivorous cyprinids, or piscivores. The abundance and condition metrics include the total number of individuals in the sample, proportion of the total number of fish that are hybrids, and proportion of the total number of individuals with signs of disease or other anomalies. Each metric is scored based on expected values for systems with various intensities of stress. The total of the scores for all 12 metrics are used to categorize the integrity of the system in question. These metrics were selected based on the assumptions listed below. (Summarized from Table 2 of Fausch et al.[126])

1. Native species richness declines with increased degradation.
2. Fish abundance decreases with increased degradation.
3. As degradation increases, the number of sensitive species drops and the number of tolerant species increases.
4. As degradation increases, species that are specialized feeders and top predators decline but the number of trophic generalists increases.
5. Siltation associated with stream degradation will decrease the availability of optimum spawning substrate with a consequent increase in species hybridization.
6. Incidences of infectious disease and abnormalities increase as degradation worsens.
7. Introduced species increase with degradation.

The overlap in metrics used in the IBI imparts a degree of robustness to the index.[126] As discussed, the IBI includes functional in addition to structural community qualities. It also combines qualities of individuals (e.g., incidence of disease), populations (e.g., incidence of hybridization), and the community (e.g., species abundance). These qualities enhance its value in reflecting overall community response.[126] Yoder[127] suggested from the relatively narrow coefficient of variation generated during its application that the IBI is as reliable as the bioassay and chemical analyses commonly used for environmental assessments. The enhanced realism associated with community level metrics offsets the necessity for considerable taxonomic and field expertise to generate the associated metrics and, if needed, to modify the index intelligently.

The IBI is modified before application to a particular ecological region.[126] Leonard and Orth[128] applied a modified IBI to a fish community of cool, high-gradient streams Karr and Dionne[129] developed an IBI for lakes and reservoirs. Karr and Kerans[130] discussed an invertebrate-based IBI, and Maxted[131] developed a similar type of index based on marine invertebrates.

V. TROPHIC EXCHANGE

In the 1960s, the conditional nature of the dilution paradigm ("the solution to pollution is dilution") became painfully clear. Fallout from atmospheric testing of nuclear weapons introduced low but detectable levels of radionuclides over vast regions of the globe. Some, like strontium-90 (^{90}Sr), accrued to discomfortingly high concentrations in foodstuffs. Epidemics of metal poisoning (Minimata and Itai-atai diseases) were also linked to accumulation of toxicants in food. After dissipation in the environment, DDT (dichloro-diphenyl-trichloroethane) reappeared at extremely high concentrations in species occupying upper trophic levels. *Silent Spring*[132] and articles published in broadly read journals (e.g., Woodwell[133,134]) presented this trophic "boomerang effect"[135] to the public as one of the first signs of a failing paradigm. The dilution paradigm was replaced as a central theme in environmental management, and the specter of toxicant transfer through trophic levels has become part of our collective knowledge. Unfortunately, as an important but favored child of our environmental conscience, the hypothesis of toxicant transfer to dangerous levels tends to be invoked dogmatically despite equally viable, alternative hypotheses.[6,136] (See ruling theories and precipitate explanation in Chapter 1.)

Because toxicants can enter aquatic organisms from both food and water, two terms are needed to distinguish between the associated processes. Bioconcentration is the increase in toxicant concentration in an organism resulting from direct uptake from water alone. As discussed in Chapter 3, bioconcentration is often quantified with the bioconcentration factor (BCF). In contrast, bioaccumulation is the increase in toxicant concentration in an organism resulting from the combined uptake from food and water. The term bioaccumulation is often applied in field surveys or other studies in which sources are poorly understood or vaguely defined. In our present discussion of community level phenomena, an additional term and associated methods must be described. Biomagnification (bioamplification) is the increase in toxicant concentration of organisms in successive trophic levels.[6] With biomagnification, the concentration of a toxicant increases with passage upward through the trophic chain or web. Often the implied consequence is damage to species at the highest trophic levels. Hunt and Bischoff[137] describe such a scenario involving DDD (dichloro-diphenyl-dichloroethane) spraying with fatal consequences for the western grebe (*Aechmophorus occidentalis*).

The degree to which a toxicant is transferred between successive trophic levels can be measured as the ratio of the toxicant concentrations in the two trophic levels of interest (the bioaccumulation factor or BF).[138,139] Implied in this term is the potential for significant accumulation from water as well as food. Other, very similar terms used in studies focusing only on food sources are biomagnification factor,[140] enrichment ratio,[141] and transfer factor.[135] Although normally presented in the following simple form, formulation of the BF can be more involved if many individuals with different weights are sampled from each trophic level.[140]

$$\text{BF} = \frac{C_{n+1}}{C_n} \tag{73}$$

where C_{n+1} = toxicant concentration in members of trophic level $n + 1$; and C_n =

toxicant concentration in members of the next lowest level, n. Biomagnification exists if BF > 1. Simple transfer occurs if BF $= 1$. Obviously, concentrations decrease (trophic dilution) if BF < 1. A BF can be estimated across several trophic levels as illustrated below with Ramade's formulations.[135] Here, accumulation from water is ignored. The concentration of a toxicant at trophic level 1 is estimated with the BF, and the concentration at trophic level 0 is estimated by rearranging Equation (73).

$$C_1 = BF_{(0,1)}C_0 \tag{74}$$

If uptake from water is assumed to be insignificant, the BF can be expressed in terms of the weight (g) of organisms in level 1 (b_1), weight (g) of individuals from level 0 consumed (a_1), fraction of the toxicant absorbed after ingestion (f_1), and daily fraction of the pollutant excreted at rate k_1.

The next transfer (trophic level 1 to trophic level 2) is described in these same terms.

$$BF_{(0,1)} = \frac{a_1 f_1}{b_1 k_1} \tag{75}$$

$$C_2 = BF_{(1,2)}C_1 \tag{76}$$

$$BF_{(1,2)} = \frac{a_2 f_2}{b_2 k_2} \tag{77}$$

The concentration of toxicant in trophic level 2 can be expressed in terms of the concentration in level 0 by simple substitution.

$$C_2 = \left[\frac{a_2 f_2}{b_2 k_2}\right]\left[\frac{a_1 f_1}{b_1 k_1}\right]C_0 \tag{78}$$

Expression of the concentration of toxicant at any trophic level based on the concentration in the lowest level can be generalized.

$$C_n = \left[\prod_{i=1}^{n} \frac{a_i f_i}{b_i k_i}\right]C_0 \tag{79}$$

Many detailed models of such trophic transfer for terrestrial (e.g., Holleman et al.[142]) and aquatic (e.g., Conover and Francis,[143] Patrick and Loutit[144]) systems exist. Equation (50) in Chapter 3 is one such simple model. Models are also common for toxicant transfers from both water and food sources (e.g., Spacie and Hamelink,[145] Thomann[159]).

$$C_{n+1} = \frac{k_{u(n+1)}}{k_{n+1}} C_w + \frac{\alpha_{n+1,n} R_{n+1,n}}{k_{n+1}} C_n \tag{80}$$

where C_w = concentration in the water; C_n = concentration in the nth trophic level (prey); C_{n+1} = concentration in the trophic level $n + 1$ (predator); $\alpha_{n+1,n}$ = assimilation efficiency (amount absorbed/amount ingested); $R_{n+1,n}$ = weight-specific ration or ingestion (amount ingested/g of total weight); $k_{n+1} = k_{e(n+1)} + g_{n+1}$; $k_{e(n+1)}$ = elimination rate constant for the predator; and g_{n+1} = growth rate constant for the predator.

Bioaccumulation to high concentrations of metals,[146,147] metalloids,[147] radionuclides,[148] and organic compounds[149] can occur at the lowest trophic levels; therefore, the potential still exists for adverse effects on consumers even in the absence of biomagnification. However, although there is reason for concern, consumption of contaminated food does not necessarily imply significant exposure because the toxicant must also be in a chemical form available for assimilation. For example, material accumulating on submerged surfaces and often defined as aufwuchs (periphyton) accumulates high concentrations of metals. But the bioavailability of the metals can be ambiguous because the metals can be associated primarily with hydrous oxides, not the periphytic microflora. Another example involves the significant incorporation of zinc into intracellular phosphate granules as a detoxification mechanism of molluscs. A significant portion of the zinc in these granules is not available for assimilation when the common periwinkle (*Littorina littorea*) is consumed by another snail (*Nassarius reticulatus*).[150]

Reichle and Van Hook[151] described biomagnification of essential elements along a terrestrial food chain and found that potassium, sodium, phosphorus, and nitrogen can exhibit biomagnification in a terrestrial food chain, but calcium is diluted at each transfer. After finding no biomagnification of cadmium, lead, or zinc in a terrestrial (arthropod) food chain, van Straalen and van Wensen[152] indicate that many other factors correlated with trophic level, including animal size and longevity, can render questionable the interpretation of biomagnification results. Predators tend to be larger and live longer than their prey. Consequently, they likely have different (size-dependent) rate constants and have longer times to accumulate metals than their prey. The complex mechanisms determining body concentration in each species can be more important than trophic status. Beyer[136] also suggested that the favored status given to biomagnification in many studies of terrestrial food chains results in inadequate consideration of alternative mechanisms such as size-related kinetics, longevity, and individual species variation in accumulation. Beyer[136] noted that the essential element zinc can display biomagnification in a zinc-deficient, terrestrial ecosystem but not in one with adequate zinc.

Davis and Foster[153] indicated that biomagnification in aquatic food chains occurs for some radionuclides of essential elements such as ^{32}P. But cadmium, copper, lead, and zinc show no biomagnification in aquatic food chains studied by Gächter and Geiger[154] or Cushing et al.[155] Bryan and Gibbs[156] indicate that tributyltin (TBT) shows no biomagnification during marine trophic exchanges. Wren et al.[157] examined a series of elements (Al, B, Ba, Be, Ca, Cd, Co, Cu, Fe, Hg, Mg, Mn, Mo, Ni, P, Pb, S, Sr, Ti, Zn) along an aquatic food chain and found that only mercury displayed biomagnification. Terhaar et al.[158] detected no biomagnification of silver in an artificial aquatic system although biomagnification of mercury was measured. Mercury biomagnification has also been documented or suggested by Potter et al.,[159] Boudou et al.,[160] and Wren and MacCrimmon.[161]

Moriarty[6] suggested that studies documenting biomagnification of organic compounds have also tended toward precipitate explanation and, as a consequence, do not represent critical evaluations of toxicant distribution among trophic levels. He indicated that species variation within a trophic level can be of sufficient magnitude to make casual documentation of biomagnification across trophic levels questionable. Connell[149] also suggested that the relative roles of bioconcentration (accumulation from water) and biomagnification are often unclear in studies of aquatic food chains.

Biomagnification of organic compounds depends on the amount of material consumed, the fraction of the ingested compound that is absorbed, the excretion rate, uptake from water, and the duration of exposure. The duration of exposure is correlated with longevity and, therefore, trophic status. Many constants can be animal size dependent, and animal size is correlated with trophic level. Generally, biomagnification is enhanced by increased species longevity and near proximity to the top of the food chain.[149] Finally, many of these qualities can be predicted with the K_{ow} of an organic compound. Connolly and Pederson[162] and Thomann[139] indicated that biomagnification would be significant if the log K_{ow} for a compound was greater than about 4 to 5.

VI. SUMMARY

In this overview of community-level effects of toxicants, commonly used and potentially useful techniques are described. Emphasis is provided in many sections in which underlying mechanisms are often neglected by ecotoxicologists. Basic models of predator-prey interactions and foraging theory are presented along with examples of their application. Models of interspecies competition and general discussion of symbiotic relationships are also presented to reinforce the importance of considering all aspects of a species niche integrity. Traditional and potentially useful measures of community structure and function are outlined along with details of their applications in ecotoxicology. Finally, the transfer of toxicants through trophic levels is briefly explored.

REFERENCES

1. Taub, F. B. "Standardized Aquatic Microcosm." *Environ. Sci. Technol.* 23: 1064–1066 (1989).
2. Hutchinson, G. E. "Concluding Remarks." *Cold Spring Harbor Symp. Quant. Biol.* 22: 415–427 (1957).
3. Wetzel, R. G. *Limnology.* (Philadelphia: Saunders College Publishing, 1983).
4. Green, R. H. "A Multivariate Statistical Approach to the Hutchinsonian Niche: Bivalve Molluscs of Central Canada." *Ecology* 52: 543–556 (1971).
5. Kettlewell, H. B. D. "Selection Experiments on Industrial Melanism in the Lapidoptera." *Heredity* 9: 323–342 (1955).
6. Moriarty, F. *Ecotoxicology: The Study of Pollutants in Ecosystems.* (New York: Academic Press, Inc., 1983).
7. Atchison, G. J., M. G. Henry, and M. B. Sanheinrich. "Effects of Metals on Fish Behavior: A Review." *Environ. Biol. Fish.* 18: 11–25 (1987).
8. Henry, M. G., and C. J. Atchison. "Metal Effects on Fish Behavior Advances in Determining the Ecological Significance of Responses." In *Metal Ecotoxicology: Concepts and Applications,* ed. M. C. Newman and A. W. McIntosh. (Chelsea, MI: Lewis Publishers, 1991).
9. Clements, W. H., J. H. Van Hassel, D. S. Cherry, and J. Cairns, Jr. "Colonization, Variability, and the Application of Substractum-Filled Trays for Biomonitoring Benthic Communities." *Hydrobiologia* 173: 45–53 (1989).
10. Schneider, M. J., S. A. Barraclough, R. G. Genoway, and M. L. Wolford. "Effects of Phenol on Predation of Juvenile Rainbow Trout *Salmo gairdneri.*" *Environ. Pollut. Ser. A. Ecol. Biol.* 23: 121–130 (1980).
11. Kania, H. J., and J. O'Hara. "Behavioral Alterations in a Simple Predator-Prey System due to Sublethal Exposure to Mercury." *Trans. Am. Fish. Soc.* 103: 134–136 (1974).
12. Goodyear, C. P. "A Simple Technique for Detecting Effects of Toxicants or Other Stresses on a Predator-Prey Interaction." *Trans. Am. Fish. Soc.* 101: 367–370 (1972).
13. Sullivan, J. F., G. J. Atchison, D. J. Kolar, and A. W. McIntosh. "Changes in the Predator-Prey Behavior of Fathead Minnows (*Pimephales promelas*) and Largemouth Bass (*Micropterus salmoides*) Caused by Cadmium." *J. Fish. Res. Board Can.* 35: 446–451 (1978).
14. Tagatz, M. E. "Effect of Mirex on Predator-Prey Interaction in an Experimental Estuarine Ecosystem." *Trans. Am. Fish. Soc.* 105: 546–549 (1976).

15. Rohlf, F. J., and R. R. Sokal. *Statistical Tables*. (New York: W. H. Freeman and Company, 1981).
16. Dixon, W. J., and F. J. Massey, Jr. *Introduction to Statistics*. (New York: McGraw-Hill Book Company, 1969).
17. Emmel, T. C. *Population Biology* (New York: Harper & Row, 1976).
18. Smith, R. L. *Ecology and Field Biology*. (New York: Harper & Row Publishers, 1980).
19. May, R. M. *Theoretical Ecology. Principles and Applications*. (Philadelphia: W. B. Saunders Company, 1976).
20. Gause, G. F. *The Struggle for Existence*. (Baltimore: William and Wilkins, 1934).
21. Huffaker, C. B. "Experimental Studies on Predation: Dispersion Factors and Predator-Prey Oscillations." *Hilgardia* 27: 343–383 (1958).
22. May, R. M. *Stability and Complexity in Model Ecosystems*. (Princeton: Princeton University Press, 1974).
23. Holling, C. S. "The Components of Predation as Revealed by a Study of Small-Mammal Predation of the European Pine Sawfly." *Can. Entomol.* 91: 293–320 (1959).
24. Poole, R. W. *An Introduction to Quantitative Ecology*. (New York: McGraw-Hill Book Company, 1974).
25. Woltering, D. M., J. L. Hedtke, and L. J. Weber. "Predator-Prey Interactions of Fishes Under the Influence of Ammonia." *Trans. Am. Fish. Soc.* 107: 500–504 (1978).
26. Stephens, D. W., and J. R. Krebs. *Foraging Theory*. (Princeton: Princeton University Press, 1986).
27. Sandheinrich, M. B., and G. J. Atchison. "Sublethal Toxicant Effects on Fish Foraging Behavior: Empirical vs. Mechanistic Approaches." *Environ. Toxicol. Chem.* 9: 107–119 (1990).
28. Gilpin, M. E., and F. J. Ayala. "Global Models of Growth and Competition." *Proc. Natl. Acad. Sci. USA* 70: 3590–3593 (1973).
29. Gause, G. F., and A. A. Witt. "Behavior of Mixed Populations and the Problem of Natural Selection." *Am. Nat.* 69: 596–609 (1935).
30. Vandermeer, J. H. "The Community Matrix and the Number of Species in a Community." *Am. Nat.* 104: 73–83 (1970).
31. Ayala, F. J. "Experimental Invalidation of the Principle of Competitive Exclusion." *Nature* 224: 1076–1079 (1969).
32. Gilpin, M. E., and K. E. Justice. "Reinterpretation of the Invalidation of the Principle of Competitive Exclusion." *Nature* 236: 273–301 (1972).
33. Vandermeer, J. H. "The Competitive Structure of Communities: An Experimental Approach with Protozoa." *Ecology* 50: 362–371 (1969).
34. Gilpin, M. E. "Limit Cycles in Competition Communities." *Am. Nat.* 109: 51–60 (1975).
35. MacArthur, R. H. "Population Ecology of Some Warblers of Northeastern Coniferous Forests." *Ecology* 39: 599–619 (1958).
36. Connell, J. H. "The Influence of Interspecific Competition and Other Factors on the Distribution of the Barnacle *Chthamalus stellatus*." *Ecology* 42: 710–723 (1961).
37. Levins, R. "Evolution in Changing Environments: Some Theoretical Explorations." *Monogr. Popul. Biol.* 2 (1968).
38. Krebs, C. J. *Ecological Methodology*. (New York: Harper Collins Publishers, 1989).
39. Hurlbert, S. H. "The Measurement of Niche Overlap and Some Relatives." *Ecology* 59: 67–77 (1978).
40. Sarokin, D., and J. Schulkin. "The Role of Pollution in Large-Scale Population Disturbances, Part 1: Aquatic Populations." *Environ. Sci. Technol.* 26: 1476–1484 (1992).
41. Schindler, D. W., K. H. Mills, D. F. Malley, D. L. Findlay, J. A. Shearer, I. J. Davies, M. A. Turner, G. A. Linsey, and D. R. Cruikshank. "Long-term Ecosystem Stress:

The Effects of Years of Experimental Acidification on a Small Lake." *Science* 228: 1395–1401 (1985).

42. Snarski, V. M. "The Response of Rainbow Trout *Salmo gairdneri* to *Aeromonas hydrophila* after Sublethal Exposures to PCB and Copper." *Environ. Pollut.* 28: 219–232 (1982).

43. Ewing, M. S., S. A. Ewing, and M. A. Zimmer. "Sublethal Copper Stress and Susceptibility of Channel Catfish to Experimental Infections with *Ichthyophthirius multifilis.*" *Bull. Environ. Contam. Toxicol.* 28: 674–681 (1982).

44. Evans, N. A. "Effect of Copper and Zinc upon the Survival and Infectivity of *Echinoparyphium recurvatum* Cercariae." *Parasitology* 85: 295–303 (1982).

45. Evans, N. A. "Effects of Copper and Zinc on the Life Cycle of *Notocotylus attenuatus* (Digenea: Notocotylidae)." *Int. J. Parasitol.* 12: 363–369 (1982).

46. Benson, A. A., and R. E. Summons. "Arsenic Accumulation in Great Barrier Reef Invertebrates." *Science* 211: 482–483 (1981).

47. Pacala, S. W., M. P. Hassell, and R. M. May "Host-Parasite Associations in Patchy Environments." *Nature* 344: 150–153 (1990).

48. Suter, G. W., III. *Ecological Risk Assessment.* (Chelsea, MI: Lewis Publishers, 1993).

49. Cairns, J., Jr., and B. R. Niederlehner. "Adaptation and Resistance of Ecosystems to Stress: A Major Knowledge Gap in Understanding Anthropogenic Perturbations." *Speculations Sci. Technol.* 12: 23–30 (1989).

50. Wilbur, H. M. "Regulation of Structure in Complex Systems: Experimental Temporary Pond Communities." *Ecology* 68: 1437–1452 (1987).

51. Crossland, N. O., and T. W. La Point. "The Design of Mesocosm Experiments." *Environ. Toxicol. Chem.* 11: 1–4 (1992).

52. Liber, K., N. K. Kaushik, K. R. Solomon, and J. H. Carey. "Experimental Designs for Aquatic Mesocosm Studies: Comparison of the 'ANOVA' and 'Regression' Design for Assessing the Impact of Tetrachlorophenol on Zooplankton Populations in Limnocorrals." *Environ. Toxicol. Chem.* 11: 61–77 (1992).

53. Kimball, K. D., and S. A. Levin. "Limitations of Laboratory Bioassays: The Need for Ecosystem Level Testing." *Bioscience* 35: 165–171 (1985).

54. Perry, J. A., and N. H. Troelstrup, Jr. "Whole Ecosystem Manipulation: A Productive Avenue for Test System Research?" *Environ. Toxicol. Chem.* 7: 941–951 (1988).

55. Odum, E. P. "Trends Expected in Stressed Ecosystems." *Bioscience* 35: 419–422 (1985).

56. Rapport, D. J., H. A. Regier, and T. C. Hutchinson. "Ecosystem Behavior under Stress." *Am. Nat.* 125: 617–640 (1985).

57. Cairns, J., Jr. "Heated Waste-Water Effects on Aquatic Ecosystems." In *Thermal Ecology II,* ed. G. W. Esch and R. W. McFarlane. (Springfield, VA: National Technical Information Center, 1976).

58. Cairns, J., Jr. "Quantification of Biological Integrity." In *The Integrity of Water,* ed. R. K. Ballentine and L. J. Guarraia. (Washington, DC: U.S. EPA. Office of Water and Hazardous Materials, 1977).

59. Sheenhan, P. J. "Effect on Community and Ecosystem Structure and Dynamics." In *Effects of Pollutants at the Ecosystem Level,* ed. P. J. Sheenan, D. R. Miller, G. C. Butler, and P. Bourdeau. (New York: John Wiley & Sons, Ltd., 1984).

60. Kersting, K. "Normalizing Ecosystem Strain: A System Parameter for Analysis of Toxic Stress in (Micro-) Ecosystems." *Ecol. Bull.* 36: 150–153 (1984).

61. Sokal, R. R., and F. J. Rohlf. *Biometry. The Principles and Practice of Statistics in Biological Research.* (New York: W. H. Freeman and Company, 1981).

62. Draper, N. R., and H. Smith. *Applied Regression Analysis.* (New York: John Wiley & Sons, Inc., 1981).

63. Neter, J., W. Wasserman, and M. H. Kutner. *Applied Linear Statistical Models.*

Regression, Analysis of Variance, and Experimental Design. (Homeland, IL: Irwin, 1990).

64. Paine, R. T. "Food Web Complexity and Species Diversity." *Am. Nat.* 100: 65–75 (1966).

65. Paine, R. T., and R. L. Vadas. "The Effects of Grazing by Sea Urchins, *Strongylocentrotus* spp., on Benthic Algal Population." *Limnol. Oceanogr.* 14: 710–719 (1969).

66. Osman, R. W., and J. A. Haugeness. "Mutualism among Sessile Invertebrates: A Mediator of Competition and Predation." *Science* 211: 846–848 (1981).

67. Pielou, E. C. *An Introduction to Mathematical Ecology.* (New York: John Wiley & Sons, 1969).

68. Smith, R. L. *Elements of Ecology.* (New York: Harper & Row Publishers, 1986).

69. Magurran, A. E. *Ecological Diversity and Its Measurement.* (Princeton: Princeton University Press, 1988).

70. Ludwig, J. A., and J. F. Reynolds. *Statistical Ecology. A Primer on Methods and Computing.* (New York: John Wiley & Sons, 1988).

71. Preston, F. W. "The Commonness, and Rarity, of Species." *Ecology* 29: 254–283 (1948).

72. Preston, F. W. "The Canonical Distribution of Commonness and Rarity: Part 1." *Ecology* 43: 185–215 (1962).

73. Cohen, A. C., Jr. "Tables for Maximum Likelihood Estimates: Singly Truncated and Singly Censored Samples." *Technometrics* 3: 535–541 (1961).

74. MacArthur, R. H. "On the Relative Abundance of Bird Species." *Proc. Natl. Acad. Sci. USA* 43: 293–295 (1957).

75. Patrick, R. "Use of Algae, Especially Diatoms in the Assessment of Water Quality." *ASTM (Am. Soc. Test. Mater.) Spec. Tech. Bull.* 528: 76–95 (1973).

76. Pielou, E. C. *Population and Community Ecology. Principles and Methods.* (New York: Gordon and Breach Science Publishers, 1974).

77. Gleason, H. A. "On the Relation between Species and Area." *Ecology* 3: 158–162 (1922).

78. James, F. C., and S. Rathbun. "Rarefaction, Relative Abundance, and Diversity of Avian Communities." *Auk* 98: 785–800 (1981).

79. Hurlbert, S. H. "The Non-concept of Species Diversity: A Critique and Alternative Parameters." *Ecology* 52: 577–586 (1971).

80. Beyer, W. H. *CRC Standard Probability and Statistics Tables and Formulae.* (Boca Raton: CRC Press, 1991).

81. Heck, K. L. J., G. van Belle, and D. Simberloff. "Explicit Calculation of the Rarefaction Diversity Measurement and the Determination of Sufficient Sample Size." *Ecology* 56: 1459–1461 (1975).

82. Peet, R. K. "The Measurement of Species Diversity." *Annu. Rev. Ecol. Syst.* 5: 285–307 (1974).

83. Heltshe, J. F., and N. E. Forrester. "Estimating Species Richness Using the Jackknife Procedure." *Biometrics* 39: 1–11 (1983).

84. Smith, E. P., and G. van Belle. "Nonparametric Estimation of Species Richness." *Biometrics* 40: 119–129 (1984).

85. Manly, B. F. J. *RT. A Program for Randomization Testing.* (Cheyenne, WY: West, Inc., 1402 S. Greeley Highway, 82007; 1992).

86. Bruce, P. C. *RESAMPLING STATS. Probability and Statistics a Radically Different Way. User Guide.* (Arlington, VA: Resampling Stats, Inc., 1992).

87. Mingoti, S. A., and G. Meeden. "Estimating the Total Number of Distinct Species Using Presence and Absence Data." *Biometrics* 48: 863–875 (1992).

88. Niederlehner, B. R., J. R. Pratt, A. L. Buikema, Jr., and J. Cairns, Jr. "Comparison of Estimates of Hazard Derived at Three Levels of Complexity." In *Community*

Toxicity Testing. ASTM STP 920, ed. J. Cairns, Jr. (Philadelphia: American Society for Testing and Materials, 1986).

89. Simpson, E. H. "Measurement of Diversity." *Nature* 163: 688 (1949).
90. Heltshe, J. F., and N. E. Forrester. "Statistical Evaluation of the Jackknife Estimate of Diversity When Using Quadrat Samples." *Ecology* 66: 107–111 (1985).
91. Kaesler, R. J., and E. E. Herricks. "Analysis of Data from Biological Surveys of Streams: Diversity and Sample Size." *Water Resour. Bull.* 12: 125–135 (1976).
92. Alatalo, R. V. "Problems in the Measurement of Evenness in Ecology." *Oikos* 37: 199–204 (1981).
93. Hill, M. O. "Diversity and Evenness: A Unifying Notation and Its Consequences." *Ecology* 54: 427–432 (1973).
94. Towner, H. *ECOSTAT. An Ecological Analysis Program. User's Manual.* (Campton, NH: Trinity Software, 1992).
95. Pontasch, K. W., E. P. Smith, and J. Cairns, Jr. "Diversity Indices, Community Comparison Indices and Canonical Discriminant Analysis: Interpreting the Results of Multispecies Toxicity Tests." *Water Res.* 23: 1229–1238 (1989).
96. Snoeijs, P. J. M. "Monitoring Pollution Affects by Diatom Community Composition. A Comparison of Sampling Methods." *Arch. Hydrobiol.* 121: 497–510 (1991).
97. Wolda, H. "Similarity Indices, Sample Size and Diversity." *Oecologia* 50: 296–302 (1981).
98. Matthews, R. A., A. L. Buikema, Jr., J. Cairns, Jr., and J. H. Rodgers, Jr. "Biological Monitoring, Part IIA: Receiving System Functional Methods. Relationships and Indices." *Water Res.* 16: 129–139 (1982).
99. Sheenhan, P. J. "Functional Changes in the Ecosystem." In *Effects of Pollutants at the Ecosystem Level,* ed. P. J. Sheenan, D. R. Miller, G. C. Butler, and P. Bourdeau. (New York: John Wiley & Sons, Ltd., 1984).
100. Vannote, R. L., G. W. Minshall, K. W. Cummins, J. R. Sedell, and C. E. Cushing. "The River Continuum Concept." *Can. J. Fish. Aquat. Sci.* 37: 130–137 (1980).
101. Kondratieff, P. F., R. A. Matthews, and A. L. Buikema, Jr. "A Stressed Stream Ecosystem: Macroinvertebrate Community Integrity and Microbial Trophic Response." *Hydrobiologia* 111: 81–91 (1984).
102. Blanck, H. "A Simple, Community Level, Ecotoxicological Test Using Samples of Periphyton." *Hydrobiologia* 124: 251–261 (1985).
103. Goldsborough, L. G., and G. G. C. Robinson. "Changes in Periphytic Algal Community Structure as a Consequence of Short Herbicide Exposures." *Hydrobiologia* 139: 177–192 (1986).
104. Amblard, C., P. Couture, and G. Bourdier. "Effects of a Pulp and Paper Mill on the Structure and Metabolism of Periphytic Algae in Experimental Streams." *Aquat. Toxicol.* 18: 137–162 (1990).
105. Day, K. E. "Short-Term Effects of Herbicides on Primary Productivity of Periphyton in Lotic Environments." *Ecotoxicology* 2: 123–138 (1993).
106. Webster, J. R. and E. F. Benfield. "Vascular Plant Breakdown in Freshwater Ecosystems." *Annu. Rev. Ecol. Syst.* 17: 567–594 (1986).
107. Giesy, J. P., Jr. "Cadmium Inhibition of Leaf Decomposition in an Aquatic Microcosm." *Chemosphere* 6: 467–475 (1978).
108. Mulholland, P. J., A. V. Palumbo, J. W. Elwood and A. D. Rosemond. "Effects of Acidification on Leaf Decomposition in Streams." *J. North Am. Benthol. Soc.* 6: 147–158 (1987).
109. Schoenberg, S. A., R. Benner, A. Armstrong, P. Sobecky, and R. E. Hodson. "Effects of Acid Stress on Aerobic Decomposition of Algal and Aquatic Macrophyte Detritus: Direct Comparison in a Radiocarbon Assay." *Appl. Environ. Microbiol.* 56: 237–244 (1990).

110. Allred, P. M., and J. P. Giesy. "Use of in Situ Microcosms to Study Mass Loss and Chemical Composition of Leaf Litter Being Processed in a Blackwater Stream." *Arch. Hydrobiol.* 114: 231–250 (1988).

111. Minderman, G. "Addition, Decomposition and Accumulation of Organic Matter in Forests." *J. Ecol.* 56: 355–362 (1968).

112. Cairns, J., Jr., and J. R. Pratt. "Developing a Sampling Strategy." In *Rationale for Sampling and Interpretation of Ecological Data in the Assessment of Freshwater Ecosystems,* ASTM STP 894 ed. B. G. Isom. (Philadelphia: American Society for Testing and Materials, 1986).

113. Elwood, J. W., J. D. Newbold, R. V. O'Neill, and W. Van Winkle. "Resource Spiraling: An Operational Paradigm for Analyzing Lotic Ecosystems.". In *Dynamics of Lotic Ecosystems,* ed., T. D. Fontaine and S. M. Bartell. (Ann Arbor, MI: Ann Arbor Science, 1983).

114. Elwood, J. W., J. D. Newbold, A. F. Trimble, and R. W. Stark. "The Limiting Role of Phosphorus in a Woodland Stream Ecosystem: Effects of P Enrichment on Leaf Decomposition and Primary Producers." *Ecology* 62: 146–158 (1981).

115. Newbold, J. D., R. V. O'Neill, J. W. Elwood, and W. Van Winkle. "Nutrient Spiraling in Streams: Implications for Nutrient Limitation and Invertebrate Activity." *Am. Nat.* 120: 628–652 (1982).

116. Cairns, J., Jr., and J. R. Pratt. "Multispecies Toxicity Testing Using Indigenous Organisms: A New Cost-Effective Approach to Ecosystem Protection." In *1985 Environmental Conference, TAPPI Proceedings,* ISSN 0272-7269. (Atlanta, GA: TAPPI Press, 1985).

117. Niederlehner, B. R., J. R. Pratt, A. L. Buikema Jr., and J. Cairns, Jr. "Laboratory Tests Evaluating the Effects of Cadmium on Freshwater Protozoan Communities." *Environ. Toxicol. Chem.* 4: 155–165 (1985).

118. Pratt, J. R., and J. Cairns, Jr. "Long-Term Patterns of Protozoan Colonization in Douglas Lake, Michigan." *J. Protozool.* 32: 95–99 (1985).

119. Cairns, J., Jr., J. R. Pratt, B. R. Niederlehner, and P. V. McCormick. "A Simple Cost-Effective Multispecies Toxicity Test Using Organisms with a Cosmopolitan Distribution." *Environ. Monit. Assess.* 6: 207–220 (1986).

120. MacArthur, R. H., and E. O. Wilson. *The Theory of Island Biogeography.* (Princeton: Princeton University Press, 1967).

121. Simberloff, D. S., and E. O. Wilson. "Experimental Zoogeography of Islands: The Colonization of Empty Islands." *Ecology* 50: 278–296 (1969).

122. Karr, J. R. "Biological Integrity: A Long-Neglected Aspect of Water Resource Management." *Ecol. Applications* 1: 66–84 (1991).

123. Gammon, J. R. "Biological Monitoring in the Wabash River and Its Tributaries." In *Water Quality Standards for the 21st Century.* (Washington, DC: U.S. EPA Criteria and Standards Division, 1989).

124. Maurer, D., G. Robertson, and I. Haydock. "Coefficient of Pollution (p): The Southern California Shelf and Some Ocean Outfalls." *Mar. Pollut. Bull.* 22: 141–148 (1991).

125. Karr, J. R., K. D. Fausch, P. L. Angermeier, P. R. Yant, and I. J. Schlosser. *Assessing Biological Integrity in Running Waters. A Method and Its Rationale,* Illinois Natural History Survey Special Publication 5. (Champagne, IL: Illinois Natural History Survey, 1986).

126. Fausch, K. D., J. Lyons, J. R. Karr, and P. L. Angermeier. "Fish Communities as Indicators of Environmental Degradation." *Am. Fish. Soc. Symp.* 8: 123–144 (1990).

127. Yoder, C. O. "Answering Some Concerns about Biological Criteria Based on Experiences in Ohio." In *Water Quality Standards for the 21st Century,* (Washington, DC : U.S. EPA Criteria and Standards Division, 1989).

128. Leonard, P. M., and D. J. Orth. "Application and Testing of an Index of Biotic Integrity in Small, Coolwater Streams." *Trans. Am. Fish. Soc.* 115: 401–414 (1986).

129. Karr, J. R., and M. Dionne. "Designing Surveys to Assess Biological Integrity in Lakes and Reservoirs." In *U.S. EPA. Biological Criteria: Research and Regulation,* EPA-440/5-91-005. (Washington, DC: Office of Water, U.S. EPA, 1991).

130. Karr, J. R., and B. L. Kerans. "Components of Biological Integrity: Their Definition and Use in Development of an Invertebrate IBI." In *Proceedings of the 1991 Midwest Pollution Control Biologists Meeting,* EPA-905/R-92/003, ed. W. S. Davis and T. P. Simon. (Chicago: U.S. EPA, Region V, Instream Biological Criteria and Ecological Assessment Committee, 1992).

131. Maxted, J. R. "The Development of Biocriteria in Marine and Estuarine Waters in Delaware." In *Water Quality Standards for the 21st Century.* (Washington, DC: U.S. EPA Criteria and Standards Division, 1989).

132. Carson, R. *Silent Spring.* (Boston, MA: Houghton Muffin, 1962).

133. Woodwell, G. M. "Toxic Substances and Ecological Cycle." *Sci. Am.* 216: 24–31 (1967).

134. Woodwell, G. M. "DDT Residues in an East Coast Estuary: A Case of Biological Concentration of a Persistent Insecticide." *Science* 156: 821–823 (1967).

135. Ramade, F. *Ecotoxicology.* (New York: John Wiley & Sons, 1987).

136. Beyer, W. N. "A Reexamination of Biomagnification of Metals in Terrestrial Food Chains." *Environ. Toxicol. Chem.* 5: 863–864 (1986).

137. Hunt, E. G., and A. I. Bischoff. "Inimical Effects on Wildlife of Periodic DDD Applications to Clear Lake." *Calif. Fish Game* 46: 91–106 (1960).

138. Walker, C. H. "Kinetic Models for Predicting Bioaccumulation of Pollutants in Ecosystems." *Environ. Pollut.* 44: 227–240 (1987).

139. Thomann, R. V. "Bioaccumulation Model of Organic Chemical Distribution in Aquatic Food Chains." *Environ. Sci. Technol.* 23: 699–707 (1989).

140. Laskowski, R. "Are the Top Predators Endangered by Heavy Metal Biomagnification?" *Oikos* 60: 387–390 (1991).

141. Paasivirta, J. *Chemical Ecotoxicology.* (Chelsea, MI: Lewis Publishers, 1991).

142. Holleman, D. F., J. R. Luick, and F. W. Whicker. "Transfer of Radiocesium from Lichen to Reindeer." *Health Phys.* 21: 657–666 (1971).

143. Conover, R. J., and V. Francis. "The Use of Radioactive Isotopes to Measure the Transfer of Materials in Aquatic Food Chains." *Mar. Biol. (Berl.)* 18: 272–283 (1973).

144. Patrick, F. M., and M. W. Loutit. "Passage of Metals to Freshwater Fish from Their Food." *Water Res.* 12: 395–398 (1978).

145. Spacie, A., and J. L. Hamelink. "Bioaccumulation." In *Fundamental of Aquatic Toxicology,* ed. G. M. Rand and S. R. Petrocelli. (New York: Hemisphere Publishing Corp., 1985).

146. Newman, M. C., A. W. McIntosh, and V. A. Greenhut. "Geochemical Factors Complicating the Use of Aufwuchs as a Biomonitor for Lead Levels in Two New Jersey Reservoirs." *Water Res.* 17: 625–630 (1983).

147. Newman, M. C., J. J. Alberts, and V. A. Greenhut. "Geochemical Factors Complicating the Use of Aufwuchs to Monitor Accumulation of Arsenic, Cadmium, Chromium, Copper, and Zinc." *Water Res.* 19: 1157–1165 (1985).

148. Neal, E. C., B. C. Patten, and C. E. DePoe. "Periphyton Growth on Artificial Substrates in a Radioactively Contaminated Lake." *Ecology* 48: 918–924 (1967).

149. Connell, D. W. *Bioaccumulation of Xenobiotic Compounds.* (Boca Raton, FL: CRC Press, Inc., 1990).

150. Nott, J. A., and A. Nicolaidou. "Bioreduction of Zinc and Manganese Along a Molluscan Food Chain." *Comp. Biochem. Physiol.* 104A: 235–238 (1993).

151. Reichle, D. E., and R. I. Van Hook, Jr., "Radionuclide Dynamics in Insect Food Chains." *Manit. Entomol.* 4: 22–32 (1970).

152. van Straalen, N. M., and J. van Wensem. "Heavy Metal Content of Forest Litter Arthropods as Related to Bodysize and Trophic Level." *Environ. Pollut.* 42: 209–221 (1986).

153. Davis, J. J., and R. F. Foster. "Bioaccumulation of Radioisotopes through Aquatic Food Chains." *Ecology* 39: 530–535 (1958).

154. Gächter, R., and W. Geiger. "MELIMEX, an Experimental Heavy Metal Pollution Study: Behavior of Heavy Metals in an Aquatic Food Chain." *Schweiz. Z. Hydrol.* 41: 277–290 (1979).

155. Cushing, C. E., D. G. Watson, A. J. Scott, and J. M. Gurtisen. "Decrease of Radionuclides in Columbia River Biota Following Closure of the Hanford Reactors." *Health Phys.* 41: 59–67 (1981).

156. Bryan, G. W., and P. E. Gibbs. "Impact of Low Concentrations of Tributyltin (TBT) on Marine Organisms: A Review." In *Metal Ecotoxicology. Concepts and Applications,* ed. M. C. Newman and A. W. McIntosh. (Chelsea, MI: Lewis Publishers, 1991).

157. Wren, C. D., H. R. MacCrimmon, and B. R. Loescher. "Examination of Bioaccumulation and Biomagnification of Metals in a Precambrian Shield Lake." *Water Air Soil Pollut.* 19: 277–291 (1983).

158. Terhaar, C. J., W. S. Ewell, S. P. Dziuba, W. W. White, and P. J. Murphy. "A Laboratory Model for Evaluating the Behavior of Heavy Metals in an Aquatic Environment." *Water Res.* 11: 101–110 (1977).

159. Potter, L., D. Kidd, and D. Standiford. "Mercury Levels in Lake Powell. Bioamplification of Mercury in Man-made Desert Reservoir." *Environ. Sci. Technol.* 9: 41–46 (1975).

160. Boudou, A., A. Delarche, F. Ribeyre, and R. Larty. "Bioaccumulation and Bioamplification of Mercury Compounds in a Second Level Consumer, *Gambusia affinis*—Temperature Effects." *Bull. Environ. Contam. Toxicol.* 22: 813–818 (1979).

161. Wren, C. D., and H. R. MacCrimmon. "Comparative Bioaccumulation of Mercury in Two Adjacent Freshwater Ecosystems." *Water Res.* 20: 763–769 (1986).

162. Connolly, J. P., and C. J. Pedersen. "A Thermodynamic-based Evaluation of Organic Chemical Accumulation in Aquatic Organisms." *Environ. Sci. Technol.* 22: 99–103 (1988).

Chapter 8

Summary

But if I abstain from giving my judgement on any thing when I do not perceive it with sufficient clearness and distinction, it is plain that I act rightly and am not deceived. But if I determine to deny or affirm, I no longer make use as I should of my free will, . . . understanding should always precede the determination of the will. And it is in the misuse of the free will that the privation which constitutes the characteristic nature of error is met with.[1]

Science without utility is intellectual vanity.[2]

I. APPLICATION TO ECOLOGICAL ASSESSMENT

How are the concepts and methods described in the previous chapters bent to assessment of ecological effects? Immediately upon posing this question, the too familiar dilemma exemplified by the quotes above presents itself. Socially mandated but logically untenable assessments must be made without an adequate understanding of the system being protected. Consequently, the best possible assessments include clear statements of uncertainties associated with conclusions, the underlying assumptions, and the probability of alternative conclusions being true. Fortunately, considerable effort is now being expended to meet these requirements. Several recent publications[3-7] have provided associated details. The very brief discussion below highlights a few of the pertinent concepts as they provide a pragmatic context for the subject of this book.

In ecological assessment, formal distinction is made between the terms, hazard and risk. Ecological hazard assessment involves "comparing the expected environmental concentration (EEC) and the estimated threshold effect (ETT) and making a judgement as to whether the proposed release is safe, hazardous, or not sufficiently characterized for a conclusion to be reached."[7] Huggett et al.[8] provided an excellent illustration of such a hazard assessment for tributyltin. As indicated by Suter,[7] Cairns was responsible for developing the paradigm of hazard assessment used today to assess the separation between EEC and ETT. A sequence of testing tiers have been established. The first tier of testing results in a certain, often low, level of confidence in the separation or lack of separation between EEC and ETT. The sequence of tiers of increasingly sensitive tests is continued until the confidence in the separation between the EEC and ETT is sufficient to provide a qualitative judgment of potential hazard. The tests in the first tier(s) are often associated with lower levels of organization, i.e., LC50 or individual-based no observed effect concentration (NOEC). Indeed, quantitative methods exist assuming that tests at low levels of ecological organization adequately predict effects at higher orders, e.g. population effects predict community effects.[9-11] Cairns[12] argued aggressively that there should be simultaneous testing of several levels of ecological organization at each tier. He argued that effects at lower levels, although grossly indicative of the probability of effects at higher levels (e.g., Slooff et al.[13]), lacked sufficient predictive potential to warrant their exclusive use in the first tiers of testing.

Ecological risk assessment is similar to hazard assessment but uses "the available toxicological and ecological information to estimate the probability of some undesired ecological event will occur."[6] [Unlike formal quantitative statements of risk that also include the severity of effects (e.g., radiological risk as described by Lindell[14]), present ecological risk assessment seems to consider only the probability of an effect.]

There are two general types of risk assessment.[5] Predictive risk assessment estimates risk of a proposed action, and retrospective risk assessment quantifies the adverse effects that have already occurred. Both use a slightly different balance of the same techniques. More than the retrospective risk assessment, the predictive risk assessment tends to follow closely the National Academy of Sciences (NAS) paradigm described below.[5] Retrospective assessments such as those associated with hazardous waste sites rely more on documentation of chemical contamination, surveys of ecological impact, and causal linkage of observed contamination and ecological impact using laboratory toxicity tests.[4]

The NAS paradigm of risk assessment has been adopted for ecological risk assessment.[3,5] This paradigm includes four basic components: hazard identification, dose-response assessment, exposure assessment, and risk characterization. In the first component, the ecological quality or entity that may be at risk is identified, the endpoints for determining effect are explicitly stated, and data requirements are formally stated.[3,5] Two types of endpoints are used. Assessment endpoints (also called receptors) are the explicitly defined ecological qualities to be protected (e.g., the rainbow trout population in a stream), and measurement endpoints are the measured qualities used to assess effect (e.g., reduction in reproductive success or growth) on the assessment endpoint.[4,5] Sometimes assessment endpoints are also measurement endpoints.[7] Incorporation of a clear system of effect qualifiers as outlined in Chapter 1 is needed during hazard identification, because statistically significant change in measurement endpoints, especially those associated with community and ecosystem levels, are used dogmatically to judge the occurrence of an adverse effect for an assessment endpoint. Assuming other factors are equal between candidates, those measurement endpoints constituting the clearest indication of adverse effect should be favored.

The methods described in this book contribute most to the next two components of ecological risk assessment (assessments of exposure and dose-response). Quantitative methods are particularly pertinent, because both exposure and dose-response assessment entail clear statements of the confidence in any associated relationships. Obviously, exposure assessment is compromised without a clear statement of chemical data quality (Chapter 2). Because exposure involves estimation of availability, sections of Chapter 3 relative to prediction errors and bioaccumulation model selection are also pertinent. Similarly, dose-response assessment would be compromised by poor experimental design or selection of methods with low statistical power. Chapters 3 through 7 present many of the most appropriate endpoints for dose-response assessment at the individual, population, community, and ecosystem levels of organization.

The final step, risk characterization, combines the data on estimated exposure and dose-response to express the likelihood or probability of an adverse effect.[3] Unfortunately, high uncertainty in data or methods can preclude expression of this risk as a probability.[5] With this information and a clear discussion of the associated uncertainties, a statement of risk is made.

II. FACILITATING GROWTH OF THE SCIENCE

The underlying purpose in the preceding chapters is to detail logical and quantitative means for accelerating the growth of ecotoxicology and, consequently, the effectiveness of ecological assessment. First, a scientific context for ecotoxicology is developed and the concept of stress qualified. The commonly held belief that progress in the field is slow because of the complexity of ecological systems is rejected. Not only is this belief inconsistent with rapid progress noted in equally complex areas of study, it imbues a hobbling passivity to our movement toward a solid knowledge base. Why strive for more rigor if ecosystems can never be understood sufficiently? Instead a tradition of weak inference is presented as a major, yet resolvable, impediment to progress in the field.

Evidence supporting this proposition has been provided by John Cairns's observations that many ecotoxicological approaches are used today more as a consequence of the history of the field than of their scientific merit,[15] and that a more vigorous falsification process is needed in the field.[16] To this end, guidelines for the growth of a strong inferential approach in the field are presented in Chapter 1.

Beginning with Chapter 2, quantitative methods for addressing ecotoxicological concerns are detailed. The topics are arranged in order of increasing ecological organization. In Chapter 2, the necessary steps to control and define the quality of measurements are presented, because inaccurate or imprecise data render results from any subsequent tests useless at best. In Chapters 3 to 7, familiar and unfamiliar quantitative methods are illustrated with numerical examples. Details for the application of these quantitative methods dominate most of this book as quantitative methods are most amenable to the falsification process,[17] i.e., foster strong inference. Each chapter aims to enhance the use of presently accepted quantitative methods and to introduce nontraditional methods. Particularly in the last chapters, the nontraditional methods include many that have been firmly established for decades in other disciplines, yet are inexplicably ignored or deemed too esoteric by ecotoxicologists. Indeed, many of the descriptions of such techniques in this volume are unelaborated composites of those outlined in basic textbooks of ecological (e.g., community level metrics in Chapter 7) or toxicological (e.g., survival time models in Chapter 4) methods.

REFERENCES

1. Descartes, R. *A Discourse on Method and Other Works.* Translated by E. S. Haldane and G. R. T. Ross. (New York: Washington Square Press, Inc., 1965).
2. Johnson, S. *The Complete Works of Samuel Johnson.* (Troy, NY: Pafraets Book Company, 1750; reprinted 1903).
3. Norton, S., M. McVey, J. Colt, J. Durda, and R. Hegner. *Review of Ecological Risk Assessment Methods,* EPA/230-10-88-041. (Washington, DC: U.S. EPA, 1988).
4. Warren-Hicks, W., B. R. Parkhurst, and S. S. Baker, Jr. *Ecological Assessments of Hazardous Waste Sites: A Field and Laboratory Reference Document,* EPA/600/3-89/013. (Washington, DC: U.S. EPA, 1989).
5. U.S. EPA. *Summary Report on Issues in Ecological Risk Assessment,* EPA/625/3-91/018. (Washington, DC: U.S. EPA, 1991).
6. Bartell, S. M., R. H. Gardner, and R. V. O'Neill. *Ecological Risk Estimation.* (Chelsea, MI: Lewis Publishers, 1992).
7. Suter, G. W., II. *Ecological Risk Assessment.* (Chelsea, MI: Lewis Publishers, 1993).
8. Huggett, R. J., M. A. Unger, P. F. Seligman, and A. O. Valkirs. "The Marine Biocide Tribucyltin. Assessing and Managing the Environmental Risks." *Environ. Sci. Technol.* 26(2): 232–237 (1992).
9. Kooijman, S. A. L. M. "A Safety Factor for LC_{50} Values Allowing for Differences in Sensitivity Among Species." *Water Res.* 21: 269–276 (1987).
10. Van Straalen, N. M., and C. A. J. Denneman. "Ecotoxicological Evaluation of Soil Quality Criteria." *Ecotoxicol. Environ. Saf.* 18: 241–251 (1989).
11. Wagner, C., and H. Løkke. "Estimation of Ecotoxicological Protection Levels from NOEC Toxicity Data." Water Res. 25: 1237–1242 (1991).
12. Cairns, J., Jr. "The Case of Simultaneous Toxicity Testing at Different Levels of Biological Organization." In *Aquatic Toxicology and Hazard Assessment: Sixth Symposium,* ASTM STP 802, ed. W. E. Bishop, R. D. Cardwell, and B. B. Heidolph. (Philadelphia, PA: American Society for Testing and Materials, 1983).

13. Slooff, W., J. A. M. van Oers, and D. De Zwart. "Margins of Uncertainty in Ecotoxicological Hazard Assessment." *Environ. Toxicol. Chem.* 5: 841–852 (1986).

14. Lindell, B. "Interpretation of Dose-Response Relationships for Risk Analyses." *Ecol. Bull.* 36: 13–16 (1984).

15. Cairns, J., Jr. "The Genesis of Biomonitoring in Aquatic Ecosystems." *Environ. Prof.* 12: 169–176 (1990).

16. Cairns, J., Jr. "Paradigms Flossed: The Coming of Age of Environmental Toxicology." *Environ. Toxicol Chem.* 11: 285–287 (1992).

17. Popper, K. R. *The Logic of Scientific Discovery.* (London: Hutchinson and Company, 1968).

Appendix

Interpolation within tables in this appendix can involve linear and harmonic interpolation. Harmonic interpolation between degrees of freedom should be used in several of the tables that have widening gaps as the degrees of freedom increase. Harmonic interpolation involves using the inverse of the degrees of freedom. Additionally, a denominator is made the highest numerical value of degrees of freedom. For example, 120 would be used in the denominator if interpolation were done between 120 and ∞ degrees of freedom in Table 12. In Table 12, linear interpolation for $df = 190$ would be between 120/120 ($= 1$) and 120/∞ ($= 0$) to estimate Dunnett's t, or 120/60 ($= 2$) and 120/120 ($= 1$) for $df = 90$.

Table 1 Values of λ used for maximum likelihood estimation of mean and standard deviation of censored data (used first in Chapter 2)

γ	h = 0.01	h = 0.02	h = 0.03	h = 0.04	h = 0.05	h = 0.06	h = 0.07	h = 0.08
0.00	0.010100	0.020400	0.030902	0.041583	0.052507	0.063627	0.074953	0.086488
0.05	0.010551	0.021294	0.032225	0.043350	0.054670	0.066189	0.077909	0.089834
0.10	0.010950	0.022082	0.033398	0.044902	0.056596	0.068483	0.080568	0.092852
0.15	0.011310	0.022798	0.034466	0.046318	0.058356	0.070586	0.083009	0.095629
0.20	0.011642	0.023459	0.035453	0.047629	0.059990	0.072539	0.085280	0.098216
0.25	0.011952	0.024076	0.036377	0.048858	0.061522	0.074372	0.087413	0.100650
0.30	0.012243	0.024658	0.037249	0.050018	0.062969	0.076106	0.089433	0.102950
0.35	0.012520	0.025211	0.038077	0.051120	0.064345	0.077756	0.091355	0.105150
0.40	0.012784	0.025738	0.038866	0.052173	0.065660	0.079332	0.093193	0.107250
0.45	0.013036	0.026243	0.039624	0.053182	0.066921	0.080845	0.094958	0.109260
0.50	0.013279	0.026728	0.040352	0.054153	0.068135	0.082301	0.096657	0.111210
0.55	0.013513	0.027196	0.041054	0.055089	0.069306	0.083708	0.098298	0.113080
0.60	0.013739	0.027649	0.041733	0.055995	0.070439	0.085068	0.099887	0.114900
0.65	0.013958	0.028087	0.042391	0.056874	0.071538	0.086388	0.101430	0.116660
0.70	0.014171	0.028513	0.043030	0.057726	0.072605	0.087670	0.102920	0.118370
0.75	0.014378	0.028927	0.043652	0.058556	0.073643	0.088917	0.104380	0.120040
0.80	0.014579	0.029330	0.044258	0.059364	0.074655	0.090133	0.105800	0.121670
0.85	0.014775	0.029723	0.044848	0.060153	0.075642	0.091319	0.107190	0.123250
0.90	0.014967	0.030107	0.045425	0.060923	0.076606	0.092477	0.108540	0.124800
0.95	0.015154	0.030483	0.045989	0.061676	0.077549	0.093611	0.109870	0.126320
1.00	0.015338	0.030850	0.046540	0.062413	0.078471	0.094720	0.111160	0.127800

Table 1 Continued

γ	h = 0.09	h = 0.10	h = 0.15	h = 0.20	h = 0.25	h = 0.30	h = 0.35	h = 0.40
0.00	0.09824	0.11020	0.17342	0.24268	0.31862	0.4021	0.4941	0.5961
0.05	0.10197	0.11431	0.17935	0.25033	0.32793	0.4130	0.5066	0.6101
0.10	0.10534	0.11804	0.18479	0.25741	0.33662	0.4233	0.5184	0.6234
0.15	0.10845	0.12148	0.18985	0.26405	0.34480	0.4330	0.5296	0.6361
0.20	0.11135	0.12469	0.19460	0.27031	0.35255	0.4422	0.5403	0.6483
0.25	0.11408	0.12772	0.19910	0.27626	0.35993	0.4510	0.5506	0.6600
0.30	0.11667	0.13059	0.20338	0.28193	0.36700	0.4595	0.5604	0.6713
0.35	0.11914	0.13333	0.20747	0.28737	0.37379	0.4676	0.5699	0.6821
0.40	0.12150	0.13595	0.21139	0.29260	0.38033	0.4755	0.5791	0.6927
0.45	0.12377	0.13847	0.21517	0.29765	0.38665	0.4831	0.5880	0.7029
0.50	0.12595	0.14090	0.21882	0.30253	0.39276	0.4904	0.5967	0.7129
0.55	0.12806	0.14325	0.22235	0.30725	0.39870	0.4976	0.6051	0.7225
0.60	0.13011	0.14552	0.22578	0.31184	0.40447	0.5045	0.6133	0.7320
0.65	0.13209	0.14773	0.22910	0.31630	0.41008	0.5114	0.6213	0.7412
0.70	0.13402	0.14987	0.23234	0.32065	0.41555	0.5180	0.6291	0.7502
0.75	0.13590	0.15196	0.23550	0.32489	0.42090	0.5245	0.6367	0.7590
0.80	0.13773	0.15400	0.23858	0.32903	0.42612	0.5308	0.6441	0.7676
0.85	0.13952	0.15599	0.24158	0.33307	0.43122	0.5370	0.6515	0.7761
0.90	0.14126	0.15793	0.24452	0.33703	0.43622	0.5430	0.6586	0.7844
0.95	0.14297	0.15983	0.24740	0.34091	0.44112	0.5490	0.6656	0.7925
1.00	0.14465	0.16170	0.25022	0.34471	0.44592	0.5548	0.6724	0.8005

Table 1 Continued

γ	h = 0.45	h = 0.50	h = 0.55	h = 0.60	h = 0.65	h = 0.70	h = 0.80	h = 0.90
0.00	0.7096	0.8368	0.9808	1.1450	1.3360	1.5610	2.1760	3.2830
0.05	0.7252	0.8540	0.9994	1.1660	1.3580	1.5850	2.2030	3.3140
0.10	0.7400	0.8703	1.0170	1.1850	1.3790	1.6080	2.2290	3.3450
0.15	0.7542	0.8860	1.0350	1.2040	1.4000	1.6300	2.2550	3.3760
0.20	0.7678	0.9012	1.0510	1.2220	1.4190	1.6510	2.2800	3.4050
0.25	0.7810	0.9158	1.0670	1.2400	1.4390	1.6720	2.3050	3.4350
0.30	0.7937	0.9300	1.0830	1.2570	1.4570	1.6930	2.3290	3.4640
0.35	0.8060	0.9437	1.0980	1.2740	1.4760	1.7130	2.3530	3.4920
0.40	0.8179	0.9570	1.1130	1.2900	1.4940	1.7320	2.3760	3.5200
0.45	0.8295	0.9700	1.1270	1.3060	1.5110	1.7510	2.3990	3.5470
0.50	0.8408	0.9826	1.1410	1.3210	1.5280	1.7700	2.4210	3.5750
0.55	0.8517	0.9950	1.1550	1.3370	1.5450	1.7880	2.4430	3.6010
0.60	0.8625	1.0070	1.1690	1.3510	1.5610	1.8060	2.4650	3.6280
0.65	0.8729	1.0190	1.1820	1.3660	1.5770	1.8240	2.4860	3.6540
0.70	0.8832	1.0300	1.1950	1.3800	1.5930	1.8410	2.5070	3.6790
0.75	0.8932	1.0420	1.2070	1.3940	1.6080	1.8580	2.5280	3.7050
0.80	0.9031	1.0530	1.2200	1.4080	1.6240	1.8750	2.5480	3.7300
0.85	0.9127	1.0640	1.2320	1.4220	1.6390	1.8920	2.5680	3.7540
0.90	0.9222	1.0740	1.2440	1.4350	1.6530	1.9080	2.5880	3.7790
0.95	0.9314	1.0850	1.2550	1.4480	1.6680	1.9240	2.6070	3.8030
1.00	0.9406	1.0950	1.2670	1.4610	1.6820	1.9400	2.6260	3.8270

Table 2 **Values of $\psi_n(t)$ used to adjust for backtransformation bias (used first in Chapter 2)**

t	$n = 10$	$n = 20$	$n = 30$	$n = 40$	$n = 50$	$n = 60$	$n = 70$	$n = 80$	$n = 90$
0.05	1.0458	1.0485	1.0494	1.0499	1.0502	1.0504	1.0505	1.0506	1.0507
0.10	1.0934	1.0992	1.1012	1.1022	1.1028	1.1032	1.1034	1.1037	1.1038
0.15	1.1427	1.1521	1.1553	1.1569	1.1579	1.1585	1.1590	1.1593	1.1596
0.20	1.1938	1.2072	1.2118	1.2142	1.2156	1.2166	1.2173	1.2178	1.2182
0.25	1.2468	1.2648	1.2710	1.2742	1.2761	1.2774	1.2784	1.2791	1.2800
0.30	1.3018	1.3248	1.3329	1.3370	1.3395	1.3412	1.3424	1.3424	1.3441
0.35	1.3587	1.3874	1.3976	1.4028	1.4060	1.4081	1.4097	1.4108	1.4117
0.40	1.4177	1.4527	1.4652	1.4716	1.4756	1.4782	1.4801	1.4816	1.4827
0.45	1.4788	1.5207	1.5359	1.5437	1.5485	1.5517	1.5540	1.5558	1.5571
0.50	1.5421	1.5917	1.6097	1.6191	1.6428	1.6287	1.6315	1.6336	1.6352
0.55	1.6076	1.6657	1.6869	1.6980	1.7048	1.7094	1.7127	1.7152	1.7172
0.60	1.6754	1.7428	1.7676	1.7806	1.7886	1.7940	1.7979	1.8008	1.8031
0.65	1.7457	1.8321	1.8519	1.8670	1.8763	1.8826	1.8871	1.8906	1.8933
0.70	1.8184	1.9068	1.9399	1.9574	1.9681	1.9754	1.9807	1.9847	1.9879
0.75	1.8936	1.9940	2.0319	2.0519	2.0643	2.0727	2.0788	2.0834	2.0870
0.80	1.9714	2.0848	2.1279	2.1508	2.1650	2.1746	2.1816	2.1869	2.1911
0.85	2.0519	2.1794	2.2283	2.2542	2.2703	2.2813	2.2893	2.2954	2.3001
0.90	2.1352	2.2779	2.3330	2.3624	2.3807	2.3932	2.4022	2.4091	2.4145
0.95	2.2214	2.3804	2.4424	2.4755	2.4962	2.5103	2.5206	2.5284	2.5345
1.00	2.3104	2.4872	2.5565	2.5938	2.6170	2.6330	2.6445	2.6534	2.6603
1.05	2.4025	2.5984	2.6757	2.7174	2.7435	2.7614	2.7745	2.7844	2.7922
1.10	2.4977	2.7141	2.8002	2.8467	2.8759	2.8959	2.9106	2.9217	2.9305
1.15	2.5961	2.8345	2.9300	2.9818	3.0144	3.0368	3.0532	3.0656	3.0755
1.20	2.6978	2.9597	3.0655	3.1231	3.1594	3.1843	3.2026	3.2165	3.2275
1.25	2.8028	3.0901	3.2069	3.2707	3.3110	3.3388	3.3591	3.3746	3.3868
1.30	2.9114	3.2257	3.3544	3.4250	3.4696	3.5005	3.5230	3.5403	3.5539
1.35	3.0235	3.3668	3.5084	3.5862	3.6356	3.6697	3.6947	3.7139	3.7290
1.40	3.1393	3.5135	3.6689	3.7547	3.8092	3.8469	3.8746	3.8958	3.9125
1.45	3.2589	3.6661	3.8364	3.9307	3.9908	4.0324	4.0630	4.0864	4.1049
1.50	3.3824	3.8247	4.0111	4.1146	4.1807	4.2266	4.2603	4.2861	4.3065
1.55	3.5099	3.9897	4.1933	4.3068	4.3793	4.4297	4.4669	4.4953	4.5178
1.60	3.6415	4.1612	4.3832	4.5074	4.5870	4.6424	4.6832	4.7145	4.7393
1.65	3.7774	4.3394	4.5813	4.7171	4.8042	4.8649	4.9097	4.9441	4.9714
1.70	3.9176	4.5247	4.7878	4.9360	5.0313	5.0978	5.1469	5.1847	5.2146
1.75	4.0623	4.7173	5.0031	5.1646	5.2687	5.3415	5.3953	5.4366	5.4694
1.80	4.2116	4.9174	5.2275	5.4034	5.5170	5.5965	5.6553	5.7005	5.7365
1.85	4.3657	5.1253	5.4614	5.6527	5.7764	5.8632	5.9275	5.9770	6.0163
1.90	4.5246	5.3413	5.7052	5.9129	6.0477	6.1423	6.2124	6.2665	6.3094
1.95	4.6885	5.5657	5.9592	6.1847	6.3312	6.4342	6.5107	6.5696	6.6165
2.00	4.8575	5.7988	6.2239	6.4684	6.6276	6.7396	6.8229	6.8871	6.9383

Table 2 *Continued*

t	n = 100	n = 110	n = 120	n = 130	n = 140	n = 150	n = 160	n = 170	n = 180
0.05	1.0507	1.0508	1.0508	1.0508	1.0509	1.0509	1.0509	1.0509	1.0510
0.10	1.1040	1.1041	1.1042	1.1042	1.1043	1.1044	1.1044	1.1045	1.1045
0.15	1.1598	1.1600	1.1602	1.1603	1.1604	1.1605	1.1606	1.1607	1.1607
0.20	1.2185	1.2188	1.2190	1.2192	1.2193	1.2195	1.2196	1.2197	1.2198
0.25	1.2800	1.2804	1.2807	1.2810	1.2812	1.2814	1.2815	1.2817	1.2818
0.30	1.3446	1.3451	1.3455	1.3458	1.3461	1.3464	1.3466	1.3468	1.3470
0.35	1.4124	1.4130	1.4135	1.4140	1.4143	1.4146	1.4149	1.4152	1.4154
0.40	1.4836	1.4843	1.4849	1.4855	1.4859	1.4863	1.4866	1.4870	1.4872
0.45	1.5582	1.5591	1.5599	1.5605	1.5611	1.5616	1.5620	1.5623	1.5627
0.50	1.6366	1.6377	1.6386	1.6393	1.6400	1.6406	1.6411	1.6415	1.6419
0.55	1.7188	1.7201	1.7211	1.7221	1.7228	1.7235	1.7241	1.7247	1.7251
0.60	1.8050	1.8065	1.8078	1.8089	1.8098	1.8106	1.8113	1.8120	1.8125
0.65	1.8955	1.8973	1.8988	1.9000	1.9011	1.9021	1.9029	1.9036	1.9043
0.70	1.9904	1.9925	1.9942	1.9957	1.9969	1.9980	1.9990	1.9999	2.0006
0.75	2.0900	2.0924	2.0944	2.0961	2.0975	2.0988	2.0999	2.1009	2.1018
0.80	2.1944	2.1972	2.1995	2.2014	2.2031	2.2046	2.2059	2.2070	2.2080
0.85	2.3040	2.3071	2.3098	2.3120	2.3139	2.3156	2.3171	2.3184	2.3196
0.90	2.4189	2.4225	2.4255	2.4281	2.4303	2.4322	2.4339	2.4353	2.4367
0.95	2.5395	2.5435	2.5469	2.5498	2.5523	2.5545	2.5564	2.5581	2.5596
1.00	2.6659	2.6705	2.6744	2.6776	2.6805	2.6829	2.6851	2.6870	2.6887
1.05	2.7985	2.8037	2.8080	2.8117	2.8149	2.8177	2.8201	2.8223	2.8242
1.10	2.9376	2.9434	2.9483	2.9525	2.9560	2.9592	2.9619	2.9643	2.9665
1.15	3.0834	3.0899	3.0954	3.1001	3.1041	3.1076	3.1107	3.1134	3.1159
1.20	3.2363	3.2436	3.2498	3.2550	3.2595	3.2634	3.2669	3.2699	3.2726
1.25	3.3967	3.4049	3.4117	3.4175	3.4226	3.4270	3.4308	3.4342	3.4372
1.30	3.5649	3.5739	3.5816	3.5881	3.5937	3.5986	3.6028	3.6066	3.6100
1.35	3.7412	3.7513	3.7597	3.7670	3.7732	3.7786	3.7834	3.7876	3.7914
1.40	3.9260	3.9372	3.9466	3.9547	3.9616	3.9676	3.9729	3.9776	3.9818
1.45	4.1199	4.1323	4.1427	4.1515	4.1592	4.1659	4.1718	4.1770	4.1816
1.50	4.3231	4.3368	4.3483	4.3581	4.3666	4.3739	4.3804	4.3862	4.3913
1.55	4.5361	4.5512	4.5639	4.5748	4.5841	4.5923	4.5994	4.6058	4.6115
1.60	4.7594	4.7761	4.7901	4.8020	4.8123	4.8213	4.8293	4.8363	4.8425
1.65	4.9935	5.0118	5.0272	5.0404	5.0518	5.0617	5.0704	5.0781	5.0850
1.70	5.2389	5.2590	5.2760	5.2904	5.3029	5.3138	5.3234	5.3320	5.3396
1.75	5.4961	5.5182	5.5368	5.5527	5.5664	5.5784	5.5890	5.5983	5.6067
1.80	5.7657	5.7899	5.8103	5.8278	5.8428	5.8560	5.8675	5.8778	5.8870
1.85	6.0482	6.0748	6.0971	6.1162	6.1327	6.1471	6.1598	6.1711	6.1812
1.90	6.3444	6.3734	6.3978	6.4188	6.4368	6.4526	6.4665	6.4789	6.4899
1.95	6.6547	6.6864	6.7132	6.7360	6.7558	6.7731	6.7883	6.8018	6.8138
2.00	6.9800	7.0146	7.0438	7.0687	7.0903	7.1092	7.1258	7.1406	7.1538

Table 2 *Continued*

t	n = 190	n = 200	n = 300	n = 400	n = 500	n = 750	n = 1000	n = 5000	n = ∞
0.05	1.0510	1.0510	1.0511	1.0511	1.0512	1.0512	1.0512	1.0513	1.0513
0.10	1.1045	1.1046	1.1048	1.1049	1.1049	1.1050	1.1050	1.1051	1.1052
0.15	1.1608	1.1608	1.1612	1.1613	1.1614	1.1616	1.1616	1.1618	1.1618
0.20	1.2199	1.2199	1.2204	1.2207	1.2208	1.2210	1.2211	1.2213	1.2214
0.25	1.2819	1.2820	1.2827	1.2830	1.2832	1.2835	1.2836	1.2839	1.2840
0.30	1.3471	1.3472	1.3481	1.3485	1.3488	1.3492	1.3493	1.3498	1.3499
0.35	1.4156	1.4157	1.4168	1.4174	1.4177	1.4182	1.4184	1.4189	1.4191
0.40	1.4875	1.4877	1.4891	1.4897	1.4902	1.4907	1.4910	1.4917	1.4918
0.45	1.5630	1.5632	1.5649	1.5658	1.5663	1.5670	1.5673	1.5681	1.5683
0.50	1.6423	1.6426	1.6446	1.6456	1.6463	1.6471	1.6475	1.6485	1.6487
0.55	1.7256	1.7259	1.7284	1.7296	1.7303	1.7313	1.7318	1.7330	1.7333
0.60	1.8130	1.8135	1.8163	1.8178	1.8186	1.8198	1.8204	1.8218	1.8221
0.65	1.9049	1.9054	1.9087	1.9104	1.9115	1.9128	1.9135	1.9151	1.9155
0.70	2.0013	2.0019	2.0058	2.0078	2.0090	2.0106	2.0114	2.0133	2.0138
0.75	2.1026	2.1033	2.1078	2.1101	2.1115	2.1133	2.1142	2.1164	2.1070
0.80	2.2089	2.2098	2.2150	2.2176	2.2192	2.2213	2.2223	2.2249	2.2255
0.85	2.3206	2.3215	2.3275	2.3305	2.3323	2.3348	2.3360	2.3389	2.3396
0.90	2.4379	2.4389	2.4457	2.4492	2.4512	2.4540	2.4554	2.4588	2.4596
0.95	2.5610	2.5622	2.7004	2.7048	2.7075	2.5794	2.5809	2.5848	2.5857
1.00	2.6902	2.6916	2.5699	2.5738	2.5762	2.7111	2.7129	2.7172	2.7183
1.05	2.8259	2.8275	2.8374	2.8424	2.8454	2.8495	2.8515	2.8564	2.8577
1.10	2.9684	2.9702	2.9814	2.9870	2.9904	2.9950	2.9973	3.0028	3.0042
1.15	3.1180	3.1200	3.1325	3.1389	3.1427	3.1478	3.1504	3.1566	3.1582
1.20	3.2751	3.2773	3.2913	3.2985	3.3027	3.3085	3.3114	3.3184	3.3201
1.25	3.4400	3.4424	3.4581	3.4661	3.4709	3.4773	3.4806	3.4884	3.4903
1.30	3.6131	3.6158	3.6333	3.6422	3.6476	3.6548	3.6584	3.6671	3.6693
1.35	3.7948	3.7978	3.8173	3.3872	3.8332	3.8412	3.8453	3.8550	3.8574
1.40	3.9855	3.9889	4.0106	4.0216	4.0282	4.0372	4.0417	4.0525	4.0552
1.45	4.1858	4.1895	4.2136	4.2258	4.2332	4.2431	4.2481	4.2601	4.2631
1.50	4.3959	4.4001	4.4268	4.4403	4.4485	4.4595	4.4650	4.4783	4.4817
1.55	4.6166	4.6212	4.6507	4.6656	4.6747	4.6868	4.6930	4.7078	4.7115
1.60	4.8482	4.8532	4.8858	4.9023	4.9124	4.9258	4.9326	4.9489	4.9530
1.65	5.0912	5.0986	5.1328	5.1510	5.1621	5.1769	5.1844	5.2024	5.2070
1.70	5.3464	5.3525	5.3921	5.4122	5.4244	5.4408	5.4490	5.4689	5.4739
1.75	5.6142	5.6209	5.6645	5.6866	5.7000	5.7180	5.7271	5.7491	5.7546
1.80	5.8952	5.9027	5.9505	5.9748	5.9896	6.0094	6.0194	6.0436	6.0496
1.85	6.1902	6.1984	6.2509	6.2776	6.2938	6.3156	6.3265	6.3531	6.3598
1.90	6.4998	6.5087	6.5663	6.5956	6.6134	6.6373	6.6493	6.6785	6.6859
1.95	6.8247	6.8345	6.8975	6.9296	6.9491	6.9754	6.9886	7.0206	7.0287
2.00	7.1657	7.1764	7.2453	7.2805	7.3019	7.3306	7.3451	7.3802	7.3891

Table 3   **Factors for estimating standard deviation and control limits for range (used first in Chapter 2)**

Number of Observations in the Subgroup	d_2	D_3	D_4
2	1.128	0	3.27
3	1.693	0	2.57
4	2.059	0	2.28
5	2.326	0	2.11
6	2.534	0	2.00
7	2.704	0.08	1.92
8	2.847	0.14	1.86
9	2.970	0.18	1.82
10	3.078	0.22	1.78
11	3.173	0.26	1.74
12	3.258	0.28	1.72
13	3.336	0.31	1.69
14	3.407	0.33	1.67
15	3.472	0.35	1.65
16	3.532	0.36	1.64
17	3.588	0.38	1.62
18	3.640	0.39	1.61
19	3.689	0.40	1.60
20	3.735	0.41	1.59

Derived from Tables B and C of Grant (reference 3 in Chapter 2) with permission from McGraw-Hill, Inc. (New York).

Table 4 **One-sample tolerance probability comparisons between n_m^* and n_m (used first in Chapter 2)**

		n_m^*			
n_m $(1 - \gamma')$	$(1 - \gamma) = 0.70$	$(1 - \gamma) = 0.80$	$(1 - \gamma) = 0.90$	$(1 - \gamma) = 0.95$	$(1 - \gamma) = 0.99$
		$\alpha = 0.10$			
5 0.33	8	9	11	12	13
10 0.39	14	15	17	19	21
15 0.41	20	21	23	25	28
20 0.42	25	27	29	32	35
25 0.43	30	33	35	38	42
30 0.44	36	38	41	44	49
35 0.44	41	44	47	50	55
40 0.45	46	49	53	56	61
45 0.45	52	55	58	62	68
50 0.45	57	60	64	67	74
55 0.45	62	65	70	73	80
60 0.46	67	71	75	79	86
65 0.46	73	76	81	85	92
70 0.46	78	81	86	90	98
75 0.46	83	87	92	96	104
80 0.46	88	92	97	102	110
85 0.46	93	97	103	107	116
90 0.46	99	103	108	113	122
95 0.46	104	108	114	119	127
100 0.47	109	113	119	124	133
		$\alpha = 0.05$			
5 0.26	9	10	11	12	14
10 0.34	15	16	18	19	22
15 0.37	20	22	24	26	29
20 0.39	26	27	30	32	36
25 0.40	31	33	36	38	43
30 0.41	36	39	42	44	49
35 0.42	42	44	48	50	55
40 0.42	47	50	53	56	62
45 0.43	52	55	59	62	68
50 0.43	57	60	65	68	74
55 0.43	63	66	70	74	80
60 0.44	68	71	76	80	86
65 0.44	73	77	81	85	92
70 0.44	78	82	87	91	98
75 0.44	84	87	92	97	104
80 0.44	89	93	98	102	110
85 0.45	94	98	103	108	116
90 0.45	99	103	109	114	112
95 0.45	104	109	114	119	128
100 0.45	110	114	120	125	134

Table 4 *Continued*

		$n_m{}^*$			
n_m $(1 - \gamma')$	$(1 - \gamma) = 0.70$	$(1 - \gamma) = 0.80$	$(1 - \gamma) = 0.90$	$(1 - \gamma) = 0.95$	$(1 - \gamma) = 0.99$
		$\alpha = 0.01$			
5 0.13	10	11	12	13	15
10 0.23	16	17	19	20	23
15 0.27	21	23	25	27	30
20 0.30	27	29	31	33	37
25 0.32	32	34	37	39	44
30 0.34	38	40	43	46	50
35 0.35	43	45	49	52	57
40 0.36	48	51	55	58	63
45 0.37	53	56	60	63	69
50 0.38	59	62	66	69	75
55 0.38	64	67	72	75	82
60 0.39	69	73	77	81	88
65 0.39	74	78	83	87	94
70 0.39	80	83	88	92	100
75 0.40	85	89	94	98	106
80 0.40	90	94	99	104	112
85 0.40	95	99	105	109	117
90 0.41	101	105	110	115	123
95 0.41	106	110	116	120	129
100 0.41	111	115	121	126	135

NOTE: $(1 - \gamma)$ is the tolerance probability using n_m^* and $(1 - \gamma')$ is the tolerance probability using n_m.

Table 5 **Critical values of *T* used to test for single outliers (used first in Chapter 2)**

Number of observations *n*	Significance Level		
	5%	**2.5%**	**1%**
3	1.15	1.15	1.15
4	1.46	1.48	1.49
5	1.67	1.71	1.75
6	1.82	1.89	1.94
7	1.94	2.02	2.10
8	2.03	2.13	2.22
9	2.11	2.21	2.32
10	2.18	2.29	2.41
11	2.23	2.36	2.48
12	2.29	2.41	2.55
13	2.33	2.46	2.61
14	2.37	2.51	2.66
15	2.41	2.55	2.71
16	2.44	2.59	2.75
17	2.47	2.62	2.79
18	2.50	2.65	2.82
19	2.53	2.68	2.85
20	2.56	2.71	2.88
21	2.58	2.73	2.91
22	2.60	2.76	2.94
23	2.62	2.78	2.96
24	2.64	2.80	2.99
25	2.66	2.82	3.01
30	2.75	2.91	
35	2.82	2.98	
40	2.87	3.04	
45	2.92	3.09	
50	2.96	3.13	
60	3.03	3.20	
70	3.09	3.26	
80	3.14	3.31	
90	3.18	3.35	
100	3.21	3.38	

Table 6 **Critical values for λ used to test for multiple outliers (used first in Chapter 2)**

n	$l + 1$	α 0.05	0.01	0.005
25	1	2.82	3.14	3.25
	2	2.80	3.11	3.23
	3	2.78	3.09	3.20
	4	2.76	3.06	3.17
	5	2.73	3.03	3.14
	10	2.59	2.85	2.95
26	1	2.84	3.16	3.28
	2	2.82	3.14	3.25
	3	2.80	3.11	3.23
	4	2.78	3.09	3.20
	5	2.76	3.06	3.17
	10	2.62	2.89	2.99
27	1	2.86	3.18	3.30
	2	2.84	3.16	3.28
	3	2.82	3.14	3.25
	4	2.80	3.11	3.23
	5	2.78	3.09	3.20
	10	2.65	2.93	3.03
28	1	2.88	3.20	3.32
	2	2.86	3.18	3.30
	3	2.84	3.16	3.28
	4	2.82	3.14	3.25
	5	2.80	3.11	3.23
	10	2.68	2.97	3.07
29	1	2.89	3.22	3.34
	2	2.88	3.20	3.32
	3	2.86	3.18	3.30
	4	2.84	3.16	3.28
	5	2.82	3.14	3.25
	10	2.71	3.00	3.11
30	1	2.91	3.24	3.36
	2	2.89	3.22	3.34
	3	2.88	3.20	3.32
	4	2.86	3.18	3.30
	5	2.84	3.16	3.28
	10	2.73	3.03	3.14
31	1	2.92	3.25	3.38
	2	2.91	3.24	3.36
	3	2.89	3.22	3.34
	4	2.88	3.20	3.32
	5	2.86	3.18	3.30
	10	2.76	3.06	3.17

Table 6   *Continued*

n	l + 1	α		
		0.05	**0.01**	**0.005**
32	1	2.94	3.27	3.40
	2	2.92	3.25	3.38
	3	2.91	3.24	3.36
	4	2.89	3.22	3.34
	5	2.88	3.20	3.32
	10	2.78	3.09	3.20
33	1	2.95	3.29	3.41
	2	2.94	3.27	3.40
	3	2.92	3.25	3.38
	4	2.91	3.24	3.36
	5	2.89	3.22	3.34
	10	2.80	3.11	3.23
34	1	2.97	3.30	3.43
	2	2.95	3.29	3.41
	3	2.94	3.27	3.40
	4	2.92	3.25	3.38
	5	2.91	3.24	3.36
	10	2.82	3.14	3.25
35	1	2.98	3.32	3.44
	2	2.97	3.30	3.43
	3	2.95	3.29	3.41
	4	2.94	3.27	3.40
	5	2.92	3.25	3.38
	10	2.84	3.16	3.28
36	1	2.99	3.33	3.46
	2	2.98	3.32	3.44
	3	2.97	3.30	3.43
	4	2.95	3.29	3.41
	5	2.94	3.27	3.40
	10	2.86	3.18	3.30
37	1	3.00	3.34	3.47
	2	2.99	3.33	3.46
	3	2.98	3.32	3.44
	4	2.97	3.30	3.43
	5	2.95	3.29	3.41
	10	2.88	3.20	3.32
38	1	3.01	3.36	3.49
	2	3.00	3.34	3.47
	3	2.99	3.33	3.46
	4	2.98	3.32	3.44
	5	2.97	3.30	3.43
	10	2.89	3.22	3.34

Table 6   *Continued*

			α	
n	*l* + 1	**0.05**	**0.01**	**0.005**
39	1	3.03	3.37	3.50
	2	3.01	3.36	3.49
	3	3.00	3.34	3.47
	4	2.99	3.33	3.46
	5	2.98	3.32	3.44
	10	2.91	3.24	3.36
40	1	3.04	3.38	3.51
	2	3.03	3.37	3.50
	3	3.01	3.36	3.49
	4	3.00	3.34	3.47
	5	2.99	3.33	3.46
	10	2.92	3.25	3.38
41	1	3.05	3.39	3.52
	2	3.04	3.38	3.51
	3	3.03	3.37	3.50
	4	3.01	3.36	3.49
	5	3.00	3.34	3.47
	10	2.94	3.27	3.40
42	1	3.06	3.40	3.54
	2	3.05	3.39	3.52
	3	3.04	3.38	3.51
	4	3.03	3.37	3.50
	5	3.01	3.36	3.49
	10	2.95	3.29	3.41
43	1	3.07	3.41	3.55
	2	3.06	3.40	3.54
	3	3.05	3.39	3.52
	4	3.04	3.38	3.51
	5	3.03	3.37	3.50
	10	2.97	3.30	3.43
44	1	3.08	3.43	3.56
	2	3.07	3.41	3.55
	3	3.06	3.40	3.54
	4	3.05	3.39	3.52
	5	3.04	3.38	3.51
	10	2.98	3.32	3.44
45	1	3.09	3.44	3.57
	2	3.08	3.43	3.56
	3	3.07	3.41	3.55
	4	3.06	3.40	3.54
	5	3.05	3.39	3.52
	10	2.99	3.33	3.46

Table 6 *Continued*

n	$l + 1$	α		
		0.05	**0.01**	**0.005**
46	1	3.09	3.45	3.58
	2	3.09	3.44	3.57
	3	3.08	3.43	3.56
	4	3.07	3.41	3.55
	5	3.06	3.40	3.54
	10	3.00	3.34	3.47
47	1	3.10	3.46	3.59
	2	3.09	3.45	3.58
	3	3.09	3.44	3.57
	4	3.08	3.43	3.56
	5	3.07	3.41	3.55
	10	3.01	3.36	3.49
48	1	3.11	3.46	3.60
	2	3.10	3.46	3.59
	3	3.09	3.45	3.58
	4	3.09	3.44	3.57
	5	3.08	3.43	3.56
	10	3.03	3.37	3.50
49	1	3.12	3.47	3.61
	2	3.11	3.46	3.60
	3	3.10	3.46	3.59
	4	3.09	3.45	3.58
	5	3.09	3.44	3.57
	10	3.04	3.38	3.51
50	1	3.13	3.48	3.62
	2	3.12	3.47	3.61
	3	3.11	3.46	3.60
	4	3.10	3.46	3.59
	5	3.09	3.45	3.58
	10	3.05	3.39	3.52
60	1	3.20	3.56	3.70
	2	3.19	3.55	3.69
	3	3.19	3.55	3.69
	4	3.18	3.54	3.68
	5	3.17	3.53	3.67
	10	3.14	3.49	3.63
70	1	3.26	3.62	3.76
	2	3.25	3.62	3.76
	3	3.25	3.61	3.75
	4	3.24	3.60	3.75
	5	3.24	3.60	3.74
	10	3.21	3.57	3.71

Table 6 *Continued*

n	*l* + 1	α 0.05	0.01	0.005
80	1	3.31	3.67	3.82
	2	3.30	3.67	3.81
	3	3.30	3.66	3.81
	4	3.29	3.66	3.80
	5	3.29	3.65	3.80
	10	3.26	3.63	3.77
90	1	3.35	3.72	3.86
	2	3.34	3.71	3.86
	3	3.34	3.71	3.85
	4	3.34	3.70	3.85
	5	3.33	3.70	3.84
	10	3.31	3.68	3.82
100	1	3.38	3.75	3.90
	2	3.38	3.75	3.90
	3	3.38	3.75	3.89
	4	3.37	3.74	3.89
	5	3.37	3.74	3.89
	10	3.35	3.72	3.87
150	1	3.52	3.89	4.04
	2	3.51	3.89	4.04
	3	3.51	3.89	4.03
	4	3.51	3.88	4.03
	5	3.51	3.88	4.03
	10	3.50	3.87	4.02
200	1	3.61	3.98	4.13
	2	3.60	3.98	4.13
	3	3.60	3.97	4.12
	4	3.60	3.97	4.12
	5	3.60	3.97	4.12
	10	3.59	3.96	4.11
250	1	3.67	4.04	4.19
	5	3.67	4.04	4.19
	10	3.66	4.03	4.18
300	1	3.72	4.09	4.24
	5	3.72	4.09	4.24
	10	3.71	4.09	4.23
350	1	3.77	4.14	4.28
	5	3.76	4.13	4.28
	10	3.76	4.13	4.28
400	1	3.80	4.17	4.32
	5	3.80	4.17	4.32
	10	3.80	4.16	4.31

Table 6 *Continued*

n	$l + 1$	\u03b1		
		0.05	**0.01**	**0.005**
450	1	3.84	4.20	4.35
	5	3.83	4.20	4.35
	10	3.83	4.20	4.34
500	1	3.86	4.23	4.38
	5	3.86	4.23	4.37
	10	3.86	4.22	4.37

Table 7 **Response metameters for proportion affected (used first in Chapter 4)**

Proportion (No. affected/ Total No.)	Arcsin $\sqrt{p}$ (radians)	Weibull	NED	Probit	Logit	Transformed Logit	Percent (%)
0.000	0.00000						0.0
0.005	0.07077	−5.29581	−2.57583	2.42417	−5.29330	2.35335	0.5
0.010	0.10017	−4.60015	−2.32635	2.67365	−4.59512	2.70244	1.0
0.015	0.12278	−4.19216	−2.17009	2.82991	−4.18459	2.90770	1.5
0.020	0.14190	−3.90194	−2.05375	2.94625	−3.89182	3.05409	2.0
0.025	0.15878	−3.67625	−1.95996	3.04004	−3.66356	3.16822	2.5
0.030	0.17408	−3.49137	−1.88079	3.11921	−3.47610	3.26195	3.0
0.035	0.18819	−3.33465	−1.81191	3.18809	−3.31678	3.34161	3.5
0.040	0.20136	−3.19853	−1.75069	3.24931	−3.17805	3.41097	4.0
0.045	0.21376	−3.07816	−1.69540	3.30460	−3.05505	3.47248	4.5
0.050	0.22551	−2.97020	−1.64485	3.35515	−2.94444	3.52778	5.0
0.055	0.23673	−2.87227	−1.59819	3.40181	−2.84385	3.57807	5.5
0.060	0.24747	−2.78263	−1.55477	3.44523	−2.75154	3.62423	6.0
0.065	0.25780	−2.69995	−1.51410	3.48590	−2.66616	3.66692	6.5
0.070	0.26776	−2.62319	−1.47579	3.52421	−2.58669	3.70666	7.0
0.075	0.27741	−2.55154	−1.43953	3.56047	−2.51231	3.74385	7.5
0.080	0.28676	−2.48433	−1.40507	3.59493	−2.44235	3.77883	8.0
0.085	0.29584	−2.42102	−1.37220	3.62780	−2.37627	3.81186	8.5
0.090	0.30469	−2.36116	−1.34076	3.65924	−2.31363	3.84318	9.0
0.095	0.31332	−2.30438	−1.31058	3.68942	−2.25406	3.87297	9.5
0.100	0.32175	−2.25037	−1.28155	3.71845	−2.19722	3.90139	10.0
0.105	0.32999	−2.19884	−1.25357	3.74643	−2.14286	3.92857	10.5
0.110	0.33807	−2.14957	−1.22653	3.77347	−2.09074	3.95463	11.0
0.115	0.34598	−2.10236	−1.20036	3.79964	−2.04066	3.97967	11.5
0.120	0.35374	−2.05703	−1.17499	3.82501	−1.99243	4.00378	12.0
0.125	0.36137	−2.01342	−1.15035	3.84965	−1.94591	4.02704	12.5
0.130	0.36886	−1.97140	−1.12639	3.87361	−1.90096	4.04952	13.0
0.135	0.37624	−1.93084	−1.10306	3.89694	−1.85745	4.07127	13.5
0.140	0.38350	−1.89165	−1.08032	3.91968	−1.81529	4.09236	14.0
0.145	0.39065	−1.85372	−1.05812	3.94188	−1.77437	4.11282	14.5
0.150	0.39770	−1.81696	−1.03643	3.96357	−1.73460	4.13270	15.0
0.155	0.40465	−1.78130	−1.01522	3.98478	−1.69591	4.15204	15.5
0.160	0.41152	−1.74667	−0.99446	4.00554	−1.65823	4.17089	16.0
0.165	0.41829	−1.71300	−0.97411	4.02589	−1.62149	4.18926	16.5
0.170	0.42499	−1.68024	−0.95417	4.04583	−1.58563	4.20719	17.0
0.175	0.43161	−1.64832	−0.93459	4.06541	−1.55060	4.22470	17.5
0.180	0.43815	−1.61721	−0.91537	4.08463	−1.51635	4.24183	18.0
0.185	0.44462	−1.58686	−0.89647	4.10353	−1.48283	4.25858	18.5
0.190	0.45103	−1.55722	−0.87790	4.12210	−1.45001	4.27499	19.0
0.195	0.45737	−1.52826	−0.85962	4.14038	−1.41784	4.29108	19.5
0.200	0.46365	−1.49994	−0.84162	4.15838	−1.38629	4.30685	20.0
0.205	0.46987	−1.47223	−0.82389	4.17611	−1.35533	4.32233	20.5
0.210	0.47603	−1.44510	−0.80642	4.19358	−1.32493	4.33754	21.0
0.215	0.48215	−1.41852	−0.78919	4.21081	−1.29505	4.35248	21.5

Table 7   *Continued*

Proportion (No. affected/ Total No.)	Arcsin $\sqrt{p}$ (radians)	Weibull	NED	Probit	Logit	Transformed Logit	Percent (%)
0.220	0.48821	−1.39247	−0.77219	4.22781	−1.26567	4.36717	22.0
0.225	0.49422	−1.36691	−0.75542	4.24458	−1.23676	4.38162	22.5
0.230	0.50018	−1.34184	−0.73885	4.26115	−1.20831	4.39584	23.0
0.235	0.50610	−1.31722	−0.72248	4.27752	−1.18029	4.40985	23.5
0.240	0.51197	−1.29303	−0.70630	4.29370	−1.15268	4.42366	24.0
0.245	0.51781	−1.26927	−0.69031	4.30969	−1.12546	4.43727	24.5
0.250	0.52360	−1.24590	−0.67449	4.32551	−1.09861	4.45069	25.0
0.255	0.52935	−1.22291	−0.65884	4.34116	−1.07212	4.46394	25.5
0.260	0.53507	−1.20030	−0.64335	4.35665	−1.04597	4.47702	26.0
0.265	0.54075	−1.17803	−0.62801	4.37199	−1.02014	4.48993	26.5
0.270	0.54640	−1.15610	−0.61281	4.38719	−0.99462	4.50269	27.0
0.275	0.55202	−1.13450	−0.59776	4.40224	−0.96940	4.51530	27.5
0.280	0.55760	−1.11321	−0.58284	4.41716	−0.94446	4.52777	28.0
0.285	0.56315	−1.09221	−0.56805	4.43195	−0.91979	4.54010	28.5
0.290	0.56868	−1.07151	−0.55338	4.44662	−0.89538	4.55231	29.0
0.295	0.57417	−1.05109	−0.53884	4.46116	−0.87122	4.56439	29.5
0.300	0.57964	−1.03093	−0.52440	4.47560	−0.84730	4.57635	30.0
0.305	0.58508	−1.01103	−0.51007	4.48993	−0.82360	4.58820	30.5
0.310	0.59050	−0.99138	−0.49585	4.50415	−0.80012	4.59994	31.0
0.315	0.59589	−0.97197	−0.48173	4.51827	−0.77685	4.61158	31.5
0.320	0.60126	−0.95279	−0.46770	4.53230	−0.75377	4.62311	32.0
0.325	0.60661	−0.93384	−0.45376	4.54624	−0.73089	4.63456	32.5
0.330	0.61194	−0.91510	−0.43991	4.56009	−0.70819	4.64591	33.0
0.335	0.61725	−0.89657	−0.42615	4.57385	−0.68566	4.65717	33.5
0.340	0.62253	−0.87824	−0.41246	4.58754	−0.66329	4.66835	34.0
0.345	0.62780	−0.86010	−0.39886	4.60114	−0.64109	4.67945	34.5
0.350	0.63305	−0.84215	−0.38532	4.61468	−0.61904	4.69048	35.0
0.355	0.63828	−0.82438	−0.37186	4.62814	−0.59713	4.70143	35.5
0.360	0.64350	−0.80679	−0.35846	4.64154	−0.57536	4.71232	36.0
0.365	0.64870	−0.78937	−0.34513	4.65487	−0.55373	4.72314	36.5
0.370	0.65389	−0.77211	−0.33185	4.66815	−0.53222	4.73389	37.0
0.375	0.65906	−0.75501	−0.31864	4.68136	−0.51083	4.74459	37.5
0.380	0.66422	−0.73807	−0.30548	4.69452	−0.48955	4.75523	38.0
0.385	0.66936	−0.72127	−0.29237	4.70763	−0.46838	4.76581	38.5
0.390	0.67449	−0.70462	−0.27932	4.72068	−0.44731	4.77634	39.0
0.395	0.67961	−0.68811	−0.26631	4.73369	−0.42634	4.78683	39.5
0.400	0.68472	−0.67173	−0.25335	4.74665	−0.40547	4.79727	40.0
0.405	0.68982	−0.65548	−0.24043	4.75957	−0.38467	4.80766	40.5
0.410	0.69490	−0.63935	−0.22754	4.77246	−0.36397	4.81802	41.0
0.415	0.69998	−0.62335	−0.21470	4.78530	−0.34333	4.82833	41.5
0.420	0.70505	−0.60747	−0.20189	4.79811	−0.32277	4.83861	42.0
0.425	0.71011	−0.59170	−0.18912	4.81088	−0.30228	4.84886	42.5
0.430	0.71517	−0.57604	−0.17637	4.82363	−0.28185	4.85907	43.0
0.435	0.72021	−0.56049	−0.16366	4.83634	−0.26148	4.86926	43.5
0.440	0.72525	−0.54504	−0.15097	4.84903	−0.24116	4.87942	44.0
0.445	0.73029	−0.52969	−0.13830	4.86170	−0.22089	4.88955	44.5

Table 7 *Continued*

Proportion (No. affected/ Total No.)	Arcsin $\sqrt{p}$ (radians)	Weibull	NED	Probit	Logit	Transformed Logit	Percent (%)
0.450	0.73531	−0.51444	−0.12566	4.87434	−0.20067	4.89966	45.0
0.455	0.74034	−0.49928	−0.11304	4.88696	−0.18049	4.90976	45.5
0.460	0.74536	−0.48421	−0.10043	4.89957	−0.16034	4.91983	46.0
0.465	0.75037	−0.46922	−0.08784	4.91216	−0.14023	4.92989	46.5
0.470	0.75538	−0.45432	−0.07527	4.92473	−0.12014	4.93993	47.0
0.475	0.76039	−0.43950	−0.06271	4.93729	−0.10008	4.94996	47.5
0.480	0.76539	−0.42476	−0.05015	4.94985	−0.08004	4.95998	48.0
0.485	0.77040	−0.41009	−0.03761	4.96239	−0.06002	4.96999	48.5
0.490	0.77540	−0.39550	−0.02507	4.97493	−0.04001	4.98000	49.0
0.495	0.78040	−0.38097	−0.01253	4.98747	−0.02000	4.99000	49.5
0.500	0.78540	−0.36651	0.00000	5.00000	0.00000	5.00000	50.0
0.505	0.79040	−0.35212	0.01253	5.01253	0.02000	5.01000	50.5
0.510	0.79540	−0.33778	0.02507	5.02507	0.04001	5.02000	51.0
0.515	0.80040	−0.32351	0.03761	5.03761	0.06002	5.03001	51.5
0.520	0.80540	−0.30929	0.05015	5.05015	0.08004	5.04002	52.0
0.525	0.81041	−0.29512	0.06271	5.06271	0.10008	5.05004	52.5
0.530	0.81542	−0.28101	0.07527	5.07527	0.12014	5.06007	53.0
0.535	0.82043	−0.26694	0.08784	5.08784	0.14023	5.07011	53.5
0.540	0.82544	−0.25292	0.10043	5.10043	0.16034	5.08017	54.0
0.545	0.83046	−0.23895	0.11304	5.11304	0.18049	5.09024	54.5
0.550	0.83548	−0.22501	0.12566	5.12566	0.20067	5.10034	55.0
0.555	0.84051	−0.21111	0.13830	5.13830	0.22089	5.11045	55.5
0.560	0.84554	−0.19726	0.15097	5.15097	0.24116	5.12058	56.0
0.565	0.85058	−0.18343	0.16366	5.16366	0.26148	5.13074	56.5
0.570	0.85563	−0.16964	0.17637	5.17637	0.28185	5.14093	57.0
0.575	0.86068	−0.15588	0.18912	5.18912	0.30228	5.15114	57.5
0.580	0.86574	−0.14214	0.20189	5.20189	0.32277	5.16139	58.0
0.585	0.87081	−0.12843	0.21470	5.21470	0.34333	5.17167	58.5
0.590	0.87589	−0.11474	0.22754	5.22754	0.36397	5.18198	59.0
0.595	0.88098	−0.10107	0.24043	5.24043	0.38467	5.19234	59.5
0.600	0.88608	−0.08742	0.25335	5.25335	0.40547	5.20273	60.0
0.605	0.89119	−0.07379	0.26631	5.26631	0.42634	5.21317	60.5
0.610	0.89631	−0.06017	0.27932	5.27932	0.44731	5.22366	61.0
0.615	0.90144	−0.04656	0.29237	5.29237	0.46838	5.23419	61.5
0.620	0.90658	0.03295	0.30548	5.30548	0.48955	5.24477	62.0
0.625	0.91174	0.01936	0.31864	5.31864	0.51083	5.25541	62.5
0.630	0.91691	0.00576	0.33185	5.33185	0.53222	5.26611	63.0
0.635	0.92209	0.00783	0.34513	5.34513	0.55373	5.27686	63.5
0.640	0.92730	0.02142	0.35846	5.35846	0.57536	5.28768	64.0
0.645	0.93251	0.03502	0.37186	5.37186	0.59713	5.29857	64.5
0.650	0.93774	0.04862	0.38532	5.38532	0.61904	5.30952	65.0
0.655	0.94299	0.06223	0.39886	5.39886	0.64109	5.32055	65.5
0.660	0.94826	0.07586	0.41246	5.41246	0.66329	5.33165	66.0
0.665	0.95355	0.08950	0.42615	5.42615	0.68566	5.34283	66.5
0.670	0.95886	0.10315	0.43991	5.43991	0.70819	5.35409	67.0
0.675	0.96418	0.11683	0.45376	5.45376	0.73089	5.36544	67.5

Table 7 *Continued*

Proportion (No. affected/ Total No.)	Arcsin $\sqrt{p}$ (radians)	Weibull	NED	Probit	Logit	Transformed Logit	Percent (%)
0.680	0.96953	0.13053	0.46770	5.46770	0.75377	5.37689	68.0
0.685	0.97490	0.14426	0.48173	5.48173	0.77685	5.38842	68.5
0.690	0.98030	0.15801	0.49585	5.49585	0.80012	5.40006	69.0
0.695	0.98571	0.17180	0.51007	5.51007	0.82360	5.41180	69.5
0.700	0.99116	0.18563	0.52440	5.52440	0.84730	5.42365	70.0
0.705	0.99663	0.19949	0.53884	5.53884	0.87122	5.43561	70.5
0.710	1.00212	0.21340	0.55338	5.55338	0.89538	5.44769	71.0
0.715	1.00764	0.22735	0.56805	5.56805	0.91979	5.45990	71.5
0.720	1.01320	0.24135	0.58284	5.58284	0.94446	5.47223	72.0
0.725	1.01878	0.25540	0.59776	5.59776	0.96940	5.48470	72.5
0.730	1.02440	0.26952	0.61281	5.61281	0.99462	5.49731	73.0
0.735	1.03004	0.28369	0.62801	5.62801	1.02014	5.51007	73.5
0.740	1.03573	0.29793	0.64335	5.64335	1.04597	5.52298	74.0
0.745	1.04144	0.31225	0.65884	5.65884	1.07212	5.53606	74.5
0.750	1.04720	0.32663	0.67449	5.67449	1.09861	5.54931	75.0
0.755	1.05299	0.34110	0.69031	5.69031	1.12546	5.56273	75.5
0.760	1.05882	0.35566	0.70630	5.70630	1.15268	5.57634	76.0
0.765	1.06470	0.37030	0.72248	5.72248	1.18029	5.59015	76.5
0.770	1.07062	0.38504	0.73885	5.73885	1.20831	5.60416	77.0
0.775	1.07658	0.39989	0.75542	5.75542	1.23676	5.61838	77.5
0.780	1.08259	0.41484	0.77219	5.77219	1.26567	5.63283	78.0
0.785	1.08865	0.42991	0.78919	5.78919	1.29505	5.64752	78.5
0.790	1.09476	0.44510	0.80642	5.80642	1.32493	5.66246	79.0
0.795	1.10093	0.46042	0.82389	5.82389	1.35533	5.67767	79.5
0.800	1.10715	0.47588	0.84162	5.84162	1.38629	5.69315	80.0
0.805	1.11343	0.49149	0.85962	5.85962	1.41784	5.70892	80.5
0.810	1.11977	0.50726	0.87790	5.87790	1.45001	5.72501	81.0
0.815	1.12617	0.52319	0.89647	5.89647	1.48283	5.74142	81.5
0.820	1.13265	0.53930	0.91537	5.91537	1.51635	5.75817	82.0
0.825	1.13919	0.55559	0.93459	5.93459	1.55060	5.77530	82.5
0.830	1.14581	0.57208	0.95417	5.95417	1.58563	5.79281	83.0
0.835	1.15250	0.58879	0.97411	5.97411	1.62149	5.81074	83.5
0.840	1.15928	0.60573	0.99446	5.99446	1.65823	5.82911	84.0
0.845	1.16614	0.62290	1.01522	6.01522	1.69591	5.84796	84.5
0.850	1.17310	0.64034	1.03643	6.03643	1.73460	5.86730	85.0
0.855	1.18015	0.65805	1.05812	6.05812	1.77437	5.88718	85.5
0.860	1.18730	0.67606	1.08032	6.08032	1.81529	5.90764	86.0
0.865	1.19456	0.69439	1.10306	6.10306	1.85745	5.92873	86.5
0.870	1.20193	0.71306	1.12639	6.12639	1.90096	5.95048	87.0
0.875	1.20943	0.73210	1.15035	6.15035	1.94591	5.97296	87.5
0.880	1.21705	0.75154	1.17499	6.17499	1.99243	5.99622	88.0
0.885	1.22482	0.77141	1.20036	6.20036	2.04066	6.02033	88.5
0.890	1.23273	0.79176	1.22653	6.22653	2.09074	6.04537	89.0
0.895	1.24080	0.81262	1.25357	6.25357	2.14286	6.07143	89.5
0.900	1.24905	0.83403	1.28155	6.28155	2.19722	6.09861	90.0

Table 7 *Continued*

Proportion (No. affected/ Total No.)	Arcsin $\sqrt{p}$ (radians)	Weibull	NED	Probit	Logit	Transformed Logit	Percent (%)
0.905	1.25747	0.85606	1.31058	6.31058	2.25406	6.12703	90.5
0.910	1.26610	0.87877	1.34076	6.34076	2.31363	6.15682	91.0
0.915	1.27495	0.90223	1.37220	6.37220	2.37627	6.18814	91.5
0.920	1.28404	0.92653	1.40507	6.40507	2.44235	6.22117	92.0
0.925	1.29339	0.95176	1.43953	6.43953	2.51231	6.25615	92.5
0.930	1.30303	0.97805	1.47579	6.47579	2.58669	6.29334	93.0
0.935	1.31300	1.00553	1.51410	6.51410	2.66616	6.33308	93.5
0.940	1.32333	1.03440	1.55477	6.55477	2.75154	6.37577	94.0
0.945	1.33407	1.06486	1.59819	6.59819	2.84385	6.42193	94.5
0.950	1.34528	1.09719	1.64485	6.64485	2.94444	6.47222	95.0
0.955	1.35704	1.13175	1.69540	6.69540	3.05505	6.52752	95.5
0.960	1.36944	1.16903	1.75069	6.75069	3.17805	6.58903	96.0
0.965	1.38260	1.20968	1.81191	6.81191	3.31678	6.65839	96.5
0.970	1.39671	1.25463	1.88079	6.88079	3.47610	6.73805	97.0
0.975	1.41202	1.30532	1.95996	6.95996	3.66356	6.83178	97.5
0.980	1.42890	1.36405	2.05375	7.05375	3.89182	6.94591	98.0
0.985	1.44801	1.43501	2.17009	7.17009	4.18459	7.09230	98.5
0.990	1.47063	1.52718	2.32635	7.32635	4.59512	7.29756	99.0
0.995	1.50003	1.66739	2.57583	7.57583	5.29330	7.64665	99.5
1.000	1.57080						100.0

Table 8 *E* values used to estimate 95% confidence intervals for LT50 using the Litchfield method (used first in Chapter 4)

x'	E
−2.5	1.00056
−2.4	1.00078
−2.3	1.00107
−2.2	1.00147
−2.1	1.00200
−2.0	1.00270
−1.9	1.00363
−1.8	1.00485
−1.7	1.00645
−1.6	1.00852
−1.5	1.01120
−1.4	1.01467
−1.3	1.01914
−1.2	1.02488
−1.1	1.03224
−1.0	1.04168
−0.9	1.05376
−0.8	1.06923
−0.7	1.08904
−0.6	1.11442
−0.5	1.14696
−0.4	1.18878
−0.3	1.24252
−0.2	1.31180
−0.1	1.40127
0.0	1.51708
0.1	1.66743
0.2	1.86310
0.3	2.11857
0.4	2.45318
0.5	2.89293
0.6	3.47293
0.7	4.24075
0.8	5.26121
0.9	6.62291
1.0	8.44766
1.1	10.90365
1.2	14.22420
1.3	18.73495
1.4	24.89192
1.5	33.33860

Table 8 *Continued*

x'	E
1.6	44.98586
1.7	61.13204
1.8	83.63826
1.9	115.18668
2.0	159.66335

Derived from Bliss (reference 37 in Chapter 4) with permission from the Association of Applied Biologists (Warwickshire, UK).

Table 9 Coefficients (a_{n-i+1}) for Shapiro-Wilk's test for normality (used first in Chapter 5)

$i\backslash n$	2	3	4	5	6	7	8	9	10
1	0.7071	0.7071	0.6872	0.6646	0.6431	0.6233	0.6052	0.5888	0.5739
2		0.0000	0.1677	0.2413	0.2806	0.3031	0.3164	0.3244	0.3291
3				0.0000	0.0875	0.1401	0.1743	0.1976	0.2141
4						0.0000	0.0561	0.0947	0.1224
5								0.0000	0.0399

$i\backslash n$	11	12	13	14	15	16	17	18	19	20
1	0.5601	0.5475	0.5359	0.5251	0.5150	0.5056	0.4968	0.4886	0.4808	0.4734
2	0.3315	0.3325	0.3325	0.3318	0.3306	0.3290	0.3273	0.3253	0.3232	0.3211
3	0.2260	0.2347	0.2412	0.2460	0.2495	0.2521	0.2540	0.2553	0.2561	0.2565
4	0.1429	0.1586	0.1707	0.1802	0.1878	0.1939	0.1988	0.2027	0.2059	0.2085
5	0.0695	0.0922	0.1099	0.1240	0.1353	0.1447	0.1524	0.1587	0.1641	0.1686
6	0.0000	0.0303	0.0539	0.0727	0.0880	0.1005	0.1109	0.1197	0.1271	0.1334
7			0.0000	0.0240	0.0433	0.0593	0.0725	0.0837	0.0932	0.1013
8					0.0000	0.0196	0.0359	0.0496	0.0612	0.0711
9							0.0000	0.0163	0.0303	0.0422
10									0.0000	0.0140

Table 9 Continued

$i\backslash n$	21	22	23	24	25	26	27	28	29	30
1	0.4643	0.4590	0.4542	0.4493	0.4450	0.4407	0.4366	0.4328	0.4291	0.4254
2	0.3185	0.3156	0.3126	0.3098	0.3069	0.3043	0.3018	0.2992	0.2968	0.2944
3	0.2578	0.2571	0.2563	0.2554	0.2543	0.2533	0.2522	0.2510	0.2499	0.2487
4	0.2119	0.2131	0.2139	0.2145	0.2148	0.2151	0.2152	0.2151	0.2150	0.2148
5	0.1736	0.1764	0.1787	0.1807	0.1822	0.1836	0.1848	0.1857	0.1864	0.1870
6	0.1399	0.1443	0.1480	0.1512	0.1539	0.1563	0.1584	0.1601	0.1616	0.1630
7	0.1092	0.1150	0.1201	0.1245	0.1283	0.1316	0.1346	0.1372	0.1395	0.1415
8	0.0804	0.0878	0.0941	0.0997	0.1046	0.1089	0.1128	0.1162	0.1192	0.1219
9	0.0530	0.0618	0.0696	0.0764	0.0823	0.0876	0.0923	0.0965	0.1002	0.1036
10	0.0263	0.0368	0.0459	0.0539	0.0610	0.0672	0.0728	0.0778	0.0822	0.0862
11	0.0000	0.0122	0.0228	0.0321	0.0403	0.0476	0.0540	0.0598	0.0650	0.0697
12			0.0000	0.0107	0.0200	0.0284	0.0358	0.0424	0.0483	0.0537
13					0.0000	0.0094	0.0178	0.0253	0.0320	0.0381
14							0.0000	0.0084	0.0159	0.0227
15									0.0000	0.0076

Table 9 Continued

i\n	31	32	33	34	35	36	37	38	39	40
1	0.4220	0.4188	0.4156	0.4127	0.4096	0.4068	0.4040	0.4015	0.3989	0.3964
2	0.2921	0.2898	0.2876	0.2854	0.2834	0.2813	0.2794	0.2774	0.2755	0.2737
3	0.2475	0.2463	0.2451	0.2439	0.2427	0.2415	0.2403	0.2391	0.2380	0.2368
4	0.2145	0.2141	0.2137	0.2132	0.2127	0.2121	0.2116	0.2110	0.2104	0.2098
5	0.1874	0.1878	0.1880	0.1882	0.1883	0.1883	0.1883	0.1881	0.1880	0.1878
6	0.1641	0.1651	0.1660	0.1667	0.1673	0.1678	0.1683	0.1686	0.1689	0.1691
7	0.1433	0.1449	0.1463	0.1475	0.1487	0.1496	0.1505	0.1513	0.1520	0.1526
8	0.1243	0.1265	0.1284	0.1301	0.1317	0.1331	0.1344	0.1356	0.1366	0.1376
9	0.1066	0.1093	0.1118	0.1140	0.1160	0.1179	0.1196	0.1211	0.1225	0.1237
10	0.0899	0.0931	0.0961	0.0988	0.1013	0.1036	0.1056	0.1075	0.1092	0.1108
11	0.0739	0.0777	0.0812	0.0844	0.0873	0.0900	0.0924	0.0947	0.0967	0.0986
12	0.0585	0.0629	0.0669	0.0706	0.0739	0.0770	0.0798	0.0824	0.0848	0.0870
13	0.0435	0.0485	0.0530	0.0572	0.0610	0.0645	0.0677	0.0706	0.0733	0.0759
14	0.0289	0.0344	0.0395	0.0441	0.0484	0.0523	0.0559	0.0592	0.0622	0.0651
15	0.0144	0.0206	0.0262	0.0314	0.0361	0.0404	0.0444	0.0481	0.0515	0.0546
16	0.0000	0.0068	0.0131	0.0187	0.0239	0.0287	0.0331	0.0372	0.0409	0.0444
17			0.0000	0.0062	0.0119	0.0172	0.0220	0.0264	0.0305	0.0343
18					0.0000	0.0057	0.0110	0.0158	0.0203	0.0244
19							0.0000	0.0053	0.0101	0.0146
20									0.0000	0.0049

Table 9 Continued

i\n	41	42	43	44	45	46	47	48	49	50
1	0.3940	0.3917	0.3894	0.3872	0.3850	0.3830	0.3808	0.3789	0.3770	0.3751
2	0.2719	0.2701	0.2684	0.2667	0.2651	0.2635	0.2620	0.2604	0.2589	0.2574
3	0.2357	0.2345	0.2334	0.2323	0.2313	0.2302	0.2291	0.2281	0.2271	0.2260
4	0.2091	0.2085	0.2078	0.2072	0.2065	0.2058	0.2052	0.2045	0.2038	0.2032
5	0.1876	0.1874	0.1871	0.1868	0.1865	0.1862	0.1859	0.1855	0.1851	0.1847
6	0.1693	0.1694	0.1695	0.1695	0.1695	0.1695	0.1695	0.1693	0.1692	0.1691
7	0.1531	0.1535	0.1539	0.1542	0.1545	0.1548	0.1550	0.1551	0.1553	0.1554
8	0.1384	0.1392	0.1398	0.1405	0.1410	0.1415	0.1420	0.1423	0.1427	0.1430
9	0.1249	0.1259	0.1269	0.1278	0.1286	0.1293	0.1300	0.1306	0.1312	0.1317
10	0.1123	0.1136	0.1149	0.1160	0.1170	0.1180	0.1189	0.1197	0.1205	0.1212
11	0.1004	0.1020	0.1035	0.1049	0.1062	0.1073	0.1085	0.1095	0.1105	0.1113
12	0.0891	0.0909	0.0927	0.0943	0.0959	0.0972	0.0986	0.0998	0.1010	0.1020
13	0.0782	0.0804	0.0824	0.0842	0.0860	0.0876	0.0892	0.0906	0.0919	0.0932
14	0.0677	0.0701	0.0724	0.0745	0.0765	0.0783	0.0801	0.0817	0.0832	0.0846
15	0.0575	0.0602	0.0628	0.0651	0.0673	0.0694	0.0713	0.0731	0.0748	0.0764
16	0.0476	0.0506	0.0534	0.0560	0.0584	0.0607	0.0628	0.0648	0.0667	0.0685
17	0.0379	0.0411	0.0442	0.0471	0.0497	0.0522	0.0546	0.0568	0.0588	0.0608
18	0.0283	0.0318	0.0352	0.0383	0.0412	0.0439	0.0465	0.0489	0.0511	0.0532
19	0.0188	0.0227	0.0263	0.0296	0.0328	0.0357	0.0385	0.0411	0.0436	0.0459
20	0.0094	0.0136	0.0175	0.0211	0.0245	0.0277	0.0307	0.0335	0.0361	0.0386
21	0.0000	0.0045	0.0087	0.0126	0.0163	0.0197	0.0229	0.0259	0.0288	0.0314
22			0.0000	0.0042	0.0081	0.0118	0.0153	0.0185	0.0215	0.0244
23					0.0000	0.0039	0.0076	0.0111	0.0143	0.0174
24							0.0000	0.0037	0.0071	0.0104
25									0.0000	0.0035

Table 10 **Percentage points of Shapiro-Wilk's W test for normality (used first in Chapter 5)**

$n\backslash\alpha$	0.01	0.02	0.05	0.10	0.50	0.90	0.95	0.98	0.99
3	0.753	0.756	0.767	0.789	0.959	0.998	0.999	1.000	1.000
4	0.687	0.707	0.748	0.792	0.935	0.987	0.992	0.996	0.997
5	0.686	0.715	0.762	0.806	0.927	0.979	0.986	0.991	0.993
6	0.713	0.743	0.788	0.826	0.927	0.974	0.981	0.986	0.989
7	0.730	0.760	0.803	0.838	0.928	0.972	0.979	0.985	0.988
8	0.749	0.778	0.818	0.851	0.932	0.972	0.978	0.984	0.987
9	0.764	0.791	0.829	0.859	0.935	0.972	0.978	0.984	0.986
10	0.781	0.806	0.842	0.869	0.938	0.972	0.978	0.983	0.986
11	0.792	0.817	0.850	0.876	0.940	0.973	0.979	0.984	0.986
12	0.805	0.828	0.859	0.883	0.943	0.973	0.979	0.984	0.986
13	0.814	0.837	0.866	0.889	0.945	0.974	0.979	0.984	0.986
14	0.825	0.846	0.874	0.895	0.947	0.975	0.980	0.984	0.986
15	0.835	0.855	0.881	0.901	0.950	0.975	0.980	0.984	0.987
16	0.844	0.863	0.887	0.906	0.952	0.976	0.981	0.985	0.987
17	0.851	0.869	0.892	0.910	0.954	0.977	0.981	0.985	0.987
18	0.858	0.874	0.897	0.914	0.956	0.978	0.982	0.986	0.988
19	0.863	0.879	0.901	0.917	0.957	0.978	0.982	0.986	0.988
20	0.868	0.884	0.905	0.920	0.959	0.979	0.983	0.986	0.988
21	0.873	0.888	0.908	0.923	0.960	0.980	0.983	0.987	0.989
22	0.878	0.892	0.911	0.926	0.961	0.980	0.984	0.987	0.989
23	0.881	0.895	0.914	0.928	0.962	0.981	0.984	0.987	0.989
24	0.884	0.898	0.916	0.930	0.963	0.981	0.984	0.987	0.989
25	0.888	0.901	0.918	0.931	0.964	0.981	0.985	0.988	0.989
26	0.891	0.904	0.920	0.933	0.965	0.982	0.985	0.988	0.989
27	0.894	0.906	0.923	0.935	0.965	0.982	0.985	0.988	0.990
28	0.896	0.908	0.924	0.936	0.966	0.982	0.985	0.988	0.990
29	0.898	0.910	0.926	0.937	0.966	0.982	0.985	0.988	0.990
30	0.900	0.912	0.927	0.939	0.967	0.983	0.985	0.988	0.990
31	0.902	0.914	0.929	0.940	0.967	0.983	0.986	0.988	0.990
32	0.904	0.915	0.930	0.941	0.968	0.983	0.986	0.988	0.990
33	0.906	0.917	0.931	0.942	0.968	0.983	0.986	0.989	0.990
34	0.908	0.919	0.933	0.943	0.969	0.983	0.986	0.989	0.990
35	0.910	0.920	0.934	0.944	0.969	0.984	0.986	0.989	0.990
36	0.912	0.922	0.935	0.945	0.970	0.984	0.986	0.989	0.990
37	0.914	0.924	0.936	0.946	0.970	0.984	0.987	0.989	0.990
38	0.916	0.925	0.938	0.947	0.971	0.984	0.987	0.989	0.990
39	0.917	0.927	0.939	0.948	0.971	0.984	0.987	0.989	0.991
40	0.919	0.928	0.940	0.949	0.972	0.985	0.987	0.989	0.991
41	0.920	0.929	0.941	0.950	0.972	0.985	0.987	0.989	0.991
42	0.922	0.930	0.942	0.951	0.972	0.985	0.987	0.989	0.991
43	0.923	0.932	0.943	0.951	0.973	0.985	0.987	0.990	0.991
44	0.924	0.933	0.944	0.952	0.973	0.985	0.987	0.990	0.991
45	0.926	0.934	0.945	0.953	0.973	0.985	0.988	0.990	0.991

Table 10   *Continued*

$n\backslash\alpha$	0.01	0.02	0.05	0.10	0.50	0.90	0.95	0.98	0.99
46	0.927	0.935	0.945	0.953	0.974	0.985	0.988	0.990	0.991
47	0.928	0.936	0.946	0.954	0.974	0.985	0.988	0.990	0.991
48	0.929	0.937	0.947	0.954	0.974	0.985	0.988	0.990	0.991
49	0.929	0.937	0.947	0.955	0.974	0.985	0.988	0.990	0.991
50	0.930	0.938	0.947	0.955	0.974	0.985	0.988	0.990	0.991

Reproduced from Shapiro and Wilk (reference 17 in Chapter 5) with permission from the *Biometrika* Trust (London).

Table 11 **Dunnett's *t* for one-sided comparisons between *p* treatment means and a control for α = 0.01 (used first in Chapter 5)**

df*p*	1	2	3	4	5	6	7	8	9
5	3.37	3.90	4.21	4.43	4.60	4.73	4.85	4.94	5.03
6	3.14	3.61	3.88	4.07	4.21	4.33	4.43	4.51	4.59
7	3.00	3.42	3.66	3.83	3.96	4.07	4.15	4.23	4.30
8	2.90	3.29	3.51	3.67	3.79	3.88	3.96	4.03	4.09
9	2.82	3.19	3.40	3.55	3.66	3.75	3.82	3.89	3.94
10	2.76	3.11	3.31	3.45	3.56	3.64	3.71	3.78	3.83
11	2.72	3.06	3.25	3.38	3.48	3.56	3.63	3.69	3.74
12	2.68	3.01	3.19	3.32	3.42	3.50	3.56	3.62	3.67
13	2.65	2.97	3.15	3.27	3.37	3.44	3.51	3.56	3.61
14	2.62	2.94	3.11	3.23	3.32	3.40	3.46	3.51	3.56
15	2.60	2.91	3.08	3.20	3.29	3.36	3.42	3.47	3.52
16	2.58	2.88	3.05	3.17	3.26	3.33	3.39	3.44	3.48
17	2.57	2.86	3.03	3.14	3.23	3.30	3.36	3.41	3.45
18	2.55	2.84	3.01	3.12	3.21	3.27	3.33	3.38	3.42
19	2.54	2.83	2.99	3.10	3.18	3.25	3.31	3.36	3.40
20	2.53	2.81	2.97	3.08	3.17	3.23	3.29	3.34	3.38
24	2.49	2.77	2.92	3.03	3.11	3.17	3.22	3.27	3.31
30	2.46	2.72	2.87	2.97	3.05	3.11	3.16	3.21	3.24
40	2.42	2.68	2.82	2.92	2.99	3.05	3.10	3.14	3.18
60	2.39	2.64	2.78	2.87	2.94	3.00	3.04	3.08	3.12
120	2.36	2.60	2.73	2.82	2.89	2.94	2.99	3.03	3.06
∞	2.33	2.56	2.68	2.77	2.84	2.89	2.93	2.97	3.00

Table 12 **Dunnett's *t* for one-sided comparisons between *p* treatment means and a control for α = 0.05 (used first in Chapter 5).**

$df\backslash p$	1	2	3	4	5	6	7	8	9
5	2.02	2.44	2.68	2.85	2.98	3.08	3.16	3.24	3.30
6	1.94	2.34	2.56	2.71	2.83	2.92	3.00	3.07	3.12
7	1.89	2.27	2.48	2.62	2.73	2.82	2.89	2.95	3.01
8	1.86	2.22	2.42	2.55	2.66	2.74	2.81	2.87	2.92
9	1.83	2.18	2.37	2.50	2.60	2.68	2.75	2.81	2.86
10	1.81	2.15	2.34	2.47	2.56	2.64	2.70	2.76	2.81
11	1.80	2.13	2.31	2.44	2.53	2.60	2.67	2.72	2.77
12	1.78	2.11	2.29	2.41	2.50	2.58	2.64	2.69	2.74
13	1.77	2.09	2.27	2.39	2.48	2.55	2.61	2.66	2.71
14	1.76	2.08	2.25	2.37	2.46	2.53	2.59	2.64	2.69
15	1.75	2.07	2.24	2.36	2.44	2.51	2.57	2.62	2.67
16	1.75	2.06	2.23	2.34	2.43	2.50	2.56	2.61	2.65
17	1.74	2.05	2.22	2.33	2.42	2.49	2.54	2.59	2.64
18	1.73	2.04	2.21	2.32	2.41	2.48	2.53	2.58	2.62
19	1.73	2.03	2.20	2.31	2.40	2.47	2.52	2.57	2.61
20	1.72	2.03	2.19	2.30	2.39	2.46	2.51	2.56	2.60
24	1.71	2.01	2.17	2.28	2.36	2.43	2.48	2.53	2.57
30	1.70	1.99	2.15	2.25	2.33	2.40	2.45	2.50	2.54
40	1.68	1.97	2.13	2.23	2.31	2.37	2.42	2.47	2.51
60	1.67	1.95	2.10	2.21	2.28	2.35	2.39	2.44	2.48
120	1.66	1.93	2.08	2.18	2.26	2.32	2.37	2.41	2.45
∞	1.64	1.92	2.06	2.16	2.23	2.29	2.34	2.38	2.42

Table 13 Dunnett's t for two-sided comparisons between p treatment means and a control for $\alpha = 0.01$ (used first in Chapter 5)

$df \backslash p$	1	2	3	4	5	6	7	8	9	10	11	12	15	20
5	4.03	$4.63^{1.8}$	$4.98^{3.0}$	$5.22^{3.9}$	$5.41^{4.6}$	$5.56^{5.2}$	$5.69^{5.7}$	$5.80^{6.1}$	$5.89^{6.5}$	$5.98^{6.9}$	$6.05^{7.2}$	$6.12^{7.4}$	$6.30^{8.1}$	$6.52^{9.0}$
6	3.71	$4.21^{1.6}$	$4.51^{2.7}$	$4.71^{3.5}$	$4.87^{4.1}$	$5.00^{4.6}$	$5.10^{5.1}$	$5.20^{5.5}$	$5.28^{5.8}$	$5.35^{6.1}$	$5.41^{6.4}$	$5.47^{6.7}$	$5.62^{7.3}$	$5.81^{8.1}$
7	3.50	$3.95^{1.5}$	$4.21^{2.4}$	$4.39^{3.1}$	$4.53^{3.7}$	$4.64^{4.2}$	$4.74^{4.6}$	$4.82^{5.0}$	$4.89^{5.3}$	$4.95^{5.6}$	$5.01^{5.8}$	$5.06^{6.1}$	$5.19^{6.7}$	$5.36^{7.4}$
8	3.36	$3.77^{1.3}$	$4.00^{2.2}$	$4.17^{2.9}$	$4.29^{3.4}$	$4.40^{3.9}$	$4.48^{4.2}$	$4.56^{4.6}$	$4.62^{4.9}$	$4.68^{5.1}$	$4.73^{5.4}$	$4.78^{5.6}$	$4.90^{6.1}$	$5.05^{6.9}$
9	3.25	$3.63^{1.2}$	$3.85^{2.1}$	$4.01^{2.7}$	$4.12^{3.2}$	$4.22^{3.6}$	$4.30^{3.9}$	$4.37^{4.2}$	$4.43^{4.5}$	$4.48^{4.8}$	$4.53^{5.0}$	$4.57^{5.2}$	$4.68^{5.7}$	$4.82^{6.4}$
10	3.17	$3.53^{1.2}$	$3.74^{1.9}$	$3.88^{2.5}$	$3.99^{3.0}$	$4.08^{3.4}$	$4.16^{3.7}$	$4.22^{4.0}$	$4.28^{4.2}$	$4.33^{4.5}$	$4.37^{4.7}$	$4.42^{4.9}$	$4.52^{5.4}$	$4.65^{6.0}$
11	3.11	$3.45^{1.1}$	$3.65^{1.8}$	$3.79^{2.4}$	$3.89^{2.8}$	$3.98^{3.2}$	$4.05^{3.5}$	$4.11^{3.8}$	$4.16^{4.0}$	$4.21^{4.2}$	$4.25^{4.4}$	$4.29^{4.6}$	$4.39^{5.1}$	$4.52^{5.7}$
12	3.05	$3.39^{1.1}$	$3.58^{1.7}$	$3.71^{2.3}$	$3.81^{2.7}$	$3.89^{3.0}$	$3.96^{3.3}$	$4.02^{3.6}$	$4.07^{3.8}$	$4.12^{4.0}$	$4.16^{4.2}$	$4.19^{4.4}$	$4.29^{4.8}$	$4.41^{5.4}$
13	3.01	$3.33^{1.0}$	$3.52^{1.7}$	$3.65^{2.2}$	$3.74^{2.6}$	$3.82^{2.9}$	$3.89^{3.2}$	$3.94^{3.4}$	$3.99^{3.6}$	$4.04^{3.8}$	$4.08^{4.0}$	$4.11^{4.2}$	$4.20^{4.6}$	$4.32^{5.2}$
14	2.98	$3.29^{1.0}$	$3.47^{1.6}$	$3.59^{2.1}$	$3.69^{2.5}$	$3.76^{2.8}$	$3.83^{3.0}$	$3.88^{3.3}$	$3.93^{3.5}$	$3.97^{3.7}$	$4.01^{3.9}$	$4.05^{4.0}$	$4.13^{4.4}$	$4.24^{5.0}$
15	2.95	$3.25^{0.9}$	$3.43^{1.5}$	$3.55^{2.0}$	$3.64^{2.4}$	$3.71^{2.7}$	$3.78^{2.9}$	$3.83^{3.2}$	$3.88^{3.4}$	$3.92^{3.6}$	$3.95^{3.7}$	$3.99^{3.9}$	$4.07^{4.3}$	$4.18^{4.8}$
16	2.92	$3.22^{0.9}$	$3.39^{1.5}$	$3.51^{1.9}$	$3.60^{2.3}$	$3.67^{2.6}$	$3.73^{2.8}$	$3.78^{3.1}$	$3.83^{3.3}$	$3.87^{3.4}$	$3.91^{3.6}$	$3.94^{3.8}$	$4.02^{4.1}$	$4.13^{4.6}$
17	2.90	$3.19^{0.9}$	$3.36^{1.5}$	$3.47^{1.9}$	$3.56^{2.2}$	$3.63^{2.5}$	$3.69^{2.7}$	$3.74^{3.0}$	$3.79^{3.2}$	$3.83^{3.3}$	$3.86^{3.5}$	$3.90^{3.6}$	$3.98^{4.0}$	$4.08^{4.5}$
18	2.88	$3.17^{0.9}$	$3.33^{1.4}$	$3.44^{1.8}$	$3.53^{2.2}$	$3.60^{2.4}$	$3.66^{2.7}$	$3.71^{2.9}$	$3.75^{3.1}$	$3.79^{3.2}$	$3.83^{3.4}$	$3.86^{3.5}$	$3.94^{3.9}$	$4.04^{4.4}$
19	2.86	$3.15^{0.9}$	$3.31^{1.4}$	$3.42^{1.8}$	$3.50^{2.1}$	$3.57^{2.4}$	$3.63^{2.6}$	$3.68^{2.8}$	$3.72^{3.0}$	$3.76^{3.2}$	$3.79^{3.3}$	$3.83^{3.4}$	$3.90^{3.8}$	$4.00^{4.3}$
20	2.85	$3.13^{0.8}$	$3.29^{1.4}$	$3.40^{1.7}$	$3.48^{2.1}$	$3.55^{2.3}$	$3.60^{2.5}$	$3.65^{2.7}$	$3.69^{2.9}$	$3.73^{3.1}$	$3.77^{3.2}$	$3.80^{3.4}$	$3.87^{3.7}$	$3.97^{4.2}$
24	2.80	$3.07^{0.8}$	$3.22^{1.3}$	$3.32^{1.6}$	$3.40^{1.9}$	$3.47^{2.1}$	$3.52^{2.4}$	$3.57^{2.5}$	$3.61^{2.7}$	$3.64^{2.8}$	$3.68^{3.0}$	$3.70^{3.1}$	$3.78^{3.4}$	$3.87^{3.8}$
30	2.75	$3.01^{0.7}$	$3.15^{1.2}$	$3.25^{1.5}$	$3.33^{1.8}$	$3.39^{2.0}$	$3.44^{2.2}$	$3.49^{2.3}$	$3.52^{2.5}$	$3.56^{2.6}$	$3.59^{2.7}$	$3.62^{2.8}$	$3.69^{3.1}$	$3.78^{3.5}$
40	2.70	$2.95^{0.7}$	$3.09^{1.1}$	$3.19^{1.4}$	$3.26^{1.6}$	$3.32^{1.8}$	$3.37^{2.0}$	$3.41^{2.1}$	$3.44^{2.3}$	$3.48^{2.4}$	$3.51^{2.5}$	$3.53^{2.6}$	$3.60^{2.8}$	$3.68^{3.2}$
60	2.66	$2.90^{0.6}$	$3.03^{1.0}$	$3.12^{1.3}$	$3.19^{1.5}$	$3.25^{1.6}$	$3.29^{1.8}$	$3.33^{1.9}$	$3.37^{2.0}$	$3.40^{2.1}$	$3.42^{2.2}$	$3.45^{2.3}$	$3.51^{2.5}$	$3.59^{2.8}$
120	2.62	$2.85^{0.6}$	$2.97^{0.9}$	$3.06^{1.1}$	$3.12^{1.3}$	$3.18^{1.5}$	$3.22^{1.6}$	$3.26^{1.7}$	$3.29^{1.8}$	$3.32^{1.9}$	$3.35^{2.0}$	$3.37^{2.1}$	$3.43^{2.2}$	$3.51^{2.5}$
∞	2.58	$2.79^{0.5}$	$2.92^{0.8}$	$3.00^{1.0}$	$3.06^{1.2}$	$3.11^{1.3}$	$3.15^{1.4}$	$3.19^{1.5}$	$3.22^{1.6}$	$3.25^{1.7}$	$3.27^{1.7}$	$3.29^{1.8}$	$3.35^{1.9}$	$3.42^{2.2}$

Reproduced from Dunnett (Reference 22 in Chapter 5) with permission from the Biometric Society.

Table 14 Dunnett's t for two-sided comparisons between p treatment means and a control for α = 0.05 (used first in Chapter 5)

df\p	1	2	3	4	5	6	7	8	9	10	11	12	15	20
5	2.57	$3.03^{2.3}$	$3.29^{3.6}$	$3.48^{4.6}$	$3.62^{5.4}$	$3.73^{5.9}$	$3.82^{6.4}$	$3.90^{6.8}$	$3.97^{7.2}$	$4.03^{7.5}$	$4.09^{7.8}$	$4.14^{8.0}$	$4.26^{8.7}$	$4.42^{9.4}$
6	2.45	$2.86^{2.1}$	$3.10^{3.4}$	$3.26^{4.3}$	$3.39^{5.0}$	$3.49^{5.6}$	$3.57^{6.0}$	$3.64^{6.4}$	$3.71^{6.8}$	$3.76^{7.1}$	$3.81^{7.4}$	$3.86^{7.6}$	$3.97^{8.2}$	$4.11^{9.0}$
7	2.36	$2.75^{2.0}$	$2.97^{3.2}$	$3.12^{4.1}$	$3.24^{4.8}$	$3.33^{5.3}$	$3.41^{5.7}$	$3.47^{6.1}$	$3.53^{6.5}$	$3.58^{6.7}$	$3.63^{7.0}$	$3.67^{7.2}$	$3.78^{7.8}$	$3.91^{8.6}$
8	2.31	$2.67^{2.0}$	$2.88^{3.1}$	$3.02^{3.9}$	$3.13^{4.5}$	$3.22^{5.1}$	$3.29^{5.5}$	$3.35^{5.9}$	$3.41^{6.2}$	$3.46^{6.5}$	$3.50^{6.7}$	$3.54^{6.9}$	$3.64^{7.5}$	$3.76^{8.2}$
9	2.26	$2.61^{1.9}$	$2.81^{3.0}$	$2.95^{3.8}$	$3.05^{4.4}$	$3.14^{4.9}$	$3.20^{5.3}$	$3.26^{5.6}$	$3.32^{5.9}$	$3.36^{6.2}$	$3.40^{6.5}$	$3.44^{6.7}$	$3.53^{7.2}$	$3.65^{7.9}$
10	2.23	$2.57^{1.8}$	$2.76^{2.9}$	$2.89^{3.6}$	$2.99^{4.2}$	$3.07^{4.7}$	$3.14^{5.1}$	$3.19^{5.4}$	$3.24^{5.7}$	$3.29^{6.0}$	$3.33^{6.2}$	$3.36^{6.5}$	$3.45^{7.0}$	$3.57^{7.7}$
11	2.20	$2.53^{1.8}$	$2.72^{2.8}$	$2.84^{3.5}$	$2.94^{4.1}$	$3.02^{4.6}$	$3.08^{4.9}$	$3.14^{5.3}$	$3.19^{5.6}$	$3.23^{5.8}$	$3.27^{6.1}$	$3.30^{6.3}$	$3.39^{6.8}$	$3.50^{7.5}$
12	2.18	$2.50^{1.7}$	$2.68^{2.7}$	$2.81^{3.4}$	$2.90^{4.0}$	$2.98^{4.4}$	$3.04^{4.8}$	$3.09^{5.1}$	$3.14^{5.4}$	$3.18^{5.7}$	$3.22^{5.9}$	$3.25^{6.1}$	$3.34^{6.6}$	$3.45^{7.3}$
13	2.16	$2.48^{1.7}$	$2.65^{2.7}$	$2.78^{3.4}$	$2.87^{3.9}$	$2.94^{4.3}$	$3.00^{4.7}$	$3.06^{5.0}$	$3.10^{5.3}$	$3.14^{5.5}$	$3.18^{5.8}$	$3.21^{6.0}$	$3.29^{6.5}$	$3.40^{7.1}$
14	2.14	$2.46^{1.7}$	$2.63^{2.6}$	$2.75^{3.3}$	$2.84^{3.8}$	$2.91^{4.2}$	$2.97^{4.6}$	$3.02^{4.9}$	$3.07^{5.2}$	$3.11^{5.4}$	$3.14^{5.6}$	$3.18^{5.8}$	$3.26^{6.3}$	$3.36^{7.0}$
15	2.13	$2.44^{1.7}$	$2.61^{2.6}$	$2.73^{3.2}$	$2.82^{3.8}$	$2.89^{4.2}$	$2.95^{4.5}$	$3.00^{4.8}$	$3.04^{5.1}$	$3.08^{5.3}$	$3.12^{5.5}$	$3.15^{5.7}$	$3.23^{6.2}$	$3.33^{6.8}$
16	2.12	$2.42^{1.6}$	$2.59^{2.5}$	$2.71^{3.2}$	$2.80^{3.7}$	$2.87^{4.1}$	$2.92^{4.4}$	$2.97^{4.7}$	$3.02^{5.0}$	$3.06^{5.2}$	$3.09^{5.4}$	$3.12^{5.6}$	$3.20^{6.1}$	$3.30^{6.7}$
17	2.11	$2.41^{1.6}$	$2.58^{2.5}$	$2.69^{3.1}$	$2.78^{3.6}$	$2.85^{4.0}$	$2.90^{4.4}$	$2.95^{4.7}$	$3.00^{4.9}$	$3.03^{5.1}$	$3.07^{5.3}$	$3.10^{5.5}$	$3.18^{6.0}$	$3.27^{6.6}$
18	2.10	$2.40^{1.6}$	$2.56^{2.5}$	$2.68^{3.1}$	$2.76^{3.6}$	$2.83^{4.0}$	$2.89^{4.3}$	$2.94^{4.6}$	$2.98^{4.8}$	$3.01^{5.1}$	$3.05^{5.3}$	$3.08^{5.4}$	$3.16^{5.9}$	$3.25^{6.5}$
19	2.09	$2.39^{1.6}$	$2.55^{2.5}$	$2.66^{3.1}$	$2.75^{3.5}$	$2.81^{3.9}$	$2.87^{4.2}$	$2.92^{4.5}$	$2.96^{4.8}$	$3.00^{5.0}$	$3.03^{5.2}$	$3.06^{5.4}$	$3.14^{5.8}$	$3.23^{6.4}$
20	2.09	$2.38^{1.6}$	$2.54^{2.4}$	$2.65^{3.0}$	$2.73^{3.5}$	$2.80^{3.9}$	$2.86^{4.2}$	$2.90^{4.5}$	$2.95^{4.7}$	$2.98^{4.9}$	$3.02^{5.1}$	$3.05^{5.3}$	$3.12^{5.7}$	$3.22^{6.3}$
24	2.06	$2.35^{1.5}$	$2.51^{2.3}$	$2.61^{2.9}$	$2.70^{3.4}$	$2.76^{3.7}$	$2.81^{4.0}$	$2.86^{4.3}$	$2.90^{4.5}$	$2.94^{4.7}$	$2.97^{4.9}$	$3.00^{5.1}$	$3.07^{5.5}$	$3.16^{6.0}$
30	2.04	$2.32^{1.5}$	$2.47^{2.3}$	$2.58^{2.8}$	$2.66^{3.2}$	$2.72^{3.6}$	$2.77^{3.9}$	$2.82^{4.1}$	$2.86^{4.3}$	$2.89^{4.5}$	$2.92^{4.7}$	$2.95^{4.8}$	$3.02^{5.2}$	$3.11^{5.8}$
40	2.02	$2.29^{1.4}$	$2.44^{2.2}$	$2.54^{2.7}$	$2.62^{3.1}$	$2.68^{3.4}$	$2.73^{3.7}$	$2.77^{3.9}$	$2.81^{4.1}$	$2.85^{4.3}$	$2.87^{4.5}$	$2.90^{4.6}$	$2.97^{5.0}$	$3.06^{5.5}$
60	2.00	$2.27^{1.4}$	$2.41^{2.1}$	$2.51^{2.6}$	$2.58^{3.0}$	$2.64^{3.3}$	$2.69^{3.5}$	$2.73^{3.7}$	$2.77^{3.9}$	$2.80^{4.1}$	$2.83^{4.2}$	$2.86^{4.4}$	$2.92^{4.7}$	$3.00^{5.1}$
120	1.98	$2.24^{1.3}$	$2.38^{2.0}$	$2.47^{2.5}$	$2.55^{2.8}$	$2.60^{3.1}$	$2.65^{3.3}$	$2.69^{3.5}$	$2.73^{3.7}$	$2.76^{3.8}$	$2.79^{4.0}$	$2.81^{4.1}$	$2.87^{4.4}$	$2.95^{4.8}$
∞	1.96	$2.21^{1.3}$	$2.35^{1.9}$	$2.44^{2.3}$	$2.51^{2.7}$	$2.57^{2.9}$	$2.61^{3.1}$	$2.65^{3.3}$	$2.69^{3.5}$	$2.72^{3.6}$	$2.74^{3.7}$	$2.77^{3.8}$	$2.83^{4.1}$	$2.91^{4.5}$

Reproduced from Dunnett (reference 22 in Chapter 5) with permission from the Biometric Society.

Table 15 **Bonferroni's adjusted *t* values for one-sided test and α = 0.01**
(Values generated as described in Chapter 5)

df\p	1	2	3	4	5	6	7	8	9	10
2	6.9646	9.9248	12.1861	14.0890	15.7639	17.2772	18.6682	19.9625	21.1778	22.3271
3	4.5407	5.8909	6.7411	7.4533	8.0526	8.5752	9.0416	9.4649	9.8538	10.2145
4	3.7470	4.6041	5.1668	5.5976	5.9514	6.2541	6.5201	6.7583	6.9746	7.1732
5	3.3649	4.0321	4.4558	4.7733	5.0302	5.2474	5.4364	5.6042	5.7554	5.8934
6	3.1427	3.7074	4.0579	4.3168	4.5241	4.6979	4.8481	4.9807	5.0996	5.2076
7	2.9980	3.4995	3.8055	4.0293	4.2071	4.3553	4.4827	4.5946	4.6947	4.7853
8	2.8965	3.3554	3.6319	3.8325	3.9910	4.1224	4.2349	4.3335	4.4214	4.5008
9	2.8214	3.2498	3.5054	3.6897	3.8345	3.9542	4.0564	4.1458	4.2252	4.2968
10	2.7638	3.1693	3.4093	3.5814	3.7162	3.8273	3.9220	4.0045	4.0778	4.1437
11	2.7181	3.1058	3.3338	3.4966	3.6238	3.7283	3.8172	3.8945	3.9631	4.0247
12	2.6810	3.0545	3.2729	3.4284	3.5495	3.6489	3.7332	3.8065	3.8714	3.9296
13	2.6503	3.0123	3.2229	3.3725	3.4887	3.5838	3.6645	3.7345	3.7965	3.8520
14	2.6245	2.9768	3.1811	3.3257	3.4379	3.5296	3.6072	3.6746	3.7341	3.7874
15	2.6025	2.9467	3.1456	3.2860	3.3948	3.4837	3.5588	3.6239	3.6814	3.7328
16	2.5835	2.9208	3.1150	3.2520	3.3579	3.4443	3.5173	3.5805	3.6363	3.6862
17	2.5669	2.8982	3.0885	3.2224	3.3259	3.4102	3.4814	3.5429	3.5972	3.6458
18	2.5524	2.8784	3.0653	3.1966	3.2979	3.3804	3.4499	3.5101	3.5631	3.6105
19	2.5395	2.8609	3.0447	3.1737	3.2731	3.3540	3.4222	3.4812	3.5330	3.5794
20	2.5280	2.8453	3.0264	3.1534	3.2512	3.3306	3.3976	3.4554	3.5063	3.5518
21	2.5177	2.8314	3.0101	3.1352	3.2315	3.3097	3.3756	3.4325	3.4825	3.5272
22	2.5083	2.8188	2.9953	3.1188	3.2138	3.2909	3.3558	3.4118	3.4610	3.5050
23	2.4999	2.8073	2.9820	3.1040	3.1978	3.2739	3.3379	3.3931	3.4416	3.4850
24	2.4922	2.7969	2.9698	3.0905	3.1832	3.2584	3.3216	3.3761	3.4240	3.4668
25	2.4851	2.7874	2.9587	3.0782	3.1699	3.2443	3.3067	3.3606	3.4080	3.4502
26	2.4786	2.7787	2.9485	3.0669	3.1577	3.2313	3.2931	3.3464	3.3933	3.4350
27	2.4727	2.7707	2.9391	3.0565	3.1465	3.2194	3.2806	3.3334	3.3797	3.4210
28	2.4671	2.7633	2.9305	3.0469	3.1362	3.2084	3.2691	3.3214	3.3673	3.4082
29	2.4620	2.7564	2.9225	3.0380	3.1266	3.1982	3.2584	3.3102	3.3557	3.3962
30	2.4573	2.7500	2.9150	3.0298	3.1177	3.1888	3.2485	3.2999	3.3450	3.3852
35	2.4377	2.7238	2.8845	2.9960	3.0813	3.1502	3.2080	3.2577	3.3013	3.3400
40	2.4233	2.7045	2.8620	2.9712	3.0545	3.1218	3.1782	3.2266	3.2691	3.3069
45	2.4121	2.6896	2.8447	2.9521	3.0340	3.1000	3.1553	3.2028	3.2445	3.2815
50	2.4033	2.6778	2.8310	2.9370	3.0177	3.0828	3.1373	3.1840	3.2250	3.2614
55	2.3961	2.6682	2.8199	2.9247	3.0045	3.0688	3.1226	3.1688	3.2092	3.2451
60	2.3901	2.6603	2.8107	2.9146	2.9936	3.0573	3.1105	3.1562	3.1962	3.2317
70	2.3808	2.6479	2.7964	2.8987	2.9766	3.0393	3.0916	3.1366	3.1759	3.2108
80	2.3739	2.6387	2.7857	2.8870	2.9640	3.0259	3.0776	3.1220	3.1608	3.1953
90	2.3685	2.6316	2.7774	2.8779	3.9542	3.0156	3.0668	3.1108	3.1492	3.1833
100	2.3642	2.6259	2.7709	2.8707	2.9464	3.0073	3.0582	3.1018	3.1399	3.1737
110	2.3607	2.6213	2.7655	2.8648	2.9401	3.0007	3.0512	3.0945	3.1324	3.1660
120	2.3578	2.6174	2.7611	2.8599	2.9348	2.9951	3.0454	3.0885	3.1261	3.1595
250	2.3414	2.5956	2.7359	2.8322	2.9051	2.9637	3.0125	3.0543	3.0908	3.1232
500	2.3338	2.5857	2.7244	2.8195	2.8916	2.9494	2.9975	3.0387	3.0747	3.1066
1000	2.3301	2.5808	2.7187	2.8133	2.8849	2.9423	2.9901	3.0310	3.0667	3.0984

Table 15 *Continued*

df\p	11	12	13	14	15	16	17	18	19
2	23.4201	24.4643	25.4657	26.4292	27.3587	28.2577	29.1290	29.9750	30.7977
3	10.5517	10.8688	11.1685	11.4532	11.7244	11.9838	12.2324	12.4715	12.7017
4	7.3571	7.5287	7.6896	7.8414	7.9851	8.1216	8.2518	8.3763	8.4957
5	6.0205	6.1384	6.2484	6.3518	6.4492	6.5414	6.6291	6.7126	6.7924
6	5.3067	5.3982	5.4834	5.5632	5.6382	5.7090	5.7761	5.8399	5.9007
7	4.8681	4.9445	5.0154	5.0815	5.1437	5.2022	5.2576	5.3101	5.3601
8	4.5732	4.6398	4.7015	4.7590	4.8129	4.8636	4.9116	4.9570	5.0001
9	4.3620	4.4219	4.4773	4.5288	4.5771	4.6224	4.6652	4.7058	4.7443
10	4.2036	4.2586	4.3094	4.3567	4.4008	4.4423	4.4814	4.5184	4.5535
11	4.0807	4.1319	4.1793	4.2232	4.2643	4.3028	4.3391	4.3735	4.4060
12	3.9825	4.0308	4.0755	4.1169	4.1555	4.1918	4.2259	4.2582	4.2887
13	3.9023	3.9484	3.9908	4.0302	4.0669	4.1013	4.1337	4.1643	4.1933
14	3.8357	3.8798	3.9205	3.9582	3.9933	4.0263	4.0572	4.0865	4.1142
15	3.7794	3.8220	3.8612	3.8975	3.9313	3.9630	3.9928	4.0209	4.0475
16	3.7313	3.7725	3.8104	3.8456	3.8783	3.9089	3.9377	3.9649	3.9906
17	3.6897	3.7297	3.7666	3.8007	3.8325	3.8623	3.8902	3.9165	3.9415
18	3.6533	3.6924	3.7283	3.7616	3.7926	3.8215	3.8487	3.8744	3.8986
19	3.6213	3.6595	3.6946	3.7271	3.7574	3.7857	3.8122	3.8373	3.8609
20	3.5929	3.6303	3.6647	3.6966	3.7262	3.7539	3.7799	3.8044	3.8275
21	3.5675	3.6043	3.6380	3.6693	3.6983	3.7255	3.7510	3.7750	3.7977
22	3.5447	3.5808	3.6141	3.6448	3.6733	3.7000	3.7251	3.7487	3.7710
23	3.5241	3.5597	3.5924	3.6226	3.6507	3.6770	3.7017	3.7249	3.7468
24	3.5053	3.5405	3.5727	3.6025	3.6302	3.6561	3.6804	3.7033	3.7249
25	3.4883	3.5230	3.5548	3.5842	3.6116	3.6371	3.6611	3.6836	3.7049
26	3.4726	3.5069	3.5384	3.5674	3.5945	3.6197	3.6433	3.6656	3.6867
27	3.4583	3.4922	3.5233	3.5520	3.5787	3.6037	3.6271	3.6491	3.6699
28	3.4450	3.4786	3.5094	3.5378	3.5643	3.5889	3.6121	3.6338	3.6544
29	3.4328	3.4660	3.4965	3.5247	3.5509	3.5753	3.5982	3.6198	3.6401
30	3.4214	3.4544	3.4846	3.5125	3.5384	3.5626	3.5853	3.6067	3.6269
35	3.3750	3.4068	3.4359	3.4628	3.4877	3.5110	3.5329	3.5534	3.5728
40	3.3409	3.3718	3.4001	3.4263	3.4506	3.4732	3.4944	3.5143	3.5331
45	3.3148	3.3451	3.3728	3.3984	3.4221	3.4442	3.4650	3.4845	3.5028
50	3.2942	3.3239	3.3512	3.3763	3.3996	3.4214	3.4417	3.4609	3.4789
55	3.2775	3.3068	3.3337	3.3585	3.3814	3.4029	3.4229	3.4418	3.4596
60	3.2637	3.2927	3.3192	3.3437	3.3664	3.3876	3.4074	3.4260	3.4436
70	3.2422	3.2707	3.2967	3.3208	3.3430	3.3638	3.3832	3.4015	3.4187
80	3.2262	3.2543	3.2800	3.3037	3.3257	3.3462	3.3653	3.3833	3.4003
90	3.2139	3.2417	3.2672	3.2906	3.3123	3.3326	3.3515	3.3693	3.3861
100	3.2041	3.2317	3.2569	3.2802	3.3017	3.3218	3.3405	3.3582	3.3748
110	3.1962	3.2235	3.2486	3.2717	3.2930	3.3130	3.3316	3.3491	3.3656
120	3.1896	3.2168	3.2417	3.2646	3.2859	3.3057	3.3242	3.3416	3.3580
250	3.1522	3.1785	3.2026	3.2248	3.2453	3.2644	3.2823	3.2991	3.3149
500	3.1352	3.1612	3.1849	3.2067	3.2269	3.2457	3.2633	3.2798	3.2954
1000	3.1268	3.1526	3.1761	3.1977	3.2178	3.2365	3.2539	3.2703	3.2857

Table 15 *Continued*

df\p	20	21	28	36	45	55	66	78
2	31.5991	32.3806	37.3965	42.4087	47.4184	52.4261	57.4326	62.4380
3	12.9240	13.1389	14.4787	15.7577	16.9858	18.1703	19.3169	20.4300
4	8.6103	8.7207	9.3983	10.0298	10.6237	11.1860	11.7211	12.2329
5	6.8688	6.9422	7.3884	7.7981	8.1783	8.5341	8.8692	9.1865
6	5.9588	6.0145	6.3510	6.6568	6.9379	7.1988	7.4428	7.6724
7	5.4079	5.4536	5.7282	5.9757	6.2019	6.4105	6.6045	6.7862
8	5.0413	5.0806	5.3162	5.5274	5.7192	5.8954	6.0586	6.2107
9	4.7809	4.8159	5.0249	5.2114	5.3801	5.5345	5.6770	5.8095
10	4.5869	4.6188	4.8087	4.9774	5.1296	5.2684	5.3962	5.5148
11	4.4370	4.4665	4.6420	4.7975	4.9374	5.0646	5.1815	5.2898
12	4.3178	4.3455	4.5099	4.6551	4.7854	4.9037	5.0123	5.1125
13	4.2208	4.2471	4.4026	4.5396	4.6624	4.7737	4.8755	4.9695
14	4.1405	4.1655	6.3138	4.4442	4.5608	4.6663	4.7628	4.8518
15	4.0728	4.0968	4.2391	3.3640	4.4756	4.5764	4.6684	4.7532
16	4.0150	4.0382	4.1754	4.2958	4.4030	4.4999	4.5882	4.6695
17	3.9651	3.9876	4.1205	4.2369	4.3406	4.4340	4.5192	4.5975
18	3.9216	3.9435	4.0727	4.1857	4.2862	4.3768	4.4593	4.5350
19	3.8834	3.9048	4.0307	4.1408	4.2386	4.3266	4.4068	4.4803
20	3.8495	3.8704	3.9935	4.1010	4.1964	4.2823	4.3603	4.4319
21	3.8193	3.8398	3.9603	4.0655	4.1588	4.2427	4.3190	4.3888
22	3.7921	3.8122	3.9306	4.0337	4.1252	4.2073	4.2819	4.3503
23	3.7676	3.7874	3.9037	4.0050	4.0948	4.1754	4.2486	4.3156
24	3.7454	3.7649	3.8794	3.9790	4.0673	4.1465	4.2184	4.2842
25	3.7251	3.7443	3.8572	3.9554	4.0423	4.1202	4.1909	4.2556
26	3.7066	3.7256	3.8369	3.9337	4.0194	4.0962	4.1658	4.2295
27	3.6896	3.7083	3.8183	3.9139	3.9984	4.0742	4.1428	4.2056
28	3.6739	3.6924	3.8012	3.8956	3.9791	4.0539	4.1216	4.1835
29	3.6594	3.6777	3.7853	3.8787	3.9612	4.0351	4.1021	4.1632
30	3.6460	3.6641	3.7706	3.8631	3.9447	4.0178	4.0839	4.1444
35	3.5911	3.6086	3.7108	3.7993	3.8774	3.9472	4.0103	4.0679
40	3.5510	3.5679	3.6670	3.7527	3.8282	3.8956	3.9565	4.0121
45	3.5203	3.5368	3.6335	3.7171	3.7907	3.8563	3.9156	3.9696
50	3.4960	3.5122	3.6071	3.6890	3.7611	3.8253	3.8833	3.9361
55	3.4764	3.4924	3.5858	3.6664	3.7372	3.8003	3.8573	3.9091
60	3.4602	3.4760	3.5682	3.6477	3.7175	3.7797	3.8358	3.8868
70	3.4350	3.4505	3.5408	3.6186	3.6869	3.7477	3.8024	3.8523
80	3.4163	3.4316	3.5205	3.5970	3.6642	3.7240	3.7778	3.8267
90	3.4019	3.4170	3.5048	3.5804	3.6467	3.7057	3.7588	3.8070
100	3.3905	3.4054	3.4924	3.5673	3.6329	3.6912	3.7437	3.7914
110	3.3812	3.3960	3.4823	3.5565	3.6216	3.6794	3.7314	3.7787
120	3.3735	3.3882	3.4739	3.5477	3.6122	3.6697	3.7213	3.7682
260	3.3299	3.3440	3.4267	3.4976	3.5596	3.6146	3.6641	3.7090
500	3.3101	3.3240	3.4052	3.4749	3.5357	3.5897	3.6382	3.6822
1000	3.3003	3.3141	3.3946	3.4636	3.5239	3.5774	3.6254	3.6689

Table 15 *Continued*

df\p	91	105	120	136	153	171	190
2	67.4426	72.4465	77.4500	82.4530	87.4557	92.4581	97.4602
3	21.5132	22.5696	23.6016	24.6113	25.6006	26.5711	27.5240
4	12.7239	13.1967	13.6531	14.0947	14.5229	14.9387	15.3433
5	9.4884	9.7768	10.0530	10.3185	10.5741	10.8210	11.0597
6	7.8894	8.0956	8.2921	8.4799	8.6601	8.8333	9.0001
7	6.9572	7.1189	7.2725	7.4188	7.5586	7.6926	7.8213
8	6.3535	6.4881	6.6155	6.7366	6.8520	6.9624	7.0681
9	5.9335	6.0500	6.1602	6.2646	6.3639	6.4586	6.5492
10	5.6255	5.7294	5.8273	5.9199	6.0079	6.0918	6.1718
11	5.3906	5.4851	5.5740	5.6580	5.7377	5.8135	5.8858
12	5.2058	5.2931	5.3751	5.4525	5.5259	5.5956	5.6620
13	5.0569	5.1385	5.2151	5.2873	5.3556	5.4205	5.4823
14	4.9343	5.0113	5.0836	5.1516	5.2160	5.2770	5.3351
15	4.8318	4.9050	4.9737	5.0383	5.0993	5.1572	5.2122
16	4.7447	4.8148	4.8805	4.9423	5.0006	5.0558	5.1083
17	4.6700	4.7374	4.8005	4.8599	4.9159	4.9689	5.0192
18	4.6051	4.6703	4.7312	4.7885	4.8424	4.8935	4.9421
19	4.5482	4.6114	4.6705	4.7259	4.7782	4.8277	4.8746
20	4.4980	4.5595	4.6169	4.6708	4.7216	4.7696	4.8151
21	4.4534	4.5133	4.5693	4.6218	4.6712	4.7180	4.7623
22	4.4134	4.4720	4.5266	4.5779	4.6262	4.6718	4.7151
23	4.3774	4.4348	4.4883	4.5385	4.5857	4.6303	4.6726
24	4.3448	4.4011	4.4536	4.5028	4.5491	4.5928	4.6342
25	4.3152	4.3705	4.4221	4.4704	4.5158	4.5587	4.5994
26	4.2882	4.3426	4.3933	4.4408	4.4855	4.5276	4.5676
27	4.2634	4.3170	4.3669	4.4137	4.4576	4.4991	4.5384
28	4.2406	4.2934	4.3426	4.3887	4.4321	4.4730	4.5116
29	4.2195	4.2717	4.3203	4.3657	4.4085	4.4488	4.4869
30	4.2000	4.2515	4.2995	3.3444	4.3866	4.4264	4.4641
35	4.1208	4.1698	4.2154	4.2581	4.2981	4.3358	4.3715
40	4.0631	4.1103	4.1541	4.1951	4.2336	4.2698	4.3041
45	4.0191	4.0649	4.1075	4.1473	4.1846	4.2197	4.2528
50	3.9845	4.0293	4.0708	4.1097	4.1460	4.1803	4.2126
55	3.9566	4.0005	4.0413	4.0793	4.1150	4.1485	4.1802
60	3.9336	3.9768	4.0169	4.0543	4.0894	4.1223	4.1535
70	3.8979	3.9400	3.9791	4.0155	4.0497	4.0818	4.1121
80	3.8715	3.9128	3.9512	3.9869	4.0204	4.0518	4.0815
90	3.8512	3.8919	3.9297	3.9649	3.9978	4.0288	4.0580
100	3.8350	3.8753	3.9126	3.9474	3.9799	4.0105	4.0394
110	3.8219	3.8618	3.8988	3.9332	3.9654	3.9957	4.0242
120	3.8111	3.8506	3.8873	3.9214	3.9534	3.9834	4.0117
250	3.7500	3.7878	3.8227	3.8553	3.8857	3.9143	3.9412
500	3.7224	3.7594	3.7936	3.8254	3.8552	3.8831	3.9095
1000	3.7087	3.7453	3.7792	3.8107	3.8401	3.8677	3.8937

Table 16 **Bonferroni's adjusted *t* values for one-sided test and α = 0.05 (values generated as described in Chapter 5)**

df\p	1	2	3	4	5	6	7	8	9	10
2	2.9200	4.3027	5.3393	6.2054	6.9646	7.6488	8.2767	8.8602	9.4076	9.9248
3	2.3534	3.1825	3.7405	4.1765	4.5407	4.8567	5.1377	5.3920	5.6251	5.8409
4	2.1319	2.7765	3.1863	3.4954	3.7470	3.9608	4.1478	4.3147	4.4657	4.6041
5	2.0151	2.5706	2.9117	3.1634	3.3649	3.5341	3.6805	3.8100	3.9263	4.0321
6	1.9432	2.4469	2.7491	2.9687	3.1427	3.2875	3.4119	3.5212	3.6190	3.7074
7	1.8946	2.3646	2.6419	2.8412	2.9980	3.1276	3.2384	3.3353	3.4216	3.4995
8	1.8596	2.3060	2.5660	2.7515	2.8965	3.0158	3.1174	3.2060	3.2846	3.3554
9	1.8331	2.2622	2.5096	2.6850	2.8214	2.9333	3.0283	3.1109	3.1841	3.2498
10	1.8125	2.2281	2.4660	2.6338	2.7638	2.8701	2.9601	3.0382	3.1073	3.1693
11	1.7959	2.2010	2.4313	2.5931	2.7181	2.8200	2.9062	2.9809	3.0468	3.1058
12	1.7823	2.1788	2.4030	2.5600	2.6810	2.7795	2.8626	2.9345	2.9978	3.0545
13	1.7709	2.1604	2.3796	2.5326	2.6503	2.7459	2.8265	2.8962	2.9575	3.0123
14	1.7613	2.1448	2.3598	2.5096	3.6245	2.7178	2.7962	2.8640	2.9236	2.9768
15	1.7531	2.1315	2.3429	2.4899	3.6025	2.6937	2.7705	2.8366	2.8948	2.9467
16	1.7459	2.1199	2.3283	2.4729	2.5835	2.6730	2.7482	2.8131	2.8700	2.9208
17	1.7396	2.1098	2.3156	2.4581	2.5669	2.6550	2.7289	2.7925	2.8484	2.8982
18	1.7341	2.1009	2.3044	2.4450	2.5524	2.6391	2.7119	2.7745	2.8295	2.8784
19	1.7291	2.0930	2.2944	2.4334	2.5395	2.6251	2.6969	2.7586	2.8127	2.8609
20	1.7247	2.0860	2.2855	2.4231	2.5280	2.6126	2.6835	2.7444	2.7978	2.8453
21	1.7207	2.0796	2.2775	2.4139	2.5177	2.6014	2.6714	2.7316	2.7844	2.8314
22	1.7171	2.0739	2.2703	2.4055	2.5083	2.5912	2.6606	2.7201	2.7723	2.8188
23	1.7139	2.0687	2.2637	2.3979	2.4999	2.5820	2.6507	2.7097	2.7614	2.8073
24	1.7109	2.0639	2.2578	2.3910	2.4922	2.5736	2.6418	2.7002	2.7514	2.7969
25	1.7081	2.0595	2.2523	2.3846	2.4851	2.5660	2.6336	2.6916	2.7423	2.7874
26	1.7056	2.0555	2.2472	2.3788	2.4786	2.5589	2.6260	2.6836	2.7340	2.7787
27	1.7033	2.0518	2.2426	2.3734	2.4727	2.5525	2.6191	2.6763	2.7263	2.7707
28	1.7011	2.0484	2.2383	2.3685	2.4671	2.5465	2.6127	2.6695	2.7192	2.7633
29	1.6991	2.0452	2.2343	2.3639	2.4620	2.5409	2.6068	2.6632	2.7126	2.7564
30	1.6973	2.0423	2.2306	2.3596	2.4573	2.5357	2.6012	2.6574	2.7064	2.7500
35	1.6896	2.0301	2.2154	2.3420	2.4377	2.5145	2.5786	2.6334	2.6813	2.7238
40	1.6839	2.0211	2.2041	2.3289	2.4233	2.4989	2.5618	2.6157	2.6627	2.7045
45	1.6794	2.0141	2.1954	2.3189	2.4121	2.4868	2.5489	2.6021	2.6485	2.6896
50	1.6759	2.0086	2.1885	2.3109	2.4033	2.4772	2.5387	2.5913	2.6372	2.6778
55	1.6730	2.0040	2.1829	2.3044	2.3961	2.4694	2.5304	2.5825	2.6280	2.6682
60	1.6707	2.0003	2.1782	2.2991	2.3901	2.4630	2.5235	2.5752	2.6203	2.6603
70	1.6669	1.9944	2.1709	2.2906	2.3808	2.4529	2.5128	2.5639	2.6085	2.6479
80	1.6641	1.9901	2.1654	2.2844	2.3739	2.4454	2.5047	2.5554	2.5996	2.6387
90	1.6620	1.9867	2.1612	2.2795	2.3685	2.4396	2.4986	2.5489	2.5928	2.6316
100	1.6602	1.9840	2.1579	2.2757	2.3642	2.4349	2.4936	2.5437	2.5873	2.6259
110	1.6588	1.9818	2.1551	2.2725	2.3607	2.4311	2.4896	2.5394	2.5829	2.6213
120	1.6577	1.9799	2.1528	2.2699	2.3578	2.4280	2.4862	2.5359	2.5792	2.6174
250	1.6510	1.9650	2.1399	2.2550	2.3414	2.4102	2.4673	2.5159	2.5582	2.5956
500	1.6479	1.9647	2.1339	2.2482	2.3338	2.4021	2.4586	2.5068	2.5487	2.5857
1000	1.6464	1.9623	2.1310	2.2448	2.3301	2.3980	2.4543	2.5022	2.5439	2.5808

Table 16   *Continued*

df\p	11	12	13	14	15	16	17	18	19
2	10.4164	10.8859	11.3359	11.7687	12.1861	12.5897	12.9808	13.3604	13.7296
3	6.0423	6.2315	6.4102	6.5797	6.7411	6.8952	7.0430	7.1849	7.3215
4	4.7320	4.8510	4.9625	5.0675	5.1668	5.2611	5.3508	5.4366	5.5187
5	4.1293	4.2193	4.3032	4.3818	4.4558	4.5257	4.5921	4.6553	4.7156
6	3.7884	3.8630	3.9323	3.9971	4.0579	4.1152	4.1694	4.2209	4.2700
7	3.5705	3.6358	3.6963	3.7527	3.8055	3.8552	3.9022	3.9467	3.9890
8	3.4198	3.4789	3.5335	3.5844	3.6319	3.6766	3.7187	3.7586	3.7965
9	3.3095	3.3642	3.4147	3.4616	3.5054	2.5465	3.5852	3.6219	3.6566
10	3.2254	3.2768	3.3242	3.3682	3.4093	3.4477	3.4839	3.5182	3.5506
11	3.1593	3.2081	3.2531	3.2949	3.3338	3.3702	3.4045	3.4368	3.4675
12	3.1058	3.1527	3.1958	3.2357	3.2729	3.3078	3.3405	3.3714	3.4007
13	3.0618	3.1070	3.1486	3.1871	3.2229	3.2565	3.2880	3.3177	3.3458
14	3.0249	3.0688	3.1091	3.1464	3.1811	3.2135	3.2440	3.2727	3.2999
15	2.9936	3.0363	3.0755	3.1118	3.1456	3.1771	3.2067	3.2346	3.2610
16	2.9666	3.0083	3.0467	3.0821	3.1150	3.1458	3.1747	3.2019	3.2276
17	2.9431	2.9840	3.0216	3.0563	3.0885	3.1186	3.1469	3.1735	3.1986
18	2.9226	2.9627	2.9996	3.0336	3.0653	3.0948	3.1225	3.1486	3.1732
19	2.9044	2.9439	2.9801	3.0136	3.0447	3.0738	3.1010	3.1266	3.1508
20	2.8882	2.9271	2.9628	2.9958	3.0264	3.0550	3.0818	3.1070	3.1308
21	2.8736	2.9121	2.9473	2.9799	3.0101	3.0382	3.0647	3.0895	3.1130
22	2.8605	2.8985	2.9334	2.9655	2.9953	3.0231	3.0492	3.0737	3.0969
23	2.8487	2.8863	2.9207	2.9525	2.9820	3.0095	3.0352	3.0595	3.0823
24	2.8379	2.8751	2.9092	2.9406	2.9698	2.9970	3.0225	3.0465	3.0691
25	2.8280	2.8649	2.8987	2.9298	2.9587	2.9856	3.0109	3.0346	3.0570
26	2.8190	2.8555	2.8890	2.9199	2.9485	2.9752	3.0002	3.0237	3.0459
27	2.8106	2.8469	2.8801	2.9107	2.9391	2.9656	2.9904	3.0137	3.0357
28	2.8029	2.8389	2.8719	2.9023	2.9305	2.9567	2.9813	3.0045	3.0263
29	2.7958	2.8316	2.8643	2.8945	2.9225	2.9485	2.9730	2.9959	3.0176
30	2.7892	2.8247	2.8572	2.8872	2.9150	2.9409	2.9652	2.9880	3.0095
35	2.7620	2.7966	2.8283	2.8575	2.8845	2.9097	2.9333	2.9554	2.9763
40	2.7419	2.7759	2.8069	2.8355	2.8620	2.8867	2.9097	2.9314	2.9519
45	2.7265	2.7599	2.7905	2.8187	2.8447	2.8690	2.8917	2.9130	2.9331
50	2.7143	2.7473	2.7775	2.8053	2.8310	2.8550	2.8774	2.8984	2.9182
55	2.7043	2.7370	2.7669	2.7944	2.8199	2.8436	2.8658	2.8866	2.9062
60	2.6961	2.7286	2.7582	2.7855	2.8107	2.8342	2.8562	2.8768	2.8962
70	2.6833	2.7153	2.7446	2.7715	2.7964	2.8195	2.8412	2.8615	2.8807
80	2.6737	2.7054	2.7344	2.7610	2.7857	2.8086	2.8301	2.8502	2.8691
90	2.6663	2.6978	2.7266	2.7530	2.7774	2.8002	2.8214	2.8414	2.8602
100	2.6605	2.6918	2.7203	2.7466	2.7709	2.7935	2.8146	2.8344	2.8530
110	2.6557	2.6868	2.7152	2.7414	2.7655	2.7880	2.8090	2.8287	2.8472
120	2.6517	2.6827	2.7110	2.7370	2.7611	2.7835	2.8044	2.8240	2.8424
250	2.6291	2.6594	2.6870	2.7124	2.7359	2.7577	2.7781	2.7972	2.8152
500	2.6188	2.6488	2.6761	2.7012	2.7244	2.7460	2.7661	2.7850	2.8028
1000	2.6137	2.6435	2.6707	2.6957	2.7187	2.7402	2.7602	2.7790	2.7966

Table 16   *Continued*

df\p	20	21	28	36	45	55	66	78
2	14.0890	14.4396	16.6883	18.9341	21.1776	23.4201	25.6613	27.9016
3	7.4533	7.5807	8.3738	9.1294	9.8538	10.5517	11.2266	11.8814
4	5.5976	5.6734	6.1380	6.5697	6.9746	7.3571	7.7207	8.0678
5	4.7733	4.8287	5.1644	5.4715	5.7554	6.0205	6.2696	6.5051
6	4.3168	4.3617	4.6317	4.8759	5.0996	5.3067	5.4998	5.6811
7	4.0293	4.0679	4.2989	4.5062	4.6947	4.8681	5.0289	5.1792
8	3.8325	3.8669	4.0724	4.2556	4.4214	4.5732	4.7133	4.8437
9	3.6897	3.7212	3.9088	4.0752	4.2252	4.3620	4.4879	4.6046
10	3.5814	3.6108	3.7852	3.9394	4.0778	4.2036	3.3191	4.4260
11	3.4966	3.5243	3.6887	3.8335	3.9631	4.0807	4.1883	4.2877
12	3.4284	3.4549	3.6112	3.7487	3.8714	3.9825	4.0840	4.1776
13	3.3725	3.3979	3.5478	3.6793	3.7965	3.9023	3.9989	4.0878
14	3.3257	3.3502	3.4949	3.6214	3.7341	3.8357	3.9283	4.0134
15	3.2860	3.3098	3.4501	3.5725	3.6814	3.7794	3.8686	3.9506
16	3.2520	3.2752	3.4116	3.5306	3.6363	3.7313	3.8177	3.8969
17	3.2224	3.2451	3.3783	3.4944	3.5972	3.6897	3.7736	3.8506
18	3.1966	3.2188	3.3492	3.4626	3.5631	3.6533	3.7352	3.8102
19	3.1737	3.1955	3.3235	3.4347	3.5330	3.6213	3.7013	3.7746
20	3.1534	3.1748	3.3006	3.4098	3.5063	3.5929	3.6713	3.7430
21	3.1352	3.1563	3.2802	3.3876	3.4825	3.5675	3.6445	3.7149
22	3.1188	3.1396	3.2618	3.3676	3.4610	3.5447	3.6204	3.6896
23	3.1040	3.1246	3.2451	3.3495	3.4416	3.5241	3.5986	3.6667
24	3.0905	3.1108	3.2300	3.3331	3.4240	3.5053	3.5789	3.6460
25	3.0782	3.0983	3.2162	3.3181	3.4080	3.4883	3.5609	3.6271
26	3.0669	3.0868	3.2035	3.3044	3.3933	3.4726	3.5444	3.6098
27	3.0565	3.0763	3.1919	3.2918	3.3797	3.4583	3.5292	3.5939
28	3.0469	3.0665	3.1811	3.2801	3.3673	3.4450	3.5152	3.5793
29	3.0680	3.0575	3.1712	3.2694	3.3557	3.4328	3.5023	3.5657
30	3.0298	3.0491	3.1620	3.2594	3.3450	3.4214	3.4903	3.5531
35	2.9960	3.0148	3.1242	3.2185	3.3013	3.3750	3.4414	3.5019
40	2.9712	2.9895	3.0964	3.1884	3.2691	3.3409	3.4055	3.4643
45	2.9521	2.9701	3.0751	3.1654	3.2445	3.3148	3.3781	3.4356
50	2.9370	2.9547	3.0582	3.1472	3.2250	3.2942	3.3564	3.4129
55	2.9247	2.9423	3.0446	3.1324	3.2092	3.2775	3.3388	3.3945
60	2.9146	2.9319	3.0333	3.1202	3.1962	3.2637	3.3243	3.3793
70	2.8987	2.9159	3.0156	3.1012	3.1759	3.2422	3.3017	3.3557
80	2.8870	2.9039	3.0026	3.0870	3.1608	3.2262	3.2849	3.3382
90	2.8779	2.8947	2.9924	3.0761	3.1492	3.2139	3.2720	3.3246
100	2.8707	2.8873	2.9844	3.0674	3.1399	3.2041	3.2617	3.3139
110	2.8648	2.8813	2.9778	3.0604	3.1324	3.1962	2.2534	3.3052
120	2.8599	2.8764	2.9724	3.0545	3.1261	3.1896	3.2464	3.2979
250	2.8322	2.8482	2.9416	3.0213	3.0908	3.1522	3.2072	3.2570
500	2.9195	2.8354	2.9276	3.0063	3.0747	3.1352	3.1894	3.2384
1000	2.8133	2.8291	2.9207	2.9988	3.0667	3.1268	3.1806	3.2291

Table 16 *Continued*

df\p	91	105	120	136	153	171	190
2	30.1413	32.3806	34.6194	36.8578	39.0960	41.3340	43.5718
3	12.5182	13.1389	13.7450	14.3379	14.9186	15.4880	16.0471
4	8.4006	8.7207	9.0294	9.3278	9.6171	9.8978	10.1708
5	6.7288	6.9422	7.1464	7.3424	7.5310	7.7129	7.8888
6	5.8522	6.0145	6.1690	6.3165	6.4578	6.5934	6.7240
7	5.3203	5.4536	5.5799	5.7001	5.8148	5.9246	6.0300
8	4.9658	5.0806	5.1892	5.2922	5.3902	5.4838	5.5735
9	4.7136	4.8159	4.9124	5.0037	5.0904	5.1730	5.2520
10	4.5256	4.6188	4.7065	4.7894	4.8680	4.9427	5.0141
11	4.3801	4.4665	4.5477	4.6242	4.6967	4.7656	4.8312
12	4.2644	4.3455	4.4215	4.4932	4.5610	4.6253	4.6865
13	4.1702	4.2471	4.3191	4.3868	4.4509	4.5116	4.5693
14	4.0921	4.1655	4.2342	4.2988	4.3597	4.4175	4.4724
15	4.0263	4.0968	4.1628	4.2247	4.2831	4.3385	4.3910
16	3.9701	4.0382	4.1018	4.1616	4.2179	4.2712	4.3217
17	3.9216	3.9876	4.0493	4.1071	4.1616	4.2131	4.2620
18	3.8793	3.9435	4.0035	4.0597	4.1126	4.1626	4.2101
19	3.8421	3.9048	3.9632	4.0180	4.0696	4.1183	4.1645
20	3.8091	3.8704	3.9276	3.9811	4.0315	4.0790	4.1241
21	3.7797	3.8398	3.8958	3.9482	3.9975	4.0441	4.0881
22	3.7532	3.8122	3.8672	3.9187	3.9670	4.0127	4.0559
23	3.7294	3.7874	3.8414	3.8920	3.9395	3.9844	4.0268
24	3.7077	3.7649	3.8181	3.8679	3.9146	3.9587	4.0004
25	3.6880	3.7443	3.7968	3.8459	3.8919	3.9354	3.9765
26	3.6699	3.7256	3.7773	3.8257	3.8712	3.9140	3.9545
27	3.6533	3.7083	3.7595	3.8073	3.8521	3.8944	3.9344
28	3.6381	3.6924	3.7430	3.7902	3.8346	3.8764	3.9159
29	3.6239	3.6777	3.7278	3.7745	3.8184	3.8597	3.8987
30	3.6108	3.6641	3.7136	3.7599	3.8033	3.8442	3.8829
35	3.5574	3.6086	3.6561	3.7005	3.7421	3.7813	3.8183
40	3.5182	3.5679	3.6140	3.6570	3.6973	3.7353	3.7710
45	3.4882	3.5368	3.5818	3.6238	3.6631	3.7001	3.7350
50	3.4646	3.5122	3.5564	3.5976	3.6362	3.6724	3.7066
55	3.4454	3.4924	3.5359	3.5764	3.6143	3.6500	3.6836
60	3.4296	3.4760	3.5189	3.5589	3.5963	3.6315	3.6646
70	3.4050	3.4505	3.4926	3.5317	3.5684	3.6028	3.6352
80	3.3868	3.4316	3.4730	3.5116	3.5476	3.5815	3.6134
90	3.3727	3.4170	3.4579	3.4960	3.5317	3.5651	3.5966
100	3.3616	3.4054	3.4460	3.4837	3.5190	3.5521	3.5832
110	3.3525	3.3960	3.4362	3.4737	3.5087	3.5415	3.5724
120	3.3449	3.3882	3.4281	3.4653	3.5001	3.5327	3.5634
250	3.3023	3.3440	3.3826	3.4184	3.4518	3.4832	3.5127
500	3.2830	3.3240	3.3619	3.3971	3.4300	3.4607	3.4897
1000	3.2734	3.3141	3.3517	3.3866	3.4191	3.4496	3.4783

Table 17 **Bonferroni's adjusted t values for two-sided test and $\alpha = 0.01$** (used first in Chapter 5)

$df\backslash p$	1	2	3	4	5	6	7	8	9	10
2	9.9248	14.0890	17.2772	19.9625	22.3271	24.4643	26.4292	28.2577	29.9750	31.5991
3	5.8409	7.4533	8.5752	9.4649	10.2145	10.8688	11.4532	11.9838	12.4715	12.9240
4	4.6041	5.5976	6.2541	6.7583	7.1732	7.5287	7.8414	8.1216	8.3763	8.6103
5	4.0321	4.7733	5.2474	5.6042	5.8934	6.1384	6.3518	6.5414	6.7126	6.8688
6	3.7074	4.3168	4.6979	4.9807	5.2076	5.3982	5.5632	5.7090	5.8399	5.9588
7	3.4995	4.0293	4.3553	4.5946	4.7853	4.9445	5.0815	5.2022	5.3101	5.4079
8	3.3554	3.8325	4.1224	4.3335	4.5008	4.6398	4.7590	4.8636	4.9570	5.0413
9	3.2498	3.6897	3.9542	4.1458	4.2968	4.4219	4.5288	4.6224	4.7058	4.7809
10	3.1693	3.5814	3.8273	4.0045	4.1437	4.2586	4.3567	4.4423	4.5184	4.5869
11	3.1058	3.4966	3.7283	3.8945	4.0247	4.1319	4.2232	4.3028	4.3735	4.4370
12	3.0545	3.4284	3.6489	3.8065	3.9296	4.0308	4.1169	4.1918	4.2582	4.3178
13	3.0123	3.3725	3.5838	3.7345	3.8520	3.9484	4.0302	4.1013	4.1643	4.2208
14	2.9768	3.3257	3.5296	3.6746	3.7874	3.8798	3.9582	4.0263	4.0865	4.1405
15	2.9467	3.2860	3.4837	3.6239	3.7328	3.8220	3.8975	3.9630	4.0209	4.0728
16	2.9208	3.2520	3.4443	3.5805	3.6862	3.7725	3.8456	3.9089	3.9649	4.0150
17	2.8982	3.2224	3.4102	3.5429	3.6458	3.7297	3.8007	3.8623	3.9165	3.9651
18	2.8784	3.1966	3.3804	3.5101	3.6105	3.6924	3.7616	3.8215	3.8744	3.9216
19	2.8609	3.1737	3.3540	3.4812	3.5794	3.6595	3.7271	3.7857	3.8373	3.8834
20	2.8453	3.1534	3.3306	3.4554	3.5518	3.6303	3.6966	3.7539	3.8044	3.8495
21	2.8314	3.1352	3.3097	3.4325	3.5272	3.6043	3.6693	3.7255	3.7750	3.8193
22	2.8188	3.1188	3.2909	3.4118	3.5050	3.5808	3.6448	3.7000	3.7487	3.7921
23	2.8073	3.1040	3.2739	3.3931	3.4850	3.5597	3.6226	3.6700	3.7249	3.7676
24	2.7969	3.0905	3.2584	3.3761	3.4668	3.5405	3.6025	3.6561	3.7033	3.7454
25	2.7874	3.0782	3.2443	3.3606	3.4502	3.5230	3.5842	3.6371	3.6836	3.7251
26	2.7787	3.0669	3.2313	3.3464	3.4350	3.5069	3.5674	3.6197	3.6656	3.7066
27	2.7707	3.0565	3.2194	3.3334	3.4210	3.4922	3.5520	3.6037	3.6491	3.6896
28	2.7633	3.0469	3.2084	3.3214	3.4082	3.4786	3.5378	3.5889	3.6338	3.6739
29	2.7564	3.0380	3.1982	3.3102	3.3962	3.4660	3.5247	3.5753	3.6198	3.6594
30	2.7500	3.0298	3.1888	3.2999	3.3852	3.4544	3.5125	3.5626	3.6067	3.6460
35	2.7238	2.9960	3.1502	3.2577	3.3400	3.4068	3.4628	3.5110	3.5534	3.5911
40	2.7045	2.9712	3.1218	3.2266	3.3069	3.3718	3.4263	3.4732	3.5143	3.5510
45	2.6896	2.9521	3.1000	3.2028	3.2815	3.3451	3.3984	3.4442	3.4845	3.5203
50	2.6778	2.9370	3.0828	3.1840	3.2614	3.3239	3.3763	3.4214	3.4609	3.4960
55	2.6682	2.9247	3.0688	3.1688	3.2451	3.3068	3.3585	3.4029	3.4418	3.4764
60	2.6603	2.9146	3.0573	3.1562	3.2317	3.2927	3.3437	3.3876	3.4260	3.4602
70	2.6479	2.8987	3.0393	3.1366	3.2108	3.2707	3.3208	3.3638	3.4015	3.4350
80	2.6387	2.8870	3.0259	3.1220	3.1953	3.2543	3.3037	3.3462	3.3833	3.4163
90	2.6316	2.8779	3.0156	3.1108	3.1833	3.2417	3.2906	3.3326	3.3693	3.4019
100	2.6259	2.8707	3.0073	3.1018	3.1737	3.2317	3.2802	3.3218	3.3582	3.3905
110	2.6213	2.8648	3.0007	3.0945	3.1660	3.2235	3.2717	3.3130	3.3491	3.3812
120	2.6174	2.8599	2.9951	3.0885	3.1595	3.2168	3.2646	3.3057	3.3416	3.3735
250	2.5956	2.8322	2.9637	3.0543	3.1232	3.1785	3.2248	3.2644	3.2991	3.3299
500	2.5857	2.8195	2.9494	3.0387	3.1066	3.1612	3.2067	3.2457	3.2798	3.3101
1000	2.5808	2.8133	2.9423	3.0310	3.0984	3.1526	3.1977	3.2365	3.2703	3.3003
∞	2.5758	2.8070	2.9352	3.0233	3.0902	3.1440	3.1888	3.2272	3.2608	3.2905

Table 17 *Continued*

df\p	11	12	13	14	15	16	17	18	19
2	33.1436	34.6194	36.0347	37.3965	38.7105	39.9812	41.2129	42.4087	43.5718
3	13.3471	13.7450	14.1214	14.4787	14.8194	15.1451	15.4575	15.7577	16.0471
4	8.8271	9.0294	9.2192	9.3983	9.5679	9.7291	9.8828	10.0298	10.1708
5	7.0128	7.1464	7.2712	7.3884	7.4990	7.6037	7.7032	7.7981	7.8888
6	6.0680	6.1690	6.2630	6.3510	6.4338	6.5121	6.5862	6.6568	6.7240
7	5.4973	5.5799	5.6565	5.7282	5.7954	5.8588	5.9188	5.9757	6.0300
8	5.1183	5.1892	5.2549	5.3162	5.3737	5.4278	5.4789	5.5274	5.5735
9	4.8494	4.9124	4.9706	5.0249	5.0757	5.1235	5.1686	5.2114	5.2520
10	4.6492	4.7065	4.7594	4.8087	4.8547	4.8980	4.9388	4.9774	5.0141
11	4.4947	4.5477	4.5966	4.6420	4.6845	4.7244	4.7620	4.7975	4.8312
12	4.3719	4.4215	4.4673	4.5099	4.5496	4.5868	4.6219	4.6551	4.6865
13	4.2721	4.3191	4.3624	4.4026	4.4401	4.4752	4.5083	4.5396	4.5693
14	4.1894	4.2342	4.2755	4.3138	4.3495	4.3829	4.4144	4.4442	4.4724
15	4.1198	4.1628	4.2024	4.2391	4.2733	4.3054	4.3355	4.3640	4.3910
16	4.0604	4.1018	4.1400	4.1754	4.2084	4.2393	4.2683	4.2958	4.3217
17	4.0091	4.0493	4.0863	4.1205	4.1525	4.1823	4.2104	4.2369	4.2620
18	3.9644	4.0035	4.0394	4.0727	4.1037	4.1327	4.1600	4.1857	4.2101
19	3.9251	3.9632	3.9983	4.0307	4.0609	4.0892	4.1157	4.1408	4.1645
20	3.8903	3.9276	3.9618	3.9935	4.0230	4.0506	4.0765	4.1010	4.1241
21	3.8593	3.8958	3.9293	3.9603	3.9892	4.0162	4.0416	4.0655	4.0881
22	3.8314	3.8672	3.9001	3.9306	3.9589	3.9854	4.0103	4.0337	4.0559
23	3.8062	3.8414	3.8738	3.9037	3.9316	3.9576	3.9820	4.0050	4.0268
24	3.7834	3.8181	3.8499	3.8794	3.9068	3.9324	3.9564	3.9790	4.0004
25	3.7626	3.7968	3.8282	3.8572	3.8842	3.9094	3.9331	3.9554	3.9765
26	3.7436	3.7773	3.8083	3.8369	3.8635	3.8884	3.9118	3.9337	3.9545
27	3.7261	3.7595	3.7900	3.8183	3.8446	3.8692	3.8922	3.9139	3.9344
28	3.7101	3.7430	3.7732	3.8012	3.8271	3.8514	3.8742	3.8956	3.9159
29	3.6952	3.7278	3.7577	3.7853	3.8110	3.8350	3.8575	3.8787	3.8987
30	3.6814	3.7136	3.7433	3.7706	3.7961	3.8198	3.8421	3.8631	3.8829
35	3.6252	3.6561	3.6845	3.7108	3.7352	3.7579	3.7792	3.7993	3.8183
40	3.5840	3.6140	3.6415	3.6670	3.6906	3.7126	3.7333	3.7527	3.7710
45	3.5525	3.5818	3.6087	3.6335	3.6565	3.6780	3.6982	3.7171	3.7350
50	3.5277	3.5564	3.5828	3.6071	3.6297	3.6508	3.6705	3.6890	3.7066
55	3.5076	3.5359	3.5618	3.5858	3.6080	3.6287	3.6481	3.6664	3.6836
60	3.4910	3.5189	3.5445	3.5682	3.5901	3.6105	3.6297	3.6477	3.6646
70	3.4652	3.4926	3.5176	3.5408	3.5622	3.5822	3.6010	3.6186	3.6352
80	3.4460	3.4730	3.4977	3.5205	3.5416	3.5613	3.5797	3.5970	3.6134
90	3.4313	3.4579	3.4823	3.5048	3.5257	3.5451	3.5633	3.5804	3.5966
100	3.4196	3.4460	3.4701	3.4924	3.5131	3.5323	3.5503	3.5673	3.5832
110	3.4100	3.4362	3.4602	3.4823	3.5028	3.5219	3.5398	3.5565	3.5724
120	3.4021	3.4281	3.4520	3.4739	3.4943	3.5132	3.5310	3.5477	3.5634
250	3.3575	3.3826	3.4055	3.4267	3.4462	3.4645	3.4815	3.4976	3.5127
500	3.3373	3.3619	3.3845	3.4052	3.4245	3.4424	3.4591	3.4749	3.4897
1000	3.3272	3.3517	3.3740	3.3946	3.4137	3.4314	3.4480	3.4636	3.4783
∞	3.3172	3.3415	3.3636	3.3840	3.4029	3.4205	3.4370	3.4524	3.4670

Table 17   *Continued*

df\p	20	21	28	36	45	55	66	78
2	44.7046	45.8094	52.9009	59.9875	67.0709	74.1519	81.2312	88.3091
3	16.3263	16.5964	18.2806	19.8889	21.4337	22.9239	24.3667	25.7675
4	10.3063	10.4367	11.2378	11.9851	12.6881	13.3540	13.9882	14.5946
5	7.9757	8.0591	8.5667	9.0332	9.4665	9.8722	10.2546	10.6168
6	6.7883	6.8500	7.2226	7.5617	7.8737	8.1636	8.4348	8.6901
7	6.0818	6.1313	6.4295	6.6987	6.9448	7.1721	7.3837	7.5819
8	5.6174	5.6594	5.9114	6.1375	6.3432	6.5323	6.7076	6.8712
9	5.2907	5.3276	5.5484	5.7458	5.9245	6.0883	6.2395	6.3803
10	5.0490	5.0823	5.2810	5.4578	3.6175	5.7634	5.8978	6.0225
11	4.8633	4.8939	5.0761	5.2378	5.3833	5.5160	5.6379	5.7509
12	4.7165	4.7450	4.9144	5.0644	5.1991	5.3216	5.4340	5.5380
13	4.5975	4.6243	4.7837	4.9244	5.0506	5.1651	5.2700	5.3670
14	4.4992	4.5247	4.6759	4.8091	4.9284	5.0364	5.1354	5.2266
15	4.4166	4.4410	4.5854	4.7125	4.8261	4.9289	5.0229	5.1094
16	4.3463	4.3698	4.5086	4.6305	4.7393	4.8377	4.9275	5.0102
17	4.2858	4.3085	4.4425	4.5600	4.6648	4.7594	4.8457	4.9251
18	4.2332	4.2551	4.3850	4.4987	4.6001	4.6915	4.7748	4.8514
19	4.1869	4.2083	4.3345	4.4450	4.5434	4.6320	4.7127	4.7868
20	4.1460	4.1669	4.2900	4.3976	4.4933	4.5795	4.6579	4.7299
21	4.1096	4.1300	4.2503	4.3554	4.4487	4.5328	4.6092	4.6794
22	4.0769	4.0969	4.2147	4.3175	4.4089	4.4910	4.5657	4.6342
23	4.0474	4.0671	4.1826	4.2835	4.3730	4.4534	4.5265	4.5935
24	4.0207	4.0400	4.1536	4.2527	4.3405	4.4194	4.4911	4.5567
25	3.9964	4.0154	4.1272	4.2246	4.3109	4.3885	4.4589	4.5233
26	3.9742	3.9929	4.1031	4.1990	4.2840	4.3602	4.4295	4.4928
27	3.9538	3.9723	4.0809	4.1755	4.2592	4.3344	4.4025	4.4649
28	3.9351	3.9533	4.0606	4.1539	4.2365	4.3106	4.3778	4.4392
29	3.9177	3.9357	4.0418	4.1339	4.2155	4.2886	4.3549	4.4155
30	3.9016	3.9195	4.0243	4.1154	4.1960	4.2683	4.3337	4.3936
35	3.8362	3.8533	3.9534	4.0403	4.1170	4.1857	4.2479	4.3047
40	3.7884	3.8049	3.9017	3.9855	4.0594	4.1256	4.1854	4.2399
45	3.7519	3.7680	3.8622	3.9437	4.0156	4.0798	4.1378	4.1907
50	3.7231	3.7389	3.8311	3.9108	3.9811	4.0438	4.1004	4.1520
55	3.6999	3.7154	3.8060	3.8843	3.9532	4.0147	4.0702	4.1208
60	3.6807	3.6960	3.7853	3.8624	3.9303	3.9908	4.0454	4.0951
70	3.6509	3.6658	3.7531	3.8284	3.8946	3.9537	4.0069	4.0553
80	3.6288	3.6435	3.7293	3.8033	3.8683	3.9262	3.9784	4.0259
90	3.6118	3.6263	3.7110	3.7839	3.8480	3.9051	3.9565	4.0032
100	3.5983	3.6127	3.6964	3.7686	3.8319	3.8883	3.9391	3.9853
110	3.5874	3.6016	3.6846	3.7561	3.8189	3.8747	3.9250	3.9707
120	3.5783	3.5924	3.6748	3.7458	3.8080	3.8634	3.9133	3.9586
250	3.5270	3.5405	3.6196	3.6875	3.7471	3.8000	3.8475	3.8907
500	3.5037	3.5170	3.5946	3.6612	3.7195	3.7713	3.8179	3.8601
1000	4.4922	3.5054	3.5822	3.6481	3.7059	3.7571	3.8032	3.8449
∞	3.4808	3.4938	3.5699	3.6352	3.6923	3.7430	3.7886	3.8299

Table 17   *Continued*

$df\backslash p$	91	105	120	136	153	171	190
2	95.3861	102.4622	109.5377	116.6126	123.6871	130.7612	137.8350
3	27.1309	28.4606	29.7598	31.0310	32.2766	33.4985	34.6984
4	15.1768	15.7375	16.2788	16.8026	17.3105	17.8040	18.2841
5	10.9616	11.2910	11.6067	11.9102	12.2025	12.4848	12.7578
6	8.9317	9.1612	9.3800	9.5893	9.7901	9.9831	10.1692
7	7.7685	7.9452	8.1130	8.2729	8.4258	8.5724	8.7132
8	7.0248	7.1696	7.3069	7.4373	7.5617	7.6806	7.7947
9	6.5121	6.6361	6.7533	6.8645	6.9073	7.0713	7.1679
10	6.1391	6.2485	6.3517	6.4495	6.5423	6.6308	6.7154
11	5.8562	5.9550	6.0480	6.1359	6.2193	6.2987	6.3745
12	5.6348	5.7254	5.8107	5.8911	5.9674	6.0399	6.1091
13	5.4571	5.5413	5.6204	5.6951	5.7658	5.8329	5.8969
14	5.3113	5.3904	5.4647	5.5347	5.6009	5.6637	5.7235
15	5.1897	5.2647	5.3350	5.4011	5.4637	5.5230	5.5794
16	5.0868	5.1583	5.2252	5.2882	5.3478	5.4042	5.4578
17	4.9986	5.0671	5.1313	5.1916	5.2486	5.3025	5.3538
18	4.9222	4.9882	5.0500	5.1080	5.1628	5.2146	5.2639
19	4.8554	4.9192	4.9789	5.0350	5.0879	5.1379	5.1854
20	4.7965	4.8584	4.9163	4.9707	5.0219	5.0704	5.1163
21	4.7442	4.8044	4.8607	4.9136	4.9633	5.0104	5.0551
22	4.6974	4.7562	4.8111	4.8625	4.9111	4.9569	5.0004
23	4.6553	4.7128	4.7664	4.8167	4.8641	4.9089	4.9513
24	4.6173	4.6736	4.7261	4.7753	4.8217	4.8654	4.9070
25	4.5828	4.6380	4.6894	4.7377	4.7831	4.8261	4.8667
26	4.5513	4.6055	4.6560	4.7034	4.7480	4.7901	4.8300
27	4.5224	4.5757	4.6255	4.6721	4.7159	4.7573	4.7965
28	4.4959	4.5484	4.5974	4.6432	4.6864	4.7271	4.7657
29	4.4714	4.5232	4.5714	4.6166	4.6591	4.6993	4.7372
30	4.4487	4.4998	4.5475	4.5921	4.6340	4.6735	4.7110
35	4.3569	4.4053	4.4503	4.4924	4.5320	4.5694	4.6047
40	4.2901	4.3365	4.3797	4.4201	4.4580	4.4937	4.5275
45	4.2393	4.2843	4.3261	4.3651	4.4018	4.4363	4.4689
50	4.1994	4.2432	4.2840	4.3220	4.3577	4.3913	4.4230
55	4.1672	4.2101	4.2500	4.2872	4.3222	4.3550	4.3860
60	4.1408	4.1829	4.2221	4.2586	4.2929	4.3252	4.3556
70	4.0997	4.1407	4.1788	4.2143	4.2476	4.2790	4.3085
80	4.0694	4.1096	4.1469	4.1816	4.2142	4.2449	4.2738
90	4.0461	4.0856	4.1223	4.1565	4.1886	4.2187	4.2471
100	4.0276	4.0666	4.1028	4.1366	4.1682	4.1979	4.2260
110	4.0126	4.0512	4.0870	4.1204	4.1517	4.1811	4.2088
120	4.0001	4.0384	4.0739	4.1070	4.1380	4.1671	4.1946
250	3.9303	3.9667	4.0004	4.0318	4.0612	4.0889	4.1149
500	3.8987	3.9343	3.9673	3.9980	4.0266	4.0536	4.0790
1000	3.8831	3.9183	3.9509	3.9812	4.0095	4.0362	4.0612
∞	3.8676	3.9024	3.9346	3.9646	3.9926	4.0189	4.0436

Table 18 **Bonferroni's adjusted *t* values for two-sided test and $\alpha = 0.05$ (used first in Chapter 5)**

$df \backslash p$	1	2	3	4	5	6	7	8	9	10
2	4.3027	6.2053	7.6488	8.8602	9.9248	10.8859	11.7687	12.5897	13.3604	14.0890
3	3.1824	4.1765	4.8567	5.3919	5.8409	6.2315	6.5797	6.8952	7.1849	7.4533
4	2.7764	3.4954	3.9608	4.3147	4.6041	4.8510	5.0675	5.2611	5.4366	5.5976
5	2.5706	3.1634	3.5341	3.8100	4.0321	4.2193	4.3818	4.5257	4.6553	4.7733
6	2.4469	2.9687	3.2875	3.5212	3.7074	3.8630	3.9971	4.1152	4.2209	4.3168
7	2.3646	2.8412	3.1276	3.3353	3.4995	3.6358	3.7527	3.8552	3.9467	4.0293
8	2.3060	2.7515	3.0158	3.2060	3.3554	3.4789	3.5844	3.6766	3.7586	3.8325
9	2.2622	2.6850	2.9333	3.1109	3.2498	3.3642	3.4616	3.5465	3.6219	3.6897
10	2.2281	2.6338	2.8701	3.0382	3.1693	3.2768	3.3682	3.4477	3.5182	3.5814
11	2.2010	2.5931	2.8200	2.9809	3.1058	3.2081	3.2949	3.3702	3.4368	3.4966
12	2.1788	2.5600	2.7795	2.9345	3.0545	3.1527	3.2357	3.3078	3.3714	3.4284
13	2.1604	2.5326	2.7459	2.8961	3.0123	3.1070	3.1871	3.2565	3.3177	3.3725
14	2.1448	2.5096	2.7178	2.8640	2.9768	3.0688	3.1464	3.2135	3.2727	3.3257
15	2.1314	2.4899	2.6937	2.8366	2.9467	3.0363	3.1118	3.1771	3.2346	3.2860
16	2.1199	2.4729	2.6730	2.8131	2.9208	3.0083	3.0821	3.1458	3.2019	3.2520
17	2.1098	2.4581	2.6550	2.7925	2.8982	2.9840	3.0563	3.1186	3.1735	3.2224
18	2.1009	2.4450	2.6391	2.7745	2.8784	2.9627	3.0336	3.0948	3.1486	3.1966
19	2.0930	2.4334	2.6251	2.7586	2.8609	2.9439	3.0136	3.0738	3.1266	3.1737
20	2.0860	2.4231	2.6126	2.7444	2.8453	2.9271	2.9958	3.0550	3.1070	3.1534
21	2.0796	2.4138	2.6013	2.7316	2.8314	2.9121	2.9799	3.0382	3.0895	3.1352
22	2.0739	2.4055	2.5912	2.7201	2.8188	2.8985	2.9655	3.0231	3.0737	3.1188
23	2.0687	2.3979	2.5820	2.7097	2.8073	2.8863	2.9525	3.0095	3.0595	3.1040
24	2.0639	2.3909	2.5736	2.7002	2.7969	2.8751	2.9406	2.9970	3.0465	3.0905
25	2.0595	2.3846	2.5660	2.6916	2.7874	2.8649	2.9298	2.9856	3.0346	3.0782
26	2.0555	2.3788	2.5589	2.6836	2.7787	2.8555	2.9199	2.9752	3.0237	3.0669
27	2.0518	2.3734	2.5525	2.6763	2.7707	2.8469	2.9107	2.9656	3.0137	3.0565
28	2.0484	2.3685	2.5465	2.6695	2.7633	2.8389	2.9023	2.9567	3.0045	3.0469
29	2.0452	2.3638	2.5409	2.6632	2.7564	2.8316	2.8945	2.9485	2.9959	3.0380
30	2.0423	2.3596	2.5357	2.6574	2.7500	2.8247	2.8872	2.9409	2.9880	3.0298
35	2.0301	2.3420	2.5145	2.6334	2.7238	2.7966	2.8575	2.9097	2.9554	2.9960
40	2.0211	2.3289	2.4989	2.6157	2.7045	2.7759	2.8355	2.8867	2.9314	2.9712
45	2.0141	2.3189	2.4868	2.6021	2.6896	2.7599	2.8187	2.8690	2.9130	2.9521
50	2.0086	2.3109	2.4772	2.5913	2.6778	2.7473	2.8053	2.8550	2.8984	2.9370
55	2.0040	2.3044	2.4694	2.5825	2.6682	2.7370	2.7944	2.8436	2.8866	2.9247
60	2.0003	2.2990	2.4630	2.5752	2.6603	2.7286	2.7855	2.8342	2.8768	2.9146
70	1.9944	2.2906	2.4529	2.5639	2.6479	2.7153	2.7715	2.8195	2.8615	2.8987
80	1.9901	2.2844	2.4454	2.5554	2.6387	2.7054	2.7610	2.8086	2.8502	2.8870
90	1.9867	2.2795	2.4395	2.5489	2.6316	2.6978	2.7530	2.8002	2.8414	2.8779
100	1.9840	2.2757	2.4349	2.5437	2.6259	2.6918	2.7466	2.7935	2.8344	2.8707
110	1.9818	2.2725	2.4311	2.5394	2.6213	2.6868	2.7414	2.7880	2.8287	2.8648
120	1.9799	2.2699	2.4280	2.5359	2.6174	2.6827	2.7370	2.7835	2.8240	2.8599
250	1.9695	2.2550	2.4102	2.5159	2.5956	2.6594	2.7124	2.7577	2.7972	2.8322
500	1.9647	2.2482	2.4021	2.5068	2.5857	2.6488	2.7012	2.7460	2.7850	2.8195
1000	1.9623	2.2448	2.3980	2.5022	2.5808	2.6435	2.6957	2.7402	2.7790	2.8133
∞	1.9600	2.2414	2.3940	2.4977	2.5758	2.6383	2.6901	2.7344	2.7729	2.8070

Table 18 *Continued*

df \p	11	12	13	14	15	16	17	18	19
2	14.7818	15.4435	16.0780	18.6883	17.2772	17.8466	18.3984	18.9341	19.4551
3	7.7041	7.9398	8.1625	8.3738	8.5752	8.7676	8.9521	9.1294	9.3001
4	5.7465	5.8853	6.0154	6.1380	6.2541	6.3643	8.4693	6.5697	6.6659
5	4.8819	4.9825	5.0764	5.1644	5.2474	5.3259	8.4005	5.4715	5.5393
6	4.4047	4.4858	4.5612	4.6317	4.6979	4.7604	4.8196	4.8759	4.9295
7	4.1048	4.1743	4.2388	4.2989	4.3553	4.4084	4.4586	4.5062	4.5514
8	3.8999	3.9618	4.0191	4.0724	4.1224	4.1693	4.2137	4.2556	4.2955
9	3.7513	3.8079	3.8602	3.9088	3.9542	3.9969	4.0371	4.0752	4.1114
10	3.6388	3.6915	3.7401	3.7852	3.8273	3.8669	3.9041	3.9394	3.9728
11	3.5508	3.6004	3.6462	3.6887	3.7283	3.7654	3.8004	3.8335	3.8648
12	3.4801	3.5274	3.5709	3.6112	3.6489	3.6842	3.7173	3.7487	3.7783
13	3.4221	3.4674	3.5091	3.5478	3.5838	3.6176	3.6493	3.6793	3.7076
14	3.3736	3.4173	3.4576	3.4949	3.5296	3.5621	3.5926	3.6214	3.6487
15	3.3325	3.3749	3.4139	3.4501	3.4837	3.5151	3.5447	3.5725	3.5989
16	3.2973	3.3386	3.3765	3.4116	3.4443	3.4749	3.5036	3.5306	3.5562
17	3.2667	3.3070	3.3440	3.3783	3.4102	3.4400	3.4680	3.4944	3.5193
18	3.2399	3.2794	3.3156	3.3492	3.3804	3.4095	3.4369	3.4626	3.4870
19	3.2163	3.2550	3.2906	3.3235	3.3540	3.3826	3.4094	3.4347	3.4585
20	3.1952	3.2333	3.2683	3.3006	3.3306	3.3587	3.3850	3.4098	3.4332
21	3.1764	3.2139	3.2483	3.2802	3.3097	3.3373	3.3632	3.3876	3.4106
22	3.1595	3.1965	3.2304	3.2618	3.2909	3.3181	3.3436	3.3676	3.3903
23	3.1441	3.1807	3.2142	3.2451	3.2739	3.3007	3.3259	3.3495	3.3719
24	3.1302	3.1663	3.1994	3.2300	3.2584	3.2849	3.3097	3.3331	3.3552
25	3.1175	3.1532	3.1859	3.2162	3.2443	3.2705	3.2950	3.3181	3.3400
26	3.1058	3.1412	3.1736	3.2035	3.2313	3.2572	3.2815	3.3044	3.3260
27	3.0951	3.1301	3.1622	3.1919	3.2194	3.2451	3.2691	3.2918	3.3132
28	3.0852	3.1199	3.1517	3.1811	3.2084	3.2339	3.2577	3.2801	3.3013
29	3.0760	3.1105	3.1420	3.1712	3.1982	3.2235	3.2471	3.2694	3.2904
30	3.0675	3.1017	3.1330	3.1620	3.1888	3.2138	3.2373	3.2594	3.2802
35	3.0326	3.0658	3.0962	3.1242	3.1502	3.1744	3.1971	3.2185	3.2386
40	3.0069	3.0393	3.0690	3.0964	3.1218	3.1455	3.1676	3.1884	3.2081
45	2.9872	3.0191	3.0482	3.0751	3.1000	3.1232	3.1450	3.1654	3.1846
50	2.9716	3.0030	3.0318	3.0582	3.0828	3.1057	3.1271	3.1472	3.1661
55	2.9589	2.9900	3.0184	3.0446	3.0688	3.0914	3.1125	3.1324	3.1511
60	2.9485	2.9792	3.0074	3.0333	3.0573	3.0796	3.1005	3.1202	3.1387
70	2.9321	2.9624	2.9901	3.0156	3.0393	3.0613	3.0818	3.1012	3.1194
80	2.9200	2.9500	2.9773	3.0026	3.0259	3.0476	3.0679	3.0870	3.1050
90	2.9106	2.9403	2.9675	2.9924	3.0156	3.0371	3.0572	3.0761	3.0939
100	2.9032	2.9327	2.9596	2.9844	3.0073	3.0287	3.0487	3.0674	3.0851
110	2.8971	2.9264	2.9532	2.9778	3.0007	3.0219	3.0417	3.0604	3.0779
120	2.8921	2.9212	2.9479	2.9724	2.9951	3.0162	3.0360	3.0545	3.0720
250	2.8635	2.8919	2.9178	2.9416	2.9637	2.9842	3.0034	3.0213	3.0383
500	2.8505	2.8785	2.9041	2.9276	2.9424	2.9696	2.9885	3.0063	3.0230
1000	2.8440	2.8719	2.8973	2.9207	2.9423	2.9624	2.9812	2.9988	3.0154
∞	2.8376	2.8653	2.8905	2.9137	2.9352	2.9552	2.9738	2.9913	3.0078

Table 18 *Continued*

df \ p	20	21	28	36	45	55	66	78
2	19.9625	20.4573	23.6326	26.8049	29.9750	33.1436	38.3112	39.4778
3	9.4649	9.6242	10.6166	11.5632	12.4715	13.3471	14.1943	15.0165
4	6.7583	6.8471	7.3924	7.8998	8.3763	8.8271	9.2558	9.6655
5	5.6042	5.6665	6.0447	6.3914	6.7126	7.0128	7.2952	7.5625
6	4.9807	5.0297	5.3255	5.5937	5.8399	6.0680	6.2810	6.4813
7	4.5946	4.6359	4.8839	5.1068	5.3101	5.4973	5.6712	5.8339
8	4.3335	4.3699	4.5869	4.7810	4.9570	5.1183	5.2675	5.4065
9	4.1458	4.1786	4.3744	4.5485	4.7058	4.8494	4.9818	5.1048
10	4.0045	4.0348	4.2150	4.3747	4.5184	4.6492	4.7695	4.8810
11	3.8945	3.9229	4.0913	4.2400	4.3735	4.4947	4.6059	4.7087
12	3.8065	3.8334	3.9925	4.1327	4.2582	4.3719	4.4761	4.5722
13	3.7345	3.7602	3.9118	4.0452	4.1643	4.2721	4.3706	4.4614
14	3.6746	3.6992	3.8448	3.9725	4.0865	4.1894	4.2833	4.3698
15	3.6239	3.6477	3.7882	3.9113	4.0209	4.1198	4.2099	4.2928
16	3.5805	3.6036	3.7398	3.8589	3.9649	4.0604	4.1473	4.2272
17	3.5429	3.5654	3.6980	3.8137	3.9165	4.0091	4.0933	4.1706
18	3.5101	3.5321	3.6614	3.7742	3.8744	3.9644	4.0463	4.1214
19	3.4812	3.5027	3.6292	3.7395	3.8373	3.9251	4.0050	4.0781
20	3.4554	3.4765	3.6006	3.7087	3.8044	3.8903	3.9683	4.0398
21	3.4325	3.4532	3.5751	3.6812	3.7750	3.8593	3.9357	4.0056
22	3.4118	3.4322	3.5522	3.6564	3.7487	3.8314	3.9064	3.9750
23	3.3931	3.4132	3.5314	3.6341	3.7249	3.8062	3.8800	3.9474
24	3.3761	3.3960	3.5126	3.6139	3.7033	3.7834	3.8560	3.9223
25	3.3606	3.3803	3.4955	3.5954	3.6836	3.7626	3.8342	3.8995
26	3.3464	3.3659	3.4797	3.5785	3.6656	3.7436	3.8142	3.8787
27	3.3334	3.3526	3.4653	3.5629	3.6491	3.7261	3.7959	3.8595
28	3.3214	3.3404	3.4520	3.5486	3.6338	3.7101	3.7790	3.8419
29	3.3102	3.3291	3.4397	3.5354	3.6198	3.6952	3.7634	3.8256
30	3.2999	3.3186	3.4282	3.5231	3.6067	3.6814	3.7489	3.8105
35	3.2577	3.2758	3.3816	3.4730	3.5534	3.6252	3.6900	3.7490
40	3.2266	3.2443	3.3473	3.4362	3.5143	3.5840	3.6468	3.7040
45	3.2028	3.2201	3.3211	3.4081	3.4845	3.5525	3.6138	3.6696
50	3.1840	3.2010	3.3003	3.3858	3.4609	3.5277	3.5878	3.6425
55	3.1688	3.1856	3.2836	3.3679	3.4418	3.5076	3.5668	3.6206
60	3.1562	3.1728	3.2697	3.3530	3.4260	3.4910	3.5494	3.6025
70	3.1366	3.1529	3.2481	3.3299	3.4015	3.4652	3.5224	3.5744
80	3.1220	3.1381	3.2321	3.3127	3.3833	3.4460	3.5024	3.5536
90	3.1108	3.1267	3.2197	3.2995	3.3693	3.4313	3.4870	3.5375
100	3.1018	3.1176	3.2099	3.2890	3.3582	3.4196	3.4747	3.5248
110	3.0945	3.1102	3.2018	3.2804	3.3491	3.4100	3.4648	3.5144
120	3.0885	3.1041	3.1952	3.2733	3.3416	3.4021	3.4565	3.5058
250	3.0543	3.0694	3.1577	3.2332	3.2991	3.3575	3.4099	3.4573
500	3.0387	3.0537	3.1406	3.2150	3.2798	3.3373	3.3887	3.4354
1000	3.0310	3.0459	3.1322	3.2059	3.2703	3.3272	3.3783	3.4245
∞	3.0233	3.0381	3.1237	3.1970	3.2608	3.3172	3.3678	3.4136

Table 18   *Continued*

df \ p	91	105	120	136	153	171	190
2	42.6439	45.8094	48.9745	52.1392	35.3037	58.4679	61.6320
3	15.8165	16.5964	17.3582	18.1035	18.8336	19.5497	20.2528
4	10.0585	10.4367	10.8016	11.1545	11.4966	11.8288	12.1519
5	7.8166	8.0591	8.2913	8.5143	8.7290	8.9362	9.1365
6	6.6705	6.8500	7.0210	7.1844	7.3410	7.4914	7.6383
7	5.9868	6.1313	6.2684	6.3990	6.5236	6.6430	6.7577
8	5.5368	5.6594	5.7755	5.8857	5.9906	6.0909	6.1869
9	5.2197	5.3276	5.4295	5.5260	5.6177	5.7051	5.7888
10	4.9849	5.0823	5.1740	5.2608	5.3431	5.4215	5.4963
11	4.8044	4.8939	4.9781	5.0576	5.1330	5.2046	5.2729
12	4.6615	4.7450	4.8233	4.8972	4.9672	5.0336	5.0969
13	4.5457	4.6243	4.6981	4.7675	4.8332	4.8956	4.9549
14	4.4500	4.5247	4.5947	4.6606	4.7228	4.7818	4.8379
15	4.3695	4.4410	4.5079	4.5708	4.6302	4.6865	4.7400
16	4.3011	4.3698	4.4341	4.4946	4.5516	4.6056	4.6568
17	4.2421	4.3085	4.3706	4.4289	4.4839	4.5360	4.5853
18	4.1907	4.2551	4.3154	4.3719	4.4251	4.4755	4.5232
19	4.1456	4.2083	4.2669	4.3218	4.3736	4.4225	4.4688
20	4.1057	4.1669	4.2240	4.2776	4.3280	4.3756	4.4208
21	4.0701	4.1300	4.1858	4.2381	4.2874	4.3339	4.3780
22	4.0382	4.0969	4.1516	4.2028	4.2510	4.2966	4.3397
23	4.0095	4.0671	4.1207	4.1710	4.2183	4.2629	4.3052
24	3.9834	4.0400	4.0928	4.1422	4.1886	4.2325	4.2739
25	3.9597	4.0154	4.0674	4.1160	4.1616	4.2047	4.2455
26	3.9380	3.9929	4.0441	4.0920	4.1370	4.1794	4.2196
27	3.9181	3.9723	4.0228	4.0700	4.1144	4.1562	4.1958
28	3.8997	3.9533	4.0032	4.0498	4.0936	4.1349	4.1739
29	3.8828	3.9357	3.9850	4.0311	4.0744	4.1151	4.1537
30	3.8671	3.9195	3.9682	4.0138	4.0566	4.0969	4.1350
35	3.8032	3.8533	3.8999	3.9434	3.9842	4.0226	4.0590
40	3.7564	3.8049	3.8499	3.8919	3.9314	3.9684	4.0035
45	3.7208	3.7680	3.8118	3.8527	3.8911	3.9271	3.9612
50	3.6926	3.7389	3.7818	3.8218	3.8594	3.8946	3.9279
55	3.6699	3.7154	3.7576	3.7969	3.8337	3.8684	3.9010
60	3.6511	3.6960	3.7376	3.7763	3.8126	3.8467	3.8789
70	3.6220	3.6658	3.7065	3.7444	3.7798	3.8131	3.8445
80	3.6004	3.6435	3.6835	3.7207	3.7555	3.7883	3.8191
90	3.5837	3.6263	3.6658	3.7025	3.7369	3.7691	3.7995
100	3.5705	3.6127	3.6517	3.6880	3.7220	3.7539	3.7840
110	3.5598	3.6016	3.6403	3.6763	3.7100	3.7416	3.7714
120	3.5509	3.5924	3.6308	3.6665	3.7000	3.7313	3.7609
250	3.5007	3.5405	3.5774	3.6117	3.6437	3.6737	3.7020
500	3.4779	3.5170	3.5532	3.5868	3.6182	3.6477	3.6754
1000	3.4666	3.5054	3.5412	3.5745	3.6056	3.6348	3.6622
∞	3.4554	3.4938	3.5293	3.5623	3.5931	3.6219	3.6491

Table 19 Dunn-Šidák's t for comparisons between p treatment means and a control for α = 0.01, 0.05, 0.10, and 0.20 (one-sided test) (values generated as described in Chapter 5)

$df\backslash p$	α	2	3	4	5	6	7	8	9	10	15	20	25	30	35	40	45	50
2	0.01	9.912	12.166	14.062	15.732	17.241	18.628	19.919	21.130	22.277	27.295	31.524	35.249	38.617	41.713	44.595	47.302	49.862
	0.05	4.273	5.292	6.144	6.892	7.566	8.185	8.760	9.300	9.810	12.040	13.918	15.571	17.064	18.437	19.715	20.914	22.049
	0.10	2.876	3.606	4.213	4.742	5.219	5.655	6.060	6.439	6.798	8.363	9.678	10.835	11.879	12.839	13.732	14.571	15.363
	0.20	1.815	2.348	2.783	3.159	3.495	3.802	4.085	4.351	4.601	5.689	6.601	7.401	8.123	8.785	9.401	9.979	10.526
3	0.01	5.836	6.733	7.444	8.041	8.563	9.028	9.451	9.839	10.199	11.706	12.903	13.913	14.795	15.583	16.299	16.958	17.569
	0.05	3.166	3.716	4.146	4.506	4.819	5.097	5.349	5.580	5.793	6.685	7.391	7.985	8.503	8.965	9.385	9.770	10.128
	0.10	2.325	2.779	3.132	3.425	3.678	3.902	4.105	4.290	4.462	5.176	5.738	6.211	6.623	6.990	7.323	7.628	7.911
	0.20	1.585	1.971	2.264	2.505	2.712	2.895	3.059	3.209	3.348	3.920	4.368	4.743	5.069	5.360	5.622	5.863	6.087
4	0.01	4.601	5.162	5.592	5.945	6.247	6.513	6.750	6.966	7.165	7.975	8.600	9.114	9.556	9.945	10.293	10.610	10.902
	0.05	2.764	3.169	3.474	3.723	3.935	4.121	4.286	4.436	4.574	5.132	5.560	5.912	6.212	6.477	6.713	6.928	7.126
	0.10	2.109	2.469	2.738	2.956	3.141	3.302	3.445	3.575	3.693	4.173	4.539	4.838	5.094	5.319	5.519	5.702	5.869
	0.20	1.487	1.817	2.059	2.254	2.417	2.558	2.684	2.797	2.900	3.314	3.628	3.885	4.103	4.294	4.464	4.618	4.760
5	0.01	4.030	4.452	4.769	5.026	5.242	5.431	5.599	5.750	5.887	6.443	6.862	7.202	7.491	7.743	7.967	8.170	8.355
	0.05	2.560	2.897	3.146	3.346	3.514	3.660	3.788	3.904	4.009	4.430	4.746	5.002	5.218	5.406	5.573	5.723	5.860
	0.10	1.995	2.309	2.538	2.721	2.874	3.006	3.123	3.227	3.322	3.700	3.982	4.210	4.402	4.568	4.716	4.849	4.970
	0.20	1.434	1.734	1.951	2.122	2.264	2.385	2.492	2.588	2.674	3.017	3.270	3.474	3.645	3.794	3.925	4.043	4.150
6	0.01	3.705	4.055	4.313	4.520	4.694	4.844	4.976	5.095	5.203	5.633	5.953	6.211	6.428	6.616	6.782	6.932	7.068
	0.05	2.438	2.736	2.954	3.127	3.270	3.394	3.503	3.600	3.688	4.037	4.295	4.501	4.674	4.824	4.956	5.074	5.182
	0.10	1.924	2.211	2.418	2.581	2.716	2.832	2.934	3.024	3.106	3.428	3.666	3.855	4.013	4.150	4.270	4.378	4.476
	0.20	1.400	1.683	1.884	2.041	2.170	2.281	2.377	2.462	2.539	2.841	3.061	3.236	3.382	3.507	3.618	3.717	3.806
7	0.01	3.498	3.803	4.026	4.204	4.352	4.479	4.591	4.691	4.781	5.139	5.403	5.614	5.791	5.943	6.077	6.197	6.306
	0.05	2.356	2.630	2.828	2.984	3.112	3.223	3.319	3.405	3.482	3.787	4.010	4.187	4.335	4.462	4.574	4.673	4.764
	0.10	1.877	2.146	2.338	2.488	2.612	2.717	2.809	2.891	2.965	3.253	3.463	3.629	3.767	3.885	3.989	4.082	4.166
	0.20	1.376	1.648	1.839	1.987	2.108	2.211	2.300	2.379	2.450	2.725	2.925	3.082	3.212	3.323	3.421	3.507	3.586

Table 19 Continued

$df\backslash p$	α	2	3	4	5	6	7	8	9	10	15	20	25	30	35	40	45	50
8	0.01	3.354	3.630	3.830	3.988	4.119	4.232	4.330	4.418	4.497	4.809	5.037	5.219	5.370	5.499	5.613	5.715	5.807
	0.05	2.298	2.555	2.739	2.883	3.002	3.103	3.191	3.269	3.340	3.615	3.815	3.973	4.104	4.217	4.315	4.403	4.482
	0.10	1.843	2.099	2.281	2.422	2.538	2.636	2.722	2.798	2.866	3.131	3.322	3.473	3.597	3.704	3.797	3.880	3.954
	0.20	1.359	1.622	1.806	1.948	2.063	2.160	2.245	2.319	2.386	2.644	2.829	2.974	3.093	3.195	3.284	3.363	3.434
9	0.01	3.248	3.503	3.687	3.832	3.951	4.054	4.143	4.222	4.294	4.574	4.777	4.939	5.072	5.187	5.287	5.376	5.457
	0.05	2.254	2.499	2.673	2.809	2.920	3.015	3.097	3.170	3.235	3.490	3.674	3.819	3.938	4.040	4.129	4.208	4.280
	0.10	1.817	2.064	2.238	2.373	2.483	2.576	2.657	2.729	2.793	3.041	3.219	3.358	3.473	3.571	3.657	3.733	3.801
	0.20	1.346	1.603	1.782	1.918	2.029	2.123	2.204	2.275	2.338	2.583	2.758	2.894	3.006	3.101	3.183	3.257	3.322
10	0.01	3.168	3.407	3.579	3.714	3.825	3.919	4.002	4.075	4.141	4.398	4.584	4.730	4.851	4.955	5.046	5.126	5.199
	0.05	2.221	2.456	2.623	2.752	2.858	2.947	3.025	3.094	3.156	3.395	3.567	3.701	3.812	3.907	3.989	4.062	4.128
	0.10	1.797	2.037	2.205	2.335	2.441	2.530	2.607	2.675	2.736	2.972	3.140	3.271	3.379	3.471	3.551	3.622	3.685
	0.20	1.336	1.588	1.762	1.895	2.003	2.094	2.172	2.240	2.302	2.537	2.703	2.833	2.939	3.028	3.107	3.176	3.238
11	0.01	3.104	3.332	3.494	3.621	3.726	3.815	3.892	3.960	4.022	4.262	4.434	4.570	4.682	4.777	4.860	4.934	5.001
	0.05	2.194	2.422	2.582	2.707	2.808	2.894	2.968	3.034	3.093	3.320	3.483	3.610	3.714	3.803	3.880	3.949	4.010
	0.10	1.780	2.015	2.179	2.304	2.407	2.493	2.567	2.633	2.692	2.918	3.078	3.203	3.305	3.392	3.468	3.535	3.595
	0.20	1.328	1.576	1.747	1.877	1.982	2.070	2.146	2.213	2.272	2.500	2.660	2.784	2.886	2.972	3.046	3.112	3.171
12	0.01	3.053	3.271	3.426	3.547	3.647	3.731	3.804	3.869	3.927	4.153	4.315	4.442	4.547	4.636	4.714	4.783	4.844
	0.05	2.172	2.394	2.550	2.670	2.768	2.851	2.922	2.986	3.042	3.260	3.415	3.536	3.635	3.720	3.793	3.858	3.916
	0.10	1.767	1.997	2.157	2.280	2.379	2.463	2.535	2.599	2.656	2.873	3.028	3.148	3.246	3.329	3.401	3.465	3.522
	0.20	1.321	1.566	1.734	1.862	1.965	2.051	2.125	2.190	2.248	2.469	2.625	2.745	2.843	2.925	2.997	3.060	3.117
13	0.01	3.011	3.221	3.371	3.487	3.582	3.662	3.732	3.794	3.850	4.064	4.218	4.339	4.437	4.522	4.595	4.660	4.718
	0.05	2.153	2.370	2.523	2.640	2.735	2.815	2.885	2.946	3.000	3.211	3.360	3.476	3.571	3.652	3.721	3.783	3.839
	0.10	1.756	1.982	2.139	2.259	2.356	2.438	2.508	2.570	2.626	2.837	2.986	3.102	3.197	3.277	3.346	3.407	3.462
	0.20	1.315	1.557	1.723	1.849	1.950	2.035	2.108	2.172	2.228	2.444	2.596	2.712	2.807	2.887	2.957	3.018	3.073

14	0.01	4.614	4.558	4.497	4.427	4.347	4.253	4.138	3.991	3.785	3.732	3.672	3.605	3.527	3.436	3.324	3.179	2.976
	0.05	3.775	3.721	3.662	3.595	3.517	3.426	3.314	3.169	2.965	2.912	2.853	2.785	2.707	2.614	2.500	2.351	2.138
	0.10	3.412	3.359	3.300	3.233	3.156	3.064	2.952	2.807	2.600	2.546	2.486	2.417	2.337	2.242	2.124	1.969	1.746
	0.20	3.035	2.982	2.923	2.856	2.778	2.685	2.571	2.423	2.212	2.156	2.093	2.022	1.938	1.839	1.714	1.550	1.310
15	0.01	4.526	4.473	4.414	4.348	4.271	4.181	4.070	3.929	3.731	3.679	3.622	3.557	3.482	3.393	3.284	3.144	2.945
	0.05	3.721	3.669	3.612	3.547	3.472	3.383	3.274	3.134	2.935	2.884	2.826	2.760	2.683	2.592	2.480	2.334	2.125
	0.10	3.370	3.319	3.261	3.196	3.121	3.032	2.922	2.781	2.579	2.526	2.467	2.399	2.321	2.227	2.111	1.959	1.738
	0.20	3.004	2.952	2.894	2.828	2.752	2.662	2.550	2.405	2.197	2.142	2.081	2.010	1.928	1.829	1.707	1.544	1.306
16	0.01	4.451	4.401	4.344	4.280	4.206	4.119	4.013	3.876	3.684	3.634	3.578	3.515	3.442	3.356	3.250	3.113	2.920
	0.05	3.674	3.624	3.569	3.506	3.433	3.346	3.240	3.104	2.910	2.859	2.802	2.738	2.663	2.573	2.463	2.320	2.113
	0.10	3.334	3.284	3.228	3.164	3.091	3.004	2.897	2.758	2.560	2.508	2.450	2.384	2.306	2.214	2.099	1.949	1.731
	0.20	2.976	2.926	2.869	2.805	2.730	2.642	2.532	2.389	2.185	2.131	2.070	2.000	1.919	1.821	1.700	1.539	1.303
17	0.01	4.387	4.338	4.284	4.222	4.150	4.066	3.963	3.830	3.644	3.595	3.541	3.479	3.408	3.324	3.221	3.087	2.897
	0.05	3.634	3.586	3.531	3.470	3.399	3.315	3.211	3.077	2.887	2.838	2.782	2.718	2.645	2.557	2.449	2.307	2.103
	0.10	3.302	3.253	3.199	3.137	3.065	2.980	2.875	2.739	2.544	2.493	2.436	2.370	2.294	2.203	2.090	1.941	1.725
	0.20	2.953	2.903	2.848	2.785	2.711	2.624	2.516	2.376	2.174	2.120	2.060	1.991	1.911	1.814	1.694	1.534	1.299
18	0.01	4.332	4.284	4.231	4.171	4.102	4.020	3.920	3.790	3.608	3.561	3.508	3.448	3.378	3.296	3.195	3.064	2.877
	0.05	3.599	3.552	3.499	3.439	3.369	3.287	3.186	3.054	2.868	2.819	2.764	2.702	2.629	2.543	2.436	2.296	2.094
	0.10	3.275	3.227	3.173	3.113	3.042	2.958	2.855	2.721	2.530	2.480	2.423	2.358	2.283	2.193	2.081	1.934	1.720
	0.20	2.932	2.883	2.829	2.767	2.695	2.609	2.503	2.364	2.164	2.111	2.052	1.984	1.904	1.808	1.689	1.530	1.297
19	0.01	4.283	4.236	4.185	4.126	4.059	3.979	3.881	3.755	3.577	3.531	3.479	3.420	3.352	3.271	3.172	3.043	2.860
	0.05	3.568	3.522	3.470	3.411	3.343	3.262	3.163	3.034	2.850	2.802	2.748	2.687	2.615	2.530	2.424	2.286	2.087
	0.10	3.250	3.204	3.151	3.091	3.022	2.940	2.838	2.706	2.517	2.468	2.412	2.348	2.273	2.184	2.073	1.927	1.715
	0.20	2.914	2.866	2.812	2.751	2.680	2.595	2.490	2.353	2.156	2.103	2.044	1.977	1.898	1.803	1.684	1.526	1.294
20	0.01	4.239	4.194	4.144	4.087	4.021	3.943	3.847	3.724	3.550	3.504	3.454	3.396	3.329	3.249	3.152	3.025	2.844
	0.05	3.541	3.496	3.445	3.387	3.320	3.240	3.143	3.016	2.835	2.788	2.734	2.673	2.603	2.518	2.414	2.277	2.080
	0.10	3.229	3.183	3.131	3.072	3.004	2.923	2.823	2.693	2.506	2.457	2.401	2.338	2.264	2.176	2.066	1.922	1.711
	0.20	2.897	2.850	2.797	2.737	2.667	2.583	2.479	2.344	2.148	2.096	2.038	1.971	1.892	1.798	1.680	1.523	1.292

Table 19 Continued

df\p	α	2	3	4	5	6	7	8	9	10	15	20	25	30	35	40	45	50
21	0.01	2.830	3.009	3.134	3.230	3.308	3.374	3.431	3.481	3.525	3.696	3.817	3.911	3.987	4.052	4.108	4.157	4.201
	0.05	2.073	2.269	2.405	2.508	2.592	2.662	2.722	2.774	2.821	3.000	3.125	3.221	3.299	3.365	3.422	3.472	3.517
	0.10	1.707	1.916	2.060	2.169	2.257	2.330	2.392	2.447	2.496	2.681	2.809	2.908	2.988	3.055	3.113	3.164	3.209
	0.20	1.290	1.520	1.676	1.794	1.887	1.965	2.032	2.090	2.141	2.335	2.470	2.572	2.655	2.724	2.784	2.836	2.883
22	0.01	2.818	2.994	3.117	3.212	3.289	3.354	3.410	3.459	3.503	3.671	3.790	3.882	3.957	4.020	4.075	4.123	4.166
	0.05	2.068	2.262	2.397	2.499	2.582	2.651	2.710	2.762	2.809	2.985	3.108	3.203	3.280	3.345	3.401	3.451	3.495
	0.10	1.703	1.912	2.055	2.163	2.250	2.322	2.384	2.439	2.487	2.670	2.797	2.894	2.974	3.040	3.097	3.147	3.192
	0.20	1.288	1.518	1.673	1.790	1.883	1.960	2.026	2.084	2.135	2.328	2.461	2.562	2.644	2.713	2.772	2.823	2.870
23	0.01	2.806	2.981	3.102	3.196	3.272	3.336	3.391	3.440	3.483	3.649	3.766	3.856	3.930	3.992	4.045	4.093	4.135
	0.05	2.062	2.256	2.389	2.491	2.572	2.641	2.700	2.751	2.797	2.972	3.094	3.187	3.264	3.328	3.383	3.431	3.475
	0.10	1.700	1.907	2.049	2.157	2.243	2.315	2.377	2.431	2.479	2.660	2.786	2.882	2.960	3.026	3.083	3.132	3.176
	0.20	1.286	1.515	1.670	1.786	1.879	1.956	2.021	2.079	2.130	2.321	2.453	2.553	2.634	2.702	2.761	2.812	2.858
24	0.01	2.796	2.968	3.089	3.182	3.257	3.320	3.374	3.422	3.465	3.628	3.743	3.832	3.905	3.966	4.019	4.065	4.107
	0.05	2.058	2.250	2.382	2.483	2.564	2.632	2.690	2.742	2.787	2.960	3.080	3.173	3.248	3.311	3.366	3.414	3.457
	0.10	1.697	1.904	2.045	2.152	2.237	2.309	2.370	2.424	2.471	2.651	2.776	2.871	2.948	3.013	3.069	3.118	3.162
	0.20	1.285	1.513	1.667	1.783	1.875	1.952	2.017	2.074	2.124	2.314	2.445	2.545	2.626	2.693	2.751	2.802	2.847
25	0.01	2.786	2.957	3.077	3.168	3.243	3.305	3.359	3.406	3.448	3.610	3.723	3.811	3.882	3.942	3.995	4.040	4.081
	0.05	2.053	2.244	2.376	2.476	2.557	2.624	2.682	2.733	2.778	2.949	3.068	3.160	3.234	3.297	3.351	3.398	3.440
	0.10	1.694	1.900	2.041	2.147	2.232	2.303	2.364	2.417	2.464	2.642	2.766	2.861	2.938	3.002	3.057	3.106	3.149
	0.20	1.283	1.511	1.664	1.780	1.872	1.948	2.013	2.070	2.120	2.308	2.439	2.538	2.617	2.684	2.742	2.792	2.837
26	0.01	2.778	2.947	3.065	3.156	3.230	3.291	3.345	3.391	3.433	3.593	3.705	3.791	3.862	3.921	3.972	4.018	4.058
	0.05	2.049	2.239	2.370	2.470	2.550	2.617	2.674	2.724	2.769	2.939	3.057	3.148	3.221	3.283	3.337	3.383	3.425
	0.10	1.692	1.897	2.037	2.142	2.227	2.298	2.358	2.411	2.458	2.635	2.758	2.852	2.928	2.991	3.046	3.094	3.137
	0.20	1.282	1.509	1.662	1.777	1.868	1.944	2.009	2.065	2.115	2.303	2.432	2.531	2.610	2.676	2.733	2.783	2.828

df	α																		
27	0.01	2.770	2.938	3.055	3.145	3.218	3.279	3.332	3.378	3.419	3.577	3.688	3.773	3.843	3.901	3.952	3.997	4.036
	0.05	2.046	2.235	2.365	2.464	2.543	2.610	2.667	2.717	2.761	2.929	3.047	3.137	3.210	3.271	3.324	3.370	3.411
	0.10	1.690	1.894	2.033	2.138	2.223	2.293	2.353	2.405	2.452	2.628	2.750	2.843	2.918	2.982	3.036	3.084	3.126
	0.20	1.281	1.507	1.660	1.774	1.865	1.941	2.006	2.062	2.111	2.298	2.427	2.524	2.603	2.669	2.726	2.775	2.819
28	0.01	2.762	2.929	3.045	3.135	3.207	3.267	3.320	3.366	3.406	3.562	3.672	3.757	3.825	3.883	3.933	3.977	4.017
	0.05	2.042	2.231	2.360	2.458	2.537	2.603	2.660	2.710	2.754	2.921	3.037	3.126	3.199	3.259	3.312	3.358	3.398
	0.10	1.687	1.891	2.030	2.134	2.218	2.288	2.348	2.400	2.447	2.621	2.743	2.835	2.910	2.973	3.026	3.074	3.116
	0.20	1.280	1.506	1.658	1.772	1.863	1.938	2.002	2.058	2.108	2.294	2.421	2.519	2.597	2.662	2.719	2.768	2.811
29	0.01	2.755	2.921	3.037	3.125	3.197	3.257	3.309	3.354	3.395	3.549	3.658	3.741	3.809	3.866	3.916	3.959	3.998
	0.05	2.039	2.227	2.355	2.453	2.532	2.597	2.654	2.703	2.747	2.913	3.028	3.117	3.189	3.249	3.301	3.346	3.387
	0.10	1.685	1.888	2.027	2.131	2.214	2.284	2.344	2.395	2.442	2.615	2.736	2.828	2.902	2.964	3.018	3.065	3.106
	0.20	1.279	1.504	1.656	1.770	1.860	1.935	1.999	2.055	2.104	2.289	2.417	2.513	2.591	2.656	2.712	2.761	2.804
30	0.01	2.749	2.914	3.028	3.116	3.187	3.247	3.298	3.343	3.383	3.537	3.644	3.727	3.794	3.851	3.900	3.943	3.981
	0.05	2.036	2.223	2.351	2.448	2.527	2.592	2.648	2.697	2.741	2.905	3.020	3.108	3.179	3.239	3.290	3.335	3.376
	0.10	1.684	1.886	2.024	2.128	2.211	2.280	2.339	2.391	2.437	2.610	2.730	2.821	2.895	2.956	3.010	3.056	3.098
	0.20	1.278	1.503	1.654	1.768	1.858	1.933	1.997	2.052	2.101	2.285	2.412	2.508	2.586	2.650	2.706	2.754	2.798
40	0.01	2.703	2.861	2.970	3.053	3.120	3.177	3.225	3.267	3.305	3.449	3.549	3.626	3.689	3.741	3.787	3.827	3.862
	0.05	2.015	2.197	2.321	2.415	2.490	2.553	2.607	2.654	2.695	2.853	2.962	3.045	3.113	3.169	3.218	3.260	3.298
	0.10	1.670	1.869	2.003	2.104	2.185	2.252	2.309	2.359	2.404	2.570	2.685	2.772	2.843	2.901	2.952	2.996	3.036
	0.20	1.271	1.493	1.642	1.753	1.841	1.914	1.977	2.031	2.078	2.257	2.380	2.472	2.547	2.609	2.662	2.709	2.750
60	0.01	2.659	2.809	2.913	2.992	3.056	3.109	3.155	3.195	3.230	3.365	3.459	3.530	3.589	3.637	3.679	3.716	3.749
	0.05	1.995	2.171	2.291	2.382	2.455	2.515	2.567	2.612	2.652	2.802	2.906	2.985	3.049	3.102	3.148	3.188	3.223
	0.10	1.658	1.851	1.983	2.081	2.160	2.225	2.280	2.328	2.371	2.531	2.642	2.725	2.792	2.848	2.896	2.938	2.976
	0.20	1.264	1.483	1.629	1.738	1.825	1.896	1.957	2.010	2.056	2.230	2.348	2.437	2.509	2.569	2.620	2.664	2.704
120	0.01	2.617	2.760	2.859	2.934	2.994	3.044	3.087	3.125	3.158	3.284	3.372	3.439	3.493	3.538	3.577	3.611	3.641
	0.05	1.974	2.146	2.262	2.350	2.420	2.478	2.528	2.571	2.609	2.753	2.852	2.927	2.987	3.037	3.080	3.118	3.152
	0.10	1.645	1.835	1.963	2.059	2.135	2.198	2.252	2.298	2.340	2.494	2.599	2.679	2.743	2.797	2.843	2.882	2.918
	0.20	1.257	1.473	1.617	1.724	1.809	1.878	1.938	1.989	2.034	2.203	2.317	2.403	2.472	2.530	2.579	2.621	2.659

Table 20 Dunn-Šidák's *t* for comparisons between *p* treatment means and a control for α = 0.01, 0.05, 0.10, and 0.20 (two-sided test) (used first in Chapter 5)

df\p	α	2	3	4	5	6	7	8	9	10	15	20	25	30	35	40	45	50
2	0.01	14.071	17.248	19.925	22.282	24.413	26.372	28.196	29.908	31.528	38.620	44.598	49.865	54.626	59.004	63.079	66.906	70.526
	0.05	6.164	7.582	8.774	9.823	10.769	11.639	12.449	13.208	13.927	17.072	19.721	22.054	24.163	26.103	27.908	29.603	31.206
	0.10	4.243	5.243	6.081	6.816	7.480	8.090	8.656	9.188	9.691	11.890	13.741	15.371	16.845	18.199	19.459	20.642	21.761
	0.20	2.828	3.531	4.116	4.428	5.089	5.512	5.904	6.272	6.620	8.138	9.414	10.537	11.552	12.484	13.351	14.166	14.936
3	0.01	7.447	8.565	9.453	10.201	10.853	11.436	11.966	12.453	12.904	14.796	16.300	17.569	18.678	19.670	20.570	21.398	22.167
	0.05	4.156	4.826	5.355	5.799	6.185	6.529	6.842	7.128	7.394	8.505	9.387	10.129	10.778	11.357	11.883	12.366	12.815
	0.10	3.149	3.690	4.115	4.471	4.780	5.055	5.304	5.532	5.744	6.627	7.326	7.914	8.427	8.886	9.301	9.683	10.038
	0.20	2.294	2.734	3.077	3.363	3.610	3.829	4.028	4.209	4.377	5.076	5.628	6.091	6.495	6.855	7.181	7.481	7.759
4	0.01	5.594	6.248	6.751	7.166	7.520	7.832	8.112	8.367	8.600	9.556	10.294	10.902	11.424	11.884	12.297	12.672	13.017
	0.05	3.481	3.941	4.290	4.577	4.822	5.036	5.228	5.402	5.562	6.214	6.714	7.127	7.480	7.790	8.069	8.322	8.554
	0.10	2.751	3.150	3.452	3.699	3.909	4.093	4.257	4.406	4.542	5.097	5.521	5.870	6.169	6.432	6.667	6.880	7.076
	0.20	2.084	2.434	2.697	2.911	3.092	3.250	3.391	3.518	3.635	4.107	4.468	4.763	5.015	5.237	5.435	5.614	5.779
5	0.01	4.771	5.243	5.599	5.888	6.133	6.346	6.535	6.706	6.862	7.491	7.968	8.355	8.684	8.971	9.226	9.457	9.668
	0.05	3.152	3.518	3.791	4.012	4.197	4.358	4.501	4.630	4.747	5.219	5.573	5.861	6.105	6.317	6.506	6.676	6.831
	0.10	2.549	2.882	3.129	3.327	3.493	3.638	3.765	3.880	3.985	4.403	4.718	4.972	5.187	5.374	5.540	5.689	5.826
	0.20	1.973	2.278	2.503	2.683	2.834	2.964	3.079	3.182	3.275	3.649	3.928	4.153	4.343	4.508	4.654	4.786	4.906
6	0.01	4.315	4.695	4.977	5.203	5.394	5.559	5.704	5.835	5.954	6.428	6.782	7.068	7.308	7.516	7.701	7.867	8.018
	0.05	2.959	3.274	3.505	3.690	3.845	3.978	4.095	4.200	4.296	4.675	4.956	5.182	5.372	5.536	5.682	5.812	5.930
	0.10	2.428	2.723	2.939	3.110	3.253	3.376	3.484	3.580	3.668	4.015	4.272	4.477	4.649	4.798	4.930	5.048	5.155
	0.20	1.904	2.184	2.387	2.547	2.681	2.795	2.895	2.985	3.066	3.385	3.620	3.808	3.965	4.100	4.220	4.327	4.424
7	0.01	4.027	4.353	4.591	4.782	4.941	5.078	5.198	5.306	5.404	5.791	6.077	6.306	6.497	6.663	6.809	6.936	7.058
	0.05	2.832	3.115	3.321	3.484	3.620	3.736	3.838	3.929	4.011	4.336	4.574	4.764	4.923	5.059	5.180	5.287	5.385
	0.10	2.347	2.618	2.814	2.969	3.097	3.206	3.302	3.388	3.465	3.768	3.990	4.167	4.314	4.441	4.552	4.651	4.741
	0.20	1.858	2.120	2.309	2.457	2.579	2.684	2.775	2.856	2.929	3.214	3.423	3.588	3.725	3.842	3.946	4.038	4.121

8	0.01	3.831	4.120	4.331	4.498	4.637	4.756	4.860	4.953	5.038	5.370	5.613	5.807	5.969	6.107	6.230	6.339	6.437
	0.05	2.743	3.005	3.193	3.342	3.464	3.569	3.661	3.743	3.816	4.105	4.316	4.482	4.621	4.740	4.844	4.937	5.021
	0.10	2.289	2.544	2.726	2.869	2.987	3.088	3.176	3.254	3.324	3.598	3.798	3.955	4.086	4.198	4.296	4.383	4.462
	0.20	1.824	2.075	2.254	2.393	2.508	2.605	2.690	2.765	2.832	3.095	3.286	3.435	3.559	3.665	3.758	3.840	3.914
9	0.01	3.688	3.952	4.143	4.294	4.419	4.526	4.619	4.703	4.778	5.072	5.287	5.457	5.598	5.720	5.826	5.921	6.006
	0.05	2.677	2.923	3.099	3.237	3.351	3.448	3.532	3.607	3.675	3.938	4.129	4.280	4.405	4.512	4.605	4.688	4.763
	0.10	2.246	2.488	2.661	2.796	2.907	3.001	3.083	3.155	3.221	3.474	3.658	3.802	3.921	4.023	4.112	4.191	4.262
	0.20	1.799	2.041	2.212	2.345	2.454	2.546	2.627	2.698	2.761	3.008	3.185	3.324	3.438	3.536	3.621	3.697	3.765
10	0.01	3.580	3.825	4.002	4.141	4.256	4.354	4.439	4.515	4.584	4.852	5.046	5.199	5.326	5.434	5.529	5.614	5.690
	0.05	2.626	2.860	3.027	3.157	3.264	3.355	3.434	3.505	3.568	3.813	3.989	4.128	4.243	4.341	4.426	4.502	4.571
	0.10	2.213	2.446	2.611	2.739	2.845	2.934	3.012	3.080	3.142	3.380	3.552	3.686	3.796	3.891	3.973	4.046	4.112
	0.20	1.779	2.014	2.180	2.308	2.413	2.501	2.578	2.646	2.706	2.941	3.108	3.239	3.346	3.438	3.517	3.588	3.651
11	0.01	3.495	3.726	3.892	4.022	4.129	4.221	4.300	4.371	4.434	4.682	4.860	5.001	5.117	5.216	5.303	5.380	5.450
	0.05	2.586	2.811	2.970	3.094	3.196	3.283	3.358	3.424	3.484	3.715	3.880	4.010	4.117	4.208	4.288	4.358	4.422
	0.10	2.186	2.412	2.571	2.695	2.796	2.881	2.955	3.021	3.079	3.306	3.468	3.595	3.699	3.788	3.865	3.933	3.995
	0.20	1.763	1.993	2.154	2.279	2.380	2.465	2.539	2.605	2.663	2.888	3.048	3.172	3.274	3.361	3.436	3.503	3.563
12	0.01	3.427	3.647	3.804	3.927	4.029	4.114	4.189	4.256	4.315	4.547	4.714	4.845	4.953	5.045	5.125	5.196	5.260
	0.05	2.553	2.770	2.924	3.044	3.141	3.224	3.296	3.359	3.416	3.636	3.793	3.916	4.017	4.103	4.178	4.244	4.304
	0.10	2.164	2.384	2.539	2.658	2.756	2.838	2.910	2.973	3.029	3.247	3.402	3.522	3.621	3.705	3.779	3.843	3.901
	0.20	1.750	1.975	2.133	2.254	2.353	2.436	2.508	2.571	2.628	2.845	2.999	3.118	3.216	3.299	3.371	3.434	3.491
13	0.01	3.371	3.582	3.733	3.850	3.946	4.028	4.099	4.162	4.218	4.438	4.595	4.718	4.819	4.906	4.981	5.048	5.108
	0.05	2.526	2.737	2.886	3.002	3.096	3.176	3.245	3.306	3.361	3.571	3.722	3.839	3.935	4.017	4.088	4.151	4.207
	0.10	2.146	2.361	2.512	2.628	2.723	2.803	2.872	2.933	2.988	3.198	3.347	3.463	3.557	3.638	3.708	3.770	3.825
	0.20	1.739	1.961	2.116	2.234	2.331	2.412	2.482	2.544	2.599	2.809	2.958	3.074	3.168	3.248	3.317	3.378	3.433
14	0.01	3.324	3.528	3.673	3.785	3.878	3.956	4.024	4.084	4.138	4.347	4.497	4.614	4.710	4.792	4.863	4.926	4.982
	0.05	2.503	2.709	2.854	2.967	3.058	3.135	3.202	3.261	3.314	3.518	3.662	3.775	3.867	3.946	4.014	4.074	4.128
	0.10	2.131	2.342	2.489	2.603	2.696	2.774	2.841	2.900	2.953	3.157	3.301	3.413	3.504	3.582	3.649	3.708	3.761
	0.20	1.730	1.949	2.101	2.217	2.312	2.392	2.460	2.520	2.574	2.779	2.924	3.036	3.128	3.205	3.272	3.331	3.384

Table 20 Continued

df\p	α	2	3	4	5	6	7	8	9	10	15	20	25	30	35	40	45	50
15	0.01	3.285	3.482	3.622	3.731	3.820	3.895	3.961	4.019	4.071	4.271	4.414	4.526	4.618	4.696	4.764	4.824	4.877
	0.05	2.483	2.685	2.827	2.937	3.026	3.101	3.166	3.224	3.275	3.472	3.612	3.721	3.810	3.885	3.951	4.009	4.060
	0.10	2.118	2.325	2.470	2.582	2.672	2.748	2.814	2.872	2.924	3.122	3.262	3.370	3.459	3.534	3.599	3.656	3.708
	0.20	1.722	1.938	2.088	2.203	2.296	2.374	2.441	2.500	2.553	2.754	2.896	3.005	3.094	3.169	3.234	3.291	3.343
16	0.01	3.251	3.443	3.579	3.684	3.771	3.844	3.907	3.963	4.013	4.206	4.344	4.451	4.540	4.614	4.679	4.737	4.788
	0.05	2.467	2.665	2.804	2.911	2.998	3.072	3.135	3.191	3.241	3.433	3.569	3.675	3.761	3.834	3.897	3.953	4.003
	0.10	2.106	2.311	2.453	2.563	2.652	2.726	2.791	2.848	2.898	3.092	3.228	3.334	3.420	3.493	3.556	3.612	3.662
	0.20	1.715	1.929	2.077	2.190	2.282	2.359	2.425	2.483	2.535	2.732	2.871	2.978	3.064	3.138	3.201	3.257	3.307
17	0.01	3.221	3.409	3.541	3.644	3.728	3.799	3.860	3.914	3.963	4.150	4.284	4.387	4.472	4.544	4.607	4.662	4.712
	0.05	2.452	2.647	2.783	2.889	2.974	3.046	3.108	3.163	3.212	3.399	3.532	3.634	3.718	3.789	3.851	3.905	3.954
	0.10	2.096	2.298	2.439	2.547	2.634	2.708	2.771	2.826	2.876	3.066	3.199	3.303	3.387	3.458	3.519	3.574	3.622
	0.20	1.709	1.921	2.068	2.179	2.270	2.346	2.411	2.468	2.519	2.713	2.849	2.954	3.039	3.111	3.173	3.227	3.276
18	0.01	3.195	3.379	3.508	3.609	3.691	3.760	3.820	3.872	3.920	4.102	4.231	4.332	4.414	4.484	4.544	4.598	4.646
	0.05	2.439	2.631	2.766	2.869	2.953	3.024	3.085	3.138	3.186	3.370	3.499	3.599	3.681	3.750	3.810	3.863	3.910
	0.10	2.088	2.287	2.426	2.532	2.619	2.691	2.753	2.808	2.857	3.043	3.174	3.275	3.358	3.427	3.487	3.540	3.587
	0.20	1.704	1.914	2.059	2.170	2.259	2.334	2.399	2.455	2.505	2.696	2.830	2.933	3.017	3.087	3.148	3.201	3.249
19	0.01	3.173	3.353	3.479	3.578	3.658	3.725	3.784	3.835	3.881	4.059	4.185	4.283	4.363	4.430	4.489	4.541	4.588
	0.05	2.427	2.617	2.750	2.852	2.934	3.004	3.064	3.116	3.163	3.343	3.470	3.569	3.649	3.716	3.775	3.826	3.872
	0.10	2.080	2.277	2.415	2.520	2.605	2.676	2.738	2.791	2.839	3.023	3.152	3.251	3.332	3.400	3.459	3.511	3.557
	0.20	1.699	1.908	2.052	2.161	2.250	2.324	2.388	2.443	2.493	2.682	2.813	2.915	2.997	3.066	3.126	3.179	3.225
20	0.01	3.152	3.329	3.454	3.550	3.629	3.695	3.752	3.802	3.848	4.021	4.144	4.239	4.317	4.383	4.441	4.491	4.536
	0.05	2.417	2.605	2.736	2.836	2.918	2.986	3.045	3.097	3.143	3.320	3.445	3.541	3.620	3.686	3.743	3.794	3.839
	0.10	2.073	2.269	2.405	2.508	2.593	2.663	2.724	2.777	2.824	3.005	3.132	3.229	3.309	3.376	3.433	3.484	3.530
	0.20	1.695	1.902	2.045	2.154	2.241	2.315	2.378	2.433	2.482	2.668	2.798	2.898	2.979	3.048	3.106	3.158	3.204

21	0.01	3.134	3.308	3.431	3.525	3.602	3.667	3.724	3.773	3.817	3.987	4.108	4.201	4.277	4.342	4.397	4.447	4.491
	0.05	2.408	2.594	2.723	2.822	2.903	2.970	3.028	3.080	3.125	3.300	3.422	3.517	3.594	3.659	3.715	3.765	3.809
	0.10	2.067	2.261	2.396	2.498	2.581	2.651	2.711	2.764	2.810	2.989	3.114	3.210	3.288	3.354	3.411	3.461	3.505
	0.20	1.691	1.897	2.039	2.147	2.234	2.306	2.369	2.424	2.472	2.656	2.785	2.884	2.964	3.031	3.089	3.140	3.185
22	0.01	3.118	3.289	3.410	3.503	3.579	3.643	3.698	3.747	3.790	3.957	4.075	4.166	4.241	4.304	4.359	4.407	4.450
	0.05	2.400	2.584	2.712	2.810	2.889	2.956	3.014	3.064	3.109	3.281	3.402	3.495	3.571	3.634	3.690	3.738	3.782
	0.10	2.061	2.254	2.387	2.489	2.572	2.641	2.700	2.752	2.798	2.974	3.098	3.193	3.270	3.334	3.390	3.440	3.484
	0.20	1.688	1.892	2.033	2.141	2.227	2.299	2.361	2.415	2.463	2.646	2.773	2.871	2.950	3.016	3.073	3.123	3.168
23	0.01	3.103	3.272	3.392	3.483	3.558	3.621	3.675	3.723	3.766	3.930	4.046	4.135	4.208	4.270	4.324	4.371	4.413
	0.05	2.392	2.574	2.701	2.798	2.877	2.943	3.000	3.050	3.094	3.264	3.383	3.475	3.550	3.613	3.667	3.715	3.757
	0.10	2.056	2.247	2.380	2.481	2.563	2.631	2.690	2.741	2.787	2.961	3.083	3.177	3.253	3.317	3.372	3.421	3.464
	0.20	1.685	1.888	2.028	2.135	2.221	2.292	2.354	2.407	2.455	2.636	2.762	2.859	2.937	3.002	3.059	3.109	3.153
24	0.01	3.089	3.257	3.375	3.465	3.539	3.601	3.654	3.702	3.744	3.905	4.019	4.107	4.179	4.240	4.292	4.339	4.380
	0.05	2.385	2.566	2.692	2.788	2.866	2.931	2.988	3.037	3.081	3.249	3.366	3.457	3.531	3.593	3.646	3.693	3.735
	0.10	2.051	2.241	2.373	2.473	2.554	2.622	2.680	2.731	2.777	2.949	3.070	3.162	3.238	3.301	3.355	3.403	3.446
	0.20	1.682	1.884	2.024	2.130	2.215	2.286	2.347	2.400	2.448	2.627	2.752	2.848	2.925	2.990	3.046	3.095	3.139
25	0.01	3.077	3.243	3.359	3.449	3.521	3.583	3.635	3.682	3.723	3.882	3.995	4.081	4.152	4.212	4.263	4.309	4.350
	0.05	2.379	2.558	2.683	2.779	2.856	2.921	2.976	3.025	3.069	3.235	3.351	3.440	3.513	3.574	3.627	3.674	3.715
	0.10	2.047	2.236	2.367	2.466	2.547	2.614	2.672	2.722	2.767	2.938	3.058	3.149	3.224	3.286	3.340	3.387	3.430
	0.20	1.679	1.881	2.020	2.125	2.210	2.280	2.341	2.394	2.441	2.619	2.743	2.838	2.914	2.979	3.034	3.083	3.126
26	0.01	3.006	3.230	3.345	3.433	3.505	3.566	3.618	3.664	3.705	3.862	3.972	4.058	4.128	4.186	4.237	4.282	4.322
	0.05	2.373	2.551	2.675	2.770	2.847	2.911	2.966	3.014	3.058	3.222	3.337	3.425	3.497	3.558	3.610	3.656	3.697
	0.10	2.043	2.231	2.361	2.460	2.540	2.607	2.664	2.714	2.759	2.928	3.047	3.137	3.211	3.273	3.326	3.373	3.415
	0.20	1.677	1.878	2.016	2.121	2.205	2.275	2.335	2.388	2.435	2.612	2.735	2.829	2.905	2.968	3.023	3.071	3.114

Table 20 Continued

$df \backslash p$	α	2	3	4	5	6	7	8	9	10	15	20	25	30	35	40	45	50
27	0.01	3.056	3.218	3.332	3.419	3.491	3.550	3.602	3.647	3.688	3.843	3.952	4.036	4.105	4.163	4.213	4.257	4.297
	0.05	2.368	2.545	2.668	2.762	2.838	2.902	2.956	3.004	3.047	3.210	3.324	3.411	3.483	3.542	3.594	3.639	3.680
	0.10	2.039	2.227	2.356	2.454	2.534	2.600	2.657	2.707	2.751	2.919	3.036	3.126	3.199	3.261	3.313	3.360	3.401
	0.20	1.675	1.875	2.012	2.117	2.201	2.270	2.330	2.383	2.429	2.605	2.727	2.820	2.896	2.959	3.013	3.061	3.103
28	0.01	3.046	3.207	3.320	3.407	3.477	3.536	3.587	3.632	3.672	3.825	3.933	4.017	4.084	4.142	4.191	4.235	4.274
	0.05	2.363	2.539	2.661	2.755	2.830	2.893	2.948	2.995	3.038	3.199	3.312	3.399	3.469	3.528	3.579	3.624	3.664
	0.10	2.036	2.222	2.351	2.449	2.528	2.594	2.650	2.700	2.744	2.911	3.027	3.116	3.188	3.249	3.301	3.347	3.388
	0.20	1.672	1.872	2.009	2.113	2.196	2.266	2.326	2.378	2.424	2.599	2.720	2.812	2.887	2.950	3.004	3.051	3.093
29	0.01	3.037	3.197	3.309	3.395	3.464	3.523	3.574	3.618	3.658	3.809	3.916	3.998	4.065	4.122	4.171	4.214	4.252
	0.05	2.358	2.534	2.655	2.748	2.823	2.886	2.940	2.987	3.029	3.189	3.301	3.387	3.457	3.515	3.566	3.610	3.650
	0.10	2.033	2.218	2.346	2.444	2.522	2.588	2.644	2.693	2.737	2.903	3.018	3.107	3.178	3.239	3.291	3.336	3.377
	0.20	1.671	1.869	2.006	2.110	2.193	2.262	2.321	2.373	2.419	2.593	2.713	2.805	2.880	2.942	2.995	3.042	3.084
30	0.01	3.029	3.188	3.298	3.384	3.453	3.511	3.561	3.605	3.644	3.794	3.900	3.981	4.048	4.103	4.152	4.194	4.232
	0.05	2.354	2.528	2.649	2.742	2.816	2.878	2.932	2.979	3.021	3.180	3.291	3.376	3.445	3.503	3.553	3.597	3.637
	0.10	2.030	2.215	2.342	2.439	2.517	2.582	2.638	2.687	2.731	2.895	3.010	3.098	3.169	3.229	3.280	3.325	3.366
	0.20	1.669	1.867	2.003	2.106	2.189	2.258	2.317	2.369	2.414	2.587	2.707	2.798	2.872	2.934	2.987	3.034	3.076
40	0.01	2.970	3.121	3.225	3.305	3.370	3.425	3.472	3.513	3.549	3.689	3.787	3.862	3.923	3.975	4.019	4.058	4.093
	0.05	2.323	2.492	2.608	2.696	2.768	2.827	2.878	2.923	2.963	3.113	3.218	3.298	3.363	3.418	3.464	3.506	3.542
	0.10	2.009	2.189	2.312	2.406	2.481	2.544	2.597	2.644	2.686	2.843	2.952	3.036	3.103	3.160	3.208	3.251	3.289
	0.20	1.656	1.850	1.983	2.083	2.164	2.231	2.288	2.338	2.382	2.548	2.663	2.751	2.821	2.880	2.931	2.975	3.015

60	0.01	2.914	3.056	3.155	3.230	3.291	3.342	3.386	3.425	3.459	3.589	3.679	3.749	3.805	3.852	3.893	3.929	3.961
	0.05	2.294	2.456	2.568	2.653	2.721	2.777	2.826	2.869	2.906	3.049	3.148	3.223	3.284	3.336	3.379	3.418	3.452
	0.10	1.989	2.163	2.283	2.373	2.446	2.506	2.558	2.603	2.643	2.793	2.897	2.976	3.040	3.093	3.139	3.179	3.214
	0.20	1.643	1.834	1.963	2.061	2.139	2.204	2.259	2.308	2.350	2.511	2.621	2.705	2.772	2.828	2.876	2.918	2.956
120	0.01	2.859	2.994	3.087	3.158	3.215	3.263	3.304	3.340	3.372	3.493	3.577	3.641	3.693	3.736	3.774	3.807	3.836
	0.05	2.265	2.422	2.529	2.610	2.675	2.729	2.776	2.816	2.852	2.987	3.081	3.152	3.209	3.257	3.298	3.334	3.366
	0.10	1.968	2.138	2.254	2.342	2.411	2.469	2.519	2.562	2.600	2.744	2.843	2.918	2.978	3.029	3.072	3.110	3.143
	0.20	1.631	1.817	1.944	2.039	2.115	2.178	2.231	2.278	2.319	2.474	2.580	2.660	2.724	2.778	2.824	2.864	2.899
∞	0.01	2.806	2.934	3.022	3.089	3.143	3.188	3.226	3.260	3.289	3.402	3.480	3.539	3.587	3.627	3.661	3.691	3.718
	0.05	2.237	2.388	2.491	2.569	2.631	2.683	2.727	2.766	2.800	2.928	3.016	3.083	3.137	3.182	3.220	3.254	3.284
	0.10	1.949	2.114	2.226	2.311	2.378	2.434	2.482	2.523	2.560	2.697	2.791	2.862	2.920	2.967	3.008	3.044	3.075
	0.20	1.618	1.801	1.925	2.018	2.091	2.152	2.204	2.249	2.289	2.438	2.540	2.617	2.678	2.729	2.773	2.811	2.844

Reprinted from Games (reference 29 in Chapter 5) with permission from the *Journal of the American Statistical Association*. Copyright (1977) by the American Statistical Association.

Table 21 Williams's $\bar{t}_{i,\alpha}$ for $w = 1$ and extrapolation β_t (superscript) for a one-sided test and $\alpha = 0.01$ (used first in Chapter 5)

$df\backslash p$	2	3	4	5	6	8	10
5	3.501[4]	3.548[6]	3.572[7]	3.586[8]	3.595[9]	3.607[9]	3.614[10]
6	3.256[4]	3.294[6]	3.313[7]	3.324[8]	3.332[8]	3.341[9]	3.346[9]
7	3.097[4]	3.130[6]	3.146[7]	3.155[7]	3.161[8]	3.169[8]	3.173[9]
8	2.985[4]	3.015[6]	3.029[6]	3.037[7]	3.042[7]	3.049[8]	3.053[9]
9	2.903[4]	2.930[5]	2.943[6]	2.950[7]	2.955[7]	2.961[8]	2.964[8]
10	2.840[3]	2.865[5]	2.877[6]	2.883[7]	2.888[7]	2.893[8]	2.896[8]
11	2.791[3]	2.814[5]	2.824[6]	2.831[7]	2.835[7]	2.840[7]	2.843[8]
12	2.750[3]	2.772[5]	2.782[6]	2.788[6]	2.792[7]	2.797[7]	2.799[7]
13	2.717[3]	2.738[5]	2.747[6]	2.753[6]	2.757[7]	2.761[7]	2.764[7]
14	2.689[3]	2.709[5]	2.718[6]	2.723[6]	2.727[7]	2.731[7]	2.733[7]
15	2.665[3]	2.684[5]	2.693[6]	2.698[6]	2.701[7]	2.705[7]	2.708[7]
16	2.644[3]	2.663[5]	2.671[6]	2.676[6]	2.680[7]	2.683[7]	2.686[7]
17	2.626[3]	2.644[5]	2.653[6]	2.658[6]	2.661[6]	2.664[7]	2.666[7]
18	2.610[3]	2.628[5]	2.636[5]	2.641[6]	2.644[6]	2.647[7]	2.650[7]
19	2.596[3]	2.614[5]	2.622[5]	2.626[6]	2.629[6]	2.633[7]	2.635[7]
20	2.584[3]	2.601[5]	2.609[5]	2.613[6]	2.616[6]	2.619[7]	2.621[7]
22	2.563[3]	2.579[5]	2.586[5]	2.591[6]	2.593[6]	2.597[7]	2.599[7]
24	2.545[3]	2.561[5]	2.568[5]	2.572[6]	2.575[6]	2.578[6]	2.580[7]
26	2.531[3]	2.546[4]	2.553[5]	2.557[6]	2.559[6]	2.562[6]	2.564[6]
28	2.518[3]	2.533[4]	2.540[5]	2.544[6]	2.546[6]	2.549[6]	2.551[6]
30	2.507[3]	2.522[4]	2.529[5]	2.533[6]	2.535[6]	2.538[6]	2.540[6]
35	2.486[3]	2.501[4]	2.507[5]	2.511[6]	2.513[6]	2.516[6]	2.517[6]
40	2.471[3]	2.484[4]	2.491[5]	2.494[5]	2.496[6]	2.499[6]	2.500[6]
60	2.453[3]	2.448[4]	2.435[5]	2.457[5]	2.459[5]	2.461[6]	2.462[6]
120	2.400[3]	2.412[4]	2.417[5]	2.420[5]	2.422[5]	2.424[5]	2.425[6]
∞	2.366[3]	2.377[4]	2.382[5]	2.385[5]	2.386[5]	2.388[5]	2.389[6]

Reproduced from Williams (reference 23 in Chapter 5) with permission from the Biometric Society.

Table 22 William's $\bar{t}_{i,\alpha}$ for $w = 1$ and extrapolation β_t (superscript) for a one-sided test and $\alpha = 0.05$ (used first in Chapter 5)

$d \backslash p$	2	3	4	5	6	8	10
5	2.142^2	2.186^4	2.209^5	2.223^5	2.232^6	2.243^6	2.250^6
6	2.058^2	2.098^4	2.119^5	2.131^5	2.139^6	2.149^6	2.154^6
7	2.002^2	2.039^4	2.058^5	2.069^5	2.076^6	2.085^6	2.091^7
8	1.962^2	1.997^4	2.014^5	2.024^5	2.031^6	2.040^6	2.045^7
9	1.931^2	1.965^4	1.981^5	1.991^5	1.998^6	2.006^6	2.010^7
10	1.908^3	1.940^4	1.956^5	1.965^5	1.971^6	1.979^6	1.983^7
11	1.889^3	1.920^4	1.935^5	1.944^5	1.950^6	1.958^6	1.962^7
12	1.873^3	1.903^4	1.918^5	1.927^5	1.933^6	1.940^6	1.944^7
13	1.860^3	1.890^4	1.904^5	1.913^5	1.919^6	1.926^6	1.930^7
14	1.849^3	1.878^4	1.892^5	1.901^5	1.906^6	1.913^6	1.917^7
15	1.840^3	1.868^4	1.882^5	1.891^5	1.896^6	1.903^6	1.907^7
16	1.831^3	1.860^4	1.873^5	1.882^5	1.887^6	1.893^6	1.897^7
17	1.824^3	1.852^4	1.866^5	1.874^5	1.879^6	1.885^6	1.889^7
18	1.818^3	1.845^4	1.859^5	1.867^5	1.872^6	1.878^6	1.882^7
19	1.812^3	1.840^4	1.853^5	1.861^5	1.866^6	1.872^6	1.876^7
20	1.807^3	1.834^4	1.847^5	1.855^5	1.860^6	1.866^6	1.870^7
22	1.798^3	1.825^4	1.838^5	1.846^5	1.851^6	1.857^6	1.860^7
24	1.791^3	1.818^4	1.830^5	1.838^5	1.843^6	1.849^6	1.852^7
26	1.785^3	1.811^4	1.824^5	1.831^5	1.836^6	1.842^6	1.846^7
28	1.780^3	1.806^4	1.819^5	1.826^5	1.831^6	1.836^6	1.840^7
30	1.776^3	1.801^4	1.814^5	1.821^5	1.826^6	1.832^6	1.835^7
35	1.767^3	1.792^4	1.804^5	1.811^5	1.816^6	1.822^6	1.825^7
40	1.761^3	1.785^4	1.797^5	1.804^5	1.809^6	1.814^6	1.818^7
60	1.746^3	1.770^4	1.781^5	1.788^5	1.792^6	1.798^6	1.801^7
120	1.731^3	1.754^4	1.765^5	1.772^5	1.776^6	1.781^6	1.784^7
∞	1.716^3	1.739^4	1.750^5	1.756^5	1.760^6	1.765^6	1.768^7

Reproduced from Williams (reference 23 in Chapter 5) with permission from the Biometric Society.

Table 23 Williams's $\bar{t}_{i,\alpha}$ for $w = 1$ and extrapolation β_t (superscript) for a two-sided test and $\alpha = 0.01$ (used first in Chapter 5)

$df\backslash p$	2	3	4	5	6	8	10
5	4.179^5	4.229^7	4.255^9	4.270^{10}	4.279^{10}	4.292^{11}	4.299^{11}
6	3.825^5	3.864^7	3.883^8	3.895^9	3.902^9	3.912^{10}	3.917^{10}
7	3.599^4	3.631^6	3.647^7	3.657^8	3.663^9	3.670^9	3.674^{10}
8	3.443^4	3.471^6	3.484^7	3.492^8	3.497^8	3.504^9	3.507^9
9	3.329^4	3.354^6	3.366^7	3.373^7	3.377^8	3.383^8	3.886^9
10	3.242^4	3.265^6	3.275^6	3.281^7	3.286^7	3.290^8	3.293^8
11	3.173^4	3.194^5	3.204^6	3.210^7	3.214^7	3.218^8	3.221^8
12	3.118^4	3.138^5	3.147^6	3.152^7	3.156^7	3.160^7	3.162^8
13	3.073^4	3.091^5	3.100^6	3.105^6	3.108^7	3.112^7	3.114^7
14	3.035^4	3.052^5	3.060^6	3.065^6	3.068^6	3.072^7	3.074^7
15	3.003^3	3.019^5	3.027^6	3.031^6	3.034^6	3.037^7	3.039^7
16	2.957^3	2.991^5	2.998^5	3.002^6	3.005^6	3.008^7	3.010^7
17	2.951^3	2.966^5	2.973^5	2.977^6	2.980^6	2.983^7	2.984^7
18	2.929^3	3.944^5	2.951^5	2.955^6	2.958^6	2.960^6	2.962^7
19	2.911^3	2.925^4	2.932^5	2.936^6	2.938^6	2.941^6	2.942^7
20	2.894^3	2.908^4	2.915^5	2.918^6	2.920^6	2.923^6	2.925^7
22	2.866^3	2.879^4	2.885^5	2.889^6	2.891^6	2.893^6	2.895^6
24	2.842^3	2.855^4	2.861^5	2.864^5	2.866^6	2.869^6	2.870^6
26	2.823^3	2.835^4	2.841^5	2.844^5	2.846^6	2.848^6	2.850^6
28	2.806^3	2.819^4	2.824^5	2.827^5	2.829^5	2.831^6	2.832^6
30	2.792^3	2.804^4	2.809^5	2.812^5	2.814^5	2.816^6	2.817^6
35	2.764^3	2.776^4	2.781^5	2.783^5	2.785^5	2.787^5	2.788^6
40	2.744^3	2.755^4	2.759^5	2.762^5	2.764^5	2.765^5	2.766^5
60	2.697^3	2.707^4	2.711^4	2.713^5	2.715^5	2.716^5	2.717^5
120	2.651^3	2.660^4	2.664^4	2.666^4	2.667^5	2.669^5	2.669^5
∞	2.607^3	2.615^4	2.618^4	2.620^4	2.621^4	2.623^5	2.623^5

Reproduced from Williams (reference 23 in Chapter 5) with permission from the Biometric Society.

Table 24   Williams's $\bar{t}_{i,\alpha}$ for $w = 1$ and extrapolation β_t (superscript) for a two-sided test and $\alpha = 0.05$ (used first in Chapter 5)

$df \backslash p$	2	3	4	5	6	8	10
5	2.699^3	2.743^5	2.766^6	2.779^6	2.788^7	2.799^7	2.806^8
6	2.559^3	2.597^5	2.617^6	2.628^6	2.635^7	2.645^7	2.650^8
7	2.466^3	2.501^5	2.518^6	2.528^6	2.535^7	2.543^7	2.548^8
8	2.400^3	2.432^5	2.448^6	2.457^6	2.436^7	2.470^7	2.475^8
9	2.351^3	2.381^5	2.395^6	2.404^6	2.410^7	2.416^7	2.421^8
10	2.313^3	2.341^5	2.355^6	2.363^6	2.368^6	2.375^7	2.379^7
11	2.283^3	2.310^5	2.323^6	2.330^6	2.335^6	2.342^7	2.345^7
12	2.258^3	2.284^5	2.297^6	2.304^6	2.309^6	2.315^7	2.318^7
13	2.238^3	2.263^5	2.275^5	2.282^6	2.286^6	2.292^7	2.295^7
14	2.220^3	2.245^5	2.256^5	2.263^6	2.268^6	2.273^7	2.276^7
15	2.205^3	2.229^5	2.241^5	2.247^6	2.252^6	2.257^7	2.260^7
16	2.193^3	2.216^5	2.227^5	2.234^6	2.238^6	2.243^7	2.246^7
17	2.181^3	2.204^4	2.215^5	2.222^6	2.226^6	2.231^7	2.234^7
18	2.171^3	2.194^4	2.205^5	2.211^6	2.215^6	2.220^7	2.223^7
19	2.163^3	2.185^4	2.195^5	2.202^6	2.205^6	2.210^7	2.213^7
20	2.155^3	2.177^4	2.187^5	2.193^6	2.197^6	2.202^7	2.205^7
22	2.141^3	2.163^4	2.173^5	2.179^6	2.183^6	2.187^7	2.190^7
24	2.130^3	2.151^4	2.161^5	2.167^6	2.171^6	2.175^7	2.178^7
26	2.121^3	2.142^4	2.151^5	2.157^6	2.161^6	2.165^7	2.168^7
28	2.113^3	2.133^4	2.143^5	2.149^6	2.152^6	2.156^7	2.159^7
30	2.106^3	2.126^4	2.136^5	2.141^6	2.145^6	2.149^7	2.151^7
35	2.093^3	2.112^4	2.122^5	2.127^6	2.130^6	2.134^7	2.137^7
40	2.083^3	2.102^4	2.111^5	2.116^6	2.119^6	2.123^6	2.126^7
60	2.060^3	2.078^4	2.087^5	2.092^6	2.095^6	2.099^6	2.101^7
120	2.037^3	2.055^4	2.063^5	2.068^6	2.071^6	2.074^6	2.076^7
∞	2.015^3	2.032^4	2.040^5	2.044^6	2.047^6	2.050^6	2.052^6

Reproduced from Williams (reference 23 in Chapter 5) with permission from the Biometric Society.

Table 25 Significant values of Steel's rank sums for a one-sided test with $\alpha = 0.05$ or 0.01 (used first in Chapter 5)

$n\backslash p$	α	2	3	4	5	6	7	8	9
4	0.05	11	10	10	10	10			
	0.01								
5	0.05	18	17	17	16	16	16	16	15
	0.01	15							
6	0.05	27	26	25	25	24	24	24	23
	0.01	23	22	21	21				
7	0.05	37	36	35	35	34	34	33	33
	0.01	32	31	30	30	29	29	29	29
8	0.05	49	48	47	46	46	45	45	44
	0.01	43	42	41	40	40	40	39	39
9	0.05	63	62	61	60	59	59	58	58
	0.01	56	55	54	53	52	52	51	51
10	0.05	79	77	76	75	74	74	73	72
	0.01	71	69	68	67	66	66	65	65
11	0.05	97	95	93	92	91	90	90	89
	0.01	87	85	84	83	82	81	81	80
12	0.05	116	114	112	111	110	109	108	108
	0.01	105	103	102	100	99	99	98	98
13	0.05	138	135	133	132	130	129	129	128
	0.01	125	123	121	120	119	118	117	117
14	0.05	161	158	155	154	153	152	151	150
	0.01	147	144	142	141	140	139	138	137
15	0.05	186	182	180	178	177	176	175	174
	0.01	170	167	165	164	162	161	160	160
16	0.05	213	209	206	204	203	201	200	199
	0.01	196	192	190	188	187	186	185	184
17	0.05	241	237	234	232	231	229	228	227
	0.01	223	219	217	215	213	212	211	210
18	0.05	272	267	264	262	260	259	257	256
	0.01	252	248	245	243	241	240	239	238
19	0.05	304	299	296	294	292	290	288	287
	0.01	282	278	275	273	271	270	268	267
20	0.05	339	333	330	327	325	323	322	320
	0.01	315	310	307	305	303	301	300	299

Reproduced from Steel (reference 32 in Chapter 5) with permission from the Biometric Society.

Table 26 Significant values of Steel's rank sums for a two-sided test with
$\alpha = 0.05$ or 0.01 (used first in Chapter 5)

$n\backslash p$	α	2	3	4	5	6	7	8	9
4	0.05	10							
	0.01								
5	0.05	16	16	15	15				
	0.01								
6	0.05	25	24	23	23	22	22	22	21
	0.01	21							
7	0.05	35	33	33	32	32	31	31	30
	0.01	30	29	28	28				
8	0.05	46	45	44	43	43	42	42	41
	0.01	41	40	39	38	38	37	37	37
9	0.05	60	58	57	56	55	55	54	54
	0.01	53	52	51	50	49	49	49	48
10	0.05	75	73	72	71	70	69	69	68
	0.01	68	66	65	64	63	62	62	62
11	0.05	92	90	88	87	86	85	85	84
	0.01	84	82	80	79	78	78	77	77
12	0.05	111	108	107	105	104	103	103	102
	0.01	101	99	97	96	95	94	94	93
13	0.05	132	129	127	125	124	123	122	121
	0.01	121	118	116	115	114	113	112	112
14	0.05	154	151	149	147	145	144	144	143
	0.01	142	139	137	135	134	133	132	132
15	0.05	179	175	172	171	169	168	167	166
	0.01	165	162	159	158	156	155	154	154
16	0.05	205	201	196	196	194	193	192	191
	0.01	189	186	184	182	180	179	178	177
17	0.05	233	228	225	223	221	219	218	217
	0.01	216	212	210	208	206	205	204	203
18	0.05	263	258	254	252	250	248	247	246
	0.01	244	240	237	235	233	232	231	230
19	0.05	294	289	285	283	280	279	277	276
	0.01	274	270	267	265	262	261	260	259
20	0.05	328	322	318	315	313	311	309	308
	0.01	306	302	298	296	293	292	290	289

Reproduced from Steel (reference 32 in Chapter 5) with permission from the Biometric Society.

Table 27 **Wilcoxon (Mann-Whitney) rank-sum test critical values with Bonferroni's adjustments: one-sided test and $\alpha = 0.05$ (first used in Chapter 5)**

P	Observations in Control (m)	Observations in Experimental Treatment (n)							
		3	4	5	6	7	8	9	10
1 ($\alpha' = 0.05$)	3	6	10	16	23	30	39	49	59
	4	6	11	17	24	32	41	51	62
	5	7	12	19	26	34	44	54	66
	6	8	13	20	28	36	46	57	69
	7	8	14	21	29	39	49	60	72
	8	9	15	23	31	41	51	63	72
	9	10	16	24	33	43	54	66	79
	10	10	17	26	35	45	56	69	82
2 ($\alpha' = 0.025$)	3			15	22	29	38	47	58
	4		10	16	23	31	40	49	60
	5	6	11	17	24	33	42	52	63
	6	7	12	18	26	34	44	55	66
	7	7	13	20	27	36	46	57	69
	8	8	14	21	29	38	49	60	72
	9	8	14	22	31	40	51	62	75
	10	9	15	23	32	42	53	65	78
3 ($\alpha' = 0.016\bar{6}$)	3				21	28	37	46	57
	4		10	16	22	30	39	48	59
	5		11	17	24	32	41	51	62
	6	6	11	18	25	33	42	53	65
	7	6	12	19	26	35	45	56	68
	8	7	13	20	28	37	47	58	70
	9	7	13	21	29	39	49	61	73
	10	8	14	22	31	41	51	63	76
4 ($\alpha' = 0.0125$)	3				21	28	37	46	56
	4			15	22	30	38	48	59
	5		10	16	23	31	40	50	61
	6	6	11	17	24	33	42	52	64
	7	6	12	18	26	34	44	55	67
	8	7	12	19	27	36	46	57	69
	9	7	13	20	28	38	48	60	72
	10	7	14	21	30	40	50	62	75

Table 27 *Continued*

P	Observations in Control (m)	Observations in Experimental Treatment (n)							
		3	4	5	6	7	8	9	10
5 ($\alpha' = 0.0100$)	3					28	36	46	56
	4			15	22	29	38	48	58
	5		10	16	23	31	40	50	61
	6		11	17	24	32	42	52	63
	7	6	11	18	25	34	44	54	66
	8	6	12	19	27	36	46	56	68
	9	7	13	20	28	37	47	59	71
	10	7	13	21	29	39	49	61	74
6 ($\alpha' = 0.008\bar{3}$)	3					28	36	45	56
	4			15	21	29	38	47	58
	5		10	16	22	30	39	49	60
	6		10	16	24	32	41	51	63
	7	6	11	17	25	33	43	54	65
	8	6	12	18	26	35	45	56	68
	9	6	12	19	27	37	47	58	70
	10	7	13	20	29	38	49	60	73
7 ($\alpha' = 0.0071$)	3						36	45	56
	4				21	29	37	47	58
	5			15	22	30	39	49	60
	6		10	16	23	32	41	51	62
	7		11	17	25	33	43	53	65
	8	6	11	18	26	35	45	55	67
	9	6	12	19	27	36	46	58	70
	10	7	13	20	28	38	48	60	72
8 ($\alpha' = 0.00625$)	3						36	45	55
	4				21	29	37	47	57
	5			15	22	30	39	49	59
	6		10	16	23	31	41	51	62
	7		11	17	24	33	42	53	64
	8	6	11	18	26	34	44	55	67
	9	6	12	19	27	36	46	57	69
	10	6	12	19	28	37	48	59	72
9 ($\alpha' = 0.0055\bar{5}$)	3							45	55
	4				21	28	37	46	57
	5			15	22	30	39	48	59
	6		10	16	23	31	40	50	62
	7		10	17	24	32	42	52	64
	8		11	18	25	34	44	55	66
	9	6	11	18	26	35	46	57	69
	10	6	12	19	28	37	47	59	71

Table 27 *Continued*

P	Observations in Control (m)	Observations in Experimental Treatment (n)							
		3	4	5	6	7	8	9	10
10 ($\alpha' = 0.0050$)	3							45	55
	4				21	28	37	46	57
	5			15	22	30	39	48	59
	6		10	16	23	31	40	50	61
	7		10	17	24	32	42	52	64
	8		11	18	25	34	44	54	66
	9	6	11	18	26	35	45	56	68
	10	6	12	19	27	37	47	58	71

Table 28 **Wilcoxon (Mann-Whitney) rank-sum test critical values with Bonferroni's adjustments: two-sided test and α = 0.05 (first used in Chapter 5)**

P	Observations in Control (m)	Observations in Experimental Treatment (n)							
		3	**4**	**5**	**6**	**7**	**8**	**9**	**10**
1 ($\alpha' = 0.0250$)	3			15	22	29	38	47	58
	4		10	16	23	31	40	50	60
	5	6	11	17	24	33	42	52	63
	6	7	12	18	26	34	44	55	66
	7	7	13	20	27	36	46	57	69
	8	8	14	21	29	38	49	60	72
	9	8	15	22	31	40	51	63	75
	10	9	15	23	32	42	53	65	78
2 ($\alpha' = 0.0125$)	3				21	28	37	46	56
	4			15	22	30	38	48	59
	5		10	16	23	31	40	50	61
	6	6	11	17	24	33	42	52	64
	7	6	12	18	26	34	44	55	67
	8	7	12	19	27	36	46	57	69
	9	7	13	20	28	38	48	60	72
	10	7	14	21	30	40	50	62	75
3 ($\alpha' = 0.008\bar{3}$)	3					28	36	45	56
	4			15	21	29	38	47	58
	5		10	16	22	30	39	49	60
	6		10	16	24	32	41	51	62
	7	6	11	17	25	33	43	54	65
	8	6	12	18	26	35	45	56	68
	9	6	12	19	27	37	47	58	70
	10	7	13	20	28	38	49	60	73
4 ($\alpha' = 0.0625$)	3						36	45	55
	4				21	29	37	47	57
	5			15	22	30	39	49	60
	6		10	16	23	31	41	51	62
	7		11	17	24	33	42	53	64
	8	6	11	18	26	34	44	55	67
	9	6	12	19	27	36	46	57	69
	10	6	12	20	28	37	48	59	72

Table 28 *Continued*

P	Observations in Control (m)	Observations in Experimental Treatment (n)							
		3	4	5	6	7	8	9	10
5 ($\alpha' = 0.005$)	3							45	55
	4				21	28	37	46	57
	5			15	22	30	39	48	59
	6		10	16	23	31	40	50	62
	7		10	17	24	32	42	52	64
	8		11	18	25	34	44	54	66
	9	6	11	18	26	35	45	57	69
	10	6	12	19	28	37	47	59	71
6 ($\alpha' = 0.0042$)	3								55
	4					28	37	46	57
	5			15	22	29	38	48	59
	6			16	23	31	40	50	61
	7		10	16	24	32	41	52	63
	8		11	17	25	33	43	54	66
	9		11	18	26	35	45	56	68
	10	6	12	19	27	36	47	58	71
7 ($\alpha' = 0.0036$)	3								55
	4					28	36	46	56
	5				21	29	38	48	58
	6			15	22	30	39	49	60
	7		10	16	23	32	41	51	63
	8		10	17	24	33	43	53	65
	9		11	18	25	34	44	55	67
	10	6	11	18	26	36	46	57	70
8 ($\alpha' = 0.0031$)	3								55
	4					28	36	46	56
	5				21	29	38	48	58
	6			15	22	30	39	49	60
	7		10	16	23	32	41	51	63
	8		10	17	24	33	43	53	65
	9		11	18	25	34	44	55	67
	10	6	11	18	26	36	46	57	70
9 ($\alpha' = 0.0028$)	3								
	4						36	45	56
	5				21	28	37	47	58
	6			15	22	30	39	49	60
	7			15	23	31	40	50	62
	8		10	16	24	32	42	52	64
	9		10	17	25	33	43	54	66
	10		11	18	26	35	45	56	68

Table 28 *Continued*

P	Observations in Control (*m*)	Observations in Experimental Treatment (*n*)							
		3	**4**	**5**	**6**	**7**	**8**	**9**	**10**
10									
($\alpha' = 0.0025$)	3								
	4						36	45	56
	5				21	28	37	47	58
	6			15	22	30	39	49	60
	7			15	23	31	40	50	62
	8		10	16	24	32	42	52	64
	9		10	17	25	33	43	54	66
	10		11	18	26	35	45	56	68

Table 29 **FORTRAN code for selection component analysis (two allele loci) (used in Chapter 6)**

```
     COMMON IDF, PROB
     BYTE NAM1(30),NAM3(30)
     WRITE(5,999)
999  FORMAT(1X, 'WHAT IS THE NAME OF YOUR DATA FILE?')
     READ (5,998) K, (NAM1(I),I=1,K)
998  FORMAT(Q,30A1)
     WRITE(5,997)
997  FORMAT(1X, 'WHAT IS THE NAME OF YOUR OUTPUT FILE?')
     READ(5,998) K, (NAM3(I),I=1,K)
     WRITE(5,996)
996  FORMAT(1X, 'HOW MANY LOCI ARE YOU ANALYZING?')
     ACCEPT *,NLOK
     OPEN(UNIT=1,NAME=NAM1,ACCESS='SEQUENTIAL',
    &   FORM='FORMATTED',TYPE='OLD')
     OPEN(UNIT=3,NAME=NAM3,ACCESS='SEQUENTIAL',
    &   FORM='FORMATTED',TYPE='NEW',DISP='KEEP')
21   NLOK=NLOK-1
     IF(NLOK.LT.0) GOTO 23
     READ(1,3)C11,C12,C21,C22,C23,C32,C33,S1,S2,S3,AM1,AM2,AM3
3    FORMAT(13F5.1)
     READ(1,5) FF1,FF2,FF3
5    FORMAT(3F6.2)
22   F1=C11+C12
     N = 0
     F2=C21+C22+C23
     F3=C32+C33
     F0=F1+F2+F3
     FM1=C11+C21+C32
     FM2=C12+C23+C33
     FM0=FM1+FM2
     S0=S1+S2+S3
     AM0=AM1+AM2+AM3
     AF1=F1+S1
     AF2=F2+S2
     AF3=F3+S3
     AF0=AF1+AF2+AF3
     A1=AM1+AF1
     A2=AM2+AF2
     A3=AM3+AF3
     A0=A1+A2+A3
     FF0=FF1+FF2+FF3
     AFF1=FF1+S1
     AFF2=FF2+S2
     AFF3=FF3+S3
     AFF0=AFF1+AFF2+AFF3
     A0A=AFF0+AM0
     A1A=AFF1+AM1
     A2A=AFF2+AM2
     A3A=AFF3+AM3
```

Table 29 *Continued*

```
     K=1
     WRITE(3,77) K
 77  FORMAT(' TEST',I5)
     K = K + 1
     X=AM1/AM0
     Z=AM3/AM0
     WRITE(3,77) K
     K = K + 1
     I = 1
  9  G=(FM0+AM1+AM2)*(X**3)-((AM1+(2*AM2))*Z*(X**2))-
   &    (FM0+AM1-AM2)*X*(Z**2)-((2*FM1)+AM1)*(X**2)+((FM2+AM1)*
   &    2*X*Z)-(FM2-FM1+AM1+AM2)*X+(AM1*(Z**3-Z**2-Z+1))
     GX=3*(FM0+AM1+AM2)*(X**2)-2*(AM1+(2*AM2))*Z*X-
   &    (FM0+AM1-AM2)*(Z**2)-2*((2*FM1)+AM1)*X+(FM2+AM1)*
   &    2*Z-(FM2-FM1+AM1+AM2)
     GY= -((AM1+(2*AM2))*(X**2))-2*(FM0+AM1-AM2)*X*Z+
   &    (FM2+AM1)*2*X+(AM1*(3*(Z**2)-(2*Z)-1))
     H=(FM0+AM3+AM2)*(Z**3)-((AM3+(2*AM2))*X*
   &    (Z**2))-(FM0+AM3-AM2)*Z*(X**2)-((2*FM2)+AM3)*(Z**2)+
   &    ((FM1+AM3)*2*X*Z)-(FM1-FM2+AM3+AM2)*Z+(AM3*
   &    (X**3-X**2-X+1))
     HY=3*(FM0+AM3+AM2)*(Z**2)-2*(AM3+(2*AM2))*
   &    Z*X-(FM0+AM3-AM2)*(X**2)-2*((2*FM2)+AM3)*Z+
   &    ((FM1+AM3)*2*X)-(FM1-FM2+AM3+AM2)
     HX= -((AM3+(2*AM2))*(Z**2))-2*(FM0+AM3-AM2)*
   &    Z*X+((FM1+AM3)*2*Z)+(AM3*(3*(X**2)-2*X-1))
     IF(GX*HY-GY*HX) 42,82,42
 82  DX = 0.0
     DY = 0.0
     GOTO 81
 42  DX=(((-G)*HY)-(GY*(-H)))/((GX*HY)-(GY*HX))
 43  DZ=(((-H)*GX)-((-G)*HX))/((GX*HY)-(GY*HX))
 81  WRITE(3,39) X,Z,DX,DZ,G,GX,GY,H,HX,HY
     X=X+DX
     Z=Z+DZ
     I = I + 1
     IF(I.GT.20) GOTO 70
     IF(ABS(DX) .GT.0. 0001 .OR. ABS(DZ) .GT. 0.0001) GOTO 9
     GOTO 71
 70  CONTINUE
     WRITE(3,77) I
 71  Y=1.0-X-Z
     WRITE(3,77) K
     K = K + 1
     U=A1/A0
     W=A3/A0
     WRITE(3,77) K
     K = K + 1
     I = 1
  8  EG=(FM0+A1+A2)*(U**3)-(A1+(2*A2))*(U**2)*W-
```

Table 29 *Continued*

```
   &   (FM0+A1−A2)*U*(W**2)−((2*FM1)+A1)*(U**2)+(FM2+A1)*
   &   2*U*W−(FM2−FM1+A1+A2)*U+(A1*(W**3−W**2−W+1))
       EGU=3*(FM0+A1+A2)*(U**2)−2*(A1+(2*A2))*U*
   &   W−(FM0+A1−A2)*(W**2)−2*((2*FM1)+A1)*U+(FM2+A1)*
   &   2*W−(FM2−FM1+A1+A2)
       EGW= −((A1+(2*A2))*(U**2))−2*(FM0+A1−A2)*
   &   U*W+(FM2+A1)*2*U+(A1* (3*(W**2) −(2*W)−1))
       EH=(FM0+A3+A2)*(W**3)−((A3+(2*A2))*U*(W**2))−
   &   (FM0+A3−A2)*W*(U**2)−((2*FM2)+A3)*(W**2)+((FM1+A3)*
   &   2*U*W)−(FM1−FM2+A3+A2)*W+(A3*((U**3)−(U**2)−U+1))
       EHU= − ((A3+(2*A2))*(W**2))−2*(FM0+A3−A2)*
   &   W*U+((FM1+A3)*2*W)+(A3*(3*(U**2)−(2*U)−1))
       EHW=3*(FM0+A3+A2)*(W**2)−2*(A3+(2*A2))*W*U−(FM0+A3−A2)*
   &   (U**2)−2*((2*FM2)+A3)*W+((FM1+A3)*2*U)−(FM1−FM2+A3+A2)
       IF(EGU*EHW−EGW*EHU) 6,80,6
  80 DU = 0.0
       DW = 0.0
       GOTO 100
   6 DU=(((−EG)*EHU)−(EGW*(−EH)))/((EGU*EHW)−(EGW*EHU))
  46 DW=(((−EH)*EGU)−((−EG)*EHU))/((EGU*EHW)−(EGW*EHU))
 100 WRITE(3,39) U,W,DU,DW,EG,EGU,EGW,EH,EHU,EHW
       U=U+DU
       W=W+DW
       I = I + 1
       IF(I .GT. 20) GOTO 60
       IF(ABS(DU) .GT. 0.0001 .OR. ABS(DW) .GT. 0.0001) GOTO 8
       GOTO 61
  60 CONTINUE
       WRITE(3,77) I
  61 V=1−U−W
       T1=2*ABC(C22,0.5,F2,1.0)
       T2=ABC(C11,F1,FM1,FM0)+ABC(C12,F1,FM2,FM0)+
   &   BAC(F2,C21,C23,FM1,FM0)+BAC(F2,C23,C21,FM2,FM0)+
   &   ABC(C32,F3,FM1,FM0)+ABC(C33,F3,FM2,FM0)
       T3=DIV((F0−0.5*F2)*((DIV(FM1,FM0)−X−0.5*Y)**2),
   &   ((X+0.5*Y)*(0.5*Y+Z)))+DIV(((AM1−AM0*X)**2),
   &   AM0*X)+DIV(((AM2−AM0*Y)**2) ,AM0*Y)+
   &   DIV(((AM3−AM0*Z)**2),AM0*Z)
       T5=DIV(F0*(1.0−0.5*V)*((X+0.5*Y−U−0.5*V)**2),
   &   ((U+0.5*V)*(0.5*V+W)))+DIV(((AF1−U*AF0)**2),
   &   (U*AF0))+DIV(((AF2−V*AF0)**2),(V*AF0))+DIV(
   &   ((AF3−W*AF0)**2),(W*AF0))+DIV(((X*AM0−U*AM0)**2),
   &   (U*AM0))+DIV(((Y*AM0−V*AM0)**2),(V*AM0))+DIV
   &   (((Z*AM0−W*AM0)**2),(W*AM0))
       WRITE(3,77) K
       K = K + 1
  35 P=(2*(FF1+S1+AM1)+FF2+S2+AM2+FM1)/(2*(FF0+S0+AM0)+FM0)
       Q=1.0−P
       UA = A1A/A0A
       WA = A3A/A0A
```

Table 29 *Continued*

```
      WRITE(3,77) K
      K = K + 1
      I = 1
   29 EG=(FM0+A1A+A2A)*(UA**3)-(A1A+(2*A2A))*(UA**2)*
     &    WA-(FM0+A1A-A2A)*UA*(WA**2)-((2*FM1)+A1A)*
     &    (UA**2)+(FM2+A1A)*2*UA*WA-(FM2-FM1+A1A+A2A)*
     &    UA+(A1A*(WA**3-WA**2-WA+1))
      EGU=3*(FM0+A1A+A2A)*(UA**2)-2*(A1A+(2*A2A))*
     &    UA*WA-(FM0+A1A-A2A)*(WA**2)-2*((2*FM1)+A1A)*
     &    UA+(FM2+A1A)*2*WA-(FM2-FM1+A1A+A2A)
      EGW=-((A1A+2*A2A))*(UA**2))-2*(FM0+A1A-A2A)*
     &    UA*WA+(FM2+A1A)*2*UA+(A1A*(3*(WA**2)-(2*WA)-1))
      EH=(FM0+A3A+A2A)*(WA**3)-((A3A+(2*A2A))*UA*
     &    (WA**2))-(FM0+A3A-A2A)*WA*(UA**2)-((2*FM2)+A3A)*
     &    (WA**2)+((FM1+A3A)*2*UA*WA)-(FM1-FM2+A3A+
     &    A2A)*WA+(A3A*((UA**3)-(UA**2)-UA+1))
      EHU=-((A3A+(2*A2A))*(WA**2))-2*(FM0+A3A-A2A)*
     &    WA*UA+((FM1+A3A)*2*WA))+(A3A*(3*(UA**2)-(2*UA)-1))
      EHW=3*(FM0+A3A+A2A)*(WA**2)-2*(A3A+(2*A2A))*WA*
     &    UA-(FM0+A3A-A2A)*(UA**2)-I*((2*FM2)+A3A)*WA+((FM1+A3A)*
     &    2*UA)-(FM1-FM2+A3A+A2A)
      IF(EGU*EHW-EGW*EHU) 7,101,7
  101 DU = 0.0
      DW = 0.0
      GOTO 102
    7 DU=(((-EG)*EHW)-(EGW*(-EH)))/((EGU*EHW)-(EGW*EHU))
   48 DW=(((-EH)*EGU)-((-EG)*EHU))/((EGU*EHW)-EGW*EHU))
  102 WRITE (3,39) UA,WA,DU,DW,EG,EGU,EGW,EH,EHU,EHW
      UA=UA+DU
      WA=WA+DW
      I = I + 1
      IF(I .GT. 20) GOTO 50
      IF(ABS(DU) .GT .0.0001 .OR. ABS(DW) .GT. 0.0001) GOTO 29
      GOTO 51
   50 CONTINUE
      WRITE(3,77) I
   51 VA=1-UA-WA
      WRITE(3,77) K
      K = K + 1
      T4=ABC(FF1,FF0,AFF1,AFF0)+ABC(FF2,FF0,AFF2,AFF0)+
     &    ABC(FF3,FF0,AFF3,AFF0)+ABC(S1,S0,AFF1,AFF0)+
     &    ABC(S2,S0,AFF2,AFF0)+ABC(S3,S0,AFF3,AFF30)
      IF (P .NE. 0.0 .AND. Q .NE. 0.0) GOTO 93
      WRITE(3,92) P,Q
   92 FORMAT('ERROR PQ',2F15.5)
      GOTO 45
   93 T6=FF0*(1.0-P*Q)*(((UA+0.5*VA-P)**2)/(P*Q))+
     &    A0A*(((UA-P**2)**2/P**2)+((VA-2*P*Q)**2/
     &    (2*P*Q))+(((WA-Q**2))**2/Q**2))
      WRITE(3,77) K
```

Table 29 *Continued*

```
      K = K + 1
      T01=T1+T2+T3+T4
      T02=T5+T6
      T0=T01+T02
      L = 100
      WRITE(3,39) FF1,FF2,FF3,FF0,AFF0,A0A,S0,UA,VA,WA
39    FORMAT (5F15.5,//,5F15.5,/)
11    H1 = (3.84/(4.0*F2))**0.5
12    HA = ABS(C22/F2−0.5)
      IF(F1) 13,86,13
13    H02= ((5.99**FM1/FM0*FM2/FM0)*(F2/(F1*(2*F1+F2))))**0.5
      GOTO 85
86    H02= 0.0
      WRITE (3,77) L
85    WRITE (3,77) K
      K = K + 1
      L = L + 1
      IF (F3) 14,87,14
14    H022= ((5.99*FM1/FM0*FM2/FM0)*(F2/(F3*(2*F3+F2))))**0.5
      GOTO 15
87    H022= 0.0
      WRITE (3,77) L
15    H03= ((3.84*FM1/FM0*FM2/FM0)*(0.5*AM0+1.0/FM0))**0.5
      L = L + 1
      IF (F1) 16,84,16
16    HB = ABS ((C11/F1)−(FM1/FM0))
      GOTO 83
84    HB = ABS(0.0−(FM1/FM0))
      WRITE (3,77) L
83    WRITE (3,77) K
      K = K + 1
      L = L + 1
      IF (F3) 17,88,17
17    HC = ABS((C32/F3)−(FM1/FM0))
      GOTO 18
88    HC = ABS(FM1/FM0)
      WRITE (6377) L
18    HD = ABS ((AM1+0.5*AM2−FM1/FM0))
19    H4 = (((5.99*AFF1*AFF2)/(AFF1+AFF2))*(1.0/FF0+1.0/S0))**0.5
      WRITE (3,77) K
      K = K + 1
31    H42 =(((5.99*AFF3*AFF2)/(AFF3+AFF2))*(1.0/FF0+1.0/S0))**0.5
32    HE = ABS(S1−FF1)
33    HF = ABS(S3−FF3)
34    H052 =((5.99*A3A*A2A/(A3A+A2A))*(1.0/AFF0+1.0/AM0))**0.5
      WRITE (3,77) K
      K = K + 1
41    H05 =((5.99*A1A*A2A/(A1A+A2A))*(1.0/AFF0+1.0/AM0))**0.5
36    HG = ABS(AFF1−AM1)
37    HH = ABS(AFF3−AM3)
```

Table 29 *Continued*

```
38  H06 = 2.0*P*Q*((3.84/A0A)**0.5)
44  HI = ABS(A2A-2.0*P*Q)
    WRITE (3,77) K
    WRITE(3,20) HA,H1,HB,H02,HC,H022,HD,H03,HE,
  &   H4,HF,H42,HG,H05,HH,H052,HI,H06
20  FORMAT (10X,2HH1,9X,2F15.5/10X,6HH02/H1,5X,2F15.5/21X,
  &   2F15.5/10X,7HH03/H02,4X,2F15.5/10X,2HH4,9X,2F15.5/
  &   21X,2F15.5/10X,7HH05/H04,4X,2F15.5/21X,2F15.5/10X,
  &   7HH06/H05,4X,2F15.5////)
    WRITE (3,2)C11,C12,F1,S1,AFF1,AM1,A1A,C21,C22,C23,F2,
  &   S2,AFF2,AM2,A2A,C32,C33,F3,S3,AFF3,AM3,A3A,F0,S0,AFF0,
  &   AM0,A0A,FM1,X,U,P,FM2,Y,V,Q,FM0,Z,W
 2  FORMAT (10X,8HGENOTYPE,5X,15HM-0 COMBINATION, 6X,33HS
  &   .FEMALES FEMALES MALES ADULTS/10X,6HMOTHER, 6X,2HMM,
  &   3X,2HMS,3X,2HSS,3X,3HSUM/12X,2HMM,6X,2F5.1,5X,F5.1,
  &   6X,F5.1,5X,F5.1,3X,F5.1,3X,F5.1/12X,2HMS,6X,4F5.1,6X,
  &   F5.1,5X,F5.1,3X,F5.1,3X,F5.1/12X,2HSS,11X,3F5.1,6X,
  &   F5.1,5X,F5.1,3X,F5.1,3X,F5.1/35X,F5.1,6X,F5.1,
  &   5X,F5.1,3X,F5.1,2X,F6.1///10X,5HM1 = , F6.1,7X,4HX =,
  &   F6.5,7X,4HU = ,F6.5,7X,4HP = ,F6.5/10X,5HM2 = ,
  &   F6.1,7X,4HY = ,F6.5,7X,4HV = ,F6.5,7X,4HQ = ,
  &   F6.5/10X,5HM0 = ,F6.1,7X,4HZ = ,F6.5,7X,4HW = ,F6.5///)
    IDF=1
    CALL PROBA(T1)
    PT1=PROB
    IDF=2
    CALL PROBA(T2)
    PT2=PROB
    IDF=1
    CALL PROBA(T3)
    PT3=PROB
    IDF=2
    CALL PROBA(T5)
    PT5=PROB
    IDF=2
    CALL PROBA(T4)
    PT4=PROB
    IDF=1
    CALL PROBA(T6)
    PT6=PROB
    IDF=6
    CALL PROBA(T01)
    PT01=PROB
    IDF=3
    CALL PROBA(T02)
    PT02=PROB
    IDF=9
    CALL PROBA(T0)
    PT0=PROB
    WRITE(3,4) T1,PT1,T2,PT2,T3,PT3,T4,PT4,T01,PT01,
```

Table 29 *Continued*

```
      &    T5,PT5,T6,PT6,T02,PT02,T0,PT0
    4 FORMAT((10X,4HTEST,5X,10HHYPOTHESIS,32X,
      &    9HDEG.FREE.,4X,6HCHI SQ,5X,5HPROB.//11X,2HT1,6X,24HFEMALE
      &    GAMETIC SELECTION,21X,1H1,8X,F8.4,5X,
      &    F7.5/11X,2HT2,6X,13HRANDOM MATING,32X,1H2,8X,F8.4,
      &    5X,F7.5/11X,2HT3,6X,27HMALE REPRODUCTIVE
      &    SELECTION,18X,1H1,8X,F9.4,5X,F7.5/11X,2HT4,6X,23HFEMALE
      &    SEXUAL SELECTION,22X,1H2,8X,F8.4,5X,F7.5/6X,
      &    11HT1+T2+T3+T4+,2X,33HALL GAMETIC AND SEXUAL
      &    COMPONENTS, 12X,1H6,8X,F9.4,5X,F7.5/11X,2HT5,6X,
      &    32HZYGOTIC SELECTION EQUAL IN SEXES,13X,1H2,
      &    8X,F8.4,5X,F7.5/11X,2HT6,6X,17HZYGOTIC SELECTION,28X,1H1,
      &    8X,F8.4,5X,F7.5/9X,5HT5+T6,5X,22HALL ZYGOTIC COMPONENTS,
      &    23X,1H3,8X,F8.4,5X,F7.5/11X,2HT0,6X,41HTOTAL FIT TO
      &    RANDOM MATING AND NEUTRALITY,4X,1H9,8X,F9.4,5X,F7.5////))
   45 GOTO 21
   23 STOP
      END
      FUNCTION ABC(A,B,C,D)
      IF(B .EQ. 0.0 .OR. C .EQ. 0.0) ABC = 0.0
      IF(B.NE.0.0.AND.C.NE.0.0) ABC=((A-(B*C/D))**2)/(B*C/D)
      IF(A .LT. 5.0) ABC=0.0
      RETURN
      END
      FUNCTION BAC(AA,BB,CC,DD,EE)
      IF(AA .EQ. 0.0 .OR. DD .EQ. 0.0) GOTO 20
      IF(BB+CC .LT. 5.0) GOTO30
      BAC=(((((0.5*AA*BB)/(BB+CC))-(0.5*AA*DD/EE))**2)/
      &    (0.5*AA*DD/EE)
      GOTO 40
   20 BAC = 0.0
      GOTO 40
   30 IF(EE.LT.5.0) BAC=0.0
      IF (EE.LT.5.0) GOTO 40
      BAC = 0.5*AA*DD/EE
   40 RETURN
      END

      FUNCTION DIV (A,B)
      IF(B.EQ.0) DIV = 0.0
      IF(B .NE. 0.0) DIV = A/B
      RETURN
      END
      SUBROUTINE PROBA(G)
      COMMON IDF,PROB
      IF(G .EQ. 0.0) PROB=1.0
      IF(G .EQ. 0.0) RETURN
      AN1=IDF
      F=G/AN1
      AN2=1.0E10
      FF=F
```

Table 29 *Continued*

```
    PROB=1.0
    IF(AN1*AN2*F .EQ. 0.0) RETURN
    IF(F .GE. 1.0) GOTO 6
    FF=1.0/F
    TEMP=AN1
    AN2=TEMP
  6 A1=2.0/AN1/9.0
    A2=2.0/AN2/9.0
    Z=ABS(((1.0-A2)*FF**(1./3.)-1.0+A1)/SQRT(A2*
  &    FF**(2./3.)+A1))
    IF(AN2.LE.3.0) Z=Z*(1.0+0.08*Z**4/AN2**3)
    FZ=EXP(-Z*Z/2.0)*0.3983423
    W=1.0/(1.0+Z*0.2316419)
    PROB=FZ*W*((((1.330274*W-1.821256)*W+1.781478)*
  &    W-0.3565638)*W+0.3193815)
    IF(F .LT. 1.0) PROB=1.0-PROB
    RETURN
    END
```

This code originated from R. Baccus and was modified by R. Chesser of the University of Georgia's Savannah River Ecology Laboratory. Electronic copies are available from M. Newman by permission of R. Baccus.

The input has the following format:

NLOK (FORMAT 12)

C11 C12 C21 C22 C23 C32 C33 NP1 NP2 NP3 AM1 AM2 AM3 (FORMAT F5.1)

PF1 PF2 PF3 (FORMAT 6.2)

Example: A data file for analyzing two loci would look like the following:

2

95. 16. 25. 12. 6. 3. 0. 463. 157. 13. 524. 148. 26.

111. 43. 3.

106. 11. 21. 18. 0. 1. 0. 549. 85. 5. 589. 107. 8.

117. 39. 1.

NLOK is the number of loci. C11 to C33 are the number of individuals in the various mother-offspring combinations.

Mothers Genotype	Offspring Genotypes		
	AA	**AB**	**BB**
AA	C11	C12	—
AB	C21	C22	C23
BB	—	C32	C33

NP1 to NP3 are the numbers of nonpregnant females of genotype AA, AB, and BB, respectively. AM1 to AM3 are the numbers of adult males of genotype AA, AB, and BB, respectively. PF1 to PF3 are the numbers of pregnant females of genotype AA, AB, and BB, respectively.

Table 30 **Values of θ used for maximum likelihood estimation of mean and standard deviation of truncated data. Used first in Chapter 7.**

γ	0.000	0.001	0.002	0.003	0.004	0.005	0.006	0.007	0.008	0.009
0.05	0.00000	0.00000	0.00000	0.00001	0.00001	0.00001	0.00001	0.00001	0.00002	0.00002
0.06	0.00002	0.00003	0.00003	0.00003	0.00004	0.00004	0.00005	0.00006	0.00007	0.00007
0.07	0.00008	0.00009	0.00010	0.00011	0.00013	0.00014	0.00016	0.00017	0.00019	0.00020
0.08	0.00022	0.00024	0.00026	0.00028	0.00031	0.00033	0.00036	0.00039	0.00042	0.00045
0.09	0.00048	0.00051	0.00055	0.00059	0.00063	0.00067	0.00071	0.00075	0.00080	0.00085
0.10	0.00090	0.00095	0.00101	0.00106	0.00112	0.00118	0.00125	0.00131	0.00138	0.00145
0.11	0.00153	0.00160	0.00168	0.00176	0.00184	0.00193	0.00202	0.00211	0.00220	0.00230
0.12	0.00240	0.00250	0.00261	0.00272	0.00283	0.00294	0.00305	0.00317	0.00330	0.00342
0.13	0.00355	0.00369	0.00382	0.00396	0.00410	0.00425	0.00440	0.00455	0.00470	0.00486
0.14	0.00503	0.00519	0.00536	0.00553	0.00571	0.00589	0.00608	0.00627	0.00646	0.00665
0.15	0.00685	0.00705	0.00726	0.00747	0.00769	0.00791	0.00813	0.00835	0.00858	0.00882
0.16	0.00906	0.00930	0.00955	0.00980	0.01006	0.01032	0.01058	0.01085	0.01112	0.01140
0.17	0.01168	0.01197	0.01226	0.01256	0.01286	0.01316	0.01347	0.01378	0.01410	0.01443
0.18	0.01476	0.01509	0.01543	0.01577	0.01611	0.01646	0.01682	0.01718	0.01755	0.01792
0.19	0.01830	0.01868	0.01907	0.01946	0.01986	0.02026	0.02067	0.02108	0.02150	0.02193
0.20	0.02236	0.02279	0.02323	0.02368	0.02413	0.02458	0.02504	0.02551	0.02599	0.02647
0.21	0.02695	0.02744	0.02794	0.02844	0.02895	0.02946	0.02998	0.03050	0.03103	0.03157
0.22	0.03211	0.03266	0.03322	0.03378	0.03435	0.03492	0.03550	0.03609	0.03668	0.03728
0.23	0.03788	0.03849	0.03911	0.03973	0.04036	0.04100	0.04165	0.04230	0.04296	0.04362
0.24	0.04429	0.04497	0.04565	0.04634	0.04704	0.04774	0.04845	0.04917	0.04989	0.05062
0.25	0.05136	0.05211	0.05286	0.05362	0.05439	0.05516	0.05594	0.05673	0.05753	0.05834
0.26	0.05915	0.05997	0.06080	0.06163	0.06247	0.06332	0.06418	0.06504	0.06591	0.06679
0.27	0.06768	0.06858	0.06948	0.07039	0.07131	0.07224	0.07317	0.07412	0.07507	0.07603
0.28	0.07700	0.07797	0.07896	0.07995	0.08095	0.08196	0.08298	0.08401	0.08504	0.08609
0.29	0.08714	0.08820	0.08927	0.09035	0.09144	0.09254	0.09364	0.09476	0.09588	0.09701
0.30	0.09815	0.09930	0.10046	0.10163	0.10281	0.10400	0.10520	0.10641	0.10762	0.10885
0.31	0.1101	0.1113	0.1126	0.1138	0.1151	0.1164	0.1177	0.1190	0.1203	0.1216
0.32	0.1230	0.1243	0.1257	0.1270	0.1284	0.1298	0.1312	0.1326	0.1340	0.1355
0.33	0.1369	0.1383	0.1398	0.1413	0.1428	0.1443	0.1458	0.1473	0.1448	0.1503
0.34	0.1519	0.1534	0.1550	0.1566	0.1582	0.1598	0.1614	0.1630	0.1647	0.1663
0.35	0.1680	0.1697	0.1714	0.1731	0.1748	0.1765	0.1782	0.1800	0.1817	0.1835
0.36	0.1853	0.1871	0.1889	0.1907	0.1926	0.1944	0.1963	0.1982	0.2001	0.2020
0.37	0.2039	0.2058	0.2077	0.2097	0.2117	0.2136	0.2156	0.2176	0.2197	0.2217
0.38	0.2238	0.2258	0.2279	0.2300	0.2321	0.2342	0.2364	0.2385	0.2407	0.2429
0.39	0.2451	0.2473	0.2495	0.2517	0.2540	0.2562	0.2585	0.2608	0.2631	0.2655
0.40	0.2678	0.2702	0.2726	0.2750	0.2774	0.2798	0.2822	0.2847	0.2871	0.2896
0.41	0.2921	0.2947	0.2972	0.2998	0.3023	0.3049	0.3075	0.3102	0.3128	0.3155
0.42	0.3181	0.3208	0.3235	0.3263	0.3290	0.3318	0.3346	0.3374	0.3402	0.3430
0.43	0.3459	0.3487	0.3516	0.3545	0.3575	0.3604	0.3634	0.3664	0.3694	0.3724
0.44	0.3755	0.3785	0.3816	0.3847	0.3878	0.3910	0.3941	0.3973	0.4005	0.4038
0.45	0.4070	0.4103	0.4136	0.4169	0.4202	0.4236	0.4269	0.4303	0.4338	0.4372

Table 30 *Continued*

γ	0.000	0.001	0.002	0.003	0.004	0.005	0.006	0.007	0.008	0.009
0.46	0.4407	0.4442	0.4477	0.4512	0.4547	0.4583	0.4619	0.4655	0.4692	0.4728
0.47	0.4765	0.4802	0.4840	0.4877	0.4915	0.4953	0.4992	0.5030	0.5069	0.5108
0.48	0.5148	0.5187	0.5227	0.5267	0.5307	0.5348	0.5389	0.5430	0.5471	0.5513
0.49	0.5555	0.5597	0.5639	0.5682	0.5725	0.5768	0.5812	0.5856	0.5900	0.5944
0.50	0.5989	0.6034	0.6079	0.6124	0.6170	0.6216	0.6263	0.6309	0.6356	0.6404
0.51	0.6451	0.6499	0.6547	0.6596	0.6645	0.6694	0.6743	0.6793	0.6843	0.6893
0.52	0.6944	0.6995	0.7046	0.7098	0.7150	0.7202	0.7255	0.7306	0.7361	0.7415
0.53	0.7469	0.7524	0.7578	0.7633	0.7689	0.7745	0.7801	0.7857	0.7914	0.7972
0.54	0.8029	0.8087	0.8146	0.8204	0.8263	0.8323	0.8383	0.8443	0.8504	0.8565
0.55	0.8627	0.8689	0.8751	0.8813	0.8876	0.8940	0.9004	0.9068	0.9133	0.9198
0.56	0.9264	0.9330	0.9396	0.9463	0.9530	0.9598	0.9668	0.9735	0.9804	0.9874
0.57	0.9944	1.001	1.009	1.016	1.023	1.030	1.037	1.045	1.052	1.060
0.58	1.067	1.075	1.082	1.090	1.097	1.105	1.113	1.121	1.129	1.137
0.59	1.145	1.153	1.161	1.169	1.177	1.185	1.194	1.202	1.211	1.219
0.60	1.228	1.236	1.245	1.254	1.262	1.271	1.280	1.289	1.298	1.307
0.61	1.316	1.326	1.335	1.344	1.353	1.363	1.373	1.382	1.392	1.402
0.62	1.411	1.421	1.431	1.441	1.451	1.461	1.472	1.482	1.492	1.503
0.63	1.513	1.524	1.534	1.545	1.556	1.567	1.578	1.589	1.600	1.611
0.64	1.622	1.634	1.645	1.657	1.668	1.680	1.692	1.704	1.716	1.728
0.65	1.740	1.752	1.764	1.777	1.789	1.802	1.814	1.827	1.840	1.853
0.66	1.866	1.879	1.892	1.905	1.919	1.932	1.946	1.960	1.974	1.988
0.67	2.002	2.016	2.030	2.044	2.059	2.073	2.088	2.103	2.118	2.133
0.68	2.148	2.163	2.179	2.194	2.210	2.225	2.241	2.257	2.273	2.290
0.69	2.306	2.322	2.339	2.356	2.373	2.390	2.407	2.424	2.441	2.459
0.70	2.477	2.495	2.512	2.531	2.549	2.567	2.586	2.605	2.623	2.643
0.71	2.662	2.681	2.701	2.720	2.740	2.760	2.780	2.800	2.821	2.842
0.72	2.863	2.884	2.905	2.926	2.948	2.969	2.991	3.013	3.036	3.058
0.73	3.081	3.104	3.127	3.150	3.173	3.197	3.221	3.245	3.270	3.294
0.74	3.319	3.344	3.369	3.394	3.420	3.446	3.472	3.498	3.525	3.552
0.75	3.579	3.606	3.634	3.662	3.690	3.718	3.747	3.776	3.805	3.834
0.76	3.864	3.894	3.924	3.955	3.986	4.017	4.048	4.080	4.112	4.144
0.77	4.177	4.210	4.243	4.277	4.311	4.345	4.380	4.415	4.450	4.486
0.78	4.52	4.56	4.60	4.63	4.67	4.71	4.75	4.79	4.82	4.86
0.79	4.90	4.94	4.99	5.03	5.07	5.11	5.15	5.20	5.24	5.28
0.80	5.33	5.37	5.42	5.46	5.51	5.56	5.61	5.65	5.70	5.75
0.81	5.80	5.85	5.90	5.95	6.01	6.06	6.11	6.17	6.22	6.28
0.82	6.33	6.39	6.45	6.50	6.56	6.62	6.68	6.74	6.81	6.87
0.83	6.93	7.00	7.06	7.13	7.19	7.26	7.33	7.40	7.47	7.54
0.84	7.61	7.68	7.76	7.83	7.91	7.98	8.06	8.14	8.22	8.30
0.85	8.39	8.47	8.55	8.64	8.73	8.82	8.91	9.00	9.09	9.18

INDEX

Z